Auto Heating & Air Conditioning

by

Chris Johanson
ASE Certified Master Technician

Publisher
The Goodheart-Willcox Company, Inc.
Tinley Park, Illinois
www.g-w.com

The Goodheart-Willcox Company, Inc. Brand Disclaimer: Brand names, company names, and illustrations for products and services included in this text are provided for educational purposes only and do not represent or imply endorsement or recommendation by the author or the publisher.

The Goodheart-Willcox Company, Inc. Safety Notice: The reader is expressly advised to carefully read, understand, and apply all safety precautions and warnings described in this book or that might also be indicated in undertaking the activities and exercises described herein to minimize risk of personal injury or injury to others. Common sense and good judgment should also be exercised and applied to help avoid all potential hazards. The reader should always refer to the appropriate manufacturer's technical information, directions, and recommendations; then proceed with care to follow specific equipment operating instructions. The reader should understand these notices and cautions are not exhaustive.

The publisher makes no warranty or representations whatsoever, either expressed or implied, including but not limited to equipment, procedures, and applications described or referred to herein, their quality, performance, merchantability, or fitness for a particular purpose. The publisher assumes no responsibility for any changes, errors, or omissions in this book. The publisher specifically disclaims any liability whatsoever, including any direct, indirect, incidental, consequential, special, or exemplary damages resulting, in whole or in part, from the reader's use or reliance upon the information, instructions, procedures, warnings, cautions, applications, or other matter contained in this book. The publisher assumes no responsibility for the activities of the reader.

Library of Congress Cataloging-in-Publication Data

Johanson, Chris.
Auto heating and air conditioning technology / by Chris Johanson.
    p. cm.
  Includes index.
  ISBN 1-59070-276-X
1. Automobiles—Heating and ventilation.
2. Automobiles—Air conditioning. I. Title.
  TL272 .j65 2004
  629.2'772—dc21

Introduction

Auto Heating & Air Conditioning Technology has been written to introduce you to the basic principles of vehicle climate control, including cooling, heating, defrosting, and air distribution. It is also intended to give you a thorough understanding of how to diagnose and repair all components used in automotive heating, cooling, and ventilating systems.

Not too long ago, cooling an automobile or truck interior was a luxury only the wealthiest buyers could afford. Today, air conditioning is standard equipment on almost all new vehicles. Heating and ventilation has been a part of the automobile for many years. The widespread use of automotive air conditioning creates a tremendous opportunity for the technician. This book will help you to take advantage of the opportunity.

Some heating and cooling concepts were discovered and put to use before the automobile was developed. Other concepts, such as electronic controls, have only recently been developed. On many vehicles, the control system for the air conditioning and heating system is part of the on-board computer. Many service procedures can only be performed with special tools and test equipment. You must know how to use these special testers and tools. For this reason, the technician must be able to take a logical approach to service, backed up by a thorough knowledge of system operation. Diagnosing air conditioning and heating systems also requires special test equipment.

Technicians must know how to find and comprehend diagnostic procedures, test specifications, and other diagnosis and repair information. In many cases, information is provided on compact discs or on the Internet. The technician must therefore know how to operate a computer. The technician must be familiar with Environmental Protection Agency (EPA) laws and related state laws governing refrigerant use and release of refrigerant into the atmosphere.

Auto Heating & Air Conditioning Technology has been carefully designed so all system parts and operating principles will be fully identified and explained before you begin work on actual systems installed in vehicles. Early chapters cover basic principles and parts. Later chapters cover service and repair operations, building on the material covered in the earlier chapters. As you work through this text, you will be introduced to basic principles and perform simple tasks involving these principles. Only then will you move on to more complex tasks. At no time will you be asked to service a system or part before you understand its construction and function.

Throughout this text, you will be directed to sources of specific service information, such as specifications and disassembly and reassembly procedures. Additionally you will be informed of safety related issues, and to proper and safe methods of performing diagnosis and service. EPA and state laws governing refrigerant handling will be covered. You will also learn about basic electrical principles and computer controls. Appendix A covers the principles and service of heated rear windows and mirrors.

Auto Heating & Air Conditioning Technology can also be used to prepare for ASE tests. It is very difficult to get a job as an automotive technician without ASE certification. This book covers all the HVAC service and repair principles you will need to know to pass the ASE Heating and Air Conditioning certification test. Many of the questions in the book are presented in the same format as those used in the actual ASE Heating and Air Conditioning certification test. ASE-type questions are included at the end of each chapter. In addition, all of Chapter 24 is devoted to information about applying for, taking, and passing the ASE tests.

A companion book to **Auto Heating & Air Conditioning Technology** is the **Workbook for Auto Heating & Air Conditioning Technology.** The workbook contains additional test questions (many in ASE format), and jobs that test your hands-on ability. The chapters in the workbook correspond to the chapters in **Auto Heating & Air Conditioning Technology.**

Chris Johanson

How to Use the Color Key

Colors are used throughout **Auto Heating & Air Conditioning Technology** to help show different refrigerant states. Other colors are used to illustrate nonrefrigerant liquids and gases, as well as electrical, mechanical, and special components. The following key shows what each color represents.

Refrigerant States

High-pressure Liquid   Low-pressure Liquid

High-pressure Gas/Vapor   Low-pressure Gas/Vapor

Components

Evaporator, Heater core, Condenser, Radiator

Compressor

Accumulator/Receiver-drier

Metering devices, Orifice tubes, Thermostats, Expansion valves

Electrical

Electrical-Computer

Supplemental Restraint System (SRS)/Air bag

Airflow

Hot air   Outside air

Cold air   Recirculated air

Blended air

Nonrefrigerant Liquids and Gases

Engine coolant   Nonrefrigerant gas, Engine vacuum

Refrigerant oil

Other

Special features or components not otherwise coded.

Table of Contents

Chapter 1

Introduction to Automotive Heating, Air Conditioning, and Ventilation

After studying this chapter, you will be able to:

❑ Explain the purpose of heating, ventilation, and air conditioning systems.
❑ Identify the components of a vehicle air conditioning system and state their purpose.
❑ Explain the basic operation of an air conditioner refrigeration system.
❑ Name the two major types of refrigerant used in automotive air conditioning.
❑ Identify the components of a vehicle heating system and state their purpose.
❑ Explain the basic operation of a heating system.
❑ Identify the components of a vehicle ventilating system and state their purpose.
❑ Explain the basic operation of a ventilation system.
❑ Trace the development of modern heating, ventilation, and air conditioning systems.
❑ Discuss the environmental effects of certain types of refrigerants.

Technical Terms

Heating, ventilating, and air conditioning (HVAC)	Ram air	Freon	Propylene glycol	Vents
	Flow restrictor	Polyalkylene glycol (PAG)	Rear window defrosters	Control panel
Bulkhead	Fixed orifice	Mineral oil	Mirror defrosters	Cables and levers
Firewall	Expansion valve	Polyol ester (POE)	Seat heaters	Vacuum diaphragms
Cowl	Accumulator	Cooling system	Intake	Vacuum valves
Refrigeration system	Receiver-drier	Water jackets	Blower	Vacuum hoses
Closed system	Desiccant	Coolant pump	Squirrel cage blowers	Drivers
Evaporator	Sight glass	Coolant	Duct	Actuators
Heat exchanger	Muffler	Radiator	Ductwork	Electrical switches
Evaporation	Metal tubing	Thermostat	Case	A/C mode
Dehumidifies	Flexible hoses	Heater core	Doors	Max A/C
Compressor	Refrigerant	Heater shutoff valve	Blend door	Vent mode
Condenser	Chemically stable	Antifreeze	Diverter doors	Heat mode
Condense	R-134a	Ethylene glycol		Defog mode
	R-12			Ozone layer

This chapter is an overview of automotive heating, ventilating, and air conditioning systems. It will familiarize you with all the mechanical and electrical components used on these systems. The basic operation of heating and ventilating, as well as the development and operation of the air conditioner refrigeration system will be covered. It previews environmental concerns addressed in later chapters. Studying this chapter will give you the foundation needed to get the most out of this text.

Heating, Ventilating, and Air Conditioning (HVAC) Systems

Vehicle heaters and ventilation systems have been around since the introduction of the closed body automobile. Crude automobile air conditioners were originally developed in the 1930s, but did not become an option on most vehicles until the 1960s. Today, heating and ventilating systems are standard equipment on every vehicle, and over 90% have air conditioning.

The purpose of the heating, ventilating, and air conditioning system is to treat the air entering the vehicle passenger compartment. The ideal system will do the following things to the air:

❑ Cool or heat, depending on the passenger's desires in relation to outside air temperature.
❑ Remove moisture to increase comfort and defog windows.
❑ Remove dust, odors, and potentially harmful or offensive gases.
❑ Circulate the treated air through the vehicle.

How the heating, ventilating, and air conditioning system performs these jobs will be discussed in later chapters.

Why Is It Called HVAC?

HVAC is an acronym (initials) for *heating, ventilating, and air conditioning.* This term is often applied to stationary air conditioning systems, and less often to automotive units. However, the interior of an automobile must be cooled, heated, dehumidified, and ventilated just as a house or factory. Therefore, the term HVAC is used where applicable in this text. Many manufacturers refer to their HVAC systems as *Climate Control, Comfort Control,* or a similar name. All of these systems, however, are HVAC systems.

At one time, all the systems making up the HVAC system were separate. There were individual sets of components to heat, ventilate, and cool the vehicle. Today, all HVAC components are combined into a single system in which all the components work together.

Major Heating, Ventilation, and Air Conditioning System Components

On modern cars and trucks, most of the HVAC components are located in the area separating the engine compartment from the passenger compartment. This area is called the *bulkhead, firewall,* or *cowl.* Other parts are mounted in the engine compartment, on the engine, or ahead of the vehicle radiator. As you read the following sections, note the relative positions of the components as shown in the accompanying figures. The HVAC system can be divided into three major parts, the refrigeration, heating, and air delivery systems.

Refrigeration System Parts and Operation

The *refrigeration system* contains four major parts as well as other components. The purpose of these components is to refrigerate, or cool air. While reading these passages, keep in mind they form a *closed system.* In a closed system, the refrigerant circulates through the refrigeration system over and over and does not escape. Refer to **Figure 1-1** as you read these next paragraphs.

Evaporator

For this discussion, flow through the refrigeration system begins at the evaporator. The *evaporator* is a large container with internal passages for refrigerant and external finned passages for air. The purpose of the evaporator is to transfer heat from the air to the refrigerant. For this reason, the evaporator is sometimes called a *heat exchanger.*

Refrigerant enters the evaporator as a liquid. As the warm air passes through the evaporator, it causes the refrigerant to turn into a vapor, or gas. This is called *evaporation.* Changing from a liquid to a gas causes the refrigerant to absorb heat from the air. The evaporating refrigerant causes the surface of the evaporator to become cold. The pressure in the evaporator is controlled for maximum cooling. Moisture in the air collects on the surface of the evaporator. This *dehumidifies* (removes moisture from) the air. Some incoming dust and particles collect on the wet evaporator. The water and dust drain out of a hole in the evaporator case. The cooled, cleaned, and dehumidified air then enters the passenger compartment.

Compressor

After absorbing heat from the air, the refrigerant gas is drawn into the *compressor.* The compressor acts as a pump, moving the refrigerant through the system. The compressor raises the refrigerant pressure by compressing the heated refrigerant. The low pressure-low temperature refrigerant vapor entering the compressor leaves as high pressure-high temperature vapor. The compressor is belt-driven by the engine. A magnetic clutch allows the compressor to be connected and disconnected from the drive belt.

Condenser

From the compressor, the refrigerant enters the **condenser.** Like the evaporator, the condenser is a heat exchanger with internal passages for refrigerant and external passages for air. However, the condenser sends heat back into the air. Air passing across the hot compressed refrigerant causes it to **condense,** or change back into a liquid. When it condenses, it gives up heat to the air passing through the condenser

A fan provides airflow through the condenser. The fan is driven by the engine or by an electric motor. This fan is also used to provide airflow through the engine cooling system. When the vehicle is moving, air is forced through the condenser. This is called **ram air.**

Flow Restrictor

The condensed refrigerant then flows to the **flow restrictor.** The flow restrictor reduces the flow of refrigerant into the evaporator. This causes the evaporator pressure to be much lower than the pressure of the condensed refrigerant coming out the condenser. The low evaporator pressures allows the refrigerant to evaporate. There are two kinds of refrigerant flow restrictors.

The fixed orifice restrictor is commonly used on modern refrigeration systems. The **fixed orifice** is a small tube which creates a narrow passage, allowing refrigerant to enter the evaporator in controlled amounts. The size of this passage is nonadjustable and controlling compressor operation controls evaporator pressure, **Figure 1-1.**

The **expansion valve** is a flow restrictor with a variable opening. An expansion valve has a small needle valve in the refrigerant opening. A diaphragm moves the needle. The diaphragm is connected to a sealed chamber containing a gas, **Figure 1-2.** The gas expands and contracts with temperature changes, moving the diaphragm and needle valve. Higher temperatures cause the valve to open allowing more refrigerant into the evaporator. Lower temperatures cause the valve to close, letting in less refrigerant.

Other Refrigeration System Components

Other components are used in the refrigerant system. Almost all systems have some of these components, while others are seldom used. Refer to **Figures 1-1** and **1-2** as you read the following sections.

To hold extra refrigerant, all systems have an **accumulator** or a **receiver-drier.** When an accumulator is used, it is located between the evaporator and the compressor. When a receiver-drier is used, it is located between the condenser and expansion valve. Besides holding refrigerant, these units contain a desiccant.

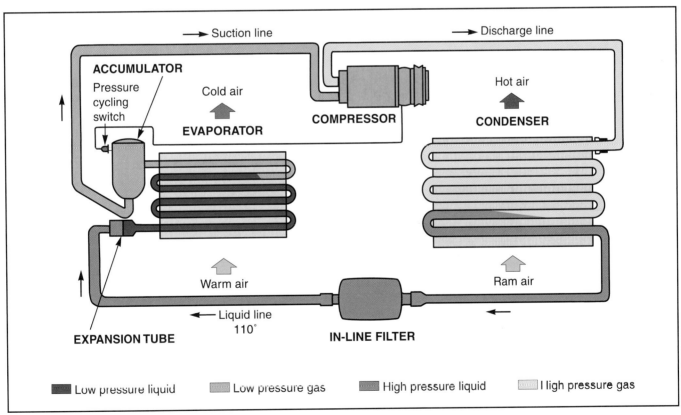

Figure 1-1. *This diagram of a refrigeration system shows all the refrigeration system parts: evaporator, compressor, condenser, and flow restrictor. The system is closed, and the refrigerant circulates over and over. Fans are used to move air through the evaporator and condenser. This is a cycling clutch orifice tube system. (Four Seasons)*

Desiccant is a substance designed to absorb moisture. Since moisture can severely damage the refrigeration system components, the desiccant must trap it as it circulates with the refrigerant.

Other components include the *sight glass.* On older systems, the technician could get a rough idea of the amount of refrigerant in the system by checking for bubbles in the sight glass. Excessive bubbles meant the system was low. Lack of bubbles meant the system was either full or completely empty. The sight glass was usually located on the receiver-drier. Sight glasses are no longer installed on vehicles.

A few vehicles have mufflers located at the compressor outlet. The *muffler* reduces noise caused by the pressurized gases leaving the compressor and is usually part of the hose assembly. *Metal tubing* and *flexible hoses* connect the components of the refrigeration system. Metal tubing is used between stationary parts. Hoses are used to connect parts that move in relation to each other, for example the engine-mounted compressor and the body-mounted condenser. Tubing and hoses must stand up under high pressures and not leak.

Refrigerant

A component common to all A/C systems is the refrigerant. *Refrigerant* is the cooling medium flowing through the refrigeration system. It must change from a liquid to a gas in the evaporator and a gas to a liquid in the condenser. It must be able to make these changes at normal outside air temperatures, and at the pressures developed by the compressor. If a substance cannot do this, it is no good as a refrigerant.

Refrigerant must also be chemically stable. A *chemically stable* compound will not break down or change into other chemicals or compounds during normal operation. Refrigerant must be compatible with the lubrication oil used in the compressor, since this oil circulates throughout the refrigeration system. Refrigerant must also be nonpoisonous and nonflammable.

The most common refrigerant used in late-model vehicles is *R-134a.* R-134a is used in all new cars and trucks built since 1993, and is used to replace other refrigerants when retrofitting air conditioners in older vehicles. *R-12,* also called *Freon,* is used in cars and trucks built before 1993. Other refrigerants are being sold as substitutes for R-12. However, all current substitutes have drawbacks, and in some cases are dangerous and illegal to use in vehicles.

Refrigerant Oils

To prevent wear in the compressor, special lubricating oil is installed in the compressor crankcase. Some of this oil also circulates with the refrigerant. It is very important the oil be compatible with the refrigerant. Systems using R-134a require a type of oil called *polyalkylene glycol,* or *PAG.* Older R-12 systems use *mineral oil.* These oils should not be mixed, as they are not compatible with each other. They are also not compatible with every refrigerant. A third type of oil, called *polyol ester,* or *POE,* can be used with either type of refrigerant.

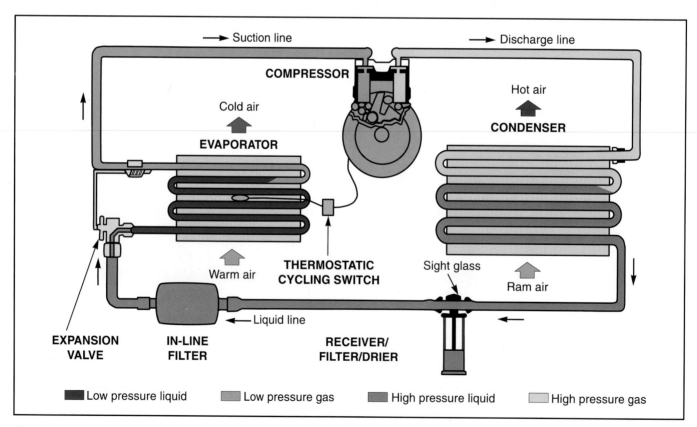

Figure 1-2. *Many of the parts in this refrigeration system are similar to those in* **Figure 1-1.** *This system uses a different type of flow restrictor. These parts will be discussed in more detail in future chapters. (Four Seasons)*

Heating System Parts and Operation

Unlike the refrigeration system, the heating system uses a medium that does not change from a liquid to a vapor. However, the heater is also a closed system with the fluid circulating over and over. To get the engine heat into the passenger compartment on cold days, the heating system uses several major parts. Refer to **Figure 1-3** as you read the following sections.

The engine's ***cooling system*** provides the needed heat. In one sense, this heat is free as only a small part of the engine's combustion heat is used to produce power. The rest must be dissipated from the engine. Some engine heat goes out with the exhaust, and some is removed by air passing over the engine. The rest of the heat must be removed by the cooling system. To do this, passages called ***water jackets*** are cast as part of the engine block. A belt-driven ***coolant pump*** circulates ***coolant*** (a mixture of antifreeze and water) through the engine where it picks up heat. The coolant then flows out of the engine and into a ***radiator,*** where the heat is transferred into the air. The lower temperature coolant then reenters the engine. Coolant constantly circulates through the engine, carrying away heat.

To keep the cooling system from removing too much engine heat, a ***thermostat*** limits coolant circulation until the engine reaches operating temperature. In very cold weather, the thermostat may re-close if the coolant temperature becomes too low. A pressure cap on the radiator pressurizes the cooling system as the engine warms up. This increases the coolant's boiling point, reducing the chance of engine overheating. To add coolant, the radiator cap can be removed when the engine and cooling system are cold.

Some of the hot coolant is diverted to the ***heater core*** through hoses. One hose is the inlet hose and the other hose is the outlet, or return hose. In most cases, the inlet hose is connected to the engine at the point where the coolant gets hot quickly, such as the cylinder head. The heater core is a sealed container with internal passages for engine coolant and external passages for air. Hot coolant passing through the heater core gives up its heat to incoming air. The heated air then enters the passenger compartment. The return hose is connected to the water jackets, allowing coolant to circulate through the cylinder heads when the thermostat is closed. This keeps the heads from overheating before the thermostat opens. The heater core can also act as a secondary radiator, removing excess heat when an engine is overheating.

On some vehicles, a ***heater shutoff valve*** is used. The heater shutoff valve is installed in the inlet hose. The driver operates it through the control system. When the heater is being used, the shutoff valve is open, and coolant can circulate through the heater core. If the heater is not being used, the shutoff valve closes. This prevents coolant circulation through the heater core. The shutoff valve is usually closed when the system is off, and during maximum air conditioner operation.

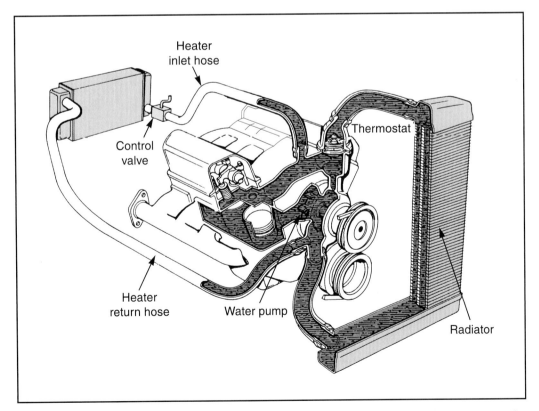

Figure 1-3. *This elementary heater system contains the hoses connecting the heater core to the engine cooling system, the core itself, and a control valve to block coolant flow to the core when necessary. A fan is used to draw air through the heater core.*

Coolant

Water could be used as an engine coolant, but it would rust the interior of the cooling system, freeze at low temperatures, and boil away at high temperatures. Modern engine coolant is a mix of water and **antifreeze.** Modern antifreeze is a chemical called **ethylene glycol** or **propylene glycol.** Ethylene and propylene glycol are used in both winter and summer, and may be called antifreeze or coolant. Antifreeze also contains corrosion inhibitors and water pump lubricants. Most antifreeze is mixed in equal parts with water for a 50-50 mix. This mixture provides the best freeze protection, corrosion protection, and heat transfer.

Other Heating Devices

On some vehicles, electrically heated units provide additional comfort and safety. Many cars have **rear window defrosters.** Rear window defrosters are wire grids placed on the rear window. Electric current is passed through the wires and causes them to heat up. The heat drives moisture from the glass. The same principle is used on **mirror defrosters. Seat heaters** also use a heated grid to warm the seat. Electrical thermostats control all these heaters. If the grid becomes too warm, the thermostat shuts off current flow.

Air Delivery System Parts and Operation

The ventilating system can be used alone, or with heating or air conditioning. The ductwork for the ventilating system is also used during the heating and cooling phases. Another job of the ventilating system is defrosting. **Figure 1-4A** shows the relationship of the ventilating, air conditioning, and heating systems.

Air enters the body at the cowl area, just ahead of the windshield, **Figure 1-4B**. This area is called the **intake.** Seals keep exhaust gases and engine vapors from entering the intake area.

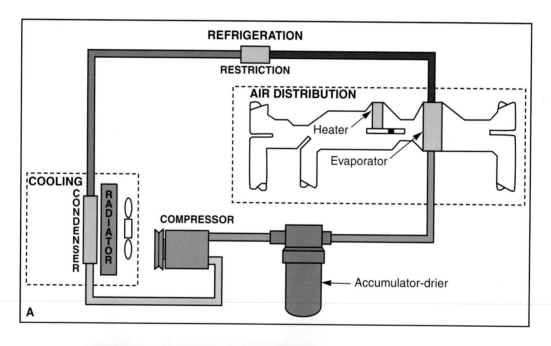

Figure 1-4. A—The relationship of the refrigeration system, heater core, and airflow system is shown here. Other systems differ in placement of the main components, but the overall design and function are the same. B—Outside air enters the HVAC system at the cowl, directly below the windshield. The screen keeps out leaves and other large debris. Note the seals that keep engine heat and fumes away from the intake. On some late-model vehicles, a cabin filter, or intake air filter, is installed under the cowl. (Delphi)

Blower

The **blower** or *fan* draws air from the intake area and pushes it through the evaporator and heater core. Air flows through the evaporator and heater even when they are not in operation.

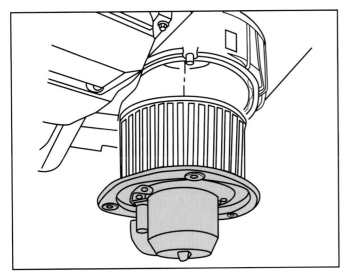

Figure 1-5. *The typical blower motor consists of a blower wheel and an electric motor. Operation of the motor is controlled by a series of electrical switches and resistors or an electronic module. (General Motors)*

Most blower wheels are referred to as **squirrel cage blowers.** The squirrel cage blower consists of a blower wheel, which resembles the wheel in a squirrel cage, driven by an electric motor. The revolving blower wheel, **Figure 1-5,** pulls air into its center as it rotates. The air is then driven outward into the passenger compartment. Electrical devices called resistors usually control motor speed. Resistors reduce the amount of current flowing into the blower, depending on which speed has been selected. A few vehicles control blower motor speed by using relays to energize different sets of wires in the blower. Modern HVAC systems usually have four blower speeds.

On all vehicles made in the last 30 years, the blower can run whenever the ignition switch is in the *On* position. This is done to provide a constant flow of air through the passenger compartment and reduce the chance of carbon monoxide poisoning.

Ductwork, Doors, and Vents

The blower output is directed through the evaporator and heater core, and into the passenger compartment, by ducts. A **duct** is a tube or passageway for a substance, in this case air. **Figure 1-6** illustrates a typical vehicle duct system. The system of ducts in a vehicle is called the **ductwork.** Most of the intake ductwork is formed by the body sheet metal and is not a separate duct assembly.

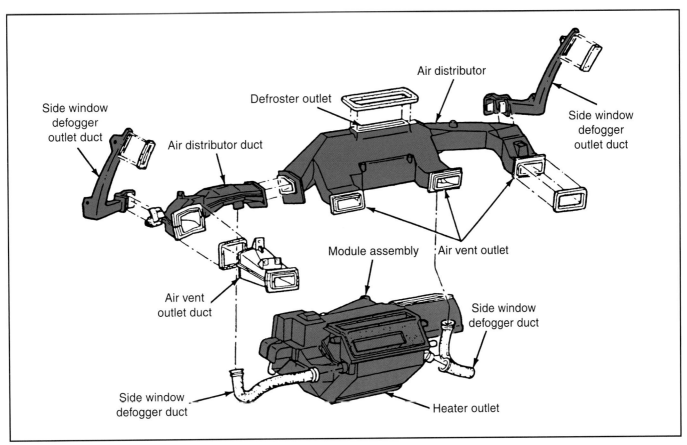

Figure 1-6. *The HVAC system vents and ducts of a common vehicle make. Air can be directed through all or some of these ducts during system operation. (Ford)*

Most of the ductwork is built into a *case,* **Figure 1-7.** The case usually contains the evaporator, heater core, and the blower. The case may be called the evaporator case, heater case, blower case, or HVAC module. The case is usually installed on the right side of the firewall, or cowl, under the dashboard. Some of the case components may extend into the engine compartment. On some vehicles, the case is a two-piece unit bolted to either side of the firewall. When assembled, the case halves and firewall form the case assembly. On other vehicles, the case is a self-contained unit bolted to the firewall, usually under the right side of the dashboard.

Movable *doors* control airflow through the ducts. These doors can be installed in the case, or in other places where they are needed. The main two types of doors are blend doors and diverter doors. The job of the *blend door* is to combine cooled air with warm air to obtain the temperature level desired by the driver. The *diverter doors* are used to direct, or divert, airflow for various operating modes. Diverter doors can direct air to the floor vents, or to the windshield defroster vents, depending on their position. Other diverter doors direct air to the dash vents for maximum cooling, or to the dash and floor vents for better circulation. Some diverter doors close off the outside air supply to recirculate the air in the passenger compartment. Most modern vehicles have one blend door and several diverter doors, **Figure 1-8.**

The outlets into the passenger compartment are called *vents.* Vents are installed in the dashboard for air conditioning, ventilation, and defrosting. These vents may have fins for directing the air. Vents on the floor provide heated air from the heater core. Modern vehicles usually have small vents at the side pillars to direct air onto the side glass. This helps to defog the side windows. The vents are often referred to in service manuals as *registers.*

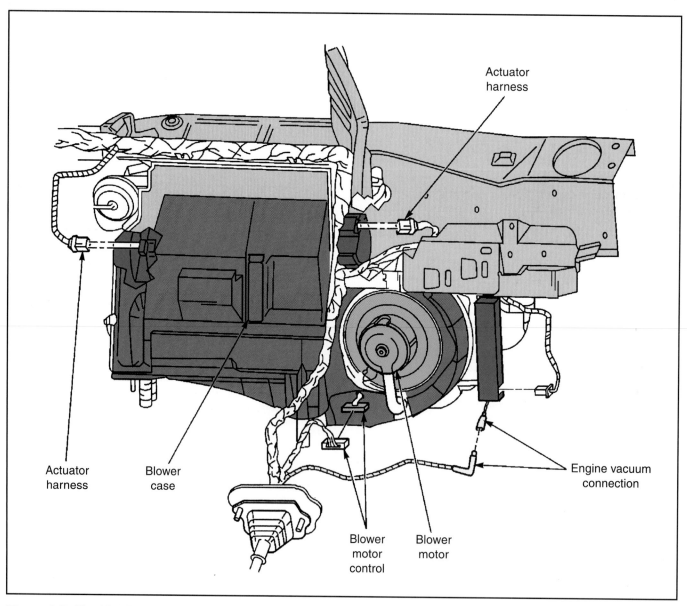

Figure 1-7. *The HVAC system case is installed under the vehicle dashboard, out of sight. This is a cutaway view from the passenger side of the vehicle. All major parts are inside the case. (General Motors)*

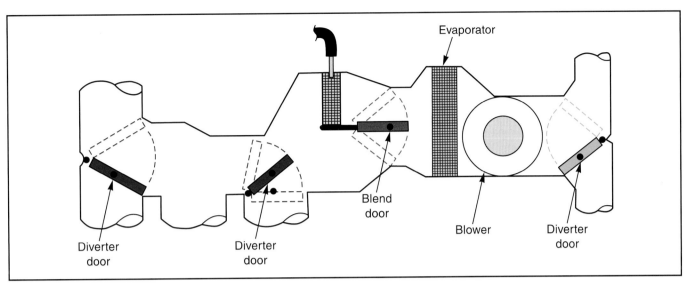

Figure 1-8. *An airflow schematic showing door movement inside an HVAC case. This schematic is an imaginary view of the case from above, with the top removed. The arrows indicate the possible movement of the air doors. (General Motors)*

HVAC System Controls

The HVAC system would be useless if the driver and passengers could not control it. To control the HVAC system, mechanical, electrical, and vacuum controls are used. Many late-model vehicles use electronic components to fully or partially control HVAC system operation.

The modern control system controls the following HVAC operations:

❑ Air conditioner compressor clutch engagement.
❑ Heater shutoff valve operation.
❑ Blower motor operation.
❑ Blower motor speeds.
❑ Blend door position.
❑ Diverter door position.

All these functions are controlled from a dashboard-mounted **control panel.** The modern control panel may also contain controls for the rear window and side mirror defoggers and the heated seat when used. The control panel connects to the HVAC system through mechanical devices, vacuum operated devices, or electrical controls. A typical control panel is shown in **Figure 1-9.**

Figure 1-9. *A control panel used on some vehicles. The controls allow the driver to select various operating conditions. Other manufacturers' panels will have different control layouts. (General Motors)*

Mechanical Devices

On some systems, especially on older vehicles, there is a direct mechanical connection between the control panel and other parts of the HVAC system. This connection is usually a series of **cables and levers, Figure 1-10.** Cables are used to open and close the blend and diverter doors. Sometimes a cable is used to operate the heater shutoff valve. The most common use of a cable connection is on the blend door. Using a cable allows the driver to control temperature by moving the dashboard lever in small steps.

On modern vehicles, **vacuum diaphragms** operate many doors and shutoff valves. A typical vacuum diaphragm is shown in **Figure 1-11.** One or more **vacuum valves** control the diaphragms by directing vacuum through **vacuum hoses.** On many modern vehicles, electric solenoids take the place of vacuum values. On other vehicles the vacuum system has been partially or completely replaced with small electric motors, sometimes called **drivers.** The diaphragms and drivers are sometimes called **actuators.**

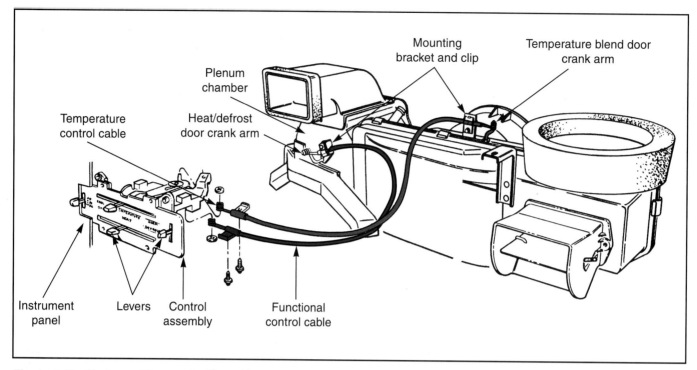

Figure 1-10. *Air door cable controls. The cable allows the driver to operate the air doors from the dashboard panel. Older vehicles used cable controls only. Cables are still used on a few vehicles to control air temperature. (Ford)*

Electrical switches in the control panel operate the blower motor, compressor clutch, and some other components. These switches are on-off electrical devices for directing current. On many modern vehicles, the control assembly contains a small computer. The computer operates the other components of the HVAC system to maintain a temperature set by the driver.

HVAC System Operation

The following sections explain how an HVAC system operates in various modes. A *mode* is a control system setting designed to achieve a desired outcome, for example heating or cooling. Every vehicle manufacturer has various names for their operating modes, and not all manufacturers provide the same number of modes. These sections will cover the main modes every vehicle will have: air conditioning, ventilating, heating, and defrosting.

Air Conditioning

In the *cooling* or *A/C mode,* the compressor clutch is engaged. The compressor pulls refrigerant from the evaporator and pumps it to the condenser. The blower motor draws air into the vehicle through the cowl area and pushes it through the evaporator. Air passing through the evaporator causes the refrigerant to vaporize and absorb heat. This cools the air and causes airborne moisture and dust to collect on the evaporator fins. The ducts deliver the cooled, dried, and cleaned air to the

passenger compartment through the dashboard vents. The vaporized refrigerant enters the compressor and is compressed. The compressed refrigerant enters the condenser. Air passing through the condenser causes the refrigerant to condense to a liquid. The condensed refrigerant returns to the evaporator through a flow restrictor. If the evaporator temperature becomes too low, the compressor clutch will turn off, or the evaporator pressure will be increased by refrigeration system valves.

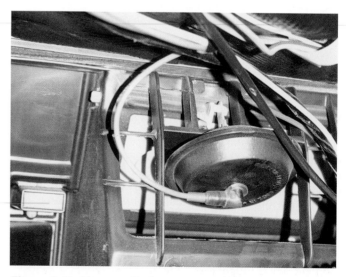

Figure 1-11. *Vacuum diaphragms are widely used on modern cars and trucks. Vacuum to move the diaphragms is developed in the engine intake manifold. Vacuum switches or electrical solenoids control the flow of vacuum to the diaphragms.*

Maximum Cooling

If *maximum cooling* or **Max A/C** is desired, a diverter door will close and prevent the entry of outside air. Already cooled air from the passenger compartment will recirculate through the evaporator. If less than maximum cooling is desired, the blend door is moved so some of the cool air is allowed to pass through the heater core before being remixed with the cooled air. On some vehicles, the heater shutoff valve is closed, and the extra heat comes from outside air only.

Ventilating

In the *ventilation* or **Vent mode,** the blower is drawing air into the vehicle through the cowl area. The air is pushed through the evaporator. The refrigeration system is off, so the evaporator does not cool the incoming air. The air is then directed by the ducts to exit from the dashboard outlets. If necessary, the blend door can be moved to allow some air to pass through the heater core. Outside air can be closed off and the inside air recirculated, however, this mode is usually not available on most cars.

Heating

In the *heating* or **Heat mode,** the heater shut off valve is open and the engine cooling system is pushing heated coolant through the heater core. The blower is drawing air into the vehicle through the cowl area. The diverter and blend doors are positioned so air flows through the heater core. Air passing through the core absorbs heat from the coolant. After giving up some of its heat to the air, the coolant returns to the engine cooling system.

The heated air is delivered through the ducts in the passenger compartment, primarily the floor ducts. If the driver desires less than full heat, the blend door is moved so cool air from outside mixes with the heated air. The outside air often passes through the evaporator, which is not operating at this time.

Defrost

In the *defrost* or **Defog mode,** the refrigeration system is operating in the same manner as in the A/C mode. The blower draws air from the cowl area and through the evaporator. The air is cooled and dehumidified, then passes through the heater core. The shutoff valve is open, and hot coolant is circulating through the heater core. The air passing through the heater core is warmed. The diverter doors are positioned so the air is then blown through the windshield vents. The warm, dry air strikes the glass and evaporates moisture from the glass.

HVAC System History

Some of the early principles upon which HVAC is based, such as heat transfer and heat absorption during a change of state, were discovered during the 17th and 18th centuries. These principles were slowly put to use in industrial processes in the 19th century. During the 20th century, the development of the automobile HVAC system began slowly. The major developments included the introduction of hot water heaters, no-draft ventilation, and air conditioning.

Heaters

In early cars, warm clothes and lap robes were the only options for cold weather driving. When closed body cars became popular in the early 1920s, the idea of using hot water from the engine to heat the passenger compartment began to be developed. Hot water heaters were first introduced in 1926 on several models. Many early cars also had rear seat heaters, using extra hoses to connect to a second heater core.

Before the advent of water heaters, and for some time after, air heaters using exhaust manifold heat were used. Exhaust manifold heat is still used to heat the passenger compartment on air-cooled cars. A few manufacturers tried electrical heating, but was abandoned since they tended to overload the low amperage electrical systems used in early vehicles. A few auto manufacturers offered gasoline powered heaters. The last gasoline heater was used in the mid-1960s on the Chevrolet Corvair.

On most cars, the heater was optional equipment until the 1960s. Many auto manufacturers and parts suppliers offered aftermarket heaters to be added to vehicles not originally equipped with them. Almost all heaters today use the heated coolant from the engine.

Ventilation

Ventilation has been around since the earliest cars. Open car bodies obviously did not have a problem with ventilation. As closed bodies evolved, there became need for passenger compartment ventilation. The earliest ventilation was the vent window. Vent windows pushed air into the passenger compartment as the vehicle moved and were placed in the front and rear of the vehicle. Rear vent windows were used well into the 1980s.

Many manufacturers also used hood or fender scoops to push air into the passenger compartment of the moving vehicle. In 1934, General Motors introduced No Draft ventilation, in which air was drawn from the outside and exited at vents on and under the dashboard. Chrysler pioneered windshield defroster vents in 1936, followed by other manufacturers soon after.

In 1938, Nash developed the Weather Eye system, which combined heating and ventilation into one unit. Air entering the passenger compartment was cleaned, heated, and blown through floor, dashboard, or defroster vents, depending on the control positions. A unique feature of this system was an air filter located ahead of the blower.

Ventilation systems began to be integrated with the heater and air conditioner. The operation of a ventilation system combined with the heater as used in an older car is illustrated in **Figure 1-12**. Ventilation systems remained basically unchanged through the 1950s and 1960s. Studebaker used a ventilation system on its last model that pulled air from just ahead of the windshield and exhausted it through vents over the taillights. In the mid 1960s, Chrysler introduced Flow Through Ventilation in which air entered at the firewall and exited through vents in the door pillars and the trunk.

In 1970, General Motors introduced Astro Ventilation, which worked in a similar fashion, but forced air to exit from vents between the rear window and trunk. Astro ventilation allowed General Motors to eliminate the side vent window. Other manufacturers developed similar systems. With air conditioning systems becoming standard equipment in almost all vehicles, side vents have largely disappeared.

Air Conditioning

The development of air conditioning was foreshadowed when the first ice making plants were developed during the 19th century. The same refrigeration principles were put to use in air conditioning systems. Willis Carrier patented the first modern air conditioner in 1906. This air conditioner was not designed to cool the air but to control moisture in textile mills and printing plants. Air conditioning was later installed in movie theaters and government and office buildings.

The refrigerant used in these early systems was usually ammonia or sulfur dioxide. Because the refrigerants were toxic, air conditioners had to be located in basements or roofs, away from people. For this reason, many of the first home refrigerators were placed on the back porch instead of in the kitchen. Automobile air conditioning was not seriously considered at this time. A few manufacturers tried to cool the vehicle using water evaporation, but this was not widely used.

In 1930, DuPont, working with General Motors, developed R-12, the first consumer safe refrigerant. R-12 was much safer than ammonia or sulfur dioxide, and allowed the refrigerator to be placed inside the home. Once a safe refrigerant was developed, mobile air conditioning became a possibility. The first car to have air conditioning was the 1940 Packard. The compressor was engine driven, but most of the other components were in the trunk. The cooled air came out of vents located behind the rear seat. This early air conditioner was installed on only a few Packards. A few experimental units were installed in Cadillacs shortly after this.

Unfortunately, World War II and its aftermath caused a 15 year delay in automotive air conditioner development. During this time, a few custom designed aftermarket air conditioners were installed in limousines and hearses. It was 1953 before factory air conditioning was offered again, this time by Cadillac, Lincoln, and Chrysler. By 1955, they were optional on most cars.

Most early air conditioners were large and inefficient. The compressor, condenser, and related hoses occupied a large amount of space in the engine compartment. System breakdowns were very common. The extra heat load of the condenser and the drag from the compressor often caused engine overheating. In many vehicles, components were still installed in the trunk. Auto manufacturers had not figured out how to integrate the air conditioner with the other heating and ventilation components, as illustrated by the air conditioner controls mounted on top of the dashboard of the 1957 car shown in **Figure 1-13**.

As late as 1955, only 1% of American cars had air conditioners. However, as sales of home air conditioning began to increase, automotive air conditioning followed closely, especially in the Southern half of the United States.

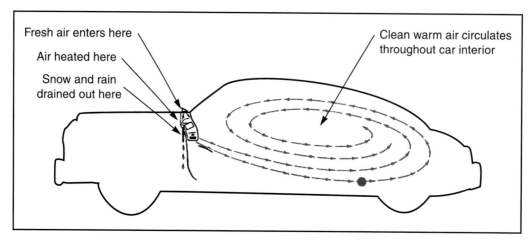

Figure 1-12. *This early heating and ventilating system cleaned and heated incoming air. While crude by modern standards, the design was a big improvement over earlier systems.*

Figure 1-13. *In the late 1950s manufacturers were still trying to figure how to integrate the heater and air conditioner. Notice the separate controls for the heater/ventilation and the air conditioner on this 1957 Pontiac.*

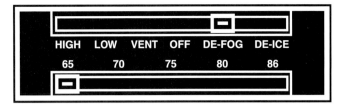

Figure 1-14. *An early automatic temperature control panel. The temperature control device was partially mechanical and relied on temperature sensitive springs to provide a temperature input. (Ford)*

Many cars were fitted with aftermarket, or add-on, air conditioners after they left the factory. Air conditioning was factory installed on 25% of 1965 cars, 70% of 1975 vehicles, and is now standard equipment in over 90% of new vehicles.

As air conditioning became more common, auto manufacturers began to integrate (combine) it with the heating and ventilating systems. The advantage of integration is the same blower, ductwork, and control panel can be used to perform all of the vehicle's HVAC functions. The engine cooling system was improved to handle the extra load of the air conditioner.

Within the last ten years the original refrigerant developed for air conditioning, R-12, has been completely replaced by R-134a. Some 1992 Ford Taurus models were the first vehicles to be factory equipped with R-134a. In the last few years some manufacturers have reintroduced the concept inspired by the Nash intake air filter, calling it the cabin filter.

Electronic Control of Air Conditioning Systems

Auto manufacturers began to replace levers and cable operated controls with vacuum diaphragms and electric solenoids. A fully automatic temperature control system appeared on the 1964 Cadillac. With the Cadillac system, the driver set the desired temperature and the control system would operate the heater and refrigeration system to maintain temperature. Updated versions of this system have been installed on many vehicles. **Figure 1-14** shows an early automatic temperature control panel. **Figure 1-15** is a later fully electronic version.

In the 1970s, evaporator pressure control valves were replaced by systems that controlled evaporator pressure by turning the compressor clutch on and off. Turning the clutch off when possible decreases the load on the engine and improves fuel mileage.

Figure 1-15. *A modern automatic temperature control panel. All functions are electronically controlled, often by the vehicle engine or body control computer. Some systems use a separate HVAC computer. The touchpad on the control panel has no mechanical or vacuum connections. (Saab)*

Figure 1-16 shows a complete modern HVAC system layout. Note how the parts fit together into a fully integrated and compact unit.

The Air Conditioner and the Ozone Layer

One of the major problems modern air conditioning is facing is ecological. Just as the automotive air conditioner was perfected, its refrigerant was discovered to be a prime cause of ozone layer destruction. The *ozone layer* is

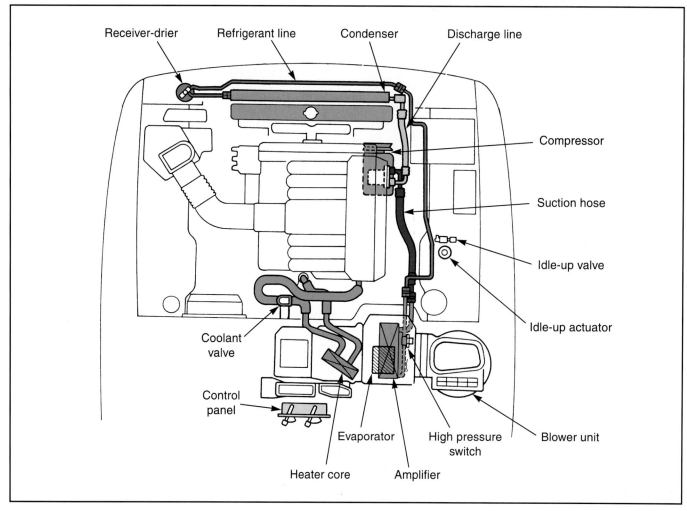

Figure 1-16. *This is an illustration of a fully integrated HVAC system. Note the relationship of the parts to form a compact efficient system. This system can cool, heat, defrost, and ventilate the interior of the vehicle. (Toyota)*

located miles above the earth and helps to protect life from ultraviolet and other harmful rays coming from the sun and space. Older refrigerants released from air conditioning and refrigeration systems (as well as spray cans and fire extinguishing systems) were found to be depleting the ozone layer. At one point, it was thought car and truck air conditioning might have to be abandoned to prevent environmental damage caused by loss of the ozone layer, **Figure 1-17.**

However, several solutions have been devised to correct many of the problems of refrigerant service. Newer air conditioning systems are better sealed to prevent leaks. Safer refrigerants have been developed and recovery equipment is in place to prevent the majority of refrigerants from entering the atmosphere. Technicians are better trained and are required to be certified in refrigerant recovery and handling.

Chapters 2, 6, 15, 16, and 23 will cover procedures for dealing with air conditioner refrigerants in ways less harmful to the environment. These procedures will actually help you to make repairs and save your customers money. In addition, they are the law.

Figure 1-17. *Satellite photo showing the depletion of the ozone layer at the South Pole. The dark area at the center is the location of the greatest area of damage. (NOAA)*

Summary

Every modern vehicle has heating and ventilating systems, and most vehicles have air conditioning. HVAC is an acronym for heating, ventilating, and air conditioning. The heating, ventilating, and air conditioning system will cool, heat, dehumidify, and clean the air entering the vehicle passenger compartment.

The major air conditioning refrigeration system parts are the evaporator, compressor, condenser, and the flow restrictor. Refrigerant turns to a gas in the evaporator and absorbs heat. The gas is drawn into the compressor where it is pressurized, and turns into a liquid in the condenser. The refrigerant then passes through the flow restrictor to begin the process again in the evaporator.

Other parts include various evaporator pressure control valves, storage for extra refrigerant, desiccant, and sight glasses. All refrigeration systems use one of two types of refrigerant. Lubricating oil is also needed.

Engine coolant circulates through the engine, carrying away heat. Some of this coolant is diverted through the heater core and gives up its heat to incoming air. Many heater systems use a shutoff valve in the inlet hose. Electrically operated heating devices include rear window and mirror defoggers and heated seats.

The ventilating system can be used alone, or with heating or air conditioning. Air flows through the evaporator and heater even when they are not operating. A blower draws in outside air and pushes it into the passenger compartment.

Air is delivered to the passenger compartment by a system of tubes called ducts. Some of the ducts are in the case, which also contains the evaporator and heater core. The blend door controls the amount of heating or cooling produced. Diverter doors direct air to the proper vents as needed for best cooling, heating, or defrosting.

Mechanical, electrical, and vacuum devices are used to control the HVAC system. Many late-model vehicles have electronic HVAC system controls. The HVAC control system operates the air conditioner compressor clutch, heater shutoff valve, blower motor, blend door, and diverter doors.

In the cooling mode, air passing through the evaporator causes the refrigerant to vaporize and absorb heat from the air. The cooled air goes to the passenger compartment, and the refrigerant is drawn into the compressor. From the compressor, the pressurized refrigerant enters the condenser. Air passing through the condenser causes the refrigerant to condense to a liquid, after which it is reused. In the heating mode, the hot coolant is circulating through the heater core. Air passing through the heater core absorbs heat from the coolant. The coolant returns through the hoses to the engine cooling system. The blend door can closely regulate cooling or heat output in any mode. In Defrost or Defog mode, incoming air is dehumidified then warmed and directed to the windshield.

HVAC principles were discovered in the last century. Over the last 50 years, heating and air conditioning have been made practical to the point where they are installed on most vehicles. A recent concern has been the effect of air conditioning refrigerant on the environment.

Review Questions—Chapter 1

Please do not write in this text. Write your answers on a separate sheet of paper.

1. List the four purposes of the HVAC system.

2. What do the initials HVAC stand for?

3. Most HVAC components are located at the _____.

4. The accumulator or receiver-drier holds extra refrigerant and a _____.

5. The heater core is warmed by _____.

6. Modern antifreeze is made of _____.

7. HVAC fans are called _____ blowers.

8. How many blend doors do most vehicles have?

9. On a late-model car, which of the following is the *most likely* to be operated by a cable?
   (A) Blend door.
   (B) Diverter door.
   (C) Heater shutoff valve.
   (D) None of the above.

10. Place the following HVAC developments in the order they occurred.
   _____ A. Automatic temperature control
   _____ B. Hot water heater
   _____ C. Flow through ventilation
   _____ D. Air conditioning
   _____ E. Replacement of R-12 by R-134a

ASE Certification-Type Questions

1. Technician A says changing refrigerant from a gas to a liquid causes it to absorb heat. Technician B says the system refrigerant is used over and over. Who is right?
   (A) A only.
   (B) B only.
   (C) Both A and B.
   (D) Neither A nor B.

2. All of the following statements about the refrigerant compressor are true, *except:*
   (A) the compressor is a pump.
   (B) the compressor moves the refrigerant through the system.
   (C) the compressor condenses the vaporized refrigerant.
   (D) refrigerant leaves the compressor as high pressure high temperature vapor.

3. The job of the desiccant is to _____.
   (A) lubricate the compressor
   (B) store extra refrigerant
   (C) quiet the compressor output
   (D) remove moisture from the refrigerant

4. All of the following statements about refrigerant are true, *except:*
   (A) refrigerant must be chemically stable.
   (B) refrigerant must not change from a gas to a liquid during normal refrigeration system operation.
   (C) refrigerant must be compatible with the compressor lubricant.
   (D) refrigerant must be nonflammable.

5. Technician A says the heater shutoff valve is closed when the engine is cold to allow quick engine warmup. Technician B says heater shutoff valve closing varies with the manufacturer. Who is right?
   (A) A only.
   (B) B only.
   (C) Both A and B.
   (D) Neither A nor B.

6. The blend door controls the _____ of the air entering the passenger compartment.
   (A) speed
   (B) moisture
   (C) temperature
   (D) direction

7. The modern HVAC control system controls of the following, *except:*
   (A) engine temperature
   (B) blower speeds
   (C) shutoff valve opening and closing
   (D) diverter door position

8. Technician A says cables are no longer used on most HVAC control systems. Technician B says many HVAC systems are electronically controlled. Who is right?
   (A) A only.
   (B) B only.
   (C) Both A and B.
   (D) Neither A nor B.

9. When the defrost or defog mode is being used, all of the following is happening, *except:*
   (A) the refrigeration system is operating.
   (B) air is flowing through the defroster vents.
   (C) the evaporator is cooling the incoming air.
   (D) the heater shutoff valve is closed.

10. At one time it was thought automotive air conditioning might have to be discontinued because of _____.
   (A) high costs
   (B) destruction of the ozone layer
   (C) depletion of sources of Freon
   (D) new and improved ventilation systems

Chapter 2

Shop Safety and Environmental Protection

After studying this chapter, you will be able to:
- ❑ Identify the major causes of accidents.
- ❑ Explain why accidents must be avoided.
- ❑ List ways to maintain a safe workplace.
- ❑ List safe work procedures.
- ❑ List refrigerant safety precautions.
- ❑ Identify and explain refrigerant first aid procedures.
- ❑ Identify types of environmental damage caused by improper shop practices.
- ❑ Identify ways to prevent environmental damage.

Technical Terms

Safety glasses

Face shield

Safety shoes

Work gloves

Material Safety Data Sheets
  (MSDS)

Carbon monoxide

Phosgene gas

Core value

Cross-contamination

Environmental Protection Agency
  (EPA)

There are many ways to do a job in an unsafe manner, and usually only one way to do it safely. This chapter covers reasons for and methods of doing things safely. The reason to make repairs in a safe manner is that it protects you, the vehicle, and the shop. Also covered in this chapter is the proper handling and disposal of waste products. Wastes must be handled properly to safeguard both you and the environment. This chapter examines various types of unsafe conditions and work practices, environmental violations, and how to correct or avoid them.

Causes of Accidents

No one usually tries to cause an accident. Accidents result in injuries that keep you from working or enjoying your time off. Some accidents can even kill. Even slight injuries are painful and annoying, and may impair your ability to work and play. Even if an accident causes no personal injury, it can result in property damage. Damage to vehicles or shop equipment can be expensive and time-consuming to fix, and could even cost you your job.

The usual cause of accidents is the attitude that it is just too much trouble to do things correctly. Often, the end result of this attitude is an accident. Unfortunately, even experienced technicians become rushed and careless. Falls, injuries to hands and feet, fires, explosions, electric shocks, and even poisonings occur in auto repair shops. Carelessness in the shop can also lead to long-term bodily harm from prolonged contact with harmful liquids, vapors, and toxic dust. Lung damage, skin disorders, and even cancer can result from contact with these substances. For these reasons, you must keep safety in mind at all times, especially when conditions tend to make it the last thing on your mind.

Personal Safety Equipment

To protect yourself from accidents and long-term harm, you should dress properly and have several types of personal safety equipment. This includes wearing proper clothing and wearing the appropriate eye, foot, and hand protection as needed.

Proper Clothing and Protective Equipment

Always dress appropriately and safely. Do not wear clothes with long, loose sleeves, open jackets, or scarves. They can get caught in moving parts and pull you into the machine or engine. Do not wear a tie unless your duty position requires it. If you must wear a tie, tuck it inside your shirt. If you wear your hair long, keep it away from moving parts by tying it up or securing it under a hat. Remove any rings or other jewelry. Not only can jewelry get caught in moving parts, it can cause a short circuit, which can result in severe burns or start a fire if caught between a positive terminal and ground.

Eye Protection

The shop environment contains many things which could be hazardous to your eyes. Flying particles can injure immediately, and exposure to dust and chemicals can cause long-term damage. You should have two types of eye protection, **safety glasses** and a **face shield.** See **Figure 2-1.** Safety glasses should be worn at all times while you are in the shop. Face shields should be worn when performing grinding or cutting operations. Other types of protective eyewear are available for specialized operations, such as welding.

Foot and Hand Protection

In any shop, there is always the danger of objects falling on your foot or having your foot crushed between two heavy objects. Therefore, you should wear foot protection when working in the shop. **Safety shoes,** preferably with steel toe inserts, should be worn at all times. Most good quality safety shoes are constructed using materials that are oil and chemical resistant. Safety shoes have soles that are not only slip resistant and insulated, but also provide support and comfort.

Always have a pair of **work gloves** available, to protect against hot objects, or parts with sharp or jagged edges. Gloves will also protect against skin rashes caused by exposure to chemicals.

Kinds of Accidents

The ways an accident can occur in the shop are many and varied. Most accidents are caused by one or a combination of two factors:
- ❑ Failure to maintain a safe workplace.
- ❑ Performing service procedures improperly.

Examples of failure to maintain a safe workplace are allowing tools and equipment to fall into disrepair, failing to dispose of old parts, containers, or other trash, and

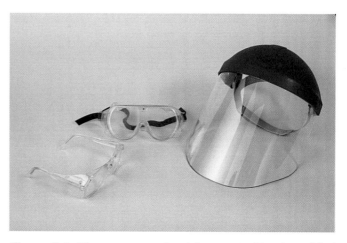

Figure 2-1. *There are very few job opportunities for a blind technician. Always wear eye protection.*

ignoring water or oil spills, **Figure 2-2.** Examples of improper service procedures are using the wrong tools or methods to perform repairs, using defective or otherwise inappropriate tools, not wearing protective equipment when necessary, and not paying close attention while performing the job.

The best way to prevent accidents is to maintain a neat workplace, use safe methods and common sense when making repairs, and wear protective equipment when needed. The following are some suggestions for reducing the possibility of accidents.

Maintaining a Safe Workplace

Return all tools and equipment to their proper storage places. This saves time as well as reducing the chance of accidents, damage, and theft. Do not leave equipment out where others could trip.

Keep workbenches clean. This reduces the chance of tools or parts falling from the bench onto the floor, where they could be lost or damaged. A falling tool or part could land on your foot. A clean workbench also reduces the possibility that critical parts will be lost in the clutter, or that a fire will start in oily debris.

Clean up spills immediately, before they get tracked around the shop. Many people are injured when they slip on floors coated with oil, antifreeze, or water. Gasoline spills can be extremely dangerous, since a flame or spark can ignite the vapors, causing a major explosion and fire, **Figure 2-3.**

Know what chemicals are stored on the shop premises. Chemicals include carburetor cleaners, hot tank solutions, parts cleaner, and even motor oil and antifreeze. Chemical manufacturers provide ***Material Safety Data***

Figure 2-2. *Oil spills like this may look like nothing much, but they can lead to serious accidents.*

Sheets, often called ***MSDS,*** for every chemical they produce. These sheets list all the known dangers of the chemical, as well as first aid procedures for skin or respiratory system contact. There should be an MSDS for every chemical in the shop. You should read the MSDS for any unfamiliar chemical before working with it. **Figure 2-4** shows a typical MSDS.

Make sure the shop is well lighted. Poor lighting makes it hard to see what you are doing, possibly leading to accidental contact with moving parts or hot surfaces. Overhead lights should be bright and centrally located. Portable lights, or drop lights, should be in operating condition and easy to use. Always use a "rough service" bulb in incandescent service lights. These bulbs are more rugged than normal light bulbs, and will not shatter if they

Figure 2-3. *This is what happens when a broken light bulb meets spilled gasoline. This fire was caused by a large gasoline spill meeting an open source of ignition.*

ACME Chemical Company

Material Safety Data Sheet

Product Name: Acetylene   **Revised 3/3/99**

24-hour Emergency Phone: Chemtrec 1-800-424-9300   Outside United States 1-905-501-0802

Trade Name/Syn: Acetylene   **NFPA Ratings**
Chemical Name/Syn: Acetylene, Ethyne, Acetylen, Ethine   Health: 0
CAS Number: 74-86-2   Flammability: 4
Formula: C_2H_2 Reactivity: 0

Hazards Identification

Simple Asphyxiant. This product does not contain oxygen and may cause asphyxia if released in a confined area. Maintain oxygen levels above 19.5%. May cause anesthetic effect. Highly flammable under pressure. Spontaneous combustion in air at pressures above 15 psig. Acetylene liquid is shock sensitive.

Effects of Exposure-Toxicity-Route of Entry

Toxic by inhalation. May cause irritation of the eyes and skin. May cause an anesthetic effect. At high concentrations, excludes an adequate oxygen supply to the lungs. Inhalation of high vapor concentrations causes rapid breathing, diminished mental alertness, impaired muscle coordination, faulty judgment, depression of sensations, emotional instability, and fatigue. Continued exposure may cause nausea, vomiting, prostration, loss of consciousness, eventually leading to convulsions, coma, and death.

Hazardous Decomposition Product

Carbon, hydrogen, carbon monoxide may be produced from burning.

Hazardous Polymerization

Can occur if acetylene is exposed to 250°F (121°C) at high pressures or at low pressures in the presence of a catalyst. Polymerization can lead to heat release, possibly causing ignition and decomposition.

Stability

Unstable—shock sensitive in its liquid form. Do not expose cylinder to shock or heat; do not allow free gas to exceed 15 psig.

Fire and Explosion Hazard

Pure acetylene can explode by decomposition above 15 psig; therefore the UEL is 100% if the ignition source is of sufficient intensity. Spontaneously combustible in air at pressures above 15 psi (207 kPa). Requires very low ignition source. Does not readily dissipate, has density similar to air. Gas may travel to source of ignition and flashback, possibly with explosive force.

Conditions to Avoid

Contact with open flame and hot surfaces, physical shock. Contact with copper, mercury, silver, brasses containing >66% copper and brazing materials containing silver or copper.

Accidental Release Measures

Evacuate all personnel from affected areas. Use appropriate protective equipment. Shut off all ignition sources. Stop leak by closing valve. Keep cylinders cool.

Ventilation, Respiratory, and Protective Equipment

General room ventilation and local exhaust to prevent accumulation and to maintain oxygen levels above 19.5%. Mechanical ventilation should be designed in accordance with electrical codes. Positive pressure air line with full face mask or SCBA. Safety goggles or glasses, PVC or rubber gloves in laboratory; and as required for cutting or welding, safety shoes.

Figure 2-4. *The MSDS shown here contains all information needed to safely use the listed chemical. It also contains information on responding to accidents involving the chemical.*

are dropped. Do not use a high wattage bulb in a portable light. Lightbulbs get very hot and can melt the light socket or cause burns. Service lights that use cool fluorescent bulbs are now available from some tool manufacturers, **Figure 2-5.**

Do not overload electrical outlets or extension cords by operating several electrical devices from one outlet. An overloaded outlet is shown in **Figure 2-6.** Do not operate high current electrical devices through extension cords. Frequently inspect electrical cords and compressed air lines to ensure they are in good condition, **Figure 2-7.** Do not close vehicle doors on electric cords or air lines. Do not run electrical cords through water puddles, or use them outside when it is raining.

Ensure all shop equipment, such as grinders, lathes, and drill presses, are equipped with safety guards, **Figure 2-8.**

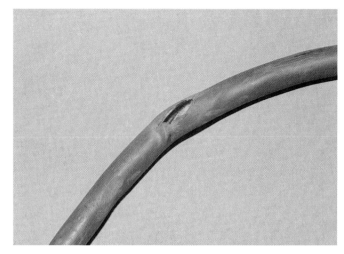

Figure 2-7. *Damaged insulation on an extension cord can cause a severe electrical shock or a short circuit that could lead to a fire.*

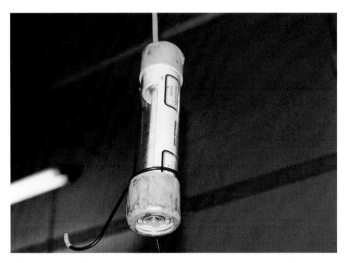

Figure 2-5. *Always use a rough service bulb in droplights, or any portable light. A rough service bulb is much cooler and reduces the chances of burns. This droplight uses a fluoroescent bulb.*

Figure 2-8. *Equipment safety guards should always be in place. Only remove a safety guard to service the equipment. The guard should be reinstalled before the equipment is reused.*

All shop equipment is equipped with guards by the manufacturer. These guards should never be removed, except for service operations such as changing the grinding wheels. When servicing any shop equipment, be sure it is turned off and unplugged. Read the equipment service literature before beginning any repairs.

Closely monitor tool and equipment condition and make repairs when necessary. This includes such varied things as replacing damaged leads on test equipment, checking and adding oil to hydraulic jacks, and regrinding the tips on screwdrivers and chisels.

Do not leave open containers of chemicals in the shop or outside. Most automotive chemicals will poison any animal (or person) who drinks it, and spills will create an extremely slippery floor.

Know where the shop fire extinguishers are located, and how to operate them. Make sure you know what type of fire extinguisher to use on what type of fire, **Figure 2-9.**

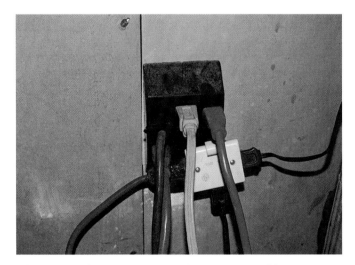

Figure 2-6. *This overloaded outlet, or "octopus," is a common cause of fires. A two socket electrical fixture cannot safely handle the electrical load caused by attaching this many electrical devices.*

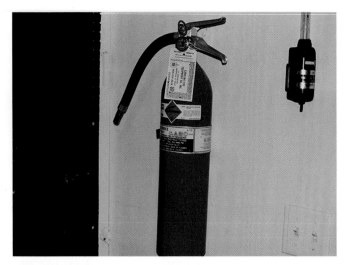

Figure 2-9. *Make sure you know where the fire extinguishers are located. Time during a fire is often lost looking for a fire extinguisher.*

Periodically check each fire extinguisher to ensure they are in working order and have them checked periodically by qualified personnel.

Performing Work Procedures Properly

Study work procedures before beginning any job that is unfamiliar. Do not assume the procedure you have used in the past will work with a different type of vehicle. Always work carefully. Speed is not nearly as important as doing the job right and avoiding injury. Avoid co-workers who will not work carefully, or who tend to engage in horseplay.

Use the right tool for the job. Using a screwdriver as a chisel or pry bar, or a wrench as a hammer, is asking for an accident or at least a broken tool. Never use a hand socket with an impact wrench. A hand socket can crack and shatter if used with an air tool. Do not use low quality tools or tools that are damaged. Use the right tool for the job.

Learn how to use new equipment before using it. This is especially true of impact wrenches, air chisels, and other air operated tools. It is also true for large electrical devices such as drill presses and brake lathes. These tools are very powerful, and can hurt you if they are used improperly. A good way to learn about new equipment is to start by reading the manufacturer's instructions.

When working on electrical systems, avoid creating a short circuit with a jumper wire or metal tool. Not only will this damage the vehicle components or wiring, it will develop enough heat to cause a severe burn or start a fire. Be careful when using a test light. Used improperly, they can cause as much or more damage than a non-fused jumper wire.

Lift safely. Make sure that you are strong enough to lift the object to be moved. Always lift with your legs, not your back. If an object is too heavy to lift by yourself, get help.

Do not smoke in the shop. You may accidentally ignite an unnoticed gasoline leak. There are other less noticeable flammable substances around every vehicle. Batteries can produce explosive hydrogen gas as part of their normal chemical reaction. A discarded cigar or cigarette can also ignite any oily rags or paper debris that may be lying around.

When using any type of vehicle lifting equipment, be sure to place the lift pads at the vehicle frame or other points that the manufacturer specifies as lifting points. Do not attempt to raise a vehicle with an unsafe or undercapacity jack. Always support a raised vehicle with good quality jackstands. Do not use a car jack as they are designed for emergency use only. Never use boards or cement blocks to support a vehicle. When using a post or hydraulic lift, be sure to place the lifting pads under the frame or on a spot that can support the vehicle's weight.

Do not leave a running vehicle unattended. The vehicle may slip into gear, or overheat while you are away. Whenever you must work on a running vehicle, set the parking brake. Do not run any engine in a closed area without good ventilation, even for a short time. **Carbon monoxide** can build up quickly, cannot be seen or smelled, and is deadly. When working on or near a running engine, keep away from all moving parts. Never reach between moving engine parts for any reason. Seemingly harmless parts such as the drive belts and fan can seriously injure you.

When road testing a vehicle, be alert, and obey all traffic laws. Do not become so absorbed in the diagnosis process that you forget to watch the road. Be alert for the actions of other drivers. If you must listen or observe a scan tool during a road test, get someone to drive the vehicle for you. Be aware of the type of tests and the effects they could have on the brake system. For example, some anti-lock brake systems are disabled whenever a scan tool is connected to the vehicle.

Accidents are not inevitable. It is up to you to notice and correct safety hazards and use safe work practices. This is the only way to prevent accidents.

Refrigerant Safety Precautions

The refrigerants used in HVAC service cause specific health dangers and require some special safety precautions. These safety precautions must *always* be taken when handling refrigerants.

Vapor and Liquid Refrigerants

Refrigerant can exist as a liquid or a vapor. Liquid refrigerants can cause skin and eye damage. Vaporized refrigerant is not hazardous in small amounts. Large amounts of vapor can cause breathing problems. The average vehicle air conditioning system does not contain enough refrigerant to cause serious breathing difficulties. A full refrigerant tank, however, does contain sufficient refrigerant that, when vaporized, can cause severe breathing difficulties. Avoid contact with any amount of liquid refrigerant, and large amounts of vaporized refrigerant.

Rules for Handling Refrigerants

Following are some general rules for refrigerant safety:
❏ If there is any possibility of contact with refrigerant, wear protective gloves.
❏ Always wear eye protection when working on any refrigeration system or cylinder containing a refrigerant.
❏ Do not discharge refrigerant in a closed shop or other enclosed area.
❏ Do not allow refrigerant to contact hot parts or an open flame. Burning refrigerant will produce poisonous **phosgene gas.**
❏ If an accidental refrigerant discharge occurs, leave the area until all danger of liquid or vapor contact is past.
❏ Be familiar with first aid for refrigerant exposure.

First Aid for Refrigerant Exposure

Three kinds of personal injury that can result from exposure to refrigerants include:
❏ Skin damage.
❏ Eye damage.
❏ Breathing difficulties.

Each type of injury must be handled by different procedures. Sometimes more than one type of injury occurs, and all symptoms must be treated. First aid procedures for these three types of injuries are given in the following paragraphs.

First Aid for Skin Exposure

Before beginning treatment, determine if any liquid refrigerant has saturated the victim's clothing. Remove any refrigerant soaked clothing immediately using rubber gloves. Allow the refrigerant to evaporate before touching the clothes again. Treatment for refrigerant exposure is the same as for frostbite—slowly warming the frozen skin. Follow these procedures:
❏ Bathe or soak the exposed skin in lukewarm water. Water temperature should be no more than 110°F (44°C). *Do not* use hot or very cold water. Skip this step if the exposure occurred more than a few minutes before treatment starts.
❏ Do not rub the affected skin, and do not cover the skin with anything other than a clean bandage.
❏ Do not break any blisters.
❏ Keep affected fingers or toes from touching, and do not allow the victim to use the appendage.
❏ Take the victim to a doctor or emergency room.

First Aid for Eye Exposure

Exposure to vaporized refrigerant, or oil suspended in the refrigerant, may cause minor eye irritation. If liquid refrigerant enters the eye, serious damage, including loss of sight, can occur. In cases of eye exposure, prompt medical attention is needed. If liquid refrigerant contacts the eye, take these steps:
❏ Immediately begin flushing the eye with running water. Tap water is okay, but do not use very hot or cold water.
❏ Continue flushing the eye for several minutes.
❏ Take the victim to a doctor or emergency room.

First Aid for Breathing Difficulty

While refrigerants are not toxic, large amounts can displace the oxygen in the air. If a large amount is directly inhaled, the refrigerant can damage lung tissue and possibly displace the air in the lungs. Refrigerant is heavier than air, and will sink to the lowest parts of a closed building, such as alignment or grease pits. If large amounts of any refrigerant are accidentally released, do the following:
❏ Immediately open doors and windows and start any building ventilation fans. If the building cannot be properly ventilated, get everyone out.
❏ Closely monitor the condition of all affected persons for signs of oxygen deprivation. Symptoms include difficult or labored breathing, headache, dizziness, incoherent speech, and signs of confused thinking. Severe cases may develop a bluish tint to the skin or fingernails.
❏ If a person shows any signs of oxygen deprivation, take them outside immediately.
❏ If anyone shows signs of severe oxygen deprivation or worsening of the above symptoms, take them to a doctor or emergency room.

Preventing Environmental Damage

Automotive repair shops cause more than their share of environmental damage. Shops often carelessly dispose of solid and liquid wastes, and cause damage to the atmosphere by failing to use proper repair procedures.

Typical solid wastes produced by automotive repair shops are scrap parts, tires, and cardboard boxes. Liquid wastes include antifreeze, brake fluid, cleaning solvents, motor oil, and transmission fluid. Improper repair procedures include allowing refrigerant gases to escape into the atmosphere and repairing engines so they emit excessive amounts of pollutants.

Toxic materials lower the air and water quality, and can even affect food supplies. The effects of poisoned air, water, and soil may be noticed almost immediately, or may take decades to become apparent. The health and financial burdens of irresponsible waste disposal, if left unchecked, will grow ever larger. Even if we escape the consequences, future generations will not.

Prevent Solid Waste Contamination

It is almost impossible to keep a shop completely spotless. Old parts and containers are an inevitable part of automotive service. You should make every effort to keep solid wastes from piling up around the shop. Not only is it unsightly, it is dangerous and expensive. The following are a few suggestions for dealing with solid wastes. In certain parts of the United States, some of these are rigidly enforced laws.

Recycle parts and scrap materials whenever possible. Do not throw away parts boxes, old tires, and salable scrap metals unless you are sure they cannot be recycled. If solid wastes cannot be recycled, dispose of them responsibly, not by illegal dumping or burning.

It makes good economic sense to recycle, since almost every rebuildable part has a return value, usually called a **core value.** The value of paper, scrap tires, and scrap metals such as aluminum, iron, and brass depends on the current market conditions, but it always has some value. Check with your local parts supplier to determine which parts can be sent back for rebuilding. Recyclers are often listed in your local telephone book, and can give you advice on what to do with recyclable materials.

Prevent Liquid Waste Contamination

Probably the most common way an automotive shop damages the environment is by allowing used motor oil, transmission fluid, antifreeze, brake fluid, or gear oil to seep into the ground. This immediately contaminates the soil. In addition, these liquids sink further into the ground with every rain, eventually contaminating the underground water table. This underground water could be your local drinking water source.

Do not attempt to solve the liquid waste problem by pouring liquid wastes into drains. Municipal waste treatment plants cannot handle antifreeze, brake fluid, or petroleum products. Most liquids also contain poisonous additives and heavy metals absorbed from the vehicle during use. In most areas, dumping any of the above liquid wastes into city drains is illegal.

In many areas, local waste management companies will accept used oil and antifreeze for recycling. Liquid wastes should be stored in a sealed above ground storage tank or in 55 gallon drums. See **Figure 2-10.** Once, collected, the used oil and antifreeze is re-refined and reused. Some used oil is burned by power plants to produce electricity, eliminating the oil and reducing the dependence on imported crude oil. Some shop heaters are powered by used motor oil. A recently developed process converts old motor oil into diesel fuel.

Figure 2-10. *A—Above ground tanks should be used to store liquid waste. B—Drums and other storage containers are used to store old oil filters.*

Prevent Gaseous Waste Contamination

Gaseous waste is the harmful vapors or gases produced as a result of automotive service operations. These gases are harmful to the atmosphere. The two most common ways automotive shops hurt the atmosphere are:
- ❑ Tampering with vehicle emission control systems.
- ❑ Discharging refrigerants into the atmosphere.

Emission Control Service and the Atmosphere

Adjusting carburetors for a richer mixture, changing the manufacturer's timing settings, and disconnecting emission controls will all increase emissions. Some seemingly harmless actions, such as installing a lower temperature cooling system thermostat or a non-stock air cleaner, can also cause a rise in engine and vehicle emissions. Not only are these actions illegal, they almost never increase power and mileage as much as hoped.

HVAC Service and the Atmosphere

In the late 1980s, the HVAC service industry came under scrutiny for its practice of venting refrigerants such as R-12 (Freon) into the atmosphere. However, R-12 was shown to contribute to ozone layer damage and is no longer manufactured.

All studies have shown refrigerants such as R-12 cause extensive damage to the ozone layer, leading to increased ultraviolet ray damage. How the ozone layer is damaged is shown in **Figure 2-11.** However, R-12 should still be used to service older refrigeration systems while it is plentiful.

It also makes good economic sense to recover and reuse refrigerants (R-12 refrigerant now costs over seven times as much as it did just a few years ago). Do not perform air conditioning repairs if you do not have a refrigerant recovery and recycling machine, **Figure 2-12.** Also read carefully the refrigerant information in Chapter 6.

Another way HVAC technicians allow refrigerant to escape is to improperly attach equipment to the refrigeration system. Before attaching or removing any hose from a refrigeration system fitting, make sure the shutoff valves are in the closed position. Be careful not to mix refrigerants in containers or in a refrigeration system. R-134a containers are always light blue, while R-12 containers are white. This minimizes the chances of *cross-contamination.*

Contaminated or unfamiliar refrigerant must be stored in a special container. Contaminated refrigerant containers are painted gray with a yellow top. Contaminated refrigerant should be shipped to a reclaiming facility for recycling or disposal. Many local refrigeration equipment and refrigerant suppliers can provide a list of the closest reclaiming facilities.

The Environmental Protection Agency

It is a federal crime, with severe penalties to:
❏ Discharge refrigerant into the atmosphere.
❏ Tamper, remove, or disable engine emission controls.

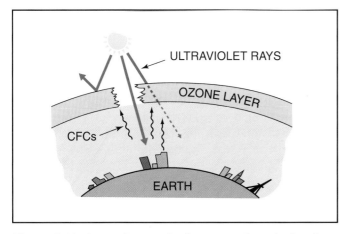

Figure 2-11. *Loss of ozone in the stratosphere is the direct result of CFC release. CFC released eventually reaches the stratosphere. Sunlight breaks up the CFC molecule forming chlorine. Chlorine breaks up the ozone molecule. Fewer ozone molecules allow more ultraviolet radiation to reach the earth. The additional ultraviolet radiation results in an increase chance for skin cancer. (Toyota Motor Corp)*

Figure 2-12. *Refrigerant recovery and recycling helps to protect a shop's profit margin, as well as the environment. (RTI Technologies Inc.)*

These emissions laws are enforced by the **Environmental Protection Agency**, usually called the **EPA.** The EPA investigates suspected violations, and often conducts "sting" operations to catch violators. In addition, some states such as California have additional laws protecting the environment.

Additional information about waste disposal and vehicle emissions can be obtained from the Environmental Protection Agency. The EPA has ten regional offices and six field offices. For the address of the nearest EPA office, write or call:

Automotive/Emissions Division
United States Environmental Protection Agency
401 M Street S. W.
Washington, DC 20460

Accidents are not inevitable. It is up to you to notice and correct safety hazards and use safe work practices. This is the only way to prevent accidents. It is also up to you to eliminate environmental damage by taking the proper steps to reduce the improper disposal of wastes. Always use common sense when working in the shop, and avoid people who do not.

Summary

People cause accidents through carelessness. Many technicians get in a hurry and forget to do the job safely. An accident may result in personal injuries, long-term bodily harm, or damage to equipment or property. No mature and competent automotive technician wants to be injured or cause property damage.

Many accidents are caused when technicians fail to correct dangerous conditions in the work area, such as oil spills or tripping hazards. Other accidents are caused when technicians try to take shortcuts instead of following proper repair procedures.

Prevent accidents by maintaining a neat work place, using proper methods of repair, and using protective equipment when needed. It is up to the technician to study the job beforehand, work safely and prevent accidents. Always use common sense when working on vehicles, and avoid people who do not.

Careless production and disposal of wastes causes environmental damage. Wastes can take the form of liquids, solids, or gases. Anyone or any shop can be a cause of pollution. The two main ways in which an automotive shop can cause environmental damage are carelessly disposing of wastes, and repairing vehicles in such a way they pollute the atmosphere.

Environmental rules should always be followed to prevent damage to the air, water, or soil. In many cases, federal and state law requires proper disposal of wastes and proper vehicle repairs.

Review Questions—Chapter 2

Please do not write in this text. Write your answers on a separate sheet of paper.

1. People cause accidents when they become _____.

2. Accidents often happen when technicians try to take _____ instead of following proper repair procedures.

3. Why should you keep workbenches clean?

4. Manufacturers provide _____ for every chemical they make.

5. There should be a separate electrical outlet for every piece of equipment that draws a lot of _____.

6. List three precautions to take when working with electrical cords and compressed air lines.

7. When can guards on shop equipment be removed?

8. Always lift heavy objects with your _____, not your _____.

9. List two results of exposure to liquid refrigerant.

10. List the following types of waste as S (solid), L (liquid), or G (gas).

    Used motor oil _____

    Antifreeze _____

    R-12 refrigerant _____

    Old coolant pump _____

    R-134a refrigerant _____

    Plastic oil jug _____

ASE Certification-Type Questions

1. Technician A says carelessness is the usual cause of accidents. Technician B says defective equipment is the usual cause of accidents. Who is right?
   (A) A only.
   (B) B only.
   (C) Both A & B.
   (D) Neither A nor B.

2. A drill press is an example of which of the following?
   (A) A totally safe piece of equipment.
   (B) Equipment that should be studied before using.
   (C) Extreme high voltage equipment.
   (D) All of the above.

3. The manufacturer provides a(n) _____ for all dangerous chemicals.
   (A) opener
   (B) Material Safety Data Sheet
   (C) glass container
   (D) salesman

4. If all tools and equipment are returned to their proper storage places, which of the following will be the *least likely* to occur?
   (A) They will be easy to find.
   (B) Someone will steal them.
   (C) No one will trip over them.
   (D) They will be broken.

5. Technician A says it is okay to use a 12-point socket with an impact wrench when the air supply to the impact wrench is weak. Technician B says a 12-point socket should be used with a hand wrench only. Who is right?
   (A) A only.
   (B) B only.
   (C) Both A & B.
   (D) Neither A nor B.

6. Using a standard light bulb in place of a "rough service" bulb in a drop light is being discussed. Technician A says this could cause a severe burn. Technician B says a standard bulb is easier to break. Who is right?
   (A) A only.
   (B) B only.
   (C) Both A & B.
   (D) Neither A nor B.

7. Ethylene glycol antifreeze should be disposed of properly because it can cause:

    (A) asphyxiation.

    (B) poisoning.

    (C) fires.

    (D) damage to paint and concrete.

8. Carbon monoxide is most dangerous at which of the following locations?

    (A) Outdoors.

    (B) In the upper atmosphere.

    (C) A closed shop.

    (D) A well-ventilated shop.

9. Technician A says you can mix refrigerants. Technician B says refrigerants must have their own separate canister, and therefore, cannot be mixed. Who is right?

    (A) A only.

    (B) B only.

    (C) Both A & B.

    (D) Neither A nor B.

10. Technician A says an automotive shop can cause damage to the atmosphere by repairing vehicles so they pollute the atmosphere. Technician B says an automotive shop can cause damage to the atmosphere by discharging refrigerants into the atmosphere. Who is right?

    (A) A only.

    (B) B only.

    (C) Both A & B.

    (D) Neither A nor B.

To diagnose air conditioning system leaks, an electronic leak detector is a must for all technicians. You should have a leak detector that can trace leaks in R-134a and R-12 systems. (Yokogawa)

Chapter 3

HVAC Tools, Equipment, and Service Information

After studying this chapter, you will be able to:
- ❏ Identify HVAC system diagnostic and test equipment.
- ❏ Identify types of refrigeration system service equipment.
- ❏ Explain the concept of dedicated refrigeration system service equipment.
- ❏ Identify engine cooling system test and service tools and equipment.
- ❏ Identify HVAC control system service tools.
- ❏ Identify HVAC and cooling system service information.

Technicial Terms

Gauge manifolds
Manifold body
Hand valves
Analog
Digital
Manifold hoses
Temperature gauges
Mechanical
 temperature gauge
Electronic
 temperature gauge
Leak detectors
Electronic leak
 detector
Dye

Halide flame
 detector
Soap solution
Refrigerant identifier
Test light
Non-powered test
 light
Powered test light
Multimeters
Voltmeter
Ohmmeter
Ammeter
Oscilloscopes
Belt tension gauge
Vacuum pump

Vacuum gauge
Snap ring pliers
Spanners
Pressure test fittings
Hose tools
Hose cutters
Crimping tools
Barb fitting
Beadlock fitting
Orifice tube tool
Oil injectors
Recovery/recycling
 equipment
Dedicated machines

Combination
 machines
Blowgun
Closed flushing
 systems
Evacuation pumps
Vacuum
Inches of mercury
 Hg
Micron
Charging scale
Charging station
Air purging
 equipment
Pressure tester

Antifreeze tester
Test strips
Combustion leak
 tester
Factory manual
General manual
Specialized manuals
Schematics
Troubleshooting charts
Technical service
 bulletins (TSB)
Telephone hotlines
Compact disc-read only
 memory (CD-ROM)

Refrigeration and general HVAC system service requires specialized tools as well as common hand and air-operated tools. This chapter covers all the specialized tools and test equipment needed to service modern HVAC systems. Studying the tools in this chapter and becoming familiar with their purposes and use will assist you when the service chapters call for their use. Make a special effort to understand the types of service literature. Almost no part of a modern vehicle can be serviced without the proper service information.

Diagnostic and Test Equipment

The following equipment must be obtained to perform any refrigeration or other HVAC system diagnosis. This equipment is also used to perform some HVAC service and replacement.

Gauge Manifolds

Gauge manifolds are the most basic of all refrigeration system tools. The gauge manifold is used as both a diagnosis and a service tool. The technician must have R-134a and R-12 manifolds to service modern vehicles. Many shops with refrigerant service machines also have one or more gauge manifolds to make pressure checks when the machine is being used on another vehicle. Manifold gauges are used to remove contaminated or unknown blend refrigerants from the air conditioning system, reducing the chance for cross-contaminating a service machine. In shops that perform a large volume of air conditioning work, technicians often have their own gauge manifolds as part of their toolset.

All gauge manifolds have the same basic parts, although there are some variations among manufacturers. The major parts of a common gauge manifold are shown in **Figure 3-1**. Refer to this figure as you read the following paragraphs.

Manifold Body and Hand Valves

The *manifold body* is made of brass or aluminum. Passages are drilled in the body to connect the other manifold parts. Some manifold bodies have a sight glass to observe the flow of refrigerant. *Hand valves* are used to control the flow of refrigerant through the passages of the manifold body. A cross-section of the manifold body and hand valves is shown in **Figure 3-2.** Note the internal passages are arranged so the gauges can read refrigeration system pressures when the valves are closed.

The hand valves used on R-12 and R-134a gauge manifolds are usually arranged in the same way, or on a slant or in front of the manifold. These different arrangements make manifold identification easier. Valve wheels for the high and low sides are identified by color. The low side hand wheel is made of blue plastic or has a blue decal in its center. The high side hand wheel is made of red plastic or has a red decal.

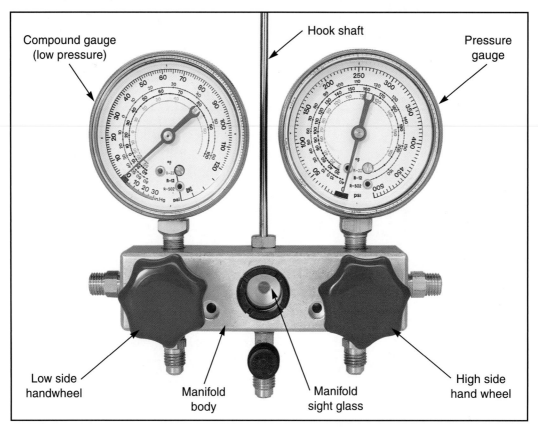

Figure 3-1. *The refrigerant pressure gauge manifold is the universal tool for HVAC service. Gauge manifolds should be carefully handled to maintain their calibration. (TIF Instruments)*

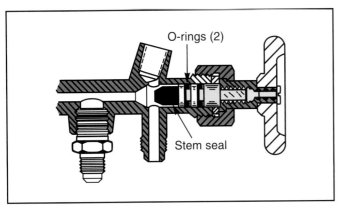

Figure 3-2. *This cross-section of a gauge manifold shows the internal passages that allow the manifold to be used to check pressure, add and remove refrigerant, and many other functions. (Robinair)*

All gauge manifolds have high and low pressure gauges. A few older gauge manifolds have three gauges. The third gauge was used to measure compressor output. Older high side gauges are calibrated from 0-500 psi (3445 kPa). Newer high side gauges may be calibrated from 0-250 psi (1723 kPa). Low side gauges are calibrated from 0 to 100-250 psi (689-1723 kPa). In addition to the pressure scale, low side gauges have a provision for measuring 0-29.9″ of vacuum (approximately 50 microns). Vacuum measurements are explained later in this chapter.

R-134a and R-12 Gauges

There are no major differences between R-134a and R-12 gauges. The internal operation of each type of gauge is the same. The difference is the calibration of each gauge, and the markings on the gauge faces. R-134a and R-12 gauge manifolds cannot be interchanged.

Gauges

The gauges used with a refrigeration gauge manifold are either *analog* (indicator needle), which resemble other pressure gauges, or *digital.* See **Figure 3-3.** In an analog gauge, the position of the needle in relation to the numbers on the gauge face indicates the pressure or vacuum in the refrigeration system. Digital gauges provide a numerical reading indicating system pressure or vacuum.

Manifold Hoses

Manifold hoses are tubes of high strength nylon or fabric cord covered by neoprene rubber. Most hoses are rated to withstand 500 psi (3445 kPa) pressures. At each end of the hose is a connector that allows it to be attached to the gauge manifold and the refrigeration system. Hoses used with a gauge manifold have connectors designed to match the refrigerant being measured by the

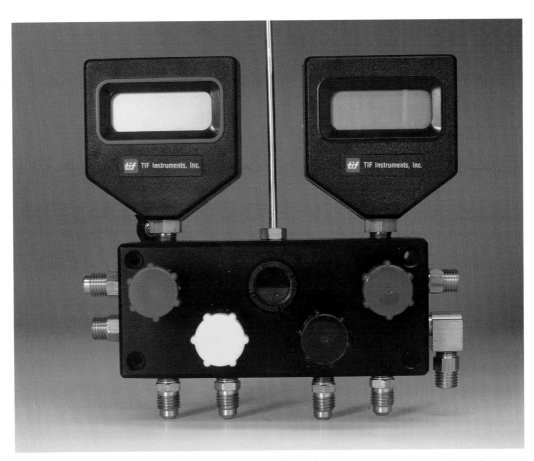

Figure 3-3. *Digital manifold gauges are available. They give accurate pressure readings, however, they do not provide some of the diagnostic advantages of analog gauges.*

gauges. Compare the R-134a and R-12 hose connectors in **Figure 3-4.** These hoses can be replaced if damaged or worn, however, they are not interchangeable. Most hoses are colored blue for the low side and red for high. The center hose is usually yellow.

Hose Fittings and Adapters

Many gauge manifolds are equipped with fitting adapters, shutoff valves, and other features. These are attached to the hoses. Typical valves are used to isolate refrigerant present in the hoses. This keeps refrigerant loss to a minimum.

R-12 refrigeration systems made since 1986 have different size service fittings to prevent the technician from accidentally crossing the high and low side connections. Adapters are used to allow the same gauge manifold to be used on these refrigeration systems. See **Figure 3-5.** Adapters are not needed on R-134a systems since they use standardized fittings.

Attaching and Reading Gauges

To attach and read the gauge manifold, first make sure the hand valves are closed. Then remove the refrigeration service fitting caps and attach the hoses. Remember the blue hose attaches to the low side of the system, and the red hose is connected to the high side.

R-134a and R-12 hose connectors have different fittings, and therefore, different attachment methods. R-12 fittings are threaded. The hose connector is also threaded and is screwed onto the fitting. R-134a fittings are somewhat similar to those used on air hoses. The hose connector is a quick disconnect with an isolation valve. Once the

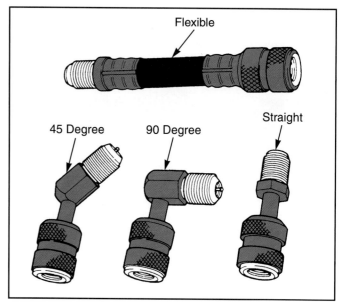

Figure 3-5. *Some R-12 systems made in the 1980s and early 90s require special adapters. These adapters are usually needed to access the high side of the refrigeration system. (Ford)*

connectors are installed, open the hose isolation valves as needed. Do not open the hand valves on the manifold body.

Once the hoses are in place, you can read the static pressures. Static pressures are the pressures in the refrigeration system when the system is not operating. If the system has been off for 30 minutes or more, the high and low side pressures should be almost the same (within a few pounds or kPa). Other uses for the gauge manifold will be covered in the chapters where they apply.

Temperature Gauges

Temperature gauges are used to test the temperature of the air exiting the HVAC system vents and to check the temperature of the engine coolant. The two main types of temperature gauges are mechanical and electronic. They are discussed in the following paragraphs.

Mechanical Temperature Gauge

The *mechanical temperature gauge,* **Figure 3-6,** relies on an internal bimetal spring to register temperature differences. The bimetal spring is a coil of wire made of two kinds of metal. Each metal expands at a different rate as its temperature changes. Therefore, changes in temperature cause the coil to tighten and loosen. The coil is attached to the gauge pointer. Movement of the pointer against the gauge face indicates the temperature.

Electronic Temperature Gauge

The *electronic temperature gauge* uses *infrared waves* to measure temperature. Infrared waves are waves similar to light waves, although they cannot be seen.

Figure 3-4. *Hose connectors vary by refrigerant type. A—R-12 connectors. B—R-134a connectors. Note the difference between the two connectors.*

Figure 3-6. *The mechanical temperature gauge shown here is useful for determining air temperatures at the outlet vents.*

Temperature changes in an object cause the infrared waves given off by the object to change. A sensor in the gauge reads the change in infrared waves as a temperature change. Internal circuitry converts sensor readings to digital temperature readouts. See **Figure 3-7.**

Leak Detectors

Leaks in the refrigeration system will cause the system to lose its refrigerant charge. Loss of refrigerant will cause the system not to work properly and could also damage the ozone layer. In many cases, it is difficult to determine the exact location of the leak as well as how severe it is. *Leak detectors* are needed to accurately locate leaks. The various kinds of leak detectors are discussed in the following sections.

> **Note: Chapter 15 contains more information on the use of leak detectors.**

Electronic Leak Detectors

Using an *electronic leak detector* is the most accurate way of locating leaks. Electronic detectors use a small solid state sensor that can detect extremely small leaks. The detector also has a probe used to draw refrigerant into the sensor. The probe tip may contain a filter to catch oil and debris. Most electronic leak detectors will make a ticking noise which increases in frequency as the probe encounters refrigerant. Large leaks raise the ticking to a high pitched squeal. Many electronic leak detectors have an LED (light emitting diode) display which indicates the leak rate. The detector may use different color LEDs or may progressively illuminate extra LEDs as the refrigerant concentration increases. Some electronic leak detectors can automatically determine the type of refrigerant in the system.

A typical electronic leak detector, **Figure 3-8,** always uses a small internal battery to power the unit. The detector will also have an on-off switch, and may contain a range selector switch to allow for checking large and small leaks.

To use an electronic leak detector, turn on the detector switch. Adjust the sensitivity to produce an occasional ticking. Then pass the detector probe end under the suspected refrigerant leak areas. Since refrigerant is heavier than air, it will flow downward from a leak. If refrigerant is leaking, the detector rate of ticking will increase. Large leaks will cause a high pitched squeal. When through using the leak detector, turn the control switch to the off position and replace the detector in its case.

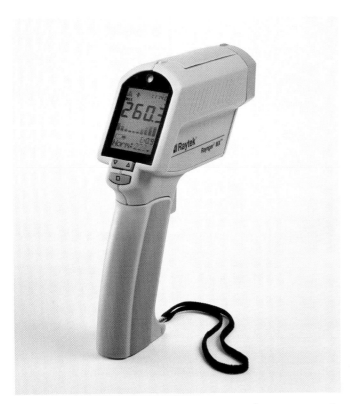

Figure 3-7. *An infrared temperature gauge gives an accurate and almost instant temperature reading. (Raytek)*

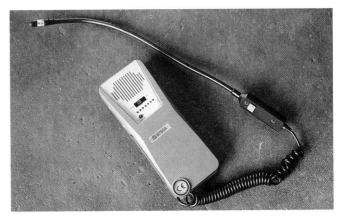

Figure 3-8. *Electronic leak detectors are needed to find small leaks on modern systems. Most detectors can locate both R-134a and R-12 leaks.*

Dye

Dye is used to locate minute leaks. The dye is injected into the refrigeration system and allowed to circulate for a few minutes. Some of the dye will leak out along with any refrigerant and stain the components at the site of the leak.

Older refrigerant dyes were colored orange and were contained in a small can resembling a one pound refrigerant can. The can was connected to the system low side through the gauge manifold. With the system operating, the dye was drawn into the system. After the dye circulated for a few minutes, the technician could look for the presence of orange dye. Dye cans are still used in some areas.

Modern dye injectors are designed to inject a fluorescent dye directly into the refrigeration system, **Figure 3-9A.** The injector is attached to one of the system service ports and the handle is turned to force the dye into the system. After the dye has circulated through the operating system for a few minutes, the technician shines an ultraviolet light, such as the one shown in **Figure 3-9B,** on the suspected leak points. If any of the dye has leaked out, it will glow under the ultraviolet light.

Halide Flame Detector

The **halide flame detector** is not as accurate as an electronic detector. It will, however, detect relatively large refrigerant leaks. The halide detector consists of a propane

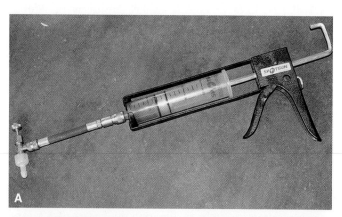

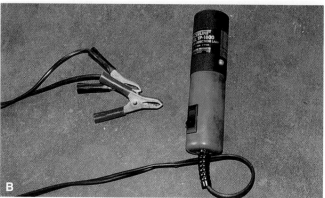

Figure 3-9. *A—Dyes are injected into the system using a device such as the one shown here. B—A black light will illuminate the dye as it leaks out with the refrigerant.*

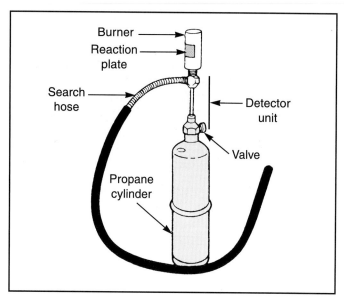

Figure 3-10. *The flame type leak detector, while seldom used today, will detect most medium sized leaks. It must be used carefully to prevent false readings and maintain shop safety. (Ford)*

cylinder attached to a burner, **Figure 3-10.** A sensing hose draws refrigerant from the suspected leak area. A copper reaction plate improves the combustion process between the refrigerant, propane, and air.

⚠ **Warning: Halide flame leak detectors are very dangerous. They should only be used if no other detection method is available. A refrigerant identifier should be used before using a halide detector to reduce the chance of fire or an explosion. They should only be used with R-12 systems.**

To use a halide leak detector, light the burner. After a few minutes operation, the reaction plate will glow dull red. Adjust the flame as necessary, then pass the free end of the sensing hose under any suspected leak areas. If refrigerant is present, the flame will change color. Small leaks will cause the flame to develop a greenish tint. A large leak will cause the flame to turn bright blue.

When the leak detecting process is finished, tightly close the propane valve and allow the tester to cool before returning it to storage. The propane valve should be closed tightly when the flame detector is not in use.

Soap Solution

The **soap solution** method will find large leaks only, and should not be relied on to locate small leaks or leaks in inaccessible locations. It is primarily used to confirm what appears to be an obvious leak. Soap solution is sometimes the only detection method available if a system has been filled with a refrigerant other than R-134a or R-12.

A soap solution can be made by mixing a small amount of dishwashing liquid or other soap with water. The solution is then sprayed or poured on the suspected leak area. Leaking refrigerant will form bubbles. The size of the bubbles and how rapidly they form will increase with the size of the leak. Slight foaming will occur at the site of a small leak, while large bubbles will be seen at a serious leak. If large bubbles form at a rate faster than one per second, the leak can be considered severe.

Refrigerant Identifiers

To avoid contaminating the recycling equipment with incorrect or contaminated refrigerant, many air conditioning specialists identify the refrigerant before beginning service. A *refrigerant identifier,* **Figure 3-11,** is used to determine what kind of refrigerant is installed in a refrigeration system or storage container. Refrigerant identifiers are usually designed to tell whether the refrigerant is R-134a, R-12, or an unknown blend. Some refrigerant identifiers can also determine the percentage of each type of refrigerant, and identify contaminated refrigerant.

Disposing of Contaminated Refrigerant

When contaminated or unfamiliar refrigerant has been found, it must be stored in special containers pending its disposal. Contaminated refrigerant containers are gray with a yellow top. Contaminated refrigerant containers should be shipped to a reclaiming facility for recycling or disposal. Storage and recycling of contaminated and unfamiliar refrigerant was discussed in Chapter 2. More information on refrigerants is located in Chapter 6.

Electrical Test Equipment

The HVAC system contains many electrical components, wire harnesses, and electrical connectors that require testing. The technician will frequently have to diagnose electrical devices and wiring.

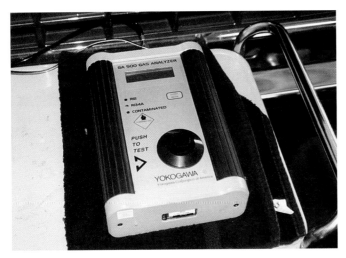

Figure 3-11. *A refrigerant identifier should always be used before recovering refrigerant.*

Note: Electrical values such as voltage and resistance will be explained in more detail in Chapter 4.

Test Lights

The ***test light*** is often used to check whether electricity is reaching a particular point in an electrical circuit, or to detect a circuit not allowing current to flow. The ***non-powered test light,*** **Figure 3-12,** can be used to probe electrical circuits to determine whether voltage is present. A ***powered test light*** resembles the non-powered light but has an internal battery. The battery supplies an electrical power source to determine whether a circuit is complete. Test lights can be useful when working on various parts of the HVAC electrical system, but must be used with care when working on any electronic system. The test light has the potential to severely damage electronic circuits. Avoid using test lights unless specifically instructed by the HVAC system manufacturers' service literature.

Figure 3-12. *Non-powered test lights are useful for determining if voltage is present in a circuit.*

Multimeters

Multimeters, such as the one in **Figure 3-13,** are devices for reading electrical values. Modern multimeters can read all common electrical values (voltage, resistance, and amperage). Many modern multimeters are able to read voltage waveforms and provide other information. Modern multimeters are *digital* types that display the electrical reading as a number. Analog multimeters use a needle, which moves against a calibrated background. Modern multimeters contain the individual meters discussed in the following paragraphs.

Note: Only digital multimeters should be used for air conditioning and other automotive work. Make sure any meter or test light has a minimum of 10 meg ohm impedance.

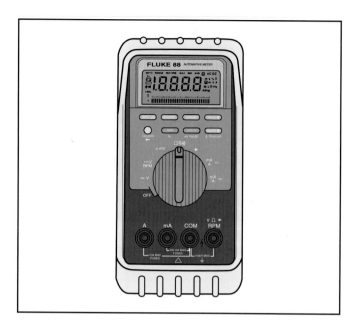

Figure 3-13. *Multimeters contain several electrical testers, such as voltmeters, ammeters, and ohmmeters. (Fluke)*

Voltmeter

The **voltmeter** section of the multimeter is connected to read voltage from a circuit. To read the voltage at an electrical connection, connect the leads to the positive side of the circuit and a ground. If necessary, select the proper voltage range, then observe the reading.

The voltmeter can also be connected to read the voltage across a connection as current flows through it. If the connection has high resistance, current will try to flow through the meter, creating a voltage reading. Voltage higher than the specified figure means the connection must be cleaned or replaced.

Ohmmeter

An **ohmmeter** can be used to check electrical resistance values. Ohmmeters can also be used to check for complete circuits. To make an ohmmeter check, turn on the multimeter and set it to ohms.

 Caution: Electronic components can be damaged by careless use of multimeters. Always check the manufacturers' literature before testing any electronic part.

Most modern digital ohmmeters will select the correct range automatically. Attach the leads to the wires or terminals to be tested. When checking wires or relay contacts for continuity, the resistance should be at or near zero. Other parts, such as motor or solenoid windings and temperature sensors should have a specific amount of resistance. If the reading is zero or infinity, the part is defective. The resistance of temperature and sunload sensors should change with changes in temperature or exposure.

Ammeter

The multimeter can usually check amperage (or amps). The amp setting of the multimeter is called an **ammeter.** Most multimeters have two separate ampere lead ports, one for low amps and one that can usually measure up to ten amps. For measuring greater amperage flows, many modern ammeters can be equipped with an inductive pickup. The pickup is clamped over the current carrying wire. The pickup reads the magnetic field created by current flowing through the wire and converts it into an amperage reading.

Oscilloscopes

Oscilloscopes have been used in diagnosing gasoline engines for years. With the addition of electronic controls on HVAC systems, they can be used as a diagnostic tool to check the waveform patterns from sensors and outputs. The newer oscilloscopes are small, hand-held devices that can be taken on road tests, **Figure 3-14.** With additional probes and adapters, they can be used for a variety of diagnostic tasks. The latest scopes have on-board memory functions, which can be used to capture and store waveform patterns for comparison to good patterns.

Belt Tension Gauges

If the compressor belt is loose, it may squeal when the compressor clutch is engaged. If the belt is too tight, it will quickly wear out and may damage the compressor and clutch bearings, water pump, or the engine. To correctly

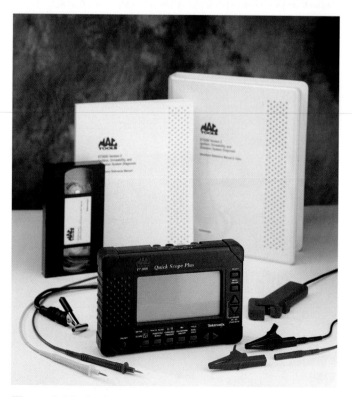

Figure 3-14. *Oscilloscopes can be used to diagnose problems in computer circuits. (MAC Tools)*

3Figure 3-15. *This belt tension gauge provides a direct reading of belt tension. (Ford)*

tighten the compressor drive belt, a ***belt tension gauge,*** **Figure 3-15,** is needed. To use the belt tension gauge, use the handle to push the center lever away from the two side levers. Slide the tool between the belt and release the handle. Then read the tension on the gauge.

Another type of belt tension gauge is shown in **Figure 3-16.** This type of tension gauge measures belt tightness by measuring the belt deflection under a certain pressure. To use this kind of tension gauge, turn the gauge sleeve until the end of the sleeve is at the proper belt tightness marked on the barrel. Then push the metal bar downward until it lines up with the line on the lower end of the barrel. Next place the gauge on a flat section of belt and push downward until the metal sleeve reaches zero. If the metal bar is pushed above the gauge line, the belt is loose.

Figure 3-16. *The belt tension gauge shown here is set to the proper reading and then the belt is deflected.*

Vacuum Pumps/Gauges

The hand-held ***vacuum pump*** and ***vacuum gauge,*** **Figure 3-17,** is used to apply and measure the effect of vacuum on HVAC system diaphragms and control valves. The pump assembly develops vacuum, which is measured on the gauge. To use the gauge, remove the hose to the vacuum operated device. Then operate the pump to apply vacuum to the device. Observe the gauge. If vacuum cannot be developed, or bleeds away rapidly, the unit is leaking. Also check for operation of the related linkage as vacuum is developed. If the linkage does not move when the vacuum is increased, the linkage or the door is stuck.

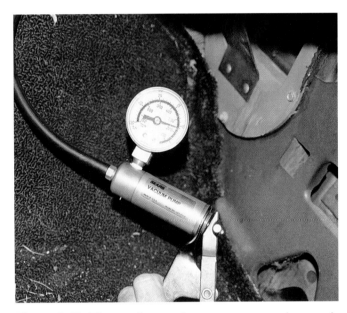

Figure 3-17. *The small manual vacuum pump and gauge is used to produce a vacuum for checking vacuum operated accessories such as air door vacuum diaphragms and vacuum heater shut off valves. It can also be used to check the operation of vacuum switches.*

HVAC System Service Tools

Once the problem has been located, the HVAC system must be repaired. Many repairs can be made with ordinary hand tools. Many service jobs require wrenches with large jaw openings. These large size wrenches are used to loosen the large fittings used on many refrigerant lines. A few HVAC service jobs require special tools.

Compressor Service Tools

Most shops prefer to replace entire compressors rather than repair them. If a compressor is being repaired, the following tools may be needed to service the seals and internal parts. Servicing older compressors may require additional special tools. The technician should consult the proper service manual for information about these tools.

Snap Ring Pliers

Special size and shape **snap ring pliers** are sometimes needed to remove compressor pressure switches, pulleys, and shaft seals. Some of these pliers are shown in **Figure 3-18.** In many cases, standard or universal snap ring pliers can also be used.

Compressor Service Tools

Compressor clutches and shaft seals cannot be removed with conventional tools. Several tools are needed to remove these and other compressor parts. They include special pliers to remove internal snap rings and holders to replace the seals and O-rings. See **Figure 3-19.**

If a replacement for an older compressor is not available, the original compressor may need to be overhauled instead of replaced. To overhaul a compressor, special internal service tools are needed.

Compressor Clutch Service Tools

Compressor clutch service often requires special tools to hold the clutch in place while other tools are used. These holding tools are sometimes called **spanners.** Other tools are used to remove pulleys or clutch hubs from the compressor. Additional tools may be needed to press new clutch pulleys or clutch hubs onto the compressor or clutch hub, as shown in **Figure 3-20.**

Pressure Test Equipment

To save unnecessary work, it is best to pressure test complex parts such as compressors after repairs are complete, but before the compressor is reinstalled. Special **pressure test fittings** allow the technician to pressurize the compressor on the bench. To prevent environmental damage, the compressor should be pressurized with nitrogen or another inert gas. **Figure 3-21A** shows a typical pressure tester using nitrogen. Nitrogen is supplied in large tanks such as the one in the left of the photo. This tester also can be used to test hoses. Hose testing adapters are shown in **Figure 3-21B.**

Hose Tools

In some cases, a refrigeration hose has to be manufactured. Sometimes the exact replacement hose is no longer available. In other cases, making a hose is much cheaper than obtaining an exact replacement. Making hoses requires the use of hose cutters and crimping tools.

Cutters

The refrigeration hoses should be cut accurately and cleanly to make a good seal and to prevent hose debris from entering the refrigeration system. **Hose cutters** are necessary to make a clean, straight cut. Typical hose cutters are shown in **Figure 3-22.**

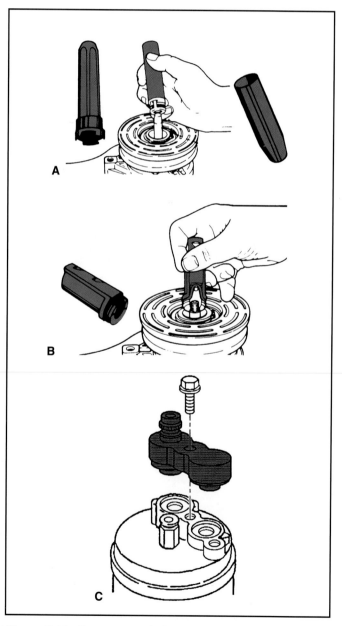

Figure 3-19. Compressor shaft seal tools include special snap ring pliers as well as O-ring and seal tools. A—Seal remover and protector. B—O-ring installer. C—Compressor leak adapter.

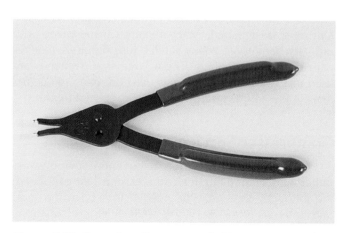

Figure 3-18. Snap ring pliers are needed to remove snap rings from inaccessible locations.

Crimping Tools

Crimping tools are made to crimp, or form, a fitting around a hose end. The major difference between types of crimping tools is the size and shape of the collets. The collets actually contact the fitting and crimp it to the hose. There are two common kinds of hose fitting types, the **barb fitting** and the **beadlock fitting.** A different crimping machine is used to make each fitting type. Each machine and its related parts are dedicated and cannot be used to make the other type of crimped fitting.

Crimping tools can be operated by hand or by hydraulic pressure. A typical hand operated crimping tool is shown in **Figure 3-23.** Hand crimping tools are inexpensive and can do a good job of making a crimped hose. The hydraulic powered crimping tool, **Figure 3-24,** uses a small hydraulic pump to create the pressure needed to operate the hydraulic piston.

To use either type of crimping tool, select the proper hose and hose fittings and lightly oil them. Then select the proper collets and place them into the crimping machine. Assemble the fitting on the hose end and place the hose and fitting in the crimping machine. Operate the machine to crimp the fitting. After the crimping operation is complete, make sure the crimp was made properly.

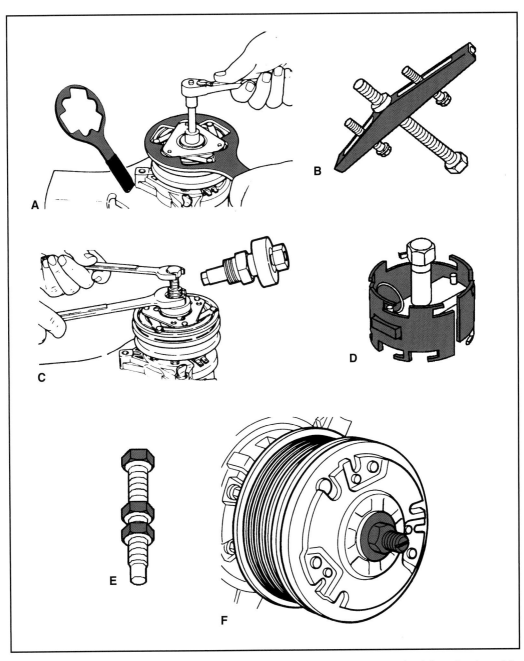

Figure 3-20. *Clutch holding tools are often needed to keep the compressor clutch from turning while the center bolt or nut is removed. A— Hub holding tool. B—Pulley puller. C—Hub and drive plate remover and installer. D—Compressor pulley puller. E—Hub and drive plate remover/installer. F—Clutch hub installation tool. (Kent Moore)*

Figure 3-21. *A—Nitrogen should be used to check for leaks. Nitrogen is an inert gas and will not harm the atmosphere. B—Adapters are used to check hoses for leaks.*

Orifice Tube Tools

Many systems require an ***orifice tube tool*** to remove the orifice tube from the evaporator inlet. A typical orifice tube tool is shown in **Figure 3-25.** To use the orifice tube removal tool, the refrigeration system must be discharged. Once the refrigerant is removed, disconnect the fitting at the evaporator inlet and insert the orifice tube tool. Slightly turn the tool to engage the tangs of the tool and tube, then

withdraw the tube. To install the new orifice tube, place it on the tool, then insert the tool and tube into the evaporator inlet. Slightly twist the tool to disengage it from the tube, then withdraw the tool.

Oil Injectors

Oil injectors are used to install lubricating oil in the refrigeration system without discharging the system. There are two kinds of oil injectors in common use. To use the type shown in **Figure 3-26A,** fill it with the proper type of compressor oil. Then with the HVAC system off, install the injector service fitting. Turn the forcing screw at the top of the injector to force the oil into the system. Then remove the injector from the fitting.

Another type of oil injector, **Figure 3-26B,** is installed in the hoses of the gauge manifold. To use this type of injector, pour the proper type of oil into the reservoir. Then attach the oil injector to one of the manifold hoses. Attach the hoses to the refrigeration system and purge the hoses and injector as necessary. Then start the engine and place the HVAC system in maximum cooling. Allow the refrigeration system pressures to stabilize. Next, slightly open the high and low side valves. The difference in pressures will force the oil into the low side of the refrigeration system. Allow the system to operate long enough for all oil to enter the system, then close the valves and remove the injector and gauge manifold from the refrigeration system.

Refrigerant Service Equipment

The following section covers large shop equipment used to service the refrigeration system. During air conditioner service, the refrigeration system may need to be emptied of refrigerant, flushed of contaminants, placed under a vacuum, and recharged.

Figure 3-22. *Bench mounted hose cutters will make a clean straight cut. Some technicians prefer to use a hand cutter.*

Figure 3-23. *Manual crimping tools can be used to make acceptable hose connections. However, most technicians prefer to use power crimping tools to ensure a leak proof seal.*

Figure 3-24. *This power crimping tool uses hydraulic pressure to crimp the hose. Power crimping tools are easier to use and provide a more positive seal.*

Recovery/Recycling Equipment

Recovery/recycling equipment is used to recover and recycle the refrigerant in a system. Refrigerant must be recovered and recycled to meet federal laws. Recycling equipment removes the refrigerant from a refrigeration system and stores it for reuse. When the refrigerant is needed, the equipment recycles (reinstalls) it into the same or another vehicle refrigeration system.

Most recovery and recycling machines also clean and dehumidify refrigerant for immediate reuse. The refrigerant may be stored in a standard 30 pound cylinder or in a separate charging tank. Most machines are designed to recycle R-134a or R-12 only. Recycling equipment is attached to the vehicle refrigeration system through hoses in the same manner as a gauge manifold.

Dedicated Machines

Different types of refrigerants cannot be mixed. Therefore, each refrigerant service machine can only be used with one type of refrigerant. These machines are called *dedicated machines.* An air conditioning service shop must have a separate dedicated machine for *each* type of refrigerant. In most shops, this means every machine used for R-134a must have a companion machine for use with R-12. If the shop uses a third refrigerant, a separate machine must be used. Some newer units are combination types, which can service systems using either type of refrigerant. Typical recovery/recycling machines for R-12 and R-134a are shown in **Figure 3-27.**

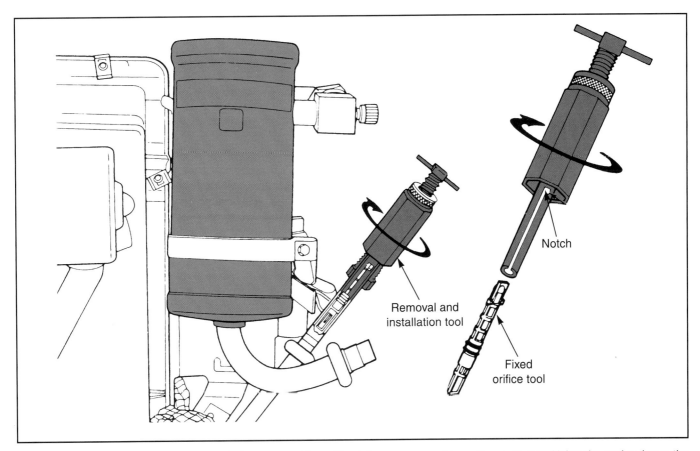

Notch

Removal and installation tool

Fixed orifice tool

Figure 3-25. *Orifice tube tools are needed to remove the orifice tube on many refrigeration systems. Using the tool reduces the chance of damage to the tube and fitting, and of leaving debris in the system. (Ford)*

Note: Vacuum pumps and charging scales are not dedicated units. Their operating procedures and settings may, however, be different for different refrigerants.

Combination Machines

Many modern machines are **combination machines.** A combination machine combines many of the components discussed in this chapter into a single unit. Most combination units have gauges, vacuum pumps, storage tanks, charging scales, and various refrigerant pressure and contamination indicators. The operation of these units is

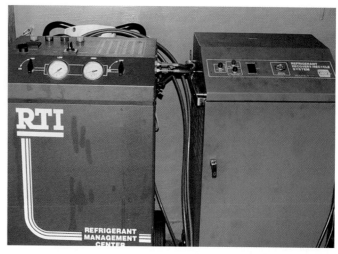

Figure 3-27. *Dedicated HVAC service machines can perform all refrigeration service operations for one type of refrigerant. The R-12 machine is red, the R-134a machine is blue.*

the same as for individual units. However, selecting switches at the combination machine control console performs all of their functions.

Flushing Equipment

Occasionally a contaminated refrigeration system must be flushed. There are two ways to flush a system, *open* and *closed loop*. Open loop flushing is done with an air-operated blowgun, while closed loop flushing is performed with special flushing equipment. Both types are explained in the following paragraphs.

Air-operated Guns

An air-operated rubber tip *blowgun*, **Figure 3-28,** is often used for open loop flushing. Some versions of the blowgun use a tank and hose design that injects the solvent into the system using air pressure, **Figure 3-29.** Open loop flushing is always done with refrigerant compatible solvents. To use an air-operated gun for flushing, recover the refrigerant and disconnect the fittings from the component to be flushed. Attach a drain hose to the inlet opening of

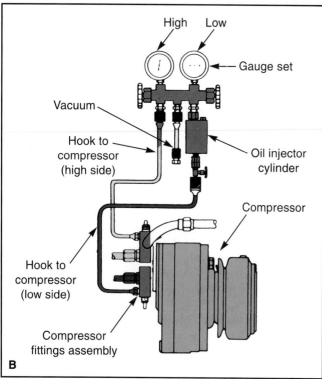

Figure 3-26. *A—This oil injector uses a screw operated piston to push oil into the refrigeration system. B—This oil injector makes use of the pressure difference between the high and low sides of the refrigeration system. (General Motors)*

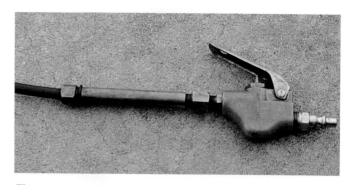

Figure 3-28. *A blowgun can be used to open flush the refrigeration system and is often used generally for removing debris from air intakes and evaporators. Always wear eye protection when using a blowgun.*

the component, then pour about one pint of solvent into the outlet opening of the component. Use the blowgun to direct compressed air or nitrogen into the outlet end of the component. Use no more than 100 psi (689 kPa) air pressure. Always blow in reverse direction to the refrigerant flow first to loosen as much debris as possible. Add solvent and repeat until only clean solvent comes out. After the component has been reverse flushed, make at least one pass in the forward direction.

Closed Flushing Systems

Closed flushing systems are machines that attach to the refrigeration system, and flush the system without taking any components loose. Closed flushing systems are special refrigerant cycling machines, or recovery and recycling machines adapted to direct the solvent through the system.

Closed loop flushing will not open completely blocked passages. It will remove oil and some contamination. To perform closed loop flushing, connect the machine to the system and follow the manufacturer's instructions.

Evacuation Pumps

Evacuation pumps are used to remove water (sometimes called moisture) from the refrigeration system.

Figure 3-29. *Tank and hose flushing equipment. Flushing compound is pushed through the refrigerant component by air pressure or nitrogen, and enters the bucket for disposal. (Robinair)*

Lowering the pressure of water allows it to evaporate, or boil, at a temperature much lower than its normal boiling point. As the pressure goes down, so does the boiling point. Creating a vacuum in the system also removes as much air as possible from the system.

How Vacuum Is Measured in the Air Conditioning System

Vacuum is a measurement of the pressure difference between two places. Any pressure below atmospheric pressure (14.7 psi or 101 kPa at sea level) is a vacuum. Vacuum is measured in two ways.

Inches of Mercury

The term **inches of mercury**, or **Hg**, refers to the ability of the pressure difference to change the level in a column of mercury. The higher the inches of mercury, the less atmospheric pressure.

Microns

A **micron** is a very small unit of pressure. One micron of pressure can move a column of mercury one millionth of a meter, or about 1/25,000th of an inch. Therefore, a micron is a unit of measure about 25,000 times smaller than an inch of mercury. While inches of mercury is a measurement of less than atmospheric pressure, microns directly measure atmospheric pressure. A chart showing the relationship between inches of mercury and microns is given in the Useful Tables section of this textbook.

Electric Pumps

The most common and most efficient type of evacuation pump is electrically operated, **Figure 3-30.** An electric pump can produce a vacuum of about 29.99" (300 microns). These pumps are equipped with fittings that allow a gauge manifold or other service equipment to be connected to the pump and then to the refrigeration system.

Figure 3-30. *Electric pumps are the best and most common method of completely evacuating the refrigeration system. (Snap-On)*

Air Pressure Pumps

The air pressure evacuation pump, **Figure 3-31,** uses what is known as a venturi effect to operate. The flow of compressed air through an air pressure operated pump creates a small vacuum that is used to draw air from the refrigeration system. Air pressure operated pumps are cheaper but less effective than electrically operated models. Most air-operated pumps produce less vacuum than an electric pump, anywhere from one half to one inch less. This type of pump must be allowed to operate for longer periods than an electric model.

Recharging Equipment

Once all repairs are completed, the refrigeration system must be recharged. The simplest method of recharging the system is to use a gauge manifold, discussed earlier in this chapter. Remember from Chapter 2 that R-134a containers are always blue, while R-12 containers are white. Be careful not to mix refrigerants.

Charging Scale

The **charging scale** allows the technician to charge a refrigeration system with the proper amount of refrigerant by weight. Charging by weight is the most accurate way to recharge a system. The charging scale somewhat resembles a bathroom scale, **Figure 3-32.** Most charging scales have digital readouts. To use a charging scale, place the refrigerant container on the scale platform and record its weight. Then add refrigerant to the system until the weight loss equals the amount of refrigerant to be added to the

Figure 3-32. *A simple charging scale enables the technician to charge the refrigeration system by weight. (TIF Instruments)*

system. Some charging scales allow you to program the amount of refrigerant to be added to a system. Once the scale is activated, it will signal when the proper weight of refrigerant has been added.

Charging Stations

A **charging station** combines the features of other refrigeration service equipment into a single unit. The typical charging station has gauges and connecting hoses, a storage area for refrigerant cylinders, and a weighing scale or other device for ensuring the proper amount of refrigerant is installed. **Figure 3-33** shows a typical charging station.

Air Purging Equipment

Most **air purging equipment** is built into refrigeration service devices such as recovery/recycling machines and charging stations. Purging is done automatically by the internal circuits of the device.

A few charging stations are equipped with a manual purging device. This device consists of a dual needle gauge, **Figure 3-34.** When both needles are in the same position, all air has been purged. When the needles are in different positions, the refrigerant cylinder must be purged. To purge the cylinder, open the cylinder valve until the needles are in the same position.

Engine Cooling System Test Equipment

The cooling system is sometimes a source of HVAC problems. Cooling system test equipment is used to check the condition of the cooling system and the coolant. Typical cooling system test equipment is discussed in the following paragraphs.

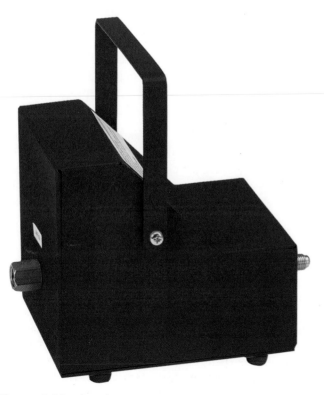

Figure 3-31. *An air pressure operated vacuum pump is cheaper but less efficient than an electric pump. (Snap-On)*

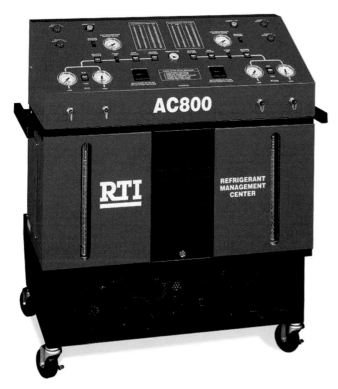

Figure 3-33. *A modern charging station contains all of the devices necessary to refill a refrigeration system. (RTI)*

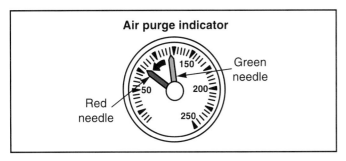

Figure 3-34. *Some charging stations use a manual air purging device. The technician vents the tank until the red and green needles line up. Most charging stations air purging devices are automatic. (Nissan)*

Pressure and Leak Testers

The quickest way to locate coolant leaks is to use a *pressure tester,* **Figure 3-35.** The pressure tester contains a pump to pressurize the cooling system. With the system pressurized, coolant will leak out. The technician can observe the leak and determine its source. The pressure tester also contains a gauge, which allows the technician to place the proper amount of pressure on the system, and helps in locating leaks in the engine or other inaccessible places.

To use the pressure tester, make sure there is no pressure in the cooling system, then remove the radiator cap.

 Warning: The cooling system must be depressurized and cool before performing this test.

Figure 3-35. *This coolant pressure tester is shown installed on a vehicle. Cooling system leaks can be observed visually, or by watching for a pressure drop. To test the radiator cap, special adapters are used attach the cap to the pressure tester.*

Add coolant if the system is low. The cooling system must be full for the leak test to work properly. Next, install the pressure tester on the radiator filler neck. If necessary, use the proper adapter to ensure a good seal. On some vehicles, the coolant reservoir is pressurized and the tester must be installed on the reservoir cap. Apply pressure until the gauge shows the pressure rating stamped on the radiator cap.

After pressurizing the cooling system, place the gauge in a location that will not bend the attaching hose. Observe the gauge for several minutes. If the gauge needle begins to drop, a leak is present in the system. Next, check the engine and cooling system for dripping coolant. Slight leaks may be located even when the gauge remains steady. After pressure testing the system, slightly bend the hose at the filler neck fitting to remove pressure. Then remove the tester from the filler neck, clean off the fitting, and store the tester it in its case.

Antifreeze Testers

If the engine coolant contains incorrect percentages of antifreeze and water, problems may occur. Too little antifreeze may cause the coolant mixture to freeze in cold weather, and will not protect the system properly against corrosion. Too much antifreeze in the mixture may actually raise the freezing point.

To check the exact percentages of antifreeze in coolant, an *antifreeze tester* is needed. There are two kinds of antifreeze testers, the float and the spectrograph. The float type, **Figure 3-36A,** makes use of the fact water and antifreeze have different weights. The weight of a liquid is usually called its specific gravity. Since antifreeze is heavier than water, a greater amount of antifreeze in the coolant will cause the float to rise higher in the mixture. The float is calibrated and the percentage of antifreeze can be read by observing how high the float rises in the mixture. Some hydrometers have a thermometer that allows the user to compensate for coolant temperature, **Figure 3-36B.**

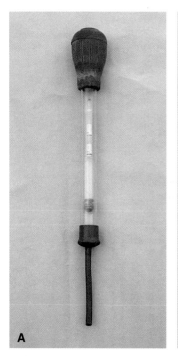 

Figure 3-36. *A—A hydrometer is the tester most commonly used to check the concentration of antifreeze in cooling systems. Hydrometers are simple to use and relatively accurate. B—Some hydrometers have a temperature-correcting device that compensates for specific gravity between hot and cold coolant.*

The spectrograph type of antifreeze tester depends on the refractive (light bending) properties of the coolant mixture. Coolant is drawn into the spectrograph and observed through a prism. The optical pattern of the coolant determines the amount of water and antifreeze in the system. **Figure 3-37** shows a typical spectrograph antifreeze tester.

Test Strips

Test strips can also be used to test antifreeze concentration. Test strips consist of a chemically treated paper strip that is dipped into the vehicle radiator filler neck, **Figure 3-38.** The color change of the test strip indicates the amount of antifreeze present.

To use a test strip, make sure the engine has thoroughly cooled off. Then remove the radiator cap or reservoir cap as necessary. Dip one of the test strips into the coolant for about one second. Remove the strip, shake off the excess antifreeze, and wait until the strip changes color. Waiting time is usually about 15-30 seconds. Then compare the color of the strip with the color chart on the strip package. As a general rule, a darker strip indicates a higher percentage of antifreeze in the coolant.

Combustion Leak Testers

Internal engine combustion leaks are caused by blown engine head gaskets or by cracked heads or blocks. Combustion leaks will affect coolant circulation and heater operation, cause coolant to be pushed out of the radiator, and will contaminate the cooling system internals with

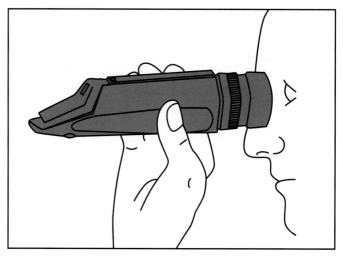

Figure 3-37. *Spectrograph type antifreeze testers are accurate and simple to use. (Leica)*

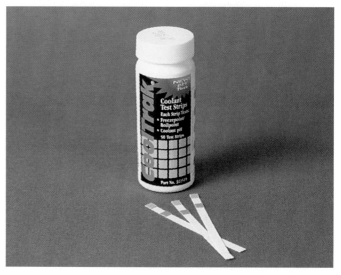

Figure 3-38. *Coolant test strips can be used to test coolant for concentration, as well as pH, which indicates the coolant's acidic level. (MVCC, Jack Klasey)*

exhaust byproducts. To prevent further damage, combustion leaks must be corrected as soon as possible.

It is sometimes difficult to tell when an engine develops an internal combustion leak. If the leak is small enough there are no telltale bubbles in the coolant, or if the leak has not allowed water to get into the engine oil, the technician must use a **combustion leak tester**, **Figure 3-39.** To use the combustion leak tester, attach it to the radiator filler neck. With the gauge attached to the filler neck, start the engine and allow it to idle. Exhaust gases will cause the chemical in the tester to change color.

Service Literature

The technician must refer to many sources of service literature to properly service HVAC systems. While modern HVAC systems operate from the same basic principles,

Figure 3-39. *A combustion leak tester may be needed to determine whether a cracked engine component or leaking gasket is causing exhaust gases to enter the cooling system. (Snap-On)*

they are complex and vary between manufacturers. Minor changes may be made between model years, often changing diagnostic and service procedures. The service literature described in the following paragraphs can simplify HVAC service by providing the latest information.

Factory Manual

The *factory manual* is published by the vehicle manufacturer or a publishing house contracted by the manufacturer. It contains all necessary service information for that one vehicle. **Figure 3-40** shows some typical factory service manuals. Most modern factory service manuals now come in volume sets for one vehicle. The major drawback to the factory manual is its relatively high cost, compared to the limited range of vehicles it can be used with.

Figure 3-40. *Factory manuals are used for one vehicle only. Most newer ones come in volume sets. (DaimlerChrysler, MVCC, Jack Klasey)*

While this type of manual is extremely detailed, it may not be the best choice if only one system, such as the brakes, is to be serviced.

General Manual

The *general manual* contains the most commonly needed service information about many different makes of vehicles, such as brake, engine, and transmission specifications, fuse replacement data, and sensor locations. General manuals also contain procedures for preventive maintenance and minor repairs.

At one time, general service information for every vehicle could be covered in one manual. Today, due to the large number of different vehicles available, this is no longer possible. Modern general manuals are divided into automobile and light truck editions. Publishers further divide their general automotive manuals into US, European, and Asian models.

The individual chapters of general manuals are grouped according to vehicle make, or several makes that are similar mechanically. Chapter subsections are devoted to particular areas of each make. General manuals also contain separate sections covering repair procedures that apply to all vehicles, such as engine overhaul, brake service, and starter/alternator overhaul. The major disadvantage of these manuals is the necessity of eliminating most of the information on specialized vehicle equipment, sheet metal, and interior.

Specialized Manual

Specialized manuals cover one common system of many types of vehicle. These manuals are often used to cover such topics as computerized engine controls, electrical systems, or brakes. They combine some of the best features of the factory and general manuals. They are often a good choice for servicing one particular system on many different makes and models of vehicles. One example of this type of manual is shown in **Figure 3-41.**

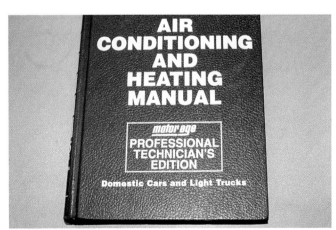

Figure 3-41. *Many service manual publishers, such as Motor and Mitchell publish overhaul manuals. In addition, overhaul manuals are available from vehicle manufacturers.*

Schematics

Schematics are pictorial diagrams which show the path of energy through a system. This energy can take the form of electricity, vacuum, air pressure, or hydraulic pressure, **Figure 3-42.** Schematics do not show an exact replica of a system, but instead indicate the flow or process within the system. Some schematics show the exact flow of a form of energy while others show the general process of a particular system. Schematics are often included as part of a service manual, or may be supplied separately.

Tracing the flow through a schematic makes diagnosis easier by showing the exact path of electricity or other form of energy. Each line represents a single wire in the vehicle's wiring harness. The schematic lines are labeled with numbers to colors to correspond with a specific color, or color and color stripe combination on the actual wires. The path can be traced by carefully following the lines from component to component. Always carefully note the color designations of the wires and any stripes or bands to ensure you are following the correct wire.

Troubleshooting Charts

Troubleshooting charts are summaries, or checklist versions, of the troubleshooting information about a particular vehicle or system. Although the information is found in a longer form elsewhere in a service manual, the troubleshooting chart allows the technician to quickly reference the problem, the possible cause, and the solution. **Figure 3-43** shows a typical troubleshooting flowchart. Some troubleshooting charts are arranged with the problem on the left-hand side of the page, the possible cause in the middle, and the corrective action on the right-hand side.

Technical Service Bulletins

Frequently, manufacturers issue **technical service bulletins (TSB),** for newer vehicles to their dealership personnel. These bulletins contain repair information that is used to describe a new service procedure, correct an unusual or frequently occurring problem, or update information in a service manual. Many of the phone

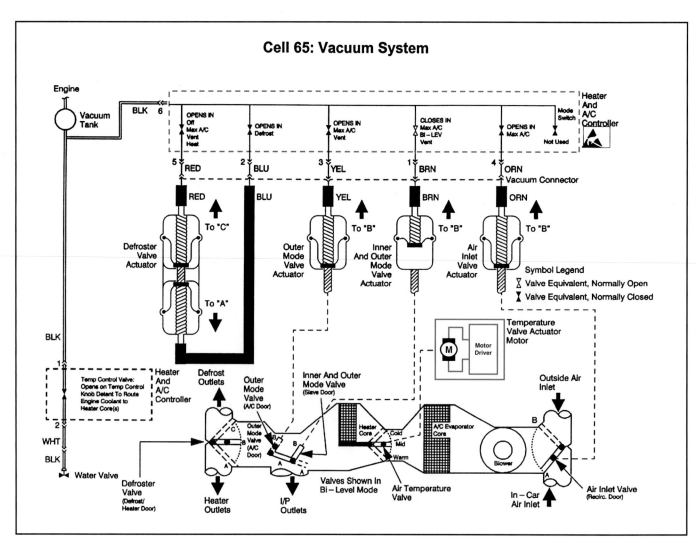

Figure 3-42. *A common electrical schematic such as the one shown above is a road map for the electricity in a circuit. As part of the job, the HVAC technician is often called on to interpret schematics. (General Motors)*

hotline and computerized assistance services receive these bulletins. They are a very good source of information to repair an unusual or frequently occurring problem. Subscriptions to these bulletins are also available through various services.

Telephone Hotlines

If all other sources of information have been exhausted, the technicians may be able to call a *telephone hotline.* Some vehicle manufacturers, part suppliers, and

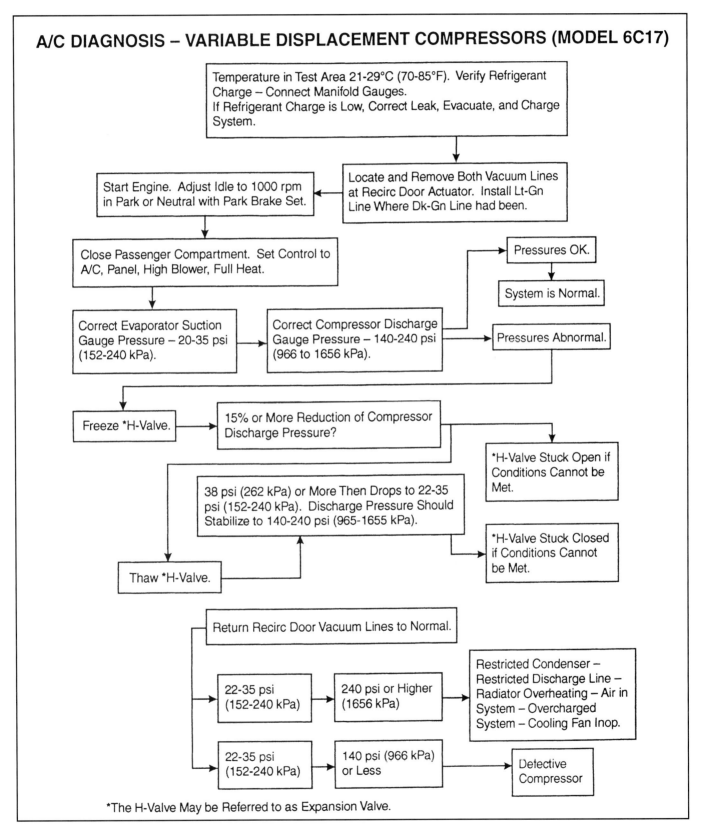

Figure 3-43. *If this troubleshooting chart is followed closely, it will almost always give you the correct diagnosis. (Four Seasons)*

service manual publishers provide technical support services over a technical hotline. Calling these hotlines will connect you with a technical support person. Hotline personnel often have information gathered from actual repair and diagnosis situations. This is a way of obtaining real life information that would otherwise not be available. Manufacturer hotlines will also have access to the latest update information from manufacturers' engineering departments.

Some vehicle manufacturers' hotlines are available only to the technicians who work for a manufacturer's dealership. Other hotlines are available by subscription. These hotlines can be accessed after a yearly fee is paid. Some parts manufacturers' hotlines are available to anyone. These hotlines are intended to help the technician who has questions about the manufacturers' parts.

Computerized Assistance

A few years after computer control came to the automobile, manufacturers began to put computer driven analyzers into the service departments of their dealerships. These were simply computers that were joined with a dedicated engine analyzer system designed to diagnose problems in that particular manufacturer's line of vehicles. Unfortunately, the first computer driven analyzers were not much better than traditional engine analyzers. Also, some technicians were very apprehensive to use the new analyzers, since many of them had little or no exposure or training in computer usage.

Improvements in software and computer technology have greatly improved the computerized analyzer. The newest computerized analyzers have user-friendly menus with touch screen capability or a standard computer mouse, which can be used to "point-and-click" on any particular menu selection. The newest analyzers can also be used on more than one manufacturer's line of vehicles. These computer driven analyzers, **Figure 3-44,** are able to communicate with the various computers on most late-model vehicles, as well as perform all standard diagnostic procedures. The analyzer's information can be updated through a computer-to-computer modem connection over the telephone to a central computer at the manufacturer's service headquarters.

One feature of these analyzers is technicians seeking service information can access the manufacturer's main computer and type in the VIN number of the vehicle in question. The computer then provides a printout of all service information, including any technical service bulletins or recall campaigns applicable to that particular vehicle. Aftermarket tool manufacturers have begun to computerize their engine analyzers, as well as their alignment equipment and exhaust gas analyzers.

Unfortunately, most small shop owners cannot afford these large, expensive computerized analyzers nor are they allowed to interface with a manufacturer's central computers. However, some software and tool companies are

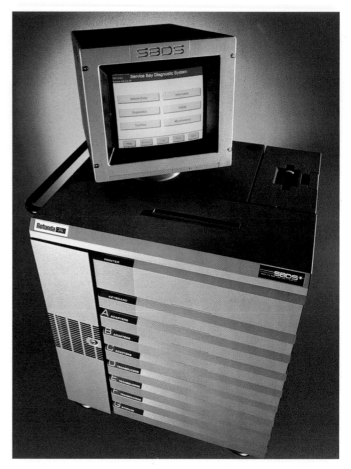

Figure 3-44. *Computerized diagnostic equipment is required to service modern vehicle systems, including the HVAC system.*

beginning to offer kits that contain harnesses to access data link connectors, along with related hardware and software that can be installed in a personal computer. These kits can turn a normal desktop or laptop computer into a computer driven analyzer, with many of the same capabilities as the large, expensive shop analyzers. In addition, many shops now write their repair and purchase orders and keep their billing and accounting in order with computers.

Manuals on CD-ROM

Service literature, including service, parts, and labor time manuals are available in compact disc format. *CD-ROM* stands for *compact disc-read only memory.* A CD-ROM is a compact disc identical in appearance to music CDs. However, this type of CD contains service manual information. One CD-ROM can provide the same amount of information found in a complete series of printed manuals. Many CD-ROM manuals cover several model years of a particular manufacturer's line of vehicles. Some of the newest CD-ROM manuals show actual, step-by-step footage of certain repair operations. The CD-ROM disc can be inserted in a computer with a CD drive. The information is then accessed and read on the computer monitor, **Figure 3-45.** The one drawback to CD-ROM manuals is they are much more expensive than a printed manual.

Figure 3-45. *Modern service literature is often provided in the form of CD-ROMs. These discs can hold the equivalent of thousands of pages of service documents.*

Internet Resources

By using a computer on-line service, small shops can access any one of several automotive central information banks over the information superhighway. These banks can offer diagnostic tips, technical service bulletins, and other service information similar to the telephone hotlines described earlier. Many of these services have interfaces that make them easy for anyone to use. Most of these on-line assistance centers are operated by aftermarket companies, private organizations, and individuals. These organizations provide a way for technicians from around the world to help each other by way of e-mail.

Summary

There are two kinds of HVAC equipment; test equipment and service equipment. Some test equipment, such as gauge manifolds, are both diagnosis and service tools. Most test equipment is dedicated, that is, it can only be used for one purpose. In addition, many refrigeration system test and service tools can be used with only one type of refrigerant.

HVAC test and service equipment can be simple (such as gauge manifolds or test lights) or extremely complex (such as charging stations and scan tools). When using any type of HVAC tools, always be careful not to damage the HVAC system or cause a release of refrigerant into the atmosphere.

Cooling system test and service equipment is used to check for leaks and pressure problems. Always allow the cooling system to cool off and remove all pressure before performing any test or service operations.

Many times the most important tools are the proper service literature or other sources of service information. Never guess at specifications or service procedures.

Review Questions—Chapter 3

Please do not write in this text. Write your answers on a separate sheet of paper.

1. A refrigeration gauge manifold body is made of _____ or _____.

2. The blue hand wheel on a gauge manifold indicates the _____ side valve.

3. Some R-12 refrigeration systems may have special service fittings. These fittings keep the technician from doing what?

4. Define refrigeration system static pressure.

5. Mechanical temperature gauges make use of a _____ spring to register temperature changes.

6. The most accurate way to check for leaks is to use a(n) _____ leak detector.

7. Older refrigerant dyes were _____ in color.

8. What color are contaminated refrigerant containers?

9. Older multimeters used a _____ that moved against a calibrated background.

10. What is the purpose of an ammeter inductive pickup?

11. To remove snap rings, the technician needs special snap ring _____.

12. The two most common kinds of hose fitting types are the _____ fitting and the _____ fitting.

13. Raising pressure in the refrigeration system causes water to boil at _____ than normal temperatures.

14. A _____ is a unit of pressure.

15. An _____ must be allowed to operate for long periods.

16. What undesirable thing will happen if too much antifreeze is placed in a cooling system?

17. Manufacturers' service manuals are often divided into _____ and _____ manuals.

18. General service manuals contain service information about many kinds of _____.

19. A specialized manual covers a specific vehicle _____.

20. Today, many manuals and other kinds of paper literature are being replaced by information that can be accessed using a _____.

ASE Certification-Type Questions

1. All of the following statements about gauge manifolds are true, *except:*
   - (A) R-134a and R-12 manifolds are needed to service modern vehicles.
   - (B) technicians often have their own gauge manifolds.
   - (C) the low side hand wheel is a different color than the low side hose.
   - (D) low side gauges can measure vacuum.

2. Technician A says that a mechanical temperature gauge may be used to determine air temperature as it leaves the HVAC system vents. Technician B says that an infrared temperature gauge may be used to check the temperature of engine coolant. Who is right?
   - (A) A only.
   - (B) B only.
   - (C) Both A and B.
   - (D) Neither A nor B.

3. An electronic leak detector is being used to check for leaks. The detector begins to squeal loudly. Which of the following is the *most likely* cause?
   - (A) Low detector battery.
   - (B) Severe refrigerant leak.
   - (C) Slight refrigerant leak.
   - (D) Presence of contaminated refrigerant.

4. Which of the following devices is used to perform open loop flushing?
   - (A) Refrigerant cycling machine.
   - (B) Recovery and recycling machine.
   - (C) Blowgun.
   - (D) Parts solvent.

5. Which of the following is *not* a dedicated refrigeration service device?
   - (A) Recycle/recovery machine.
   - (B) Charging station.
   - (C) Storage cylinder.
   - (D) Charging scale.

6. A charging station combines all of the following components, *except:*
   - (A) flushing unit.
   - (B) gauges.
   - (C) hoses.
   - (D) weighing scale.

7. Technician A says that a spectrograph can measure coolant temperature. Technician B says that a hydrometer can measure coolant freezing point. Who is right?
   - (A) A only.
   - (B) B only.
   - (C) Both A and B.
   - (D) Neither A nor B.

8. All of the following statements about service literature are true, *except:*
   - (A) overhaul manuals contain comprehensive disassembly and reassembly information.
   - (B) a schematic is a graphic representation of an electrical or vacuum system.
   - (C) troubleshooting charts contain a series of logical steps.
   - (D) vehicle manufacturers publish general service manuals.

9. A service manual CD-ROM contains which of the following?
   - (A) Sales and public relations information.
   - (B) Troubleshooting and other service information.
   - (C) Parts and labor prices.
   - (D) Addresses of paper manual providers.

10. Technician A says that a web site containing vehicle service information is called a hotline. Technician B says that web sites can be accessed through the Internet. Who is right?
    - (A) A only.
    - (B) B only.
    - (C) Both A and B.
    - (D) Neither A nor B.

Chapter 4

HVAC Electrical and Electronic Fundamentals

After studying this chapter, you will be able to:

❏ Explain the electron theory of electricity.
❏ Identify basic vehicle electrical circuits.
❏ Identify basic electrical measurements.
❏ Identify and explain the purpose of vehicle wiring and connectors.
❏ Explain how to diagnose indicator light problems.
❏ Explain how to diagnose heated glass problems.
❏ Identify and explain the purpose of common vehicle electrical devices.
❏ Explain the construction and operation of the automotive computer.
❏ Identify the major parts of vehicle computers.
❏ Explain the operation of control loops.

Technical Terms

Protons	Series circuit	Plug-in connector	Potentiometers	Central processing unit
Neutrons	Complete circuit	Schematic	Capacitors	(CPU)
Electrons	Parallel circuit	Wiring diagram	Dielectric	Memory
Conductors	Series-parallel	Circuit protection	Chip capacitors	Read only memory
Insulators	circuit	devices	Semiconductors	(ROM)
Current	Short circuit	Fuses	Diodes	Random access memory
Amperes	Open circuit	Fuse block	Rectified	(RAM)
Amps	Magnetic	Fusible link	Rectifiers	Serial data
Voltage	Polarity	Circuit breakers	Transistor	Universal asynchronous
Volts	Electromagnet	Relay	Power transistors	receive and transmit
Resistance	Induction	Solenoid	Integrated circuit (IC)	(UART)
Ohms	Electromagnetism	Magnetic clutch	Chips	Class 2 serial
Ohm's law	Harnesses	Motors	Microprocessors	communications
Direct current (dc)	Color coded	Armature	Control loop	Diagnostic output
Alternating current	Ground	Field coils	Input sensors	Trouble code
(ac)	wire	Resistors	Output devices	Scan tool
Waveform	Wire gage	Variable resistors		Self-diagnosis

You cannot diagnose problems on modern HVAC systems without a thorough knowledge of electrical and electronic theory. This chapter covers the basic electrical components used in all HVAC systems. Electronic components are also covered when they are used in modern HVAC systems. This chapter contains a brief review of the most important principles of electricity, electronics, and computer operation.

Electrical Basics

The following sections outline the fundamental principles governing the operation of all electrical equipment. All modern vehicles use 12-volt electrical systems. This information applies to any make of vehicle.

Conductors and Insulators

Everything is made of atoms. Every atom has a center of **protons** and **neutrons.** The neutrons have no charge and the protons have a positive charge, making the center of the atom positively charged. Revolving around this center are negatively charged **electrons.** See **Figure 4-1.**

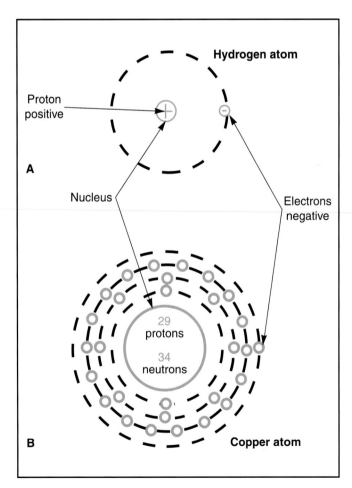

Figure 4-1. *The fundamental atom has a central positive nucleus surrounded by negatively charged electrons. All atoms are variations of this arrangement.*

Some atoms easily give up or receive electrons. These atoms make up elements that are good electrical **conductors.** Examples are copper and aluminum. Materials whose atoms resist giving up or accepting atoms are **insulators.** Glass and plastic are good insulators. Some materials can alternate between conducting and insulating. These materials are discussed in the computer section later in this chapter.

Electrical Measurements

The flow of electricity through a circuit depends on three electrical properties, all of which can be measured. These electrical properties are explained in the following paragraphs.

Current

Current is the number of electrons flowing past any point in the circuit. Current is measured in **amperes,** usually shortened to **amps.** The higher the amps rating, the more electrons are moving in the circuit.

Voltage

Voltage is electrical pressure, created by the difference in the number of electrons between two terminals. It provides the push that makes electrons flow and is measured in **volts.**

Resistance

Resistance is the opposition of the atoms in a conductor to the flow of electrons. All conductors, even copper and aluminum, have some resistance to giving up their electrons. Resistance is measured in **ohms.**

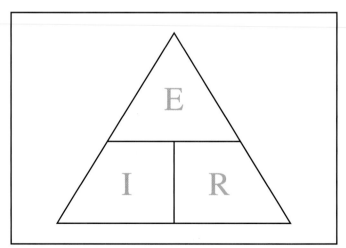

Figure 4-2. *This illustration is called the Ohm's Law Triangle. To calculate an electrical value, cover the letter representing the unknown value and multiply or divide the known values. Voltage (E) is always divided by another value to get value. Amperage (I) and resistance (R) are always multiplied to determine voltage.*

Ohm's Law

Sometimes all of the above electrical properties are not known. However, if you know two of the properties, you can calculate the third using **Ohm's law. Figure 4-2** is a graphic representation of Ohm's law. It is sometimes called the Ohm's law triangle. Using Ohm's law requires no more than simple multiplication or division. For example, you want to know the total amperage draw of one of the resistors in a resistor assembly. If the resistor has a resistance of 6 ohms and the vehicle has a 12-volt electrical system, you can use Ohm's law to calculate the amperage draw:

$$\frac{12 \text{ volts}}{6 \text{ ohms}} = 2 \text{ amps}$$

If you know the amperage and voltage and want to calculate the resistance in the previous example, use the following calculation:

$$\frac{12 \text{ volts}}{2 \text{ amps}} = 6 \text{ ohms}$$

If you know the amperage and resistance, calculate the voltage by the following formula:

$$2 \text{ amps} \times 6 \text{ ohms} = 12 \text{ volts}$$

These formulas make up Ohm's law. When applying Ohm's law to find an unknown electrical property, remember the following:

❑ When amperage or resistance is unknown, divide voltage by the other known value to obtain the unknown value.

❑ When voltage is unknown, multiply amperage and resistance to get voltage.

Direct and Alternating Current

The vehicle battery and alternator always have two terminals: positive and negative. There is always a shortage of electrons at the positive terminal, and an excess of electrons at the negative terminal. This current flows in only one direction, from negative to positive. This is called a **direct current (dc)** system.

In the system used in home, schools, and offices, the flow of electrons changes direction many times every second. These are **alternating current (ac)** systems. Since the vehicle battery cannot be charged by alternating current, automotive electrical systems are always direct current systems. Sometimes an ac voltage is used as a speed signal.

Waveforms

Depending on the circuit, each will produce a pattern. This pattern of current is known as a **waveform.** Waveforms are very useful in diagnosing vehicle problems. They are a very good indicator of system condition. Waveforms have somewhat limited use in HVAC work.

Types of Electrical Circuits

There are three types of automotive circuits. Every wire in a car or truck is part of one of these types of circuits. The three types of circuits are series, parallel, and series-parallel.

The simplest type of automotive circuit is the **series circuit.** The series circuit consists of a power source, switch, a load, and connecting wiring. Electrons flow through the wiring from the power source, through the switch and load, and back to the battery. The load could be a motor, lightbulb, or complex circuit. The other wiring is made of metals that allow the electrons to pass with very little resistance. The same current (number of electrons) flows through every part of the series circuit.

A typical series circuit is the simplified blower fan circuit shown in **Figure 4-3.** In the circuit, the electrical path is from the battery through the negative battery cable, blower motor, control switch, ignition switch, and positive cable, returning to the battery. This is called a **complete circuit,** since the electricity makes a loop from the battery, through the motor and control switches, and back to the battery.

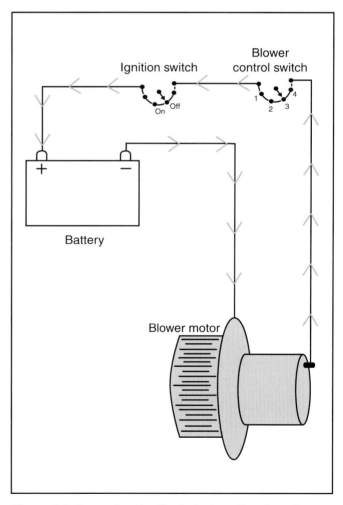

Figure 4-3. *In a series circuit, all electrons flow through every part of the circuit. Therefore, the current flow is the same throughout the circuit.*

The circuit in **Figure 4-4** is a *parallel circuit*. In a parallel circuit, current flow is split so each electrical component has its own current path. Different amounts of current will flow depending on the resistance of each part of the circuit.

The *series-parallel circuit* has some components wired in series and some wired in parallel, as shown in **Figure 4-5.** All current flows through some parts of the circuit, while the current path is split in other parts. An example of the series-parallel circuit is the compressor clutch electromagnet and blower motor. All clutch and blower current passes through the ignition switch and HVAC main switch, but only part of the current passes through the electromagnet and motor.

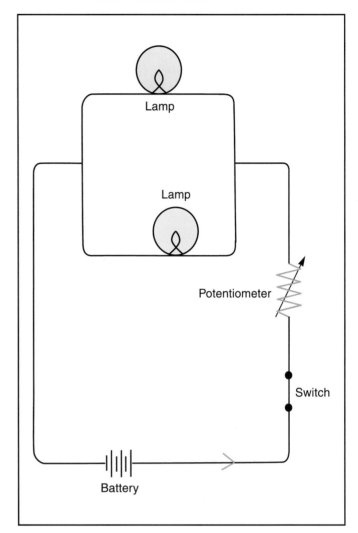

Figure 4-5. *In a series-parallel circuit, all current flows through some parts of the circuit, while the current, path is split in other parts. Some components are wired in series, and some are wired in parallel.*

Circuit Defects

There are two types of wiring defects: shorts and opens. A *short circuit* is caused when the wire insulation fails or is removed, and the wire contacts the frame, body, or another grounded part of the vehicle, as in **Figure 4-6.** If a fuse or circuit breaker does not protect the circuit, a short could lead to damaged wiring or components, or a fire. Fuses and circuit breakers are discussed later in this chapter.

An *open circuit* is a circuit that is not complete. Current cannot flow in an open circuit, as in **Figure 4-7.** Common causes of open circuits are loose or corroded connections, disconnected wires, and defects in electrical components such as switches, bulbs, and fuses. A related problem is a high resistance electrical connection, usually caused by corrosion or overheating. Current still flows, but at a reduced rate because of the high resistance. The high resistance may cause the circuit to stop operating, or possibly catch fire at the connection.

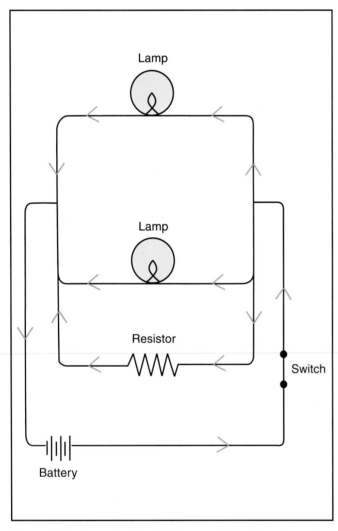

Figure 4-4. *In a parallel circuit, current flow through various parts of the circuit will be different, depending on the resistance of each part. Current flow is split so each electrical component has its own current path.*

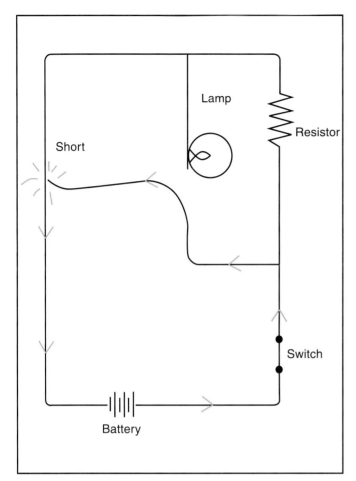

Figure 4-6. *A short circuit occurs when some part of the circuit is bypassed. A short circuit almost always results in excessive current flow.*

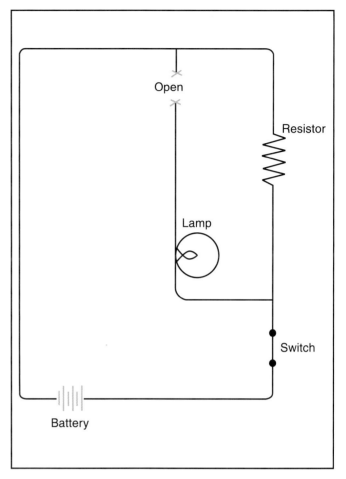

Figure 4-7. *An open circuit is incomplete. Current cannot flow in an open circuit. Any circuit turned off by opening a switch is an open circuit until the switch is closed.*

Electromagnetism

There is a well-defined relationship between electricity and magnetism. When a material is *magnetic,* the electrical charges of its electrons are aligned to create a force extending outward from the material. Magnetic fields have definite North and South poles. This property of magnetic fields is called *polarity.* Like poles repel each other and unlike poles attract. It is usually not necessary for the technician to determine the North and South poles when servicing a magnetic unit.

A magnetic material attracts iron and metals containing iron. Some materials are naturally magnetic, while others can be magnetized by electricity. When current flows in a wire, the electrons start moving in the same direction. This alignment of electrons creates a magnetic field around the wire as long as current is flowing. When the wire is wound into a coil, the magnetic fields of each wire loop combine to create a very strong magnetic field. The combination of the coil

winding and a metal core is called an *electromagnet,* **Figure 4-8.** The iron core helps to increase field strength, and may be movable. The magnetic field created may be used to move linkages or open electrical contacts. Electromagnets are the basic component of solenoids, relays, and motors. An electromagnet is used to engage the HVAC compressor clutch.

When a wire moves through a magnetic field, or a magnetic field moves through a wire, the electrons in the wire begin to move, causing current to flow in the wire. This process is called *induction,* and it is how current is produced in wheel speed sensors. In wheel speed sensors, the sensor wires are stationary, and teeth on the axle move. The teeth create a magnetic fluctuation, or movement, in a permanent magnet built into the sensor. This fluctuation induces current in the sensor wiring.

The relationship between electricity and magnetism is called *electromagnetism.* Electromagnetism is used to operate many electrical devices on the vehicle. These devices will be discussed later in this chapter and throughout this book.

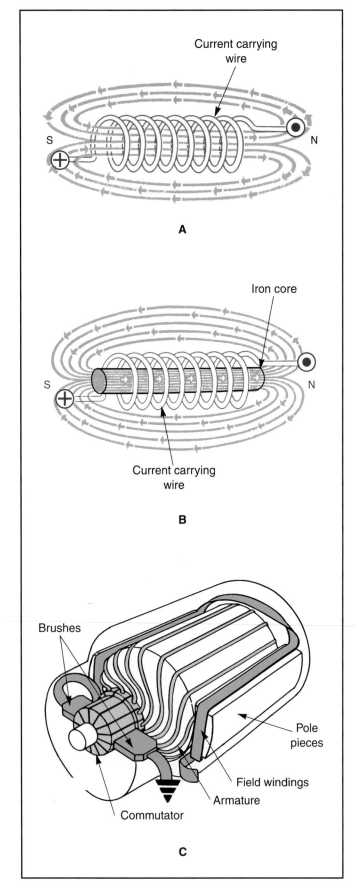

Figure 4-8. *A—The conductor is wrapped by wire to make an electromagnet. B—Placing an iron core within the wire strengthens the field. C—A motor uses the principles of electromagnetism.*

Vehicle Wiring and Electrical Components

The modern automotive electrical system is a complex arrangement of wiring and electrical components. The electrical system must produce electricity and deliver it to the proper places; protect circuits from damage; reduce or increase voltage; change electricity into light, motion, or heat; and use electricity to control the movement of liquids and gases. The construction, operation, and use of devices to accomplish this are discussed in the following paragraphs. The operation and function of electronic devices will be discussed later in this chapter.

Wiring

Most automotive wiring is made of copper, aluminum, or aluminum coated with copper. Wires are plastic coated and installed into wrapped assemblies called **harnesses,** **Figure 4-9.** For easier circuit tracing, automotive wiring is **color coded.** Color coding means giving the insulation of every vehicle wire a specific color. Modern vehicles have many wires, and it is necessary to increase the number of available colors by adding a stripe or stripes of contrasting colors to the original insulator color. Most forms of service literature contain schematics that show the diagram of the circuit's wiring and their colors.

On most vehicles, there is no return wiring from various electrical units back to the battery. The vehicle frame forms the return, or **ground wire.** On all modern vehicles, the negative terminal is the ground terminal. This means the battery, charging system, and all other electrical devices have their negative terminal connected to a common negative ground connection, usually the vehicle frame and body. On most vehicles, the battery negative cable is attached to the engine block. The body and chassis may be grounded directly to the battery through a smaller ground cable attached to the negative post, or grounded to the engine block by one or more ground straps, **Figure 4-10.**

Wire Size

All wires, no matter how well they conduct current, have some resistance and lose some electrical power as heat. It is important the wire in a circuit be large enough to carry the rated amperage without overheating. At the same time, the wire should not be unnecessarily large as to increase bulk and cost. **Wire gage** is the rating system for wire diameter. The larger the gage number, the smaller the wire. Wire gage can be measured in AWG (American Wire Gage) or metric sizes. The gage refers only to the thickness of the wire itself, without insulation. Typical wire gages are shown in **Figure 4-11.** The wire gage shows the highest amperage that can be carried by a particular gage wire without overheating. The

Figure 4-9. *Instead of installing many single wires, manufacturers combine them into a wire harness. The wire harness protects the individual wires and makes electrical connections easier.*

Figure 4-10. *Ground straps are commonly used to ensure a good electrical connection between the vehicle body and frame and the engine block.*

METRIC WIRE SIZES (mm²)	AWG SIZES (AMERICAN WIRE GAUGE)
.22	24
.35	22
.5	20
.8	18
1.0	16
2.0	14
3.0	12
5.0	10
8.0	8
13.0	6
19.0	4
32.0	2

Figure 4-11. *The right wire gage must be used to prevent wire overheating and possible fires. This chart shows the relationship of American Wire Gage (AWG) and metric sizes. It is OK to use a larger wire gage than the original wire. (General Motors)*

largest gages are used to connect the battery to the vehicle starter. To further reduce resistance losses, wires carrying high amperages are designed to be as short as possible.

Plug-in Connectors

The majority of connectors on modern vehicles are plug-in types, **Figure 4-12.** A **plug-in connector** has male and female ends. The ends are plugged into each other. Connectors having more than one wire are called *multiple connectors.* Many service manuals refer to a specific connector by the number of wires it contains, such as 12-wire connectors, 23-wire connectors, and so on. Modern plug-in connectors cannot be assembled incorrectly, due to the shape of the connector, or by the use of special aligning lugs and slots on each side of the connector.

Modern plug-in connectors are sealed to keep out dirt, moisture, and corrosion, **Figure 4-13.** Since many vehicle electronic components operate on very low voltages, a small increase in resistance due to contamination can affect circuit operation.

A few connectors may be screw terminals, or bolts which pass through a terminal eye to form a connector, **Figure 4-14.** These connectors are usually used to connect a ground strap to the vehicle body, or on circuits with very high current loads, such as a junction block or the starter.

Wiring Schematics

A *schematic,* or *wiring diagram,* is a drawing showing electrical units and the wires connecting them. Schematics also show wire colors and terminal types. Use of the schematic allows the technician to trace out defective components in the wiring system. Many vehicle manufacturers break down the overall vehicle wiring into separate circuit diagrams, as shown in **Figure 4-15.**

Figure 4-12. *Plug-in connectors can be found in almost every part of the vehicle.*

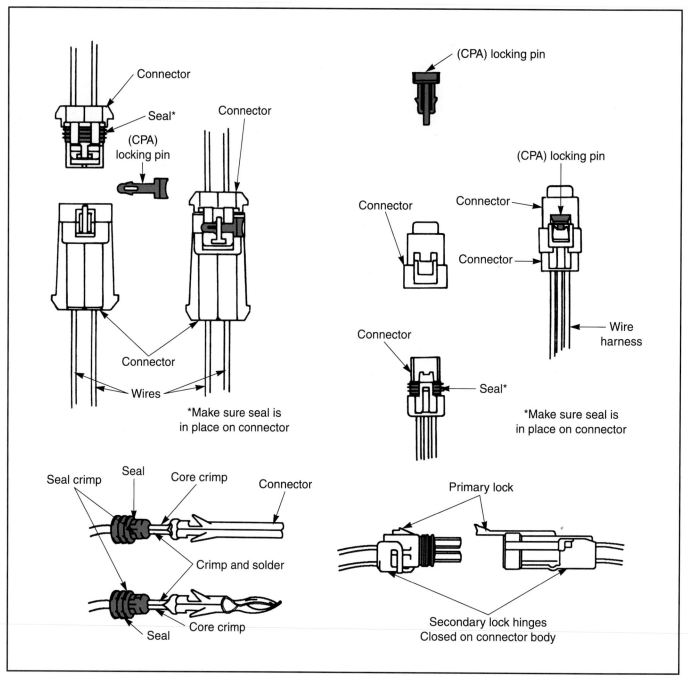

Figure 4-13. *Plug-in connectors are installed in various parts of the wiring harness to simplify the removal of electrical equipment.* *(General Motors)*

Schematics use symbols to represent electrical devices. There is some variation in the use of these symbols. Some schematics have a combination of company-specific and standardized symbols. **Figure 4-16** illustrates some symbols widely used in automotive electrical diagrams.

Circuit Protection Devices

To protect vehicle circuits from damage due to excessive current flows, *circuit protection devices* are used. These include fuses, fusible links, and circuit breakers.

Short circuits or defective components can cause excessive current flows. All electrical circuits except the starter and alternator output will have a circuit protection device.

Fuses

Fuses are made of a soft metal that melts when excess current flows through them, before the current can damage other components or circuit wiring. The types of fuses used in modern vehicles are shown in **Figure 4-17**. The majority of fuses are installed in a *fuse block,* which is usually located under the dashboard or in the glove compartment. A melted or blown fuse must be replaced.

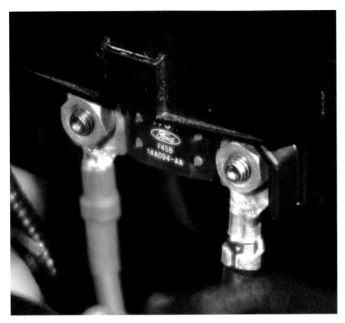

Figure 4-14. *Screw terminals are often used to connect the ends of wires carrying heavy electrical loads. The threaded connector ensures a good electrical connection. (Jack Klasey)*

Fusible Links

A *fusible link* is a length of wire made of soft metal. It operates in the same manner as a fuse, melting when excess current flows, **Figure 4-18.** Fusible links are usually installed in the main wiring leading from the battery or starter solenoid to the main electrical circuits. Both a fuse in the fuse box and a fusible link ahead of the box may be used to protect a circuit.

Circuit Breakers

Circuit breakers consist of a contact point set attached to a bimetallic strip. The bimetallic strip will bend as it heats up. When it becomes hot enough, it bends enough to open the point set and break the circuit. When the strip cools off, it straightens out and allows the points to close. The advantage of the circuit breaker is it can reset itself.

Switches

To control the flow of electricity through a circuit, some sort of switch is used. Some of these switches have

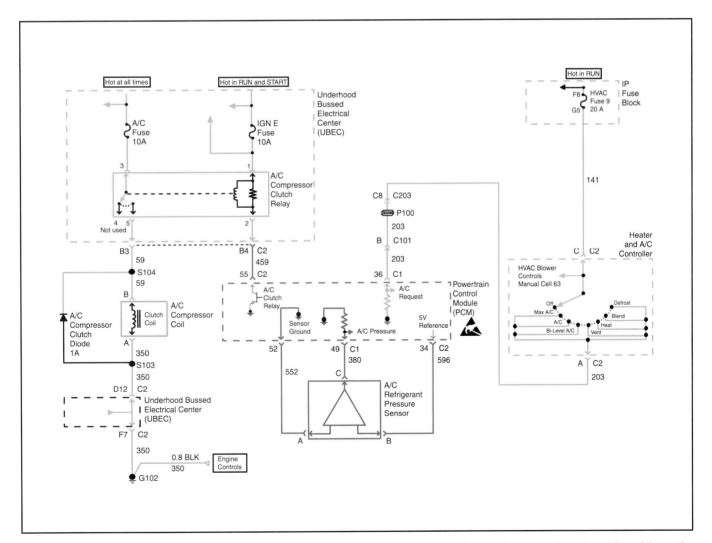

Figure 4-15. *This schematic represents the electrical wiring of a single electrical unit. Other schematics show the wiring of the entire vehicle, or a selected area. Schematics make circuit tracing easier. (General Motors)*

	Legend of Symbols Used on Wiring Diagrams		
+	Positive	⟫—	Connector
−	Negative	⟶	Male connector
⏚	Ground	⟩	Female connector
	Fuse		Denotes wire continues elsewhere
	Gang fuses with buss bar		Denotes wire goes to one of two circuits
	Circuit breaker		Splice
	Capacitor	J2 2	Splice identification
Ω	Ohms		Thermal element (Bi-metallic)
	Resistor	TIMER	Timer
	Variable resistor		Multiple connector
	Series resistor		Optional Wiring with / Wiring without
	Coil		"Y" windings
	Step up coil	88:88	Digital readout
	Open contact		Single filament lamp
	Closed contact		Dual filament lamp
	Closed switch		L.E.D. — light emitting diode
	Open switch		Thermistor
	Closed ganged switch		Gauge
	Open ganged switch		Sensor
	Two-pole single-throw switch		Fuel injector
	Pressure switch	#36	Denotes wire goes through 40 way disconnect
	Solenoid switch	#19 STRG COLUMN	Denotes wire goes through 25 way steering column connector
	Mercury switch	INST PANEL #14	Denotes wire goes through 25 way instrument panel connector
	Diode or rectifier	ENG #7	Denotes wire goes through grommet to engine compartment
	Bi-directional zener diode		Denotes wire goes through grommet
	Motor		Heated grid elements
	Armature and brushes		

Figure 4-16. *Symbols are commonly used to designate various electrical devices. Most of these symbols are the same for all vehicle manufacturers. Many manufacturers have their own symbols. (Dodge)*

Autofuse

Current Rating	Color
3	Violet
5	Tan
7.5	Brown
10	Red
15	Blue
20	Yellow
25	Natural
30	Green

Autofuse

Maxifuse

Current Rating	Color
20	Yellow
30	Green
40	Amber
50	Red
60	Blue
70	Brown
80	Natural

Maxifuse

Minifuse

Current Rating	Color
5	Tan
7.5	Brown
10	Red
15	Blue
20	Yellow
25	White
30	Green

Minifuse

Top

Side

Pacific fuse element

Pacific Fuse Element

Current Rating	Color
30	Pink
40	Green
50	Red
60	Yellow

Figure 4-17. *Fuse types used in modern automobiles. Most fuses and circuit breakers are installed in a junction block. (General Motors)*

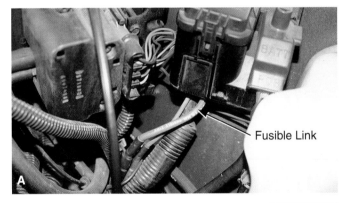

Fusible Link

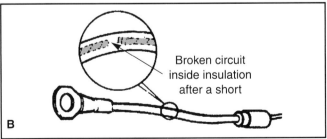

Broken circuit inside insulation after a short

Figure 4-18. *A—Fusible link is used to power vehicle systems. It is simply a wire a couple of gage sizes smaller than the actual current-carrying wire. B—When a fusible link blows, it creates an open area, much like a fuse. The insulation for the link will be burned and may come apart when touched. (General Motors)*

simple on-off positions, while some, such as windshield wiper and blower motor switches, have several positions to place the circuit in varying operating modes. The vehicle driver manually operates many switches. Examples are vehicle headlight, ignition, windshield wiper, and air conditioner/heater switches. Other switches are operated as a byproduct of other driver actions, such as the brake and backup light switches. Engine or transmission operating conditions activate some switches. Examples are oil pressure and coolant temperature switches.

Solenoids, Relays, and Magnetic Clutches

Some electrical components are electromagnetic control devices. Electricity creates a magnetic field. The magnetic field causes movement of a metal part. See **Figure 4-19.** In a *relay,* the magnetic field closes one or more sets of contacts, causing electrical flow in a circuit. This is useful when switching high current devices, such as motors or resistance heaters, without excessive lengths of heavy wire. In a *solenoid,* the magnetic field performs a mechanical task, such as opening or closing a vacuum valve or moving linkage.

The compressor clutch is a *magnetic clutch.* When the circuit is complete, current flows through the electromagnet in the clutch assembly. This creates a magnetic

Figure 4-19. *A typical electromechanical relay. When energized, the wire coil creates a magnetic field that pulls a set of points into contact against spring pressure. When current is removed from the coil, the spring opens the point set.*

field, pulling the two clutch halves together. When the two clutch halves are in contact, the engine turns the compressor. Compressor clutches are discussed in more detail in Chapter 8.

Motors

To turn electricity into rotation, **motors** are needed. The most commonly used motors consist of a central **armature** made of many loops of wire. Surrounding the armature are **field coils.** The field coils produce a magnetic field which causes current to flow through the armature windings. By controlling the direction of current flow, the armature can be made to turn inside the field windings. Current direction, and therefore motor direction, is usually controlled by the use of a commutator and brushes. Refer to **Figure 4-20.** Electric motors are found throughout the vehicle, including the starter, HVAC blower, windshield wiper, power window, seat, and antenna motors, and ABS system.

Figure 4-20. *This HVAC blower is a simple electric motor.*

Resistors

Resistors are placed in a circuit to reduce current flow. They are made of carbon or various metals. These materials cause extra resistance to the flow of electrons. This reduces current flow. Resistors are commonly used to reduce motor speeds and protect other circuit components, **Figure 4-21.** *Variable resistors* are often used in HVAC systems to control blower motor speed. A variable resistor is a resistor with a sliding contact. The total resistance can be adjusted by moving the sliding contact.

Special resistors called *potentiometers* are sometimes used in automatic temperature control systems. A potentiometer is a resistor with three terminals. The amount of resistance can be divided between the terminals by manually adjusting the potentiometer control.

Capacitors

To damp out voltage fluctuations and control electronic frequencies, *capacitors* are often used. The capacitor serves as a trap for voltage surges (sometimes called spikes) by attracting excess electrons. Capacitors consist of a *dielectric* (non-conducting) material between a positive and a negative pole. The dielectric material absorbs electrons by becoming polarized; the electrons in the dielectric material rearrange themselves into positive and negative areas similar to those in a magnet. This causes the capacitor to filter out voltage surges before they can affect radio and computer circuits. Almost all electrical devices which consume large amounts of current contain capacitors to reduce voltage surges. Modern capacitors are very small and have the same capacity as older models. They are called *chip capacitors.*

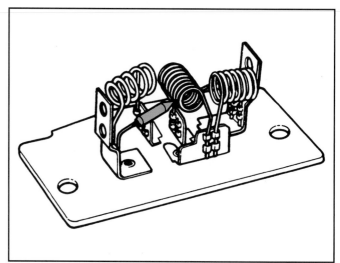

Figure 4-21. *The blower resistor shown here is used to vary the HVAC blower speeds. Each coil has a different resistance value. Directing the blower current through different resistors varies the blower speed. The thermal limiter is a type of fuse. (Ford)*

Basic Electrical Diagnosis

In many cases, HVAC system electrical diagnosis consists of checking for failure of the clutch to engage or a malfunctioning blower motor. On other occasions, you may be called on to find out why a heated rear window is not working or why a control panel light is out. The sections below contain procedures for general electrical diagnosis. More detailed electrical and electronic checking procedures are given in the chapters where they apply.

Before proceeding to check any electrical problem, remember from the information given earlier in this chapter what is necessary for any electrical device to operate properly:

❑ There must be a source of electricity (voltage).
❑ The electrical device must be in operating condition.
❑ The electrical device must be properly grounded.

Electrical Device Not Working

When any electrical device is not working, begin by checking for power at the positive terminal using a non-powered test light. If power is reaching the device, check the ground. Also check for grease or rust in the connectors leading to the device. If power and ground are available and the electrical device does not work, you can safely assume the device itself is bad. You may be able to use an ohmmeter or powered test light to make further checks of the device itself. Substituting with a known good part is often the only way to check some electrical parts.

If there is no power to the electrical device, begin by checking the system fuse or circuit breaker. If the fuse is blown, you must determine what caused the electrical overload. Do not simply replace the fuse and let the vehicle go. If some of the components controlled by the fuse are working, the fuse is good.

If the fuse is good, check the switches and relays the electricity must pass through on its way to the inoperative device. Switches or relays may be defective, disconnected, or misadjusted. Use an ohmmeter or a powered test light to check operation. The switch or relay should show infinite resistance in the *off* position, and low resistance in the *on* position. If the switch or relay does not give these readings, it should be adjusted or replaced. Also check all wiring connectors and look for places where the wire may have broken or rubbed through. In some cases you may have to unwrap wire harnesses to get a good look at wires and connectors.

Electrical Device on at All Times

If an electrical device is on at all times, the most likely causes are a stuck or melted switch or a short in a wiring harness. Begin by determining how the device is operated. It may be normal for the device to be energized under certain conditions. Once you determine the device should be off, check the switch by unplugging it. If the device is de-energized, the problem is located. The switch has probably failed internally. If the light remains on after all switches and relays have been unplugged, check for a shorted wire.

Vehicle Computers

The use of small inexpensive computers has transformed the automotive industry. The modern automotive computer is a complex device containing a collection of electronic parts and circuits that operate many vehicle systems. All modern cars and trucks have at least one computer, and most have several. The following information applies to all vehicle computers. This information will be most useful when diagnosing the computer that controls the operation of some HVAC systems. All vehicle computers have some common operating features.

Semiconductor Materials

All modern automotive computers and other electronic devices depend on the use of materials known as **semiconductors**. Semiconductors are made of silicon or germanium with small impurities to cause them to be either conductors or insulators, depending on how voltage is directed into them. Semiconductors change at the atomic level, depending on how their electrons are affected. Common semiconductors are diodes and transistors. These devices are extremely small and can be combined into compact computers, which can perform the same number of calculations as a room full of older electronic devices.

Diodes

Diodes are semiconductors that allow current to flow in one direction only. Diodes are one-way check valves for electricity, becoming a conductor or an insulator, depending on which way the current tries to flow. Diodes are a familiar part of alternators, where the alternating current produced must be **rectified,** or changed, to direct current to charge the battery. For this reason, diodes are sometimes called **rectifiers.** Diodes are used in the HVAC system at the compressor to prevent reverse voltage surges to ground when the clutch is turned on.

Transistors

A **transistor** is a semiconductor device used as a switch or amplifier. Transistors can carry heavy current, but are operated by very low currents. When a small amount of current is sent to one part of the transistor, it becomes a conductor, and much heavier current can flow through another part. When this small amount of current is removed, the transistor becomes an insulator and heavy current flow stops. A transistor with no moving parts is much more reliable than a set of mechanical points.

Transistors used in computers are usually too small to be seen. Large transistors designed to carry heavy current loads are called **power transistors.**

Computer Circuits

Many diodes and transistors, as well as chip capacitors, resistors, and other parts are combined onto a large complex electronic circuit by etching the circuitry on small pieces of semiconductor material. A circuit made in this manner is called an **integrated circuit,** or **IC.** The IC also contains resistors, capacitors, and other electronic devices. Modern ICs, sometimes called **chips,** or **microprocessors,** can control vehicle functions formerly done by mechanical devices, such as controlling engine timing. Microprocessors have made some systems, such as anti-lock brakes, not only possible but also available in even low-priced vehicles.

Control Loops

A modern computer's microprocessors can perform control operations based on their ability to quickly process input messages and issue commands. This ability allows them to operate as part of a control loop, **Figure 4-22.** A **control loop** can be thought of as an endless circle of causes and effects, which are used to operate many vehicle systems. When the control loop is operating, the **input sensors** furnish information to the computer, which makes decisions and sends commands to the **output devices.** The operation of the output devices affects the operation of the vehicle system, causing changes in the readings furnished by the input sensors. There is a continuous loop of information, from the input sensors, to the computer, to the output devices, to the vehicle system, and back to the input sensors. Some computers operate many control loops at the same time.

The most obvious example of an HVAC control loop is a temperature sensor input that tells the HVAC control computer the passenger compartment is becoming too hot. The computer then issues output commands, energizing the compressor clutch and blower motor to cool down the passenger compartment. When the sensor input tells the computer the passenger compartment temperature has been cooled off, the computer turns off the compressor clutch and blower motor.

Main Computer Sections

All automotive computers contain two main sections: the central processing unit (CPU), and the memory. The CPU and memory are built into the computer. The entire module must be replaced if either is defective. The construction and operation of these two sections is explained in the following paragraphs.

Central Processing Unit

The **central processing unit (CPU)** is the section of the computer that receives the input sensor information, compares this information with the information stored in memory, performs calculations, and makes output decisions. The CPU may contain several microprocessors, as well as other electronic parts.

Memory

The two basic types of computer **memory** used in the control module are **read only memory (ROM)** and **random access memory (RAM).** The ROM and RAM also contain microprocessors and other parts. The ROM contains permanent programs that tell the computer what to do under various operating conditions. The ROM also contains the operating standard for the system. Information is installed in ROM at the time of manufacture and cannot be modified in the field. Since it has a permanent, or nonvolatile memory, information stored in ROM will remain if the battery is disconnected or the system fuse is removed.

The random access memory (RAM) is a temporary storage place for data from the input sensors. As information is received from the sensors, it is temporarily stored in RAM, overwriting any old information. The computer constantly receives new signals from the sensors as the HVAC system operates. When a trouble code is generated by a system defect, it is stored in RAM. The RAM has a temporary, or volatile memory. If the battery cable or system fuse is removed, all data in the RAM will be erased including trouble codes.

Serial Data

On newer vehicles, the control module receives and transmits **serial data** to other on-board computers. Serial data is simply sensor and actuator information that is shared by the various vehicle on-board computers. By sharing data, it eliminates the need for duplicate sensors for each system. This information is transmitted over wiring referred to as a *data bus.* In the near future, vehicles may use fiber optics as the data bus.

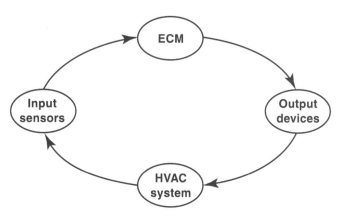

Figure 4-22. *A basic control loop. In addition to the inputs, controller and outputs, vehicle operation itself is part of the loop.*

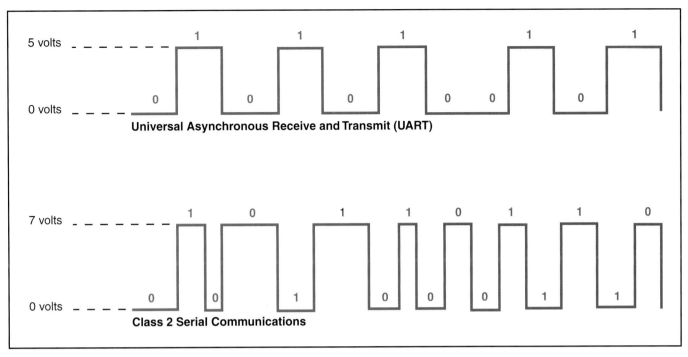

Figure 4-23. *UART and Class 2 data communication protocols are used in HVAC and other vehicle systems.*

ECM External Communications

There are two types of external serial communication used in automotive computer systems. Most modern vehicles use a system called **universal asynchronous receive and transmit (UART).** This signal is used for communication between the ECM, off-board diagnostic equipment, and other control modules. UART is a data line that varies voltage between 0-5 volts at a fixed pulse width rate. Some of the newest vehicles use UART, but also depend on the use of **Class 2 serial communications.** Class 2 data is transferred by toggling the line voltage from 0-7 volts, with 0 being the rest voltage, and by varying the pulse width. The variable pulse width and higher voltage allows Class 2 data communications to better utilize the data bus, **Figure 4-23.**

The first type of bus used with computer controls was called a *serial bus.* Vehicle manufacturers are now using more sophisticated busses called **Controller Area Networks,** or **CANs.**

Computer Diagnostic Outputs

In addition to operating the output devices, the computer provides a **diagnostic output** to the technician. This consists of two major components:

❑ A dashboard mounted light or LED display that indicates a problem is present.

❑ A diagnostic connector for retrieving diagnostic information.

When a system defect occurs, the control module saves information about the defect in memory. This information takes the form of a **trouble code.** On vehicles up to 1995, trouble codes are 2- or 3-digit numbers corresponding to a specific defect, **Figure 4-24.** On newer vehicles with the

OBD II system, the trouble code format consists of one letter and four numbers. To determine whether any trouble codes are present, the technician must use a **scan tool** to cause the module to enter the **self-diagnosis** mode and retrieve trouble codes. Code retrieval and other diagnosis procedures will be discussed in more detail in Chapter 22.

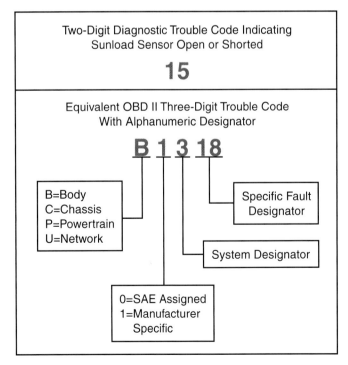

Figure 4-24. *Trouble codes for older and newer vehicles are different. Older vehicles used a two or three digit code. 1996 and newer vehicles use the alphanumeric code.*

Summary

Every atom has a positively charged center. Negatively charged electrons revolve around the center. Electricity is the movement of electrons from atom to atom. Materials whose atoms easily give up or receive electrons are conductors. Materials whose atoms resist giving up or accepting electrons are insulators. The path through which the electrons move is called a circuit. Electrons will not flow in a circuit unless there are more electrons in one place than another and if there is a path between the two places. The three basic electrical properties are amperage, voltage, and resistance. Unknown electrical properties can be calculated using Ohm's law.

The two types of current flow are alternating and direct. Automotive electrical systems are always direct current systems. Alternating current is used in some sensors. The three types of automotive circuits are series, parallel, and series-parallel. Magnetism can be used to make electricity and electricity can be used to create magnetism. This relationship between electricity and magnetism is called electromagnetism.

Automotive wiring can be copper, aluminum, or copper coated aluminum, plastic coated, color coded, and installed into harnesses. Wire gage is the rating system for wire diameter. Most electrical connectors are plug-in types. A connector with more than one wire is called a multiple connector. A schematic, or wiring diagram, is a drawing of electrical units and connecting wires. Schematics allow the technician to trace out defective components in the wiring system.

Circuit protection devices include fuses, fusible links, and circuit breakers. Switches control the flow of electricity through a circuit. Relays, solenoids, and magnetic clutches are electromagnetic control devices. Relays control electrical flow, while solenoids and magnetic clutches cause physical movement. Motors turn electricity into rotation. Resistors are used to reduce current flow through circuits.

All modern automotive computers depend on the use of materials known as semiconductors. Semiconductors can be conductors or insulators, depending on how voltage is directed into them. Common semiconductors are diodes and transistors. Diodes allow current to flow in one direction only. Transistors are used as switches or amplifiers. Transistors carry large amounts of current, but are operated by low currents. Transistors can perform switching operations with no moving parts. Transistors, diodes, capacitors, and resistors are etched onto small pieces of semiconductor material to form an integrated circuit, or IC. ICs are also called chips or microprocessors.

Computers are collections of microprocessors and other components. They form control loops, using the inputs and outputs of the system and internal controls. The computer processes inputs from sensors and issues commands to output devices. Computer internal controls are the central processing unit, or CPU, and the memory circuits. The two types of memory are read only memory or ROM, which consists of permanent settings, and random access memory, or RAM, which changes as the vehicle operates. Computers, in addition to controlling the operation of a system, can produce diagnostic codes for aid in troubleshooting problems.

Review Questions—Chapter 4

Please do not write in this text. Write your answers on a separate sheet of paper.

1. A good conductor easily gives up _____.

2. State whether the following materials are good conductors or good insulators by placing a C or I next to them.
   Glass _____
   Plastic _____
   Aluminum _____
   Copper _____

3. A difference in the number of electrons on each side of a circuit is called _____.

4. Resistance to the flow of electrons is measured in _____.

5. If voltage and resistance are known, the technician can use Ohm's law to find _____.

6. In a _____ circuit, current flows through every part of the circuit.

7. Motors turn electricity into _____.

8. A _____ is a kind of electromagnet.

9. _____ control large amounts of current.

10. _____ can have a letter and four numbers.

ASE Certification-Type Questions

1. Technician A says an excess of electrons in one part of a complete circuit causes current flow. Technician B says current can flow through some incomplete circuits. Who is right?
   (A) A only.
   (B) B only.
   (C) Both A & B.
   (D) Neither A nor B.

2. Amperage is the number of _____ flowing past a point in a circuit.
   (A) volts
   (B) protons
   (C) electrons
   (D) atoms

3. Technician A says in a series circuit, only some of the current flows through parts of the circuit. Technician B says a series-parallel circuit is a combination of series and parallel sections. Who is right?

   (A) A only.

   (B) B only.

   (C) Both A & B.

   (D) Neither A nor B.

4. Wire gage is the measurement of wire _____.

   (A) length

   (B) material

   (C) insulation

   (D) diameter

5. Which of the following is *not* a circuit protection device?

   (A) Fuses.

   (B) Circuit breakers.

   (C) Resistors.

   (D) Fusible links.

6. A blower motor does not operate. Technician A says the problem could be a switch. Technician B says the problem could be a bad motor ground. Who is right?

   (A) A only.

   (B) B only.

   (C) Both A & B.

   (D) Neither A nor B.

7. None of the components of an HVAC system operate when the main panel switch is turned on. Which of the following is the *least likely* cause?

   (A) Defective panel switch.

   (B) Blown HVAC fuse.

   (C) Disconnected HVAC panel connector.

   (D) Blower motor burned out.

8. Technician A says semiconductors are insulators. Technician B says semiconductors are conductors. Who is right?

   (A) A only.

   (B) B only.

   (C) Both A & B.

   (D) Neither A nor B.

9. All of the following statements about control loops are true, *except:*

   (A) a control loop is a circle of causes and effects.

   (B) control loops are used to operate vehicle systems.

   (C) the input sensors make decisions and send commands to the output devices.

   (D) the computer can operate several control loops at the same time.

10. The computer diagnostic output system consists of all of the following, *except:*

    (A) dashboard warning light.

    (B) scan tool.

    (C) data link connector.

    (D) internal computer circuits.

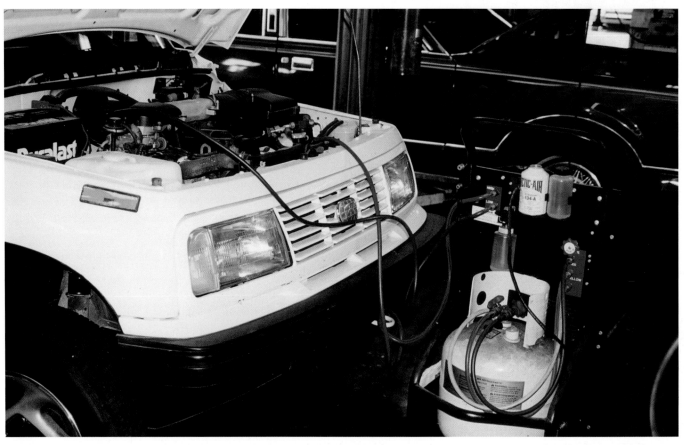

Always recover refrigerant. Never intentionally vent refrigerant to the atmosphere.

Chapter 5

Principles of Refrigeration

After studying this chapter, you will be able to:
- ❑ Identify and define heat and cold.
- ❑ Identify and explain heat transfer methods.
- ❑ Identify and explain change of state.
- ❑ Identify latent heat and explain how it is absorbed and released from a substance.
- ❑ Explain the basic refrigeration cycle.
- ❑ Identify common refrigeration system components and the purpose of each.
- ❑ Explain how heat load affects refrigeration system operation.

Technical Terms

Humidity	Vaporizing	Compressor	Fixed orifice tube (FOT)	Pilot operated absolute (POA)
Wind chill	Boiling	Reed valves	Ford orifice tube cycling clutch (FOTCC)	Valves in receiver (VIR)
Cold	Condensing	Radial compressors		
Hot	Sensible heat	Axial compressor	Variable displacement orifice tube (VDOT)	Evaporator pressure regulator (EPR)
Absolute zero	Latent heat	High pressure relief valve	Expansion valve	Metal tubing
Heat transfer	Superheat		Thermostatic expansion valve (TXV)	Flexible hoses
Convection	Superheated steam	Muffler		
Conduction	Pressure-temperature relationship	Condenser	Block valve	Service fittings
Radiation		Receiver-drier	Evaporator pressure controls	Refrigerant
States	Cycling clutch orifice tube (CCOT) system	Filter		Heat load
Change of state		Sight glass	Suction throttling valve (STV)	Ambient air
Freezing	Evaporator	Flow restrictor		Relative humidity
Melting	Accumulator	Fixed orifice		Sunload
	Desiccant			

You cannot repair a modern refrigeration system unless you know how it works. Knowledge of what is actually happening in a working refrigeration system is essential to determining what is going on when the system is not working. In this chapter, you will learn the underlying principles and components of the refrigeration system.

The Purpose of Refrigeration

Before studying the operation of the refrigeration system, we must review its purpose. Remember from Chapter 1, the main job of the air conditioner is to cool the incoming air and remove moisture. The other purposes of the system are to remove dust and airborne gases, and to circulate air through the vehicle.

Removing heat is an obvious benefit of the air conditioner. Less obvious is the need to remove humidity. **Humidity** is the amount of water vapor in the surrounding air. The basic reason for humidity control is humans are warm-blooded animals. We produce heat as a byproduct of muscle movement. In warm weather, we produce too much heat, and the body tries to remove this heat by perspiring, or sweating. As perspiration forms on our bodies, it removes heat as it evaporates. However, if the humidity is high, the air will not immediately absorb additional moisture, and perspiration will remain on the body for a much longer time. This is why humid air feels more uncomfortable than non-humid air at the same temperature. When the air conditioner removes humidity, your perspiration mechanism becomes more effective.

Air movement also affects heat removal. The more air moving over your body, the more heat is removed by convection. Airflow also removes moisture from the skin. In cold weather, this effect is called **wind chill.** Air movement through the passenger compartment when the air conditioner is operating removes more heat than would be possible if the air was still.

Basic Principles of Refrigeration

To understand how the air conditioner works, you must first learn about some of the basic principles of refrigeration. The principles the refrigeration system uses are:
- ❑ Heat transfer, especially by convection.
- ❑ Change of state from liquid to gas and back again.
- ❑ Sensible and latent heat of vaporization and condensation.
- ❑ The effects of pressure and temperature.

The underlying principles of refrigeration deal with some things you may have thought about. For instance, you may have wondered what causes things to be hot or cold. You may have also wondered why pressurizing the cooling system keeps the coolant from boiling out of the radiator. This part of the chapter addresses these questions.

Heat and Cold

When something feels cool to the touch, we say it is **cold.** When something feels hot to the touch, we say it is **hot.** What we refer to as heat and cold are actually varying levels of heat. In science, there is no such thing as cold. An object we call cold merely contains less heat than a warmer object. The only temperature point in which an object is considered to have no heat is referred to as **absolute zero,** which is -459°F (-273°C).

Heat Transfer

All substances are made up of atoms. Atoms combine to create molecules. The bonds between the atoms in a molecule cause it to move, or vibrate. The vibrations of an individual molecule are too small to be felt as movement. However, the vibration of billions of molecules can be felt as heat. The more molecular vibration, the more heat.

Heat transfer is actually the movement of these vibrations from an object with more heat to an object with less heat. Heat always travels from a warmer object to a cooler one. There are three methods of heat transfer, convection, conduction, and radiation. If all objects in the vicinity are at the same temperature, no heat transfer can occur. In actual practice, this seldom occurs and some movement of heat is always taking place.

Convection

Convection is the transfer of heat by air. The hot object heats the air, which then heats any surrounding objects. You have probably noticed the air several feet above the surface of an operating stove or barbecue grill feels warm. This is caused by convection of the heat from the cooking surface. The most common example of convection is the transfer of heat from a hot water heater to air passing over the heater's fins, **Figure 5-1.**

Figure 5-1. *Hot water radiators, such as this baseboard heater, transfer heat to the surrounding air, which then heats the room. The fins increase the air contact surface to allow quicker heat transfer.*

Conduction

Conduction is heat transfer by direct contact between two objects. For instance, if you pick up a cold soft drink can, you immediately feel the heat leaving your hand for the can metal. Another example is the heating of a cooking pot by direct contact with the burner, **Figure 5-2.** In the refrigeration system, the most common example of conduction is the transfer of heat between the air and the metal of the evaporator.

Radiation

Radiation occurs when heat travels from an object by infrared rays. These rays are not visible and do not contain heat in themselves. However, when the infrared rays strike an object, they cause its molecules to vibrate, heating it. The most obvious example of radiation is how the sun heats an object through millions of miles of space, **Figure 5-3.** Radiation is a factor in automotive air conditioning as sunlight has an effect on system efficiency.

Change of State

Everything in the universe is in one of three **states,** or physical conditions. The three states are *solid, liquid,* and *gas.* All matter can switch between solid, liquid, and gas if conditions are right. The switching of the physical condition of a substance is called **change of state.**

The usual way to cause a change of state is by adding or removing heat. Changing from a solid to a liquid or from a liquid to a gas is called *changing to a higher state,* and requires the addition of heat. Changing from a gas to a liquid or from a liquid to a solid is called *changing to a lower state,* and requires the removal of heat. See **Figure 5-4.**

Pressure also affects the change of state. The temperature at which a substance will move to a higher or lower state will increase if its pressure is increased, or decrease if its pressure is decreased.

Any kind of matter can change its state from gas to liquid to solid if the temperature is low enough. Matter can also switch from solid to liquid to gas if the temperature is high enough. No matter how much heat is removed from a solid, it remains a solid. A gas, no matter how much heat is added, always remains a gas. A liquid can change state up or down, depending on its temperature.

Note: Do not be confused with chemical changes brought on by heat. If something becomes so hot it begins burning, this is not a change of state, but a chemical reaction. Changing gasoline into a vapor to make it burn is a change of state, but the actual burning is not.

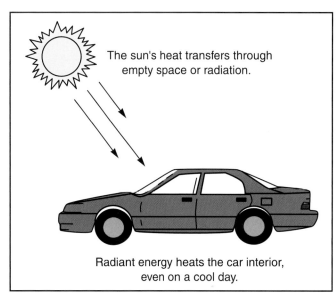

The sun's heat transfers through empty space or radiation.

Radiant energy heats the car interior, even on a cool day.

Figure 5-3. *Radiation is the transfer of heat through waves of radiant energy, such as when sunlight travels through millions of miles of empty space to heat a car interior.*

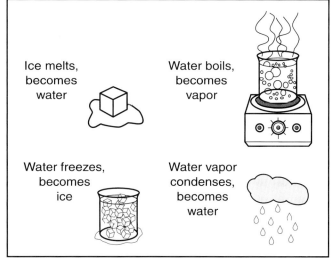

Ice melts, becomes water

Water boils, becomes vapor

Water freezes, becomes ice

Water vapor condenses, becomes water

Figure 5-4. *Water, like most substances, can occur in three states: solid (ice), liquid (water), or gas (water vapor or steam). Change of state occurs when heat is either added or removed.*

Figure 5-2. *This electric stove burner transfers heat to the pot and water by direct contact.*

Freezing and Melting

Freezing and *melting* are changes of state from a liquid to a solid, and from a solid to a liquid. The most common example of this change of state is water freezing or melting. When water freezes (sometimes called *fusion*), it goes from a liquid to solid ice. When ice melts, it turns to liquid water.

Vaporization and Condensation

The changes of state most vital to understanding the refrigeration system is the change between liquid and gas. The change of state of water from liquid to water vapor, or steam, is called **vaporizing.** Vaporizing is sometimes called **boiling.** When water vapor changes to liquid, the process is called **condensing.**

Sensible and Latent Heat

All liquids and gases contain two kinds of heat, sensible heat and latent heat. **Sensible heat** is heat that can be felt, or sensed. A drink from a soda can be removed from a refrigerator feels colder than a drink taken from a cabinet. This is because the refrigerated soda contains less heat than the other soda.

 Note: Solids contain only sensible heat since there is no further change of state possible once a substance reaches the solid state.

Latent heat is often called *hidden heat,* since it is hidden in the liquid or gas. The reason only liquids and gases contain latent heat is because latent heat is absorbed to produce a change to the next higher state. Solids contain no latent heat, but must absorb heat to change to a liquid. To change from a liquid to a gas, more heat must be absorbed. This heat does not change the temperature, just the state. For instance, it is possible to have 32°F (0°C) ice and 32°F (0°C) water. However, the water contains more heat; the latent heat added to change it from ice into water.

The types of latent heat are:

❑ *Latent heat of freezing or fusion.* When a substance gives up enough heat to go from a liquid to a solid.
❑ *Latent heat of melting.* When a substance absorbs enough heat to go from a solid to a liquid.
❑ *Latent heat of vaporization.* When a substance absorbs enough heat to go from a liquid to a gas.
❑ *Latent heat of condensation.* When a substance gives up enough heat to go from a gas to a liquid.

A great deal of latent heat is absorbed and released from substances when they change state. See **Figure 5-5.** The potential to remove vast quantities of heat by changing state is what makes latent heat so important to the operation of the refrigeration system.

Superheat

Many air conditioning manuals refer to superheat. **Superheat** is any heat added after the change to a higher state. For instance, once water is turned to steam at 212°F (100°C), more heat can be added to bring the temperature of the steam even higher. Steam with a temperature of more than 212°F is called **superheated steam.** Superheat must be removed before the substance can change state. Therefore, steam at 232°F (111°C), must have enough heat removed to lower its temperature by 20°F (11°C) before removing additional heat will cause it to condense back into water.

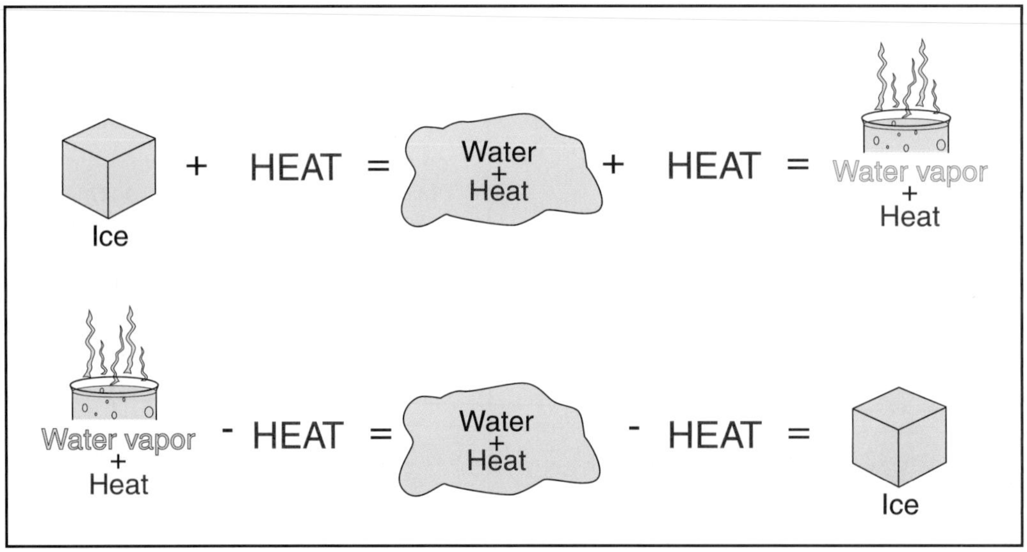

Figure 5-5. *An upward change of state occurs when ice absorbs heat and turns to water, and when water absorbs heat and becomes water vapor. A downward change of state occurs when water vapor gives off heat and becomes water, and when water gives off heat and becomes ice.*

The Pressure-Temperature Relationship

Pressure has a great effect on the temperature at which a substance changes state. The **pressure-temperature relationship** is different for every substance. An increase in pressure always results in an increase in the temperature at which a change of state will take place. A decrease in pressure always causes a decrease in the temperature at which a change of state takes place. For instance, water boils at 212°F (100°C). However, if the pressure of water is raised, such as in a pressurized engine cooling system, the boiling point is raised. For every 1 psi (6.9 kPa) rise in temperature, the boiling point of the water goes up by 3°F (1.6°C). Therefore, in a cooling system containing only water, the boiling point will be raised to 257°F (124°C) when a 15 psi (104 kPa) pressure cap is used.

Figure 5-6 shows the relationship of pressure to change of state for one particular refrigerant. As you study the chart, note that something will turn to a gas at a higher temperature when it is pressurized, it will also condense at a higher temperature.

To make the pressure-temperature relationship work in a refrigeration system, we must have a substance that can change from a liquid to a gas, and from a gas to a liquid, at the pressures developed by the refrigeration system. Various types of refrigerants have been developed, but not all of them will work in an automotive air conditioning system. Later in this chapter, we will discuss the action of refrigerants in more detail.

Pressure-Temperature Relationship		
Pressure	Water	Refrigerant
0	212°F (100°C)	−15°F (−26°C)
5	227°F (108°C)	−6°F (−21°C)
10	242°F (116°C)	+5°F (−15°C)
15	257°F (124°C)	+15°F (−9.4°C)
20	272°F (133°C)	+20°F (−6.6°C)

Figure 5-6. *Pressure affects the temperature at which a change of state occurs. Notice how the boiling points of water and a common refrigerant increase when the pressure is increased.*

Refrigeration Components

The following sections contain a brief overview of the components of the refrigeration system. These sections concentrate on the effect each component has on heat transfer, change of state, and pressure development. Detailed discussions of each component will be given in later chapters.

The Two Main Types of Refrigeration Systems

There are two major types of automotive refrigeration systems. Notice both systems use an evaporator, compressor, and condenser. The type of evaporator pressure control device used identifies the type of refrigeration system. The most common modern refrigeration system is the **cycling clutch orifice tube (CCOT) system** shown in **Figure 5-7.** It uses a fixed flow restrictor. This system always has a way to control the amount of refrigerant drawn into the compressor. Many orifice tube systems use a cycling compressor clutch. The compressor control may be a pressure switch installed on the low side of the system, or a temperature switch mounted on the evaporator outlet.

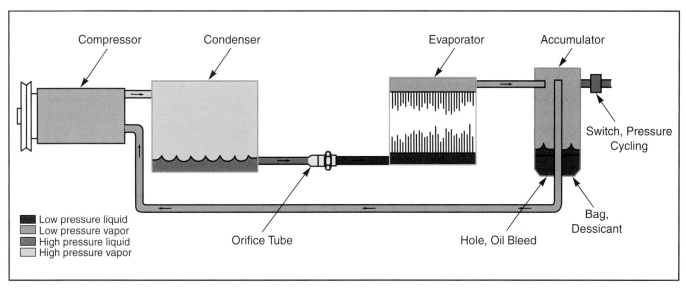

Figure 5-7. *The system shown here is a modern refrigeration system using a fixed orifice tube. (General Motors)*

Many late-model systems have a valve in the compressor that allows it to vary the amount of refrigerant pumped. These systems always have an accumulator between the evaporator and compressor.

A variation of the CCOT air conditioning system is shown in **Figure 5-8.** This type of system is used on many minivans and full-size vans. In addition to the normal cycling clutch system used to cool the front of the vehicle, it has an extra cooling circuit with an evaporator installed in the rear of the passenger compartment.

The other refrigeration system uses a thermostatic expansion valve, **Figure 5-9.** The compressor on these systems operates at all times while the system is on. Older versions of these systems use an evaporator pressure control device, located at the evaporator outlet. Some very old expansion valve systems did not have a pressure control device, and cycled the compressor clutch to control evaporator pressure. A temperature switch resembling an expansion valve controlled the clutch on these early systems.

Evaporator

An *evaporator* is a heat exchanger made from one or more lengths of tubing, folded into a series of loops, or coils. See **Figure 5-10.** This makes the evaporator as small as possible so it does not take up a significant amount of interior room. Fins are attached to the coils to aid heat transfer into the tubing. The fins are thin and flat to allow

maximum airflow. The refrigerant flows through every coil in the evaporator, absorbing heat as it travels. As moisture collects on the surface of the evaporator, it runs to the bottom of the evaporator case and drips out of the drain. Other evaporators are of different designs.

Accumulator

Most modern refrigeration systems have an *accumulator* installed between the evaporator and the compressor. The accumulator is a small tank, usually made of aluminum, **Figure 5-11.** It holds refrigerant and contains a *desiccant.* The desiccant absorbs any moisture left in the refrigeration system during service, and any small amounts of moisture that enter through the rubber hoses. The design of the internal tubes of the accumulator separates the liquid and vaporized refrigerant. The accumulator may have a return line which allows any oil in the bottom of the accumulator to return to the compressor.

Compressor

The *compressor* is a pump used to compress the vaporized refrigerant and move it through the system, **Figure 5-12.** Most modern compressors are piston types, which draw in refrigerant through one-way valves called *reed valves.* The intake reed valve is pulled open when the piston is moving downward. This allows refrigerant to enter the cylinder. When the piston begins to move

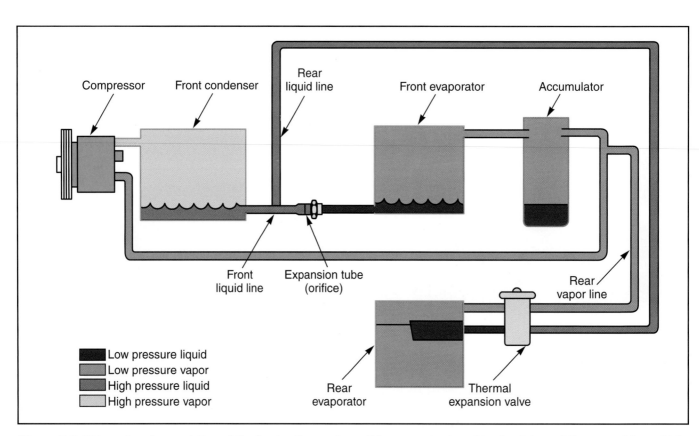

Figure 5-8. *This system is a variation of the fixed orifice system. It has an extra low side circuit to cool the rear of the vehicle. (General Motors)*

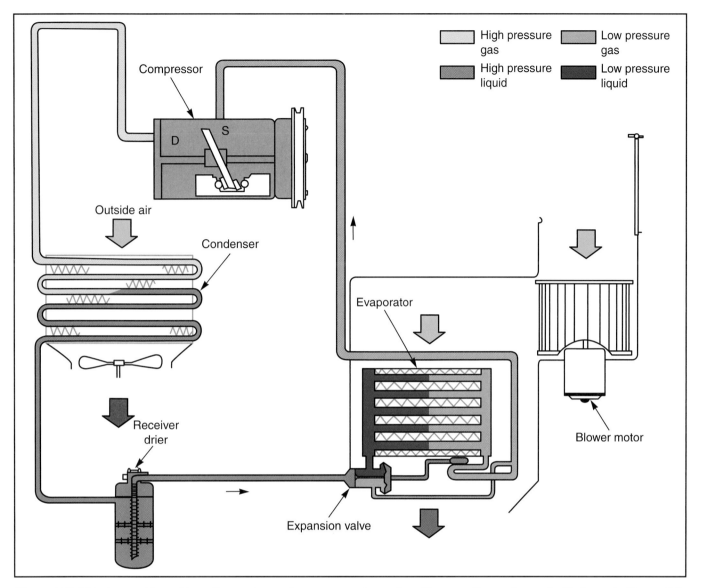

Figure 5-9. *This is an older refrigeration system using an expansion valve. Compare the parts used with those in* **Figure 5-8.** *(Nissan)*

Figure 5-10. *In the evaporator, refrigerant changes from liquid to gas and absorbs heat. The evaporator is made of a material with good heat transfer characteristics. Most evaporators are made of aluminum and copper.*

Figure 5-11. *Modern refrigeration systems use accumulators to store refrigerant and prevent liquid refrigerant from reaching the compressor. Other functions of the accumulator are to trap debris, remove moisture, and ensure that oil returns to the compressor.*

upward, the intake reed valve closes and a reed valve connected to the outlet opens. The upward movement of the piston drives the refrigerant out through the open reed valve. This cycle repeats every time the piston moves.

Some piston compressors resemble small engines, with the pistons at right angles to the crankshaft. These are called **radial compressors.** Many modern vehicles use **axial compressor** designs, with the pistons parallel to the crankshaft. The pistons are moved back and forth by the action of an off-center place called a *swash* or *wobble plate.* A few compressors are rotary vane or scroll types. They use a rotating action, similar to an engine oil pump to move the refrigerant.

Figure 5-12. *A typical axial compressor. The compressor is equipped with a magnetic clutch and pulley assembly.*

A magnetic clutch allows the compressor to be connected and disconnected from the drive belt. The compressor clutch is often cycled on and off to control evaporator pressure. Most compressors are equipped with a **high pressure relief valve.** If the compressor outlet pressure becomes too high, the valve opens to relieve pressure.

Muffler

To reduce noise caused by the compressed refrigerant leaving the compressor, a **muffler** may be used. When used, the muffler is located at the compressor outlet, as in **Figure 5-13.** Like an exhaust system muffler, it consists of a series of baffles that reverse the refrigerant flow, causing the sound impulses to cancel each other out. In some cases, the muffler is incorporated as part of the hose assembly. To save weight, many modern refrigeration systems do not use a muffler.

Condenser

Like the evaporator, the **condenser** is a heat exchanger. Its job is to transfer heat from the refrigerant to the outside air. Like the evaporator, it is one or more tubes coiled

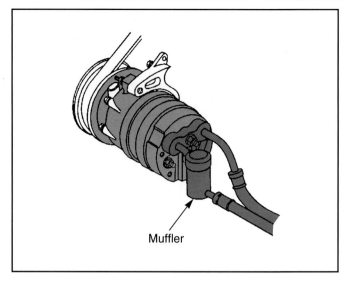

Figure 5-13. *When a muffler is used, it is installed on the outlet of the compressor to reduce compressor pulsation and noise. Some systems do not use a muffler. Other systems have two small mufflers on the compressor inlet and outlet lines. (General Motors)*

to save space. The coils are surrounded with flat fins for better heat transfer. The condenser is always installed at the front of the vehicle, ahead of the radiator. A typical condenser installation is shown in **Figure 5-14.**

Receiver-drier

Some older vehicles use a **receiver-drier,** located between the condenser and expansion valve. The receiver-drier holds extra refrigerant, and separates the liquid and vaporized refrigerant. The receiver-drier also contains desiccant to absorb any moisture that has entered the refrigeration system. Some receiver-driers contain a sight glass. See **Figure 5-15.**

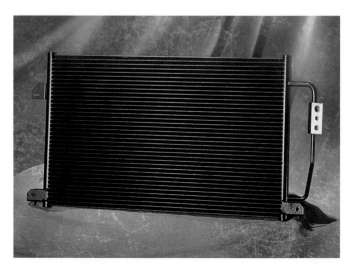

Figure 5-14. *The condenser is installed ahead of the radiator where air can be pulled through by a fan or forced through by vehicle movement. (Modine)*

Figure 5-15. *This older type receiver drier is equipped with a sight glass. Newer receivers have no sight glass.*

Filter

Some systems have a **filter** installed in the line between the condenser and flow restrictor. The filter traps metal and rubber particles to keep them from damaging other refrigeration system parts. A typical filter is shown in **Figure 5-16.**

Aftermarket filters are sometimes installed when the system has suffered a major compressor failure, and there is no economical way to remove all the debris from the refrigeration system. Aftermarket filters are also installed in the liquid line between the condenser and flow restrictor.

Sight Glass

On older R-12 systems a **sight glass** was always used, **Figure 5-17.** The technician could get a general idea of the amount of refrigerant in the system by checking for bubbles in the sight glass. A few bubbles meant the system was slightly low. Many bubbles or foam meant the system was severely low. If no bubbles were present, the system was either full or completely empty. The sight glass was always installed between the condenser and

Figure 5-16. *When a refrigeration system becomes contaminated due to catastrophic compressor failure, a filter should be installed between the condenser and the flow restrictor.*

the flow restrictor. Some sight glasses were installed at the top of the receiver-drier, while others were in the line to the evaporator. Cycling clutch type refrigeration systems do not have a sight glass. Sight glasses are sometimes installed in R-134a systems, but they should not be used to check the state of charge. Bubbles will appear in the sight glass of an R-134a system, even when the system is properly charged.

Flow Restrictor

The **flow restrictor** is installed at the entrance to the evaporator. It reduces the flow of refrigerant into the evaporator, lowering evaporator pressure. As mentioned earlier, the system is often identified by the type of flow restrictor. There are two kinds of refrigerant flow restrictors.

Fixed Orifice Tube

The **fixed orifice** is used on many modern refrigeration systems. The fixed orifice is a non-adjustable opening, usually installed in the tubing leading into the evaporator. See **Figure 5-18.** The word *orifice* means opening. The orifice allows refrigerant to enter the evaporator at a controlled rate. The size of this hole is not adjustable. Evaporator pressure is controlled by controlling the amount of refrigerant drawn into the compressor.

Figure 5-17. *This sight glass is located in the receiver-drier. Other sight glasses are located in the line between the receiver-drier and the expansion valve, or in the expansion valve assembly.*

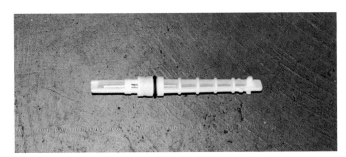

Figure 5-18. *The opening of an orifice tube flow restrictor does not change. Varying the compressor output controls refrigerant flow. The orifice is usually located at the inlet tube of the evaporator.*

Some systems using orifice tubes are called *cycling clutch orifice tube (CCOT)*, *fixed orifice tube (FOT)*, *Ford orifice tube cycling clutch (FOTCC)*, *variable displacement orifice tube (VDOT)*, thermostatic switch systems, pressure switch systems, or accumulator systems.

Expansion Valve

A flow restrictor with a variable opening is called an **expansion valve.** An expansion valve is sometimes called a **thermostatic expansion valve (TXV).** The expansion valve has a small shaft with a large end. The large end section can partially block or seal off the opening. The needle is attached to a diaphragm. On the other side of the diaphragm is a chamber. The chamber is attached to a tube, which extends into a cavity at the outlet end of the evaporator. The chamber and tube are sealed and contain a gas. **Figure 5-19A** shows the placement of an expansion valve in relation to the evaporator. A variation of the expansion valve, called a **block valve,** is shown in **Figure 5-19B.**

If the evaporator temperature becomes too low, the gas contracts. This causes the diaphragm to move and close the valve. Closing the valve cuts off refrigerant flow into the evaporator. Once the evaporator temperature begins to rise, the gas expands and causes the diaphragm to open the needle valve. Refrigerant can once again enter the evaporator. This cycle repeats as necessary.

 Note: Both types of flow restrictors will have a mesh screen just ahead of the restrictor in the high pressure (liquid) line. This screen catches any debris that might clog the restrictor.

Evaporator Pressure Controls

Some older vehicles have **evaporator pressure controls** installed between the evaporator and the compressor. The purpose of any evaporator pressure control is to keep condensed moisture from freezing on the evaporator. These devices are all variations of a pressure regulating valve. As long as the evaporator pressure is above 28-30 psi (193-207 kPa), the valve is open. When the evaporator pressure drops below this figure, the valve closes, keeping the compressor from drawing any more refrigerant from the evaporator.

When more refrigerant enters through the flow restrictor, evaporator pressure goes up and the valve opens. Keeping the evaporator pressure at 28-30 psi (193-207 kPa) or above keeps the evaporator temperature above 32°F (0°C) and the condensed moisture cannot freeze. **Figure 5-20** shows the location of a typical evaporator pressure control. Some names for these pressure controls are **suction throttling valve (STV), pilot operated absolute (POA), valves in receiver (VIR),** and **evaporator pressure regulator (EPR).**

Connecting Lines

Metal tubing and *flexible hoses* connect the components of the refrigeration system. Metal tubing is used between parts that do not move. Hoses connect parts that

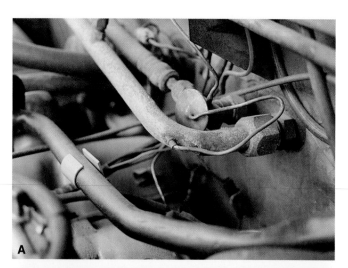

Figure 5-19. *A—This type of expansion valve is located at the entrance to the evaporator. The sensing bulb is attached to the evaporator outlet line. B—All of the parts are contained inside of the block expansion valve. Note the attachment of inlet and outlet tubes of the evaporator.*

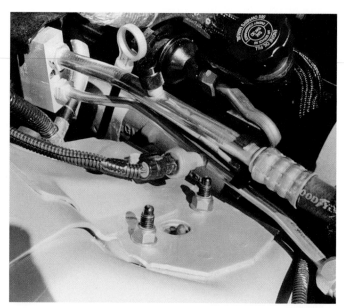

Figure 5-20. *The clutch cycling switch can be installed anywhere on the low pressure side of the refrigeration system. This switch is installed on the line between the evaporator and accumulator. Other clutch cycling switches are installed on the accumulator body.*

move in relation to each other, for example, the engine-mounted compressor and the body-mounted condenser. Tubing and hoses must stand up under high pressures and not leak.

Sometimes the lines used on a refrigeration system are called vapor or liquid lines. Vapor lines carry refrigerant vapor when the system is operating normally. Examples would be the lines from the evaporator to the compressor, and from the compressor to the condenser. Liquid lines carry liquid refrigerant during normal system operation. The line from the condenser to the flow restrictor is an example of a liquid line.

Service Fittings

Service fittings, sometimes called *service valves* or *service ports,* allow the technician to gain access to the refrigeration system. Service fittings allow pressure testing, refrigerant removal, recharging, and other service operations. Every type of refrigerant requires the use of a different type of service fitting. Service fittings will be discussed in more detail in later chapters.

Seals and Gaskets

To hold in pressurized refrigerant, seals and gaskets are used through the refrigeration system. There are many variations of seals and gaskets. Seals and gaskets will be discussed in more detail in Chapter 7.

Which Parts Are Used on Which Systems?

Not all of the parts described in this chapter are used on every refrigeration system. There is considerable variation as to what parts are used on which system. A few general rules do apply to refrigeration systems made since the 1960s. For instance:

❑ Most refrigeration systems will never have both an accumulator and a receiver-drier. A few medium-duty trucks have both an accumulator and receiver-drier.
❑ An evaporator pressure regulator valve is never used with a fixed orifice flow restrictor.
❑ Usually a system with an expansion valve type of flow restrictor will not have a cycling clutch compressor.
❑ Systems with a cycling clutch will not have an evaporator pressure control valve.
❑ Most systems with a cycling clutch will not have a sight glass.
❑ Only a few systems have a filter.

These rules apply in most cases. Always consult the service literature and make a careful search of the system to determine just what is being used.

Refrigerants

The substance circulating inside of the refrigeration system is called **refrigerant.** Using the right refrigerant is critical to the proper operation of the refrigeration system,

Figure 5-21. Most importantly, the refrigerant must be able to evaporate and condense at the pressures and temperatures developed by the refrigeration system.

The two refrigerants in common use, R-134a and R-12, will evaporate at –14.7°F (–26°C) and –21.7°F (–30°C) at sea level. To allow refrigerants to evaporate at a temperature that will not make the air too cold, they must be pressurized. Evaporator pressures are usually held at about 21-30 psi (145-207 kPa), depending on the type of refrigerant and other system design factors. Controlling the pressure raises the refrigerant boiling point to slightly over 32°F (0°C). To allow the refrigerant to condense to a liquid, pressures in the condenser are raised to about 200 psi, (1380 kPa), the temperature at which the refrigerant will condense is raised to about 135°F (57°C), which is much higher than outside air temperature. Condenser pressures may go much higher on very hot days. Note these pressures can easily be produced by the refrigeration system.

The Refrigeration Cycle

In this section, we will discuss how refrigeration principles are used with physical components to make the air conditioning system work. As you read, concentrate on the overall process rather than on individual components. Refer to **Figure 5-22** as you read these sections.

Cooling Phase—Evaporator

We will begin the basic refrigeration cycle at the evaporator. The evaporator is the place where heat is transferred from the air to the refrigerant.

The refrigerant enters the evaporator as a low pressure, low temperature liquid. Warm air passes through the evaporator. Heat from the air is transferred to the evaporator metal by convection. The heat passes from the evaporator metal to the refrigerant by conduction.

Figure 5-21. *Refrigerant is the medium used to transfer heat from inside the car.*

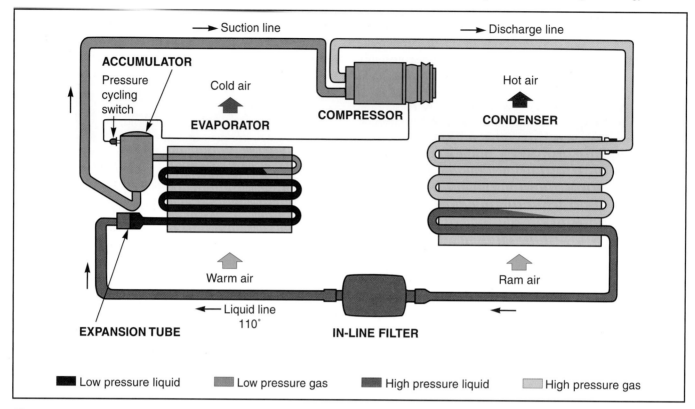

Figure 5-22. *A typical cycling clutch orifice tube air conditioning system. (Four Seasons)*

Since the pressure is low, the heat causes the refrigerant to turn to a gas, or vaporize. As the refrigerant vaporizes, it absorbs heat from the air. This heat becomes latent heat in the refrigerant vapor. The cooled air enters the passenger compartment. The refrigerant leaves the evaporator as a low pressure, low temperature vapor. The lines and accumulator will be cold to the touch. The vaporized refrigerant travels to the accumulator, on its way to the compressor.

Refrigerant Superheat in the Evaporator

Earlier in this chapter, superheat was discussed as it applies to steam. This section applies the principle of superheat to refrigerant. You know from previous sections that refrigerant evaporates (turns to vapor) when it absorbs enough heat. The refrigerant vapor can also absorb additional heat, which is called superheat. For example, assume that in a particular evaporator, the refrigerant evaporates at 21°F (11.6°C). The temperature of the refrigerant as it leaves the evaporator is 31°F (17°C). This means that the refrigerant has 10°F (5.5°C) of superheat. The refrigerant has absorbed extra heat from the air passing over the evaporator.

A small amount of superheat increases evaporator efficiency. Superheat also ensures that no liquid refrigerant reaches the compressor, where it could damage the compressor valves. An evaporator operating at maximum efficiency will have about 8–12°F (4.5–6.6°C) of superheat at the evaporator outlet.

Too much superheat, however, can damage the compressor. Excessive superheat can cause oil to collect in the evaporator. This occurs because the heated refrigerant molecules expand and cannot hold the heavier oil

molecules in suspension. The oil molecules fall to the bottom of the evaporator and do not return to the compressor. If the oil cannot return to the compressor, the compressor will be damaged from lack of lubrication. Excessive superheat can occur because of a refrigerant undercharge. For this reason, it is very important to make sure the refrigeration system has the right amount of both refrigerant and oil.

Pressure Increase Phase—Compressor

The refrigerant vapor is drawn from the accumulator into the compressor. In the compressor, the refrigerant is packed, or compressed, into a smaller space. Forcing the refrigerant vapor to occupy a smaller space does two things:
- ❏ It increases the pressure of the refrigerant.
- ❏ It forces the heat in the vapor into a smaller area.

The refrigerant enters the compressor as a low temperature, low pressure vapor, and leaves as a high temperature, high pressure vapor. Except on mild days, the vapor will be superheated. This means the vapor is hotter than the temperature at which it will condense. The lines containing the superheated refrigerant will be very hot to the touch. From the compressor, the high pressure vapor travels to the condenser.

Heat Transfer Phase—Condenser

From the compressor, the refrigerant enters the condenser as a high temperature, high pressure vapor. Air is constantly passing through the condenser, either through vehicle movement or pulled by the engine fan.

Although this air is the same temperature as the air passing through the evaporator, it is not as hot as the compressed refrigerant.

Heat in the refrigerant is transferred to the condenser metal by conduction. The heat passes from the condenser metal to the outside air by convection. At this time, any superheat is removed. Although the refrigerant is under high pressure in the condenser, the loss of heat causes the refrigerant to condense, or turn to a liquid. As the refrigerant condenses, it gives up the latent heat it absorbed in the evaporator. This heat migrates out through the condenser metal to the atmosphere. The heated air passes through the vehicle radiator and exits out of the bottom of the engine compartment.

The refrigerant leaves the condenser as a high pressure, high temperature liquid. Even though the liquid refrigerant is still hot, it has changed to a lower state and is now ready to be reused in the evaporator. On some systems, the refrigerant will go through a receiver-drier, on its way to the flow restrictor.

Refrigerant Subcooling in the Condenser

The main purpose of the condenser is to remove heat from the refrigerant vapor until it condenses back to a liquid. An efficient condenser and fan will lower the refrigerant temperature at least 8°F (4.4°C) below its condensation point. This is called **subcooling.** Subcooling is important because it ensures that all refrigerant vapor is condensed back to a liquid and that only liquid refrigerant is present at the expansion valve or orifice tube.

Since liquid refrigerant removes heat by changing from liquid to vapor in the evaporator, any refrigerant vapor entering the evaporator reduces system efficiency. Extra subcooling does not present a problem, since the system becomes more efficient as more heat is removed from the refrigerant.

Proper subcooling is especially important when the vehicle uses an orifice tube system. Systems with expansion valves have a receiver-drier between the condenser and expansion valve. The receiver is designed to separate vapor from liquid refrigerant. Orifice tube systems have no receiver-drier, and vapor could reach the evaporator if there is no subcooling. Some large vehicles with orifice tube systems have a tank between the condenser and expansion valve to perform the vapor separation once done by the receiver-drier.

Pressure Decrease Phase—Flow Restrictor

The liquid refrigerant tries to enter the evaporator but is stopped by the flow restrictor. The action of a flow restrictor is similar to the nozzle of a water hose. It sprays the liquid refrigerant in a sheet or mist of atomized particles. Breaking the refrigerant into small particles helps it to vaporize easily.

The flow restrictor also works with the compressor to control the evaporator pressure. The compressor is always trying to draw refrigerant from the evaporator. The restrictor controls evaporator pressure by controlling the amount of refrigerant entering the evaporator. Since the pressure is low, the refrigerant in the evaporator vaporizes, beginning the cycle again.

Factors That Affect System Performance

Until now we have simply referred to conditioning warm humid air passing through the evaporator. However, the amount of heat and humidity present in the air, the amount of sunlight, and other factors have a major effect on the efficiency of a vehicle's refrigeration system, and therefore the cooling power of the air conditioner.

Temperature

The **heat load** is the total effect of heat and humidity of the surrounding, or **ambient air.** The higher the temperature of the air entering the evaporator, the harder the refrigerant in the evaporator must work to remove the heat. Also, the hotter the outside air passing through the condenser, the harder it is for the refrigerant in the condenser to give up its heat.

Humidity

The effect of humidity on refrigeration system operation is often overlooked. Water vapor is turned into liquid water as it passes through the evaporator. Changing state from vapor to liquid causes the water to release latent heat. This heat, which is considerable, must be absorbed by the refrigerant.

Humidity is always referred to as **relative humidity.** The amount of humidity is always relative (proportional) to the air temperature. Warm air can hold more water vapor than cooler air. For example, on a typical summer morning, the relative humidity might be 40% at 70°F (21°C). Later in the day, the relative humidity may change to 20% at 90°F (32°C). At both temperatures the air has the same amount of water vapor, but the amount of water the air could hold has increased with temperature.

Sunlight

Earlier in this chapter we discussed heat transfer by radiation. When the sun is directly overhead, infrared radiation is striking the vehicle directly, causing a considerable heating effect. This is referred to as *sunlight load* or simply **sunload.** When the sun is low in the sky, the infrared radiation is striking the vehicle at an angle, and heating is less. Overcast skies reduce, but do not totally eliminate the effect of infrared radiation. At night, the infrared radiation from the sun is not a factor.

All of these factors change the heat load and greatly affect the operation of the refrigeration system. As an example, an air conditioner can produce 45°F (7°C) air when the outside temperature is 80°F (27°C) with low sunlight and humidity. However, it may be able to lower the temperature only to 60°F (15°C) when the outside air is at 95°F (35°C) with high humidity and the sun overhead.

Vehicle Design

Refrigeration system performance can vary from vehicle to vehicle, depending on two factors. The first is passenger compartment space. The larger the passenger compartment, the harder the refrigeration system must work to cool the area. Obviously, it is much easier to cool the passenger compartment of a pickup truck or compact car than a minivan.

The second design factor that can affect system performance is the vehicle's interior and exterior colors. Darker colors tend to attract and retain more heat than lighter colors. In some areas, you can see a notable difference in system performance between two identical model cars with the only difference being their color.

Electrothermal Cooling

Some small cooling units rely on *electrothermal cooling.* This type of cooling uses a material called *electrothermal ceramic.* Electrothermal ceramics react to electricity in an unusual way. Electrical current passing through an electrothermal ceramic carries heat away from one side of the ceramic material. The heat is then released on the other side of the material. See **Figure 5-23.**

Using electrothermal ceramic allows small areas to be cooled by current flow only. There is no need for a conventional refrigeration system. The disadvantage of electrothermal ceramics is that they draw large amounts of current for the amount of cooling produced. For this reason they are impractical for cooling large areas. However, they work well to cool small areas. On some new vehicles electrothermal ceramics are used to create a *beverage cooler* located in the dashboard. Beverage coolers usually hold one or two soft drinks.

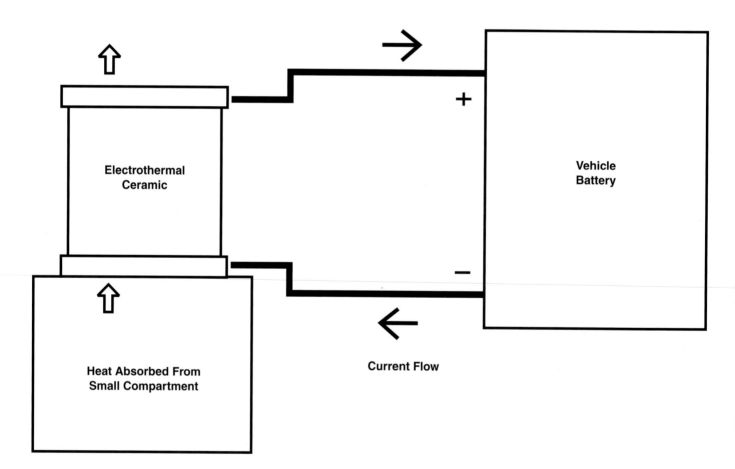

Cooling a Small Area With an Electrothermal Ceramic

Figure 5-23. *Electric current is flowing through the electrothermal ceramic. Heat is being carried away from one side of the ceramic, lowering the temperature in the enclosed area. On the other side of the electrothermal ceramic, heat is being released into the surrounding atmosphere.*

Summary

Heat and cold do not exist as such, but are actually varying levels of heat. Heat transfer is the movement of molecular vibrations from an object with more heat to an object with less heat. The three methods of heat transfer are radiation, conduction, and convection. Conduction and convection are the main heat transfer means used by the refrigeration system.

The three states of matter are solids, liquids, and gases. Changing from a solid to a liquid or from a liquid to a gas is called changing to a higher state, and requires the addition of heat. Changing from a gas to a liquid or liquid to a solid is called changing to a lower state, and requires the removal of heat. The refrigeration system makes the most use of vaporization and condensation.

All liquids and gases contain sensible heat and latent heat. Sensible heat is heat that can be felt. Latent heat is hidden in the liquid or gas. Only liquids and gases contain latent heat. Solids contain no latent heat, but must absorb heat to change to a liquid. To change from a liquid to a gas, more heat must be absorbed. This heat does not change the temperature, just the state. The types of latent heat are latent heat of freezing, latent heat of melting, latent heat of vaporization, and latent heat of condensation. Latent heat is important to the operation of the refrigeration system because change of state removes a large quantity of heat. Superheat is any heat added after the change of state.

Pressure affects the change of state. The more the pressure is increased, the higher the temperature at which a change of state takes place. The most important principles in the operation of the refrigeration system are heat transfer, change of state, latent heat, and the effect of pressure on change of state. Removing heat is an obvious benefit of the air conditioner. Reducing the humidity also helps us feel more comfortable. Air movement removes more heat than would occur when the air is still.

The substance circulating inside the refrigeration system is called refrigerant. Using the right refrigerant is critical to the proper operation of the refrigeration system. The most important property of refrigerant is it is able to evaporate and condense at the pressures produced by the refrigeration system.

There are four major parts to the refrigeration cycle. The first phase is cooling, which takes place at the evaporator. The next phase is pressure increase, which occurs inside of the compressor. The third phase is the heat transfer phase, which takes place in the condenser. The fourth and last phase is the pressure decrease phase that occurs at the flow restrictor

The main refrigeration components are the evaporator, compressor, condenser, and flow restrictor. Other refrigeration system parts are the accumulator, the evaporator pressure controls, the receiver drier, the sight glass, and filter, when used.

There are two kinds of flow restrictor. The fixed orifice is used with a cycling clutch. The expansion valve is usually found on older vehicles. Metal tubing and flexible hoses connect the other components of the refrigeration system. Vapor lines carry refrigerant vapor when the system is operating normally. Liquid lines carry liquid refrigerant during normal system operation. The actual use of parts depends on the design, the manufacturer, and the age of the system.

The heat load is the total effect of heat and humidity of the outside, or ambient, air. Other factors affecting system performance are temperature, humidity, sunlight, and vehicle design.

Review Questions—Chapter 5

Please do not write in this text. Write your answers on a separate sheet of paper.

1. In science, there is no such thing as _____.

2. Heat always travels to an object _____ than its surroundings.

3. Match the heat transfer method with the descriptions.

   (A) Convection

   (B) Conduction

   (C) Radiation

   How the sun heats the earth. ___

   Why a hot exhaust manifold could burn you. ___

   How refrigerant cools the evaporator metal. ___

   How a chicken is cooked over a barbecue pit. ___

   Makes use of infrared rays. __

   Heat transfer by direct contact. ___

4. The three states of matter are _____, _____, and _____.

5. To change water to its next highest state, it should be _____.

6. Sensible heat can be _____.

7. Define superheat.

8. Raising pressure will _____ the temperature at which a change of state takes place.

9. High humidity makes it hard for the body to remove perspiration by _____.

10. Which of the following is a heat exchanger?

    (A) Evaporator.

    (B) Condenser.

    (C) Compressor.

    (D) Both A & B.

11. In an operating refrigeration system, the refrigerant in the evaporator is under _____ pressure.

12. In an operating refrigeration system, the refrigerant in the condenser is under _____ pressure.

13. Which of the following is *always* used on an automotive refrigeration system?
    (A) Muffler.
    (B) Receiver-drier.
    (C) Flow restrictor.
    (D) Accumulator.

14. Is the refrigerant line between the condenser and flow restrictor a liquid line or a vapor line?

15. What three things make up the heat load on the refrigeration system?

ASE Certification-Type Questions

1. All of the following statements about heat are true, *except:*
   (A) all substances are made up of atoms.
   (B) all molecules vibrate.
   (C) the vibration of molecules can be felt as heat.
   (D) the less molecular vibration, the more heat.

2. Technician A says heat always travels from a warmer object to a colder one. Technician B says when all objects in an area are at the same temperature, no heat transfer will occur. Who is right?
   (A) A only.
   (B) B only.
   (C) Both A & B.
   (D) Neither A nor B.

3. The usual way to cause a change of state is by adding or removing _____.
   (A) pressure
   (B) refrigerant
   (C) air flow
   (D) heat

4. Technician A says only a liquid can change state up or down. Technician B says a chemical change can also be a change of state. Who is right?
   (A) A only.
   (B) B only.
   (C) Both A & B.
   (D) Neither A nor B.

5. All of the following statements about latent heat are true, *except:*
   (A) latent heat is often called hidden heat.
   (B) solids contain no latent heat.
   (C) changes of state release a small amount of latent heat.
   (D) adding or removing latent heat does not change the temperature of a substance.

6. Technician A says if the pressure is held constant, a substance will have different vaporizing and condensing temperatures. Technician B says higher pressures cause higher boiling points. Who is right?
   (A) A only.
   (B) B only.
   (C) Both A & B.
   (D) Neither A nor B.

7. The amount of water vapor in the surrounding air is called the _____.
   (A) humidity
   (B) latent heat
   (C) sensible heat
   (D) boiling point

8. Which of the following refrigeration system devices restricts the flow of refrigerant into the evaporator?
   (A) Accumulator.
   (B) Expansion valve.
   (C) Receiver-drier.
   (D) Condenser.

9. What two devices are *never* found on the same refrigeration system?
   (A) Compressor and condenser.
   (B) Accumulator and flow restrictor.
   (C) Accumulator and condenser.
   (D) Accumulator and receiver-drier.

10. When water vapor condenses on the evaporator, which of the following happens?
    (A) It turns to ice.
    (B) It releases latent heat.
    (C) It absorbs latent heat.
    (D) The water drips on the condenser.

Chapter 6

Refrigerants, Refrigerant Oils, and Related Chemicals

After studying this chapter, you will be able to:

❑ Identify the purposes of refrigerants.
❑ List regulations involving the use and handling of refrigerants.
❑ Identify current refrigerants and identify systems where they are used.
❑ Compare advantages and disadvantages of various refrigerants.
❑ Identify modern refrigerant oils and the refrigerants they are used with.
❑ List the purposes of refrigeration desiccants.
❑ Explain the purposes of refrigeration system flushing compounds.
❑ Identify refrigeration system flushing compounds and explain why some are no longer used.
❑ Explain the purposes of refrigeration leak detector dyes.

Technical Terms

Chloroflurocarbons (CFCs)
Chlorine (Cl)
Montreal Protocol
Clean Air Act
Environmental Protection
   Agency (EPA)
R-134a
HFC-134a
Tetrafluoroethane
   (CF$_3$CH$_2$F)
One pound can
R-12

CFC-12
Freon
Dichlorodifluoromethane
   (CCl$_2$F$_2$)
Retrofitting
Hydrochlorofluorocarbon
   (HCFC)
American Society of
   Refrigerating and Air
   Conditioning Engineers
   (ASHRAE)
Blended refrigerants

Blend
HCFC-22
R-22
Azeotropic blends
Zeotropic blend
Drop-in refrigerant
Significant New
   Alternatives Policy
   SNAP program
Barrier hoses
Refrigerant oil
Miscible

Polyalkylene glycol (PAG)
Mineral oil
Polyol ester (POE)
Hydroscopic
Alkylbenzene (AB)
Desiccants
Flushing agents
R-11
R-113
R-141b
Leak detector dyes
Biocide

The following chapter identifies and gives the uses of modern refrigerants. It also provides information about technical and legal aspects of modern refrigerants. After studying this chapter, you will be able to select the proper refrigerant for a particular system.

Refrigerant Purposes

Before identifying the various kinds of refrigerants, it is important to discuss just what a refrigerant does. The most obvious job of a refrigerant is to absorb and release heat. It does this by changing state (vaporizing and condensing) at pressures developed by the refrigeration system.

Another purpose of the refrigerant is to carry small amounts of compressor lubricating oil without chemically reacting with it. The refrigerant should not react chemically with the metals, rubber parts, or desiccants in the system. Chemical reactions would damage the components and change the chemical makeup of the refrigerant, possibly reducing its ability to transfer heat.

Because of the danger of leaks caused by accidents, any refrigerant used in a motor vehicle should be non-flammable and nonpoisonous. It also should not be so expensive as to make accidental loss a large financial burden.

Modern refrigerants must also be environmentally friendly. They should not harm the earth's environment by destroying the ozone layer or contributing to global warming.

Refrigerants and Governmental Regulation

At one time, the refrigerant situation was very clear. The only refrigerant in use was R-12. However, after many years of use, R-12 was declared unfit for use in refrigeration systems. The reason R-12 is no longer used had nothing to do with its effectiveness as a refrigerant.

Two documents that have greatly affected refrigerants, automotive air conditioning, and the HVAC industry as a whole are the Montreal Protocol and the Clean Air Act.

Montreal Protocol

In the 1970s, scientists discovered that chemicals called *chloroflurocarbons* or *CFCs* were damaging the ozone layer, which protects the earth from harmful solar radiation. It was found that CFCs were breaking down into their component parts, including the element *chlorine (Cl)*. As the chlorine reached the upper atmosphere, it was combining to deplete the ozone layer, **Figure 6-1.**

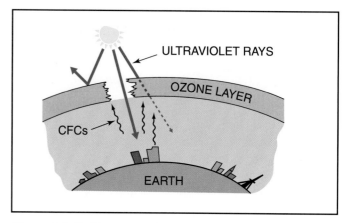

Figure 6-1. *Refrigerants and other compounds have contributed to the depletion of the Earth's ozone layer.*

At the time, CFCs were used in aerosol spray cans as a propellant. The United States and many other nations quickly banned CFC usage in spray cans. However, CFC compounds continued to be used by a few nations in spray cans, and throughout the world in many other applications, including automotive refrigerants.

In 1987, leaders from the major industrialized nations reached and signed an international agreement called the *Montreal Protocol.* This agreement called for the reduction and elimination of all ozone depleting chemicals, including CFCs. Since 1987, more than 90 nations have ratified the Montreal Protocol.

Clean Air Act

While the Montreal Protocol works at the international level, a second document enforces the Protocol's standards in the United States at the national level. This document is called the *Clean Air Act.* While the Clean Air Act regulates more than refrigerants, sections of the Act outline restrictions regarding the manufacture, import, release, and service of refrigerants and refrigerant systems.

The Clean Air Act provides severe penalties for any infractions. A section of the Act charges the *Environmental Protection Agency (EPA)* with implementation and enforcement of the regulations. It also provides for substantial rewards, encouraging individuals and businesses to turn in violators.

 Warning: Businesses or technicians who violate any provision of the Clean Air Act may be fined, lose their certifications, and possibly face criminal charges.

In cases of conflict between the Montreal Protocol, the Clean Air Act, and any other regulatory law regarding refrigerants, the more stringent law is enforced.

Types of Refrigerants

The major issue facing the automotive air conditioning industry is the type of refrigerants to be used in the future. After the ban on CFCs was announced, various alternative refrigerants were considered and almost all were rejected. Serious questions arose about the effect of using other than the original refrigerant in existing systems. Replacement refrigerants have become available, but using them required overcoming some problems. Other environmentally friendly refrigerants are available and acceptable for stationary air conditioners but not for vehicles.

R-134a

There are many refrigerants on the market. However, only one refrigerant meets all manufacturer's requirements for motor vehicle use. That refrigerant is ***R-134a,*** sometimes called ***HFC-134a.*** Its chemical name and formula is ***tetrafluoroethane*** (***CF_3CH_2F***).

R-134a has been installed in all OEM vehicle air conditioners beginning with the 1994 model year. The molecular structure of R-134a does not include a chlorine atom, so there is no danger of ozone layer depletion. While R-134a does have some effect on global warming, it is many times less damaging than R-12.

If a container of liquid R-134a is opened to atmospheric pressure, it will boil at –14.7°F (-26°C). To keep condensed moisture (water vapor) from freezing on the evaporator, its surface temperature must be kept at about 32°F (0°C). To maintain this temperature, the R-134a in the evaporator must be under pressure. Most R-134a systems use a cycling compressor clutch. The clutch is disengaged when evaporator pressure reaches about 21 psi (150 kPa).

To allow the vaporized R-134a to condense, the compressor raises the pressure of the refrigerant entering the condenser. Raising the pressure to 200 psi (1380 kPa) will allow the R-134a to condense at any temperature below 130°F (54°C). Since the outside air is always cooler than this, the refrigerant will condense easily if enough air is drawn through the condenser. On very hot days, an air conditioner may develop condenser pressures of over 400 psi (2760 kPa). **Figure 6-2** is a pressure-temperature chart for R-134a.

R-134a is supplied in 30 pound (13.5 kg) cylinders, or one pound cans, **Figure 6-3.** All R-134a cylinders are painted light blue for easy identification.

Note: The term *one pound can* is used to describe small cans of refrigerant. However, a one pound can almost never contains one pound of refrigerant. In terms of actual content, a "one pound can" of R-134a contains 12 ounces (340 g) of refrigerant while R-12 cans contain 14 ounces (397 g).

R-134a Temperature-Pressure Chart

Temperature °F (°C)		Pressure psi (kPa)	Temperature °F (°C)		Pressure psi (kPa)
16 (–9)		15 (106)	100 (38)		124 (857)
18 (–8)		17 (115)	102 (39)		129 (887)
20 (–7)		18 (124)	104 (40)		133 (917)
22 (–6)	EVAPORATOR RANGE	19 (134)	106 (41)		137 (948)
24 (–4)		21 (144)	108 (42)		142 (980)
26 (–3)		22 (155)	110 (43)	CONDENSER RANGE	147 (1012)
28 (–2)		24 (166)	112 (44)		152 (1045)
30 (–1)		26 (177)	114 (46)		157 (1079)
32 (0)		27 (188)	116 (47)		162 (1114)
34 (1)		29 (200)	118 (48)		167 (1149)
36 (2)		31 (212)	120 (49)		172 (1185)
38 (3)		33 (225)	122 (50)		177 (1222)
40 (4)		35 (238)	124 (51)		183 (1260)
45 (7)		40 (272)	126 (52)		188 (1298)
50 (10)		45 (310)	128 (53)		194 (1337)
55 (13)		51 (350)	130 (54)		200 (1377)
60 (16)		57 (392)	135 (57)		215 (1481)
65 (18)		64 (438)	140 (60)		231 (1590)
70 (21)		71 (487)	145 (63)		247 (1704)
75 (24)		78 (540)	150 (66)		264 (1823)
80 (27)		88 (609)	155 (68)		283 (1948)
85 (30)		95 (655)	160 (71)		301 (2079)
90 (32)		104 (718)	165 (74)		321 (2215)
95 (35)		114 (786)	170 (77)		342 (2358)

Figure 6-2. *A chart of R-134a pressures at different temperatures. Notice how pressure rises with temperature. This relationship remains the same for all pure R-134a in any location. (General Motors)*

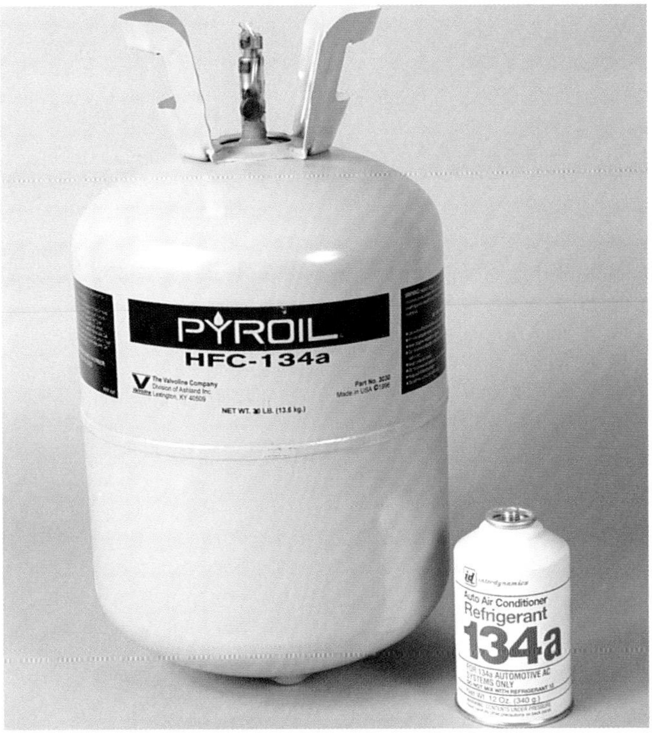

Figure 6-3. *Types of R-134a containers can vary. R-134a is sold in 30 pound and larger canisters as well as one pound cans. (Jack Klasey)*

R-12 (Freon)

R-12, or *CFC-12,* usually called *Freon,* was the only automotive refrigerant for many years. It was totally phased out of new vehicles starting with the 1994 model year. Its chemical name and formula is *dichlorodifluoromethane (CCl_2F_2).*

At atmospheric pressure, R-12 will boil at –21.7°F (-30°C). To keep condensed moisture from freezing on the evaporator, the R-12 in the evaporator is pressurized to raise its boiling point. Most R-12 systems with cycling clutches have evaporator pressure switches. These switches disengage the clutch at about 25 psi (172 kPa). On older systems the evaporator pressure control valves maintained evaporator pressure at roughly 28-30 psi (193-207 kPa).

The compressor raises the pressure of the R-12 entering the condenser to 180 psi (1235 kPa). This allows the R-12 to condense at any temperature below 130°F (54°C). The refrigerant can condense as cooler outside air is drawn through the condenser. A pressure-temperature chart for R-12 is shown in **Figure 6-4.** On very hot days, condenser pressures may be much higher.

R-12 is supplied in 30 pound (13.5 kg) and larger cylinders, painted white. One pound cans may still be found in some areas. See **Figure 6-5.**

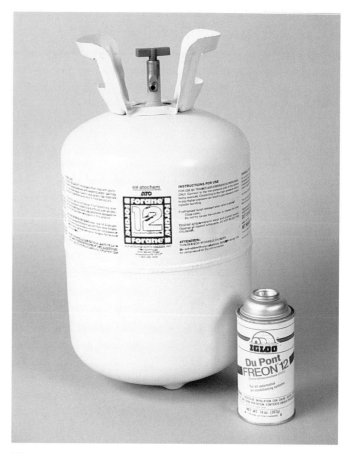

Figure 6-5. *Most commercially available R-12 is sold in 30 pound (13.4 kg) containers. R-12 was originally used in the automotive repair industry by the one pound can. (Jack Klasey)*

R-12 Temperature-Pressure Chart

Temperature °F (°C)	Pressure psi (kPa)	Temperature °F (°C)	Pressure psi (kPa)
16 (–9)	18 (127)	100 (38)	117 (808)
18 (–8)	20 (136)	102 (39)	121 (833)
20 (–7)	21 (145)	104 (40)	125 (859)
22 (–6)	22 (155)	106 (41)	129 (893)
24 (–4)	24 (165)	108 (42)	133 (917)
26 (–3)	25 (175)	110 (43)	136 (940)
28 (–2)	27 (185)	112 (44)	140 (969)
30 (–1)	28 (196)	114 (46)	145 (997)
32 (0)	30 (207)	116 (47)	149 (1027)
34 (1)	32 (219)	118 (48)	153 (1057)
36 (2)	33 (230)	120 (49)	158 (1087)
38 (3)	36 (249)	122 (50)	162 (1118)
40 (4)	37 (255)	124 (51)	167 (1150)
45 (7)	42 (287)	126 (52)	171 (1182)
50 (10)	47 (322)	128 (53)	176 (1215)
55 (13)	52 (359)	130 (54)	181 (1248)
60 (16)	58 (398)	135 (57)	194 (1334)
65 (18)	64 (440)	140 (60)	207 (1425)
70 (21)	70 (484)	145 (63)	220 (1519)
75 (24)	77 (531)	150 (66)	235 (1618)
80 (27)	84 (580)	155 (68)	250 (1721)
85 (30)	92 (633)	160 (71)	265 (1828)
90 (32)	100 (688)	165 (74)	281 (1940)
95 (35)	108 (746)	170 (77)	298 (2057)

The shaded area on the left side is labeled EVAPORATOR RANGE; the shaded area on the right side is labeled CONDENSER RANGE.

Figure 6-4. *A chart of R-12 pressures at different temperatures. R-12 pressure also rises with temperature. (General Motors)*

R-12 Supply

Manufacturing of R-12 in the United States ceased after December 31, 1995, in accordance with the Montreal Protocol and the Clean Air Act. At the same time, the importing of R-12 was also banned. However, several developing countries were granted an extension on the R-12 manufacturing ban. This led to the smuggling of R-12 refrigerant.

Since the halt of production at the beginning of 1996, the supply and price of R-12 in the United States and North America had depended on three factors:

❏ The amount of R-12 that was originally stockpiled.
❏ The amount of R-12 recovered from systems serviced.
❏ The amount of R-12 illegally imported.

Estimates indicate the supply of R-12 is decreasing rapidly. Some areas have already begun to experience shortages of R-12. In areas where R-12 is in low supply, shops are experiencing increased business in converting R-12 systems to R-134a or another refrigerant. Today, most of the R-12 in older vehicles is recovered and reused. Some chemical manufacturers are offering recycled R-12.

Demand is determined by the number of operating R-12 systems. This number should decrease as more and more vehicles with R-12 systems are either retired or converted to R-134a or another refrigerant.

Caution: Some confiscated caches of smuggled R-12 have been found to be contaminated with water and other impurities. Be sure any R-12 you use comes from reputable sources.

R-134a versus R-12

If you compare pressure and temperature figures, you might note R-12 is a little more efficient than R-134a. R-12 evaporates at a slightly higher pressure and condenses at a slightly lower pressure than R-134a.

The compressor in an R-12 system would not need to work as hard as the compressor in an R-134a system to produce the same cooling effect. Since the pressure-temperature differences are small, the two refrigerants can be assumed to operate in the same manner. If the environmental damage and expense of the two refrigerants are considered, R-134a is the better choice. Most manufacturers compensate for the use of R-134a by using a slightly larger condenser or by increasing airflow through the condenser with dual electric fans.

Note: In this text, unless reference is made to a specific refrigerant, the term "refrigerant" when used by itself, should be taken to mean R-134a.

Retrofitting

As mentioned earlier, an increasingly common practice today is retrofitting. **Retrofitting** is the process of converting an R-12 refrigeration system to operate on another type of refrigerant. R-134a is ideal for retrofitting since its pressure-temperature relationship is similar to R-12. In addition, hoses, O-rings, and other rubber parts usually do not have to be changed unless they are already leaking. The decision to retrofit a vehicle is the vehicle owner's. Retrofitting will be covered in Chapter 23.

Refrigerant Classification

Each refrigerant is classified by its chemical composition. You have already been introduced to two of the three types; CFCs, and HFCs. A third one is **hydrochlorofluorocarbon** or **HCFC.** Each of these stands for:

❑ CFC—Chlorofluorocarbons.
❑ HFC—Hydrofluorocarbons.
❑ HCFC—Hydrochlorofluorocarbons.

The abbreviated terms for refrigerant chemical composition are primarily used by the EPA. To prevent confusion, since many companies make the same refrigerant, the **American Society of Refrigerating and Air Conditioning Engineers (ASHRAE)** created a universal identification system. This identification systems designates all refrigerants to begin with the letter R. The number after the refrigerant identifies the molecular composition.

Blended Refrigerants (Azeotropic and Zeotropic Blends)

A fourth classification is **blended refrigerants.** Many refrigerants are blends of various compounds. A **blend** is a combination of two or more different refrigerants and sometimes other gases to create a new refrigerant. Some blended refrigerants contain significant amounts of R-134a, often between 59 and 80%. Other refrigerant blends contain **HCFC-22,** usually called **R-22.** R-22 is a hydrochlorofluorocarbon refrigerant and can damage R-12 and R-134a system hoses and O-rings. If a refrigerant blend containing R-22 is used, all of the system hoses, O-rings, and other rubber parts must be replaced with R-22 compatible components. Most technicians feel it is simpler to use pure R-134a.

Note: R-22 may be phased out over the next 20-30 years. Hydrochloroflurocarbon refrigerants have some ozone depletion and global warming potential.

Some refrigerant compounds combine better than others. Refrigerants that will not separate once blended are referred to as **azeotropic blends.** Azeotropic refrigerants are not normally used in automotive systems and are in the R-500 series of refrigerant numbering.

Other blends can separate into their various component refrigerants. The name for this type of refrigerant is a **zeotropic blend.** Many replacement refrigerants used in automotive air conditioning are zeotropic blends. They fall under the R-400 series of refrigerant numbering.

Performance of zeotropic blends can be unpredictable, as they usually do not have a specific temperature at which they change. Some zeotropic blend refrigerants contain chlorine, and therefore cause harm to the ozone layer. Some blends contain small amounts of flammable gases such as propane, butane, and isobutane. Many are not compatible with refrigerant oils or common desiccants. Since the component refrigerants may separate out if they are charged as a vapor, blended refrigerants can only be charged as a liquid.

Drop-in Refrigerants

The term **drop-in refrigerant** has been used to mean a refrigerant substitute that can be added to a partially charged refrigeration system with no modifications. Currently, there is no such thing as a "drop-in" refrigerant. Not only is it illegal to mix refrigerants, no replacement refrigerant is compatible with R-12 or R-134a.

A refrigeration system using R-12 must either have R-12 added or be retrofitted. A system with R-134a must have R-134a added. Any system with a blended refrigerant must have that blend added. Blended refrigerants can be recycled and reused in the original vehicle *only*. If the refrigerant is unknown or no longer available, it must be completely reclaimed and destroyed according to specific EPA guidelines, **Figure 6-6.** Only then can the shop proceed to repair and recharge the air conditioning system.

> **Note: The EPA does not use the term "drop-in" to describe any refrigerant.**

Figure 6-6. *Contaminated or unknown refrigerant blends must be disposed of properly. The tank on this recovery machine is designated for contaminated or unknown refrigerants.*

Replacement Refrigerants

The advantage of R-134a and R-12 is they are efficient and safe. Many other refrigerants have been tried in the past, but almost all of them have drawbacks. Sulfur dioxide and ammonia are poisonous, and are used only in stationary refrigeration systems, located away from people. Propane and butane will work in an automotive refrigeration system, but are extremely flammable and explosive.

Since the manufacture of R-12 was banned, many replacement refrigerants have been developed, and some are being marketed. All replacement refrigerants other than R-134a have drawbacks.

Choosing a Replacement Refrigerant

While replacement refrigerants have been approved by the Environmental Protection Agency (EPA), their approval does not mean the refrigerant is an efficient

substitute for R-12 or R-134a. It also does not mean the refrigerant will not damage automotive refrigeration system parts. EPA approval means only the refrigerant will not damage the ozone layer, contribute to global warming, and is nonflammable.

SNAP Program

Under the Clean Air Act, the EPA examines any and all R-12 substitutes. The procedure used to review potential replacement refrigerants is called the **Significant New Alternatives Policy** or **SNAP program.**

Using the guidelines outlined in the SNAP program, the EPA examines replacement refrigerants for the following characteristics:

❑ Ozone depletion.
❑ Global warming.
❑ Flammability.
❑ Toxicity.

Acceptable refrigerants are subject to fitting, labeling, and shutoff switch conditions. The EPA has determined several refrigerants are acceptable for use as R-12 replacements. **Figure 6-7** is a list of refrigerants approved by the SNAP program for use in R-12 auto air conditioners.

Any refrigerants found to have a high global warming potential, or is ozone depleting, flammable, or toxic, are classified as unacceptable. Using refrigerants found unacceptable under SNAP is a violation of the Clean Air Act.

> **Note: The EPA updates the SNAP list regularly. Be sure to keep a copy of the most current list.**

However, SNAP's evaluation does not determine how the replacement refrigerant will perform. Some SNAP approved refrigerants are actually approved only for use in stationary equipment, such as room air conditioners, vending machines, and water coolers. EPA law allows these refrigerants to be sold. They can even be advertised for use in vehicles. However, it is *illegal* to use them in a vehicle refrigeration system.

Manufacturer Approval

All vehicle manufacturers and most aftermarket parts suppliers do not allow the use of replacement refrigerants with their systems or parts. Use of any refrigerant other than R-134a or R-12 will void the vehicle or part warranty. This applies to replacement refrigerants approved by the EPA. Remember: EPA approval *does not* mean manufacturer approval. So far, the only refrigerant approved by most manufacturers for retrofitting R-12 systems is R-134a.

Refrigerant Cost and Availability

Most alternative refrigerants are less expensive than R-12, and in some cases, R-134a. However, the cost of the refrigerant should not be the only factor. The EPA

Alternative Refrigerants to R-12 (From EPA SNAP list)			
Name	**Trade Name**	**Oil Compatible**	**Comments**
HFC-134a	R-134a	PAG/POE	Acceptable for all systems.
R-406A	GHG	MO	Contains R-22, must be used with barrier hoses.
GHG-X4 R-414A	GHG-X4 Autofrost, Chill-it	MO	Contains R-22, must be used with barrier hoses.
Hot Shot R-414B	Hot Shot Kar Kool	MO/POE	Contains R-22, must be used with barrier hoses.
FRIGC FR-12 (HCFC Blend Beta) R-416A	FRIGC FR-12	POE	Contains R-143a and butane, but not flammable.
Free Zone (HCFC Blend Delta)	Free Zone RB-276	MO	Contains a lubricant (approx. 2%).
Freeze 12	Freeze 12	MO	Approx. 80% R-134a.
GHG-X5	GHG-X5	MO	Contains R-22, must be used with barrier hoses.
Ikon-12	Same	MO or POE	Rubber seals made with Buna N (NBR, XNBR, HNBR) must be changed; barrier hoses must be used.
SP34E	Same	MO or PAG	Approx. 99% R-134a.

*Subject to fitting, label, no drop-in, and compressor shutoff switch use conditions
MO–Mineral oil. PAG–Polyalkylene glycol. POE–Polyol ester.

Figure 6-7. *This list shows the R-12 substitute refrigerants currently approved or not approved by the Environmental Protection Agency. Keep in mind approval means only it meets EPA standards for ozone, global warming, and flammability protection, not that it is guaranteed to work well in a refrigeration system. (Environmental Protection Agency)*

recommends the availability of any R-12 refrigerant substitute should be weighed against the availability of R-134a. Some blend refrigerants may not be available in all areas. If the vehicle leaves the area and needs air conditioning service, the blend refrigerant may not be immediately available.

Dedicated Service and Recovery/Recycling Equipment

Since it is illegal to mix refrigerants, every type of refrigerant must have its own dedicated service tools and recovery and recycling equipment. Every HVAC service shop must have two dedicated recovery and recycling units, one for R-12 and one for R-134a. These may be combined into a dual function machine.

Gauge hoses also must comply with regulations set forth in the Clean Air Act. Manifold gauge and service machine hoses must have a unique fitting permanently attached to the hose for each refrigerant type.

To use another refrigerant, the shop must purchase or convert another dedicated recovery/recycling unit. If the refrigeration system was in use for a long time, certain components of the blend may have leaked out. This means the system no longer contains the same refrigerant blend that was originally installed. This refrigerant must be reprocessed or destroyed according to EPA guidelines.

Service Fittings and Hoses

If a substitute refrigerant is used, the EPA requires the original service fittings be converted to fittings specific to the refrigerant. This is to keep other technicians from cross-contaminating service equipment or accidentally filling a system with the wrong refrigerant.

The components of some refrigerant blends, for example R-22, can seep through conventional hoses. When these refrigerants are used, the technician must replace all hoses with **barrier hoses** if the vehicle was not originally equipped with them.

Refrigerant Oils

Refrigerant oil is needed to lubricate the moving parts of the compressor. On systems with expansion valves, the oil lubricates the valve. The proper oil reduces compressor internal friction. This increases compressor life and reduces the engine horsepower consumed by the compressor. The oil must be compatible with the refrigerant, since it is carried through the system by the refrigerant. If the oil is not compatible with the refrigerant, it may not mix well enough to stay in the refrigerant stream. The oil may then collect at the bottom of the condenser or evaporator, and the compressor will be starved for oil. Any refrigerant oil that will mix well with refrigerant is said to be **miscible.**

Incompatible oils may collect into large droplets. These droplets can damage the compressor reed valves. The wrong type of oil may also react chemically with the refrigerant or system metals and damage parts. For this reason, different types of oils should not be mixed, since at least one of them will not be compatible with the refrigerant. When retrofitting, a small amount of the old oil will be left in the system. Usually, this does not affect the efficiency of the new oil.

Types of Refrigerant oil

Refrigerant oil is not simply a lighter version of motor oil. It is formulated so it can be carried through the refrigeration system without settling out in low spots or breaking down under the temperature extremes of the refrigeration system. In addition, refrigerant oil is manufactured under extremely clean conditions to contain no moisture or impurities. Three types of refrigerant oil are used today, **Figure 6-8.**

Figure 6-8. *Using the right kind of refrigerant oil is critical to refrigeration system life. The red label PAG oil is for use in General Motors air conditioning systems. Of the three major kinds of refrigerant oil, only POE (ester) will work with both R-134a and R-12.*

Polyalkylene Glycol (PAG)

Systems using R-134a require a type of oil called *polyalkylene glycol*, or *PAG*. This oil is synthetic (non-petroleum) oil similar in chemical makeup to cooling system antifreeze. PAG oil is usually light blue in color. It is intended for R-134a systems, and cannot be used in an R-12 system. It has a viscosity rating of about 150. This viscosity applies only to refrigerant oils, and cannot be compared to motor or gear oils.

 Note: A special 164 viscosity PAG should be used in General Motors refrigeration systems.

Mineral Oils

Petroleum based *mineral oil* is used in R-12 systems. They are often called 500 or 525 viscosity refrigerant oils. This viscosity is a standard for refrigerant oil only. Mineral oil is usually clear to light yellow in color. Mineral oil is

also used with some replacement refrigerants. However, mineral oil should never be used in an R-134a system, as it will not mix with the refrigerant. The oil will collect at the bottom of the evaporator and condenser. The compressor then becomes starved for oil and will be damaged.

Polyol Ester (POE)

A third type of oil, called *Polyol ester,* or *POE,* can be used with R-134a and R-12. It is an alcohol-based oil that is compatible with small amounts of PAG or mineral oil. POE is usually clear with no tint. It has a slight odor similar to brake fluid. A feature of POE oil is its ability to change viscosity in relation to temperature changes. POE oil is usually used when an R-12 system is retrofitted to R-134a.

 Caution: Some manufacturers do not allow the use of POE in their R-134a systems.

All refrigerant oils are **hydroscopic.** A hydroscopic material will absorb water. Some oils are more hydroscopic than others. For instance, POE oil is many times more hydroscopic than mineral oil. No matter what type of compressor oil is being used, containers should be kept closed at all times to keep the oil from absorbing water.

Alkylbenzene

A refrigerant oil primarily designed for stationary units is **alkylbenzene (AB).** It is mixed with mineral oil in units that are being retrofitted to accept another refrigerant. Alkylbenzene should never be used in an automotive system.

Other Refrigeration System Chemicals

Other refrigeration system chemicals include desiccants, flushes, and dyes. Desiccants are used in the refrigeration system during normal operation. Flushes and dyes are used during system diagnosis and service.

Desiccants

When servicing an air conditioning refrigeration system, it is almost impossible to avoid some moisture entry. It is also impossible to completely remove moisture from the system. **Desiccants** are drying agents used in every refrigeration system to minimize possible damage from moisture.

Refrigerants and refrigerant oils are manufactured carefully to remove all water vapor. However, when they are exposed to outside air, they absorb water quickly. If a system has been in use for many years, even if it has never been serviced, some water has entered the system. This is because every gas, including water vapor, wants to equalize its pressure.

Any water in the refrigeration system will react with the refrigerant to form acids. The acids will attack the metal in the system. On systems with an expansion valve, water can freeze at the expansion valve, blocking refrigerant flow. If water droplets freeze into ice crystals in the evaporator, the crystals will be drawn into the compressor. Ice entering the compressor may damage the reed valves or pistons.

The chemical makeup of desiccant causes it to hold any moisture left in the refrigeration system. Whatever desiccant is used must be better at holding moisture than the refrigerants and oils. Desiccant is placed in a porous bag installed in the accumulator or receiver-drier.

Flushing Compounds

A compressor failure often sends metal particles and oxidized oil throughout the refrigeration system. Occasionally, a desiccant bag will break and the loose desiccant will become lodged in system parts. Rubber hoses sometimes break down and send rubber particles into the system. To remove this debris from the refrigeration system, various *flushing agents* are used.

In the past two refrigerants, *R-11* and *R-113* were used to flush refrigeration systems. However, both contain chlorine and have been phased out. It is illegal to vent R-11 or R-113 to the atmosphere, so they should always be used in a closed (recycling) flushing machine. In a closed system, the flush recirculates back to the flushing machine. The flushing refrigerant must be disposed according to EPA guidelines.

Other refrigerants that can be used for flushing are R-134a and *R-141b.* R-134a will remove only small amounts of oil and debris, but can be recycled after use. R-141b works well as a flush, and is inexpensive. However, it is flammable and toxic in large doses. Therefore, R-141b must be used carefully. Most technicians prefer to use R-141b in a closed flushing machine. In addition to refrigerants, dry nitrogen is sometimes used to flush slightly dirty systems. However, nitrogen tanks are usually under high pressure and care must be taken not to rupture refrigeration system parts.

Liquid flushing agents include mineral sprits, light oils, and various commercial solvents. Many liquid flushing agents leave deposits in the refrigeration system, and may affect refrigerating oil viscosity. Some flushing agents are not compatible with R-134a and R-12 refrigerants. For these reasons, you must carefully check the manufacturer's recommendations before using a flushing agent.

Leak Detector Dyes

Leak detector dyes are used to locate leaks in the refrigeration system, **Figure 6-9.** Modern dyes are injected into one of the service ports. The dye will circulate with

 Note: Some manufacturers of R-134a add dye to the refrigerant.

Figure 6-9. *Dyes are used to detect leaks. Most modern dyes are fluorescent types used with a black light. Some older dyes were red, orange, or another color and could be seen in ordinary light. This dye can also contains a chemical which can help slow or stop minor leakage. (Jack Klasey)*

the refrigerant and stain the area around any leaks. Most dyes are red, orange, or another color for easier spotting. Some dyes are fluorescent and require the use of a black light. Modern dyes are compatible with R-12 and R-134a systems. Devices for injecting dyes were discussed in Chapter 3.

Evaporator Surface Cleaners

A damp evaporator surface is an excellent place for various microorganisms to grow. These microorganisms, usually mold or mildew, can become so numerous they cause a musty smell inside the vehicle. To remove these microorganisms it may be necessary to spray the surface of the evaporator with a *biocide* cleaner that will kill the organisms and dislodge them from the evaporator surface. Most of these cleaners are supplied as spray cans, **Figure 6-10.** The technician must remove enough components to expose the surface of the evaporator. Then the cleaner is sprayed onto the evaporator and allowed to sit for a specified time. Then the cleaner and mold can be washed off with a light water spray. In some cases, you may need to remove the evaporator core and soak its surface in the cleaner.

Refrigeration System Sealers

Sealers can be used to fix minor refrigeration system leaks. Sealers work in one of two ways. Some are stop-leak products that seal pinholes caused by corrosion or punctures. Others are designed to swell leaking O-rings and other seals. Many sealers perform both functions.

Figure 6-10. *Evaporator surface cleaners are often needed to remove mold from evaporator surfaces. Most of these products are sold in pressurized cans. (Jack Klasey)*

Refrigeration system sealers should be used cautiously. Sealers may cause excessive seal swelling, creating new leaks. Too much sealer in a refrigeration system can restrict the expansion valve or orifice tube, and may damage the compressor. Large amounts of sealer have plugged hoses on refrigeration service equipment. The most common use of sealers is to plug a small leak when a vehicle is not valuable enough to justify a major repair.

Summary

The refrigerant is designed to absorb and release heat by changing state inside of the refrigeration system. The refrigerant also carries small amounts of the compressor lubricating oil. Refrigerant should not react chemically with the oil, metals, rubber parts, or desiccants in the system. Modern refrigerant must also be environmentally friendly.

The refrigerants in use today are R-134a and R-12. R-134a has replaced R-12 in all original equipment air conditioners and is used to retrofit earlier air conditioners. R-134a does not pose a danger to the ozone layer.

R-12 was the only refrigerant used in automotive air conditioners until recently. It damages the ozone layer and is no longer manufactured, although there is still plenty around to service older systems. R-12 is slightly more efficient than R-134a, but is more expensive. All replacement refrigerants other than R-134a have drawbacks. Some refrigerants are approved by the EPA, but may not work in every system. All replacement refrigerants are blends of various compounds, and some contain R-134a, or R-22. R-22 can damage rubber parts and new hoses and O-rings must installed when R-22 is used.

There is no such thing as a drop-in refrigerant. It is illegal to mix refrigerants and no replacement refrigerant is compatible with R-12 or R-134a. Since it is illegal to mix refrigerants, every type of refrigerant, including blends, must have its own dedicated recovery and recycling equipment.

Refrigerant oils lubricate the compressor and the expansion valve. Refrigerant oil must be compatible with the refrigerant, or it may not mix well enough to stay in the refrigerant stream and be carried back to the compressor. The wrong type of oil may also react chemically with the refrigerant or system parts.

R-134a systems should be filled with polyalkylene glycol (PAG). Mineral oils are used with R-12 only. POE oil can be used in R-134a or R-12 systems.

To minimize damage from moisture, desiccants are used in every refrigeration system. In the past R-11 and R-113 refrigerants could be used to flush refrigeration systems. Modern flushing refrigerants include R-134a and R-141b. Dry nitrogen is sometimes used as a flushing agent. Some flushing agents are not compatible with R-134a and R-12 refrigerants. Leak detector dyes are used to locate leaks in the refrigeration system. Modern dyes are compatible with R-12 and R-134a systems.

Review Questions—Chapter 6

Please do not write in this text. Write your answers on a separate sheet of paper.

1. Because of the danger of leaks, refrigerants used in vehicles should be _____ and _____.

2. What is meant when we say a refrigerant should be environmentally friendly?

3. Which refrigerant boils at a higher temperature, R-12 or R-134a?

4. On hot days, condenser pressures can be as high as _____.
   (A) 30 psi (207 kPa)
   (B) 200 psi (1380 kPa)
   (C) 300 psi (2070 kPa)
   (D) 400 psi (2760 kPa)

5. How do HVAC manufacturers compensate for the use of R-134a instead of R-12?

6. Some _____ is carried along with the refrigerant.

7. _____ can be used with R-134a and R-12.

8. Some flushing agents are _____.
   (A) toxic
   (B) flammable
   (C) damaging to the ozone layer
   (D) All of the above.

9. Leak detector dye is put into the refrigeration systems through one of the _____.

10. What is the purpose of evaporator surface cleaners?

ASE Certification-Type Questions

1. Technician A says R-134a has no effect on global warming. Technician B says R-134a has no effect on the ozone layer. Who is right?
   - (A) A only.
   - (B) B only.
   - (C) Both A and B.
   - (D) Neither A nor B.

2. All of the following statements about R-134a and R-12 are true, *except:*
   - (A) R-12 evaporates at a slightly higher pressure than R-134a.
   - (B) R-12 condenses at a slightly higher pressure than R-134a.
   - (C) an R-12 system compressor does not work as hard as an R-134a system compressor.
   - (D) if environmental impact and expense are considered, R-134a is the better choice than R-12.

3. Technician A says some refrigerant blends contain R-134a. Technician B says some refrigerant blends contain R-22. Who is right?
   - (A) A only.
   - (B) B only.
   - (C) Both A and B.
   - (D) Neither A nor B.

4. A refrigeration system using R-12 is one pound (.45 kg) low. Which of the following could be legally added to raise the refrigerant level to normal?
   - (A) R-12.
   - (B) R-134a.
   - (C) An approved blended refrigerant.
   - (D) All of the above.

5. Refrigerant blends may contain percentages of the following compounds, *except:*
   - (A) propane.
   - (B) chlorine.
   - (C) butane.
   - (D) nitrogen.

6. Technician A says PAG oil can be used with R-134a. Technician B says PAG oil cannot be used with R-12. Who is right?
   - (A) A only.
   - (B) B only.
   - (C) Both A and B.
   - (D) Neither A nor B.

7. Desiccant is placed in a porous bag inside the _____.
   - (A) accumulator
   - (B) receiver-drier
   - (C) evaporator housing
   - (D) Both A and B.

8. All of the following is used to flush a refrigeration system, *except:*
   - (A) R-11.
   - (B) R-12.
   - (C) R-134a.
   - (D) R-141b.

9. Leak detector dye is usually colored _____.
   - (A) red
   - (B) green
   - (C) orange
   - (D) All of the above

10. A musty smell inside of the vehicle may be caused by _____ on the evaporator.
    - (A) mold
    - (B) refrigerant
    - (C) lubricating oil
    - (D) desiccant

Hoses and lines are used to connect the stationary refrigeration system parts. (Gates)

Chapter 7

Hoses, Lines, Fittings, and Seals

After studying this chapter, you will be able to:
- ❏ Explain the purposes and construction of refrigerant hoses.
- ❏ Explain the purposes and construction of refrigerant lines.
- ❏ Explain the purposes of refrigeration system fittings.
- ❏ Identify types of refrigeration system fittings.
- ❏ Explain the purposes of refrigeration system seals.
- ❏ Identify types of refrigeration system seals.

Technical Terms

Hoses	Crimp fittings	Schrader valve
Barrier hose	Spring lock coupling	Manual service valves
Impermeable	Garter spring	Lip seals
Lines	Screw type hose clamps	Carbon ring seal
Captured O-ring	Service fittings	O-rings
Headers	Push-on service fittings	Sealing washers
Muffler	Screw-on service fittings	Gaskets
Compression fittings		

This chapter identifies and gives the locations of refrigeration system lines, hoses, fittings, and seals. It also provides technical aspects of hoses and seals as they interact with refrigerants. Studying this chapter will enable you to identify hoses, lines, fittings, and seals used in all parts of various refrigeration systems.

Hoses and Lines

The obvious purpose of hoses and lines is to connect the major parts of the refrigeration system. They must be made to withstand the high pressures and extreme changes in temperature in the engine compartment. Hoses also absorb some of the vibration caused by compressor pulses and changes in operating pressure. They must also resist exposure to dirt, salt, oil, antifreeze, and other chemicals.

Hose and line sizes vary between sides of the refrigeration system, depending on whether the line carries liquid or vapor, **Figure 7-1.** The small line is the high pressure liquid line. The larger line is the low pressure vapor return line.

The refrigeration system lines and hoses are sometimes overlooked. Also overlooked are the fittings and sealing devices used with the lines and hoses. However, if any of these minor parts leak, clog, or break, the rest of the refrigeration system is useless.

Hoses

Flexible **hoses** are made of a type of synthetic rubber, usually Buna. Flexible hoses are intended for parts that move in relation to each other and where temperature changes may cause expansion or contraction of the parts connected by the hose. The rubber construction of the hose allows it to compensate for changes in length and flexing between parts.

Figure 7-1. *The evaporator inlet line carries liquid refrigerant, and therefore, can be smaller than the outlet line that carries vaporized refrigerant.*

Hoses that carry vapor must have a large internal diameter. Liquid lines, such as the line between the condenser and flow restrictor, are usually smaller since they carry liquid refrigerant only. This difference in hoses can sometimes be used to identify the high and low sides of the system for attaching gauges. The interior of the hose is made to be as smooth as possible to reduce resistance and friction between the hose and refrigerant.

The center section of a refrigeration hose is made of several plies of polyester. Each ply is wrapped around the hose interior in a spiral pattern. The plies cross each other at an angle, strengthening the hose while allowing movement. The fabric is covered with various blends of synthetic nitrile or Buna rubber. The synthetic rubber is selected for its resistance to heat and oil. Some high pressure hoses are coated with hard plastic for added protection. Other hoses are covered with metal mesh for abrasion resistance. The interior of the hose is made of dense rubber compounds that resist refrigerant leakage and provide a smooth surface.

Barrier Hoses

All newer vehicles are equipped from the factory with barrier hoses. A **barrier hose** has an inner lining, usually made of nylon, **Figure 7-2.** The lining is **impermeable,** which means refrigerant molecules cannot pass through it. Barrier hoses are used on all R-134a systems, and must be installed as part of some retrofit operations.

Hoses for Retrofitting

Since the R-134a refrigerant molecule is smaller than the R-12 molecule, it was first thought that all hoses would have to be replaced during a retrofit. Subsequent tests have shown most R-12 to R-134a retrofits do not require hose replacement. The mineral oil used in R-12 systems coats and penetrates the inside of the hoses. This oil coating forms a barrier which keeps R-134a from leaking out. However, if a hose must be changed during an R-134a retrofit, a barrier hose must be used. This is because an unused R-12 hose does not have the oil coating, and the R-134a will leak out through the rubber.

All R-12 hoses *must* be replaced with barrier hoses in any system converted to use blended refrigerants containing R-22. R-22 blends can damage an R-12 hose, causing it to swell and break up. When an R-22 blend is used, the EPA requires all hoses be changed to barrier types. Some new hoses will seal R-134a, R-12, and R-22.

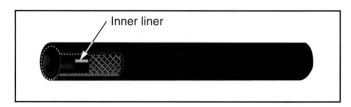

Figure 7-2. *Hoses are constructed of several layers of rubber, reinforced by fabric or plastic fibers. (Gates)*

Metal Lines

Metal refrigerant *lines* connect parts that do not move in relation to each other. Lines are made of steel or aluminum alloys formed into tubes. The line is shaped to make the shortest possible connection between parts without creating excessive bends or kinks. Metal lines are also used on large vehicles and vehicles with rear mounted evaporators.

The advantage of using metal lines is they are cheaper and can make tighter bends than hoses. Line diameter varies depending on where the line is used in the system. Vapor lines are always larger than liquid lines. Some steel lines have an internal coating, but aluminum lines are not coated. Metal lines are not replaced during retrofitting unless they are leaking or restricted. Most metal lines have the fittings formed into the end of the line, **Figure 7-3.** Some vehicles have a special depression designed to accept the O-ring. This is called a *captured O-ring* design. **Figure 7-4** illustrates the difference between noncaptured and captured O-ring fittings.

Combination Hose and Line Assemblies

Most refrigeration systems have at least one set of lines and hoses formed into a single assembly, **Figure 7-5.** Most hose and line assemblies are replaced as a unit.

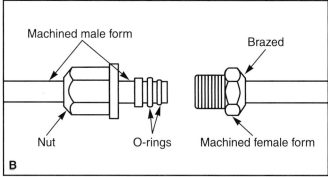

Figure 7-3. *A—Some line fittings make use of threaded nuts and mating fittings. B—The fitting shoulder is placed against the mating part or line and the nut is tightened to compress the O-ring(s). (General Motors)*

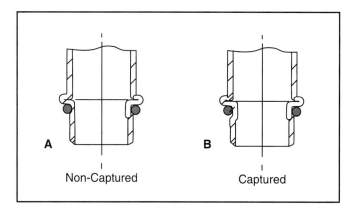

Figure 7-4. *A—Some fittings are designed so the O-ring is compressed against the shoulder and the mating part. B—The captured type of fitting has a small groove that holds the O-ring securely. (General Motors)*

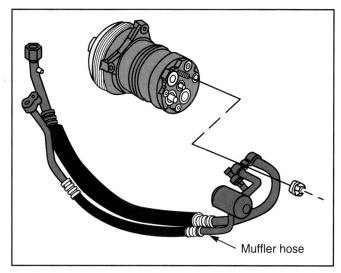

Figure 7-5. *This single hose assembly contains all of the hoses connecting the compressor to the evaporator and compressor. A two outlet connector, or header, is fastened to the rear of the compressor. (General Motors)*

Special fittings are used to make the connection between the metal lines and rubber hoses. Many hose and line assemblies have special connectors that attach to the rear of the compressor or other major units. These special connectors are called *headers.* Both low and high pressure lines are formed into the header used on compressors.

Many line and hose assemblies have a *muffler* installed in the high pressure line at the compressor outlet. The muffler reduces noises caused by pulsation of the refrigerant leaving the compressor. See **Figure 7-6.** On late-model vehicles, the muffler has been incorporated into the discharge hose or eliminated to save weight.

Hose and Line Attachments

To reduce vibration that could cause noise and wear, most hoses and lines are held in position by clips or brackets. The clips and brackets usually provide a loose fit

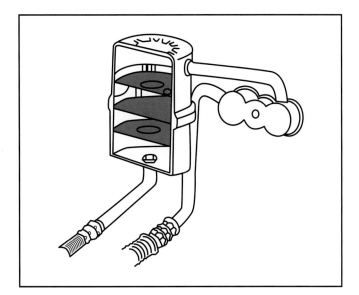

Figure 7-6. *The muffler, when used, reduces noise and vibration caused by the compressed refrigerant leaving the compressor. The internal baffles reverse refrigerant flow causing pressure and noise impulses to cancel each other. (General Motors)*

to allow for movement and expansion during operation. Metal lines may be isolated from the clips and the body by foam rubber sleeves.

Hose and Line Size Specifications

Depending on the system capacity, different size hoses and lines are used. Low pressure lines are the largest, since they carry vaporized refrigerant. High pressure liquid lines are smaller as they carry liquid refrigerant. Also by making the lines smaller, a pressure differential is created, which makes the system operate more efficiently.

Refrigeration System Fittings

Since they are required to hold a gas rather than a liquid, refrigeration system fittings must be able to seal very small openings. Several types of fittings are used on modern refrigeration systems.

Compression Fittings

Refrigeration system *compression fittings* are similar to the compression fittings used on fuel systems and other liquid and air piping. However, refrigeration system compression fittings use additional sealing in the form of an O-ring. All compression fittings consist of a nut with internal threads tightened against a seat with external threads.

When the nut is tightened, the O-ring is compressed between a shoulder formed into the line and the seat, **Figure 7-7A.** These fittings are always used to connect a metal line to another metal part such as the compressor,

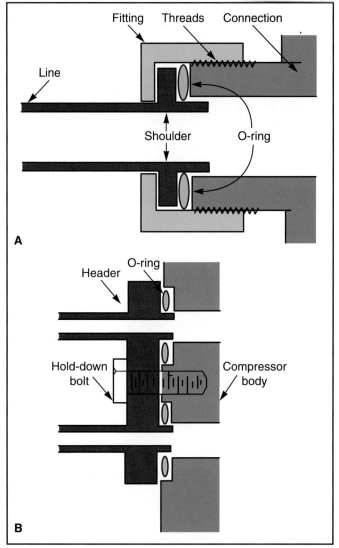

Figure 7-7. *This cutaway view shows the sealing effect caused by compressing the O-rings. A—This shows a nut tightened against the connection. B—Header tightened against the compressor body.*

condenser, or evaporator. Another design, **Figure 7-7B,** uses a central bolt to connect a header with two lines to the rear of the compressor.

Crimp Fittings

Crimp fittings are metal sleeves that are collapsed around a hose and metal line. All factory new and replacement hoses are crimped. Collapsing the sleeve causes the hose to seal against the metal line, **Figure 7-8.** The metal line always has external ribs to keep the hose from sliding off. A special machine is used to perform the crimping operation. There are two types of crimp shapes, **Figure 7-9.** Some crimped fittings are formed completely around the metal, **Figure 7-10.** Other crimped fittings consist of a metal ring formed near the end of the hose over the ribs of the metal line. Barb crimps are used on older R-12 hoses. Beadlock crimps are used on newer R-134a and with barrier hoses.

Figure 7-8. *Two crimped fittings connecting the flexible hoses to a compressor header.*

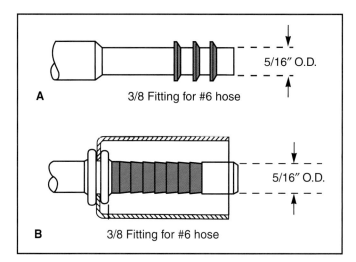

Figure 7-9. *The two main types of crimp fittings are the barb fitting and the beadlock fitting. R-134a systems must use beadlock fittings. R-12 systems can use either type. A—Barb style. B—Beadlock style. (Four Seasons)*

Figure 7-10. *A complete crimp surrounds the hose and fitting. Done properly it eliminates the chance of leaks.*

Spring Lock Couplings

The **spring lock coupling,** sometimes called the *quick connect coupling,* is found on Ford, DaimlerChrysler, and some Asian and European made vehicles. It consists of a male fitting with a circular cage about 1″ (25.4 mm) from its end. The two O-rings seal in refrigerant. See **Figure 7-11.** To hold the coupling together, a spring called a garter spring is installed inside the cage. A **garter spring** is a small spring formed into a circle, **Figure 7-12.** The cage is tapered to allow the garter spring to expand when it is pushed rearward.

When a female fitting, **Figure 7-13,** is pressed into the cage, it pushes the spring toward the back, allowing it to expand. The female fitting slips past the spring and bottoms at the rear of the cage. The garter spring then moves past the edge of the female fitting and returns to its original size. Together with the cage, the garter spring keeps the female fitting from coming out. The connection can only be taken apart by using a special tool. To further hold the spring lock coupling, a clip or clamp may be installed over the connection, **Figure 7-14.**

Figure 7-11. *Spring lock coupling used on some late-model vehicles. The garter spring holds the coupling together and the dual O-rings provide sealing.*

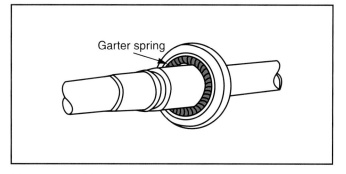

Figure 7-12. *The garter spring and spring cage are shown in the disassembled view of a spring lock coupling. (Ford)*

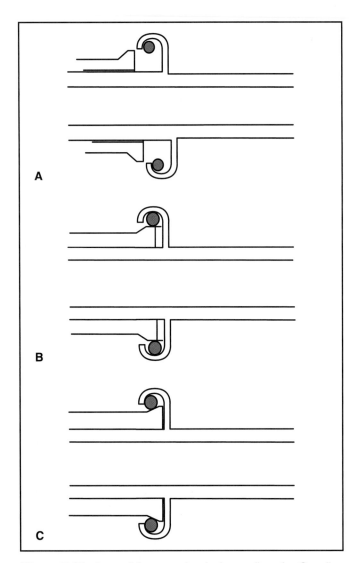

Figure 7-13. *Assembling a spring lock coupling. A—Coupling disassembled. B—Coupling being pushed together. The garter spring is expanded and pushed to the outside of the spring cage. C—Coupling assembled. The garter spring returns to its original position and holds the coupling together. (Ford)*

When a screw type hose clamp is used on an R-134a system, the smaller molecule may leak through the fitting, which would hold R-12. In addition, sliding a barrier hose over the ridges of the metal fitting will damage the barrier material past the clamp sealing point. Other types of fittings have replaced screw type fittings.

> **Caution: Do not use screw type hose clamps on barrier hose.**

Service Fittings

To perform various service operations, ***service fittings*** are used. All service fittings connect the refrigeration system to the service hoses and through them to the service

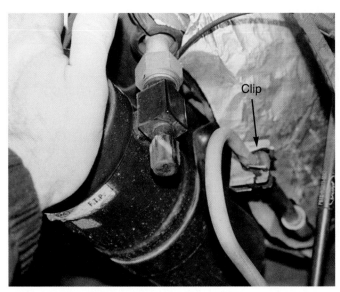

Figure 7-14. *Many spring lock couplings use a clip to further hold the coupling together. The clip reduces vibration and provides a better seal. Aftermarket clips are also available.*

Screw Type Hose Clamps

In the past, ***screw type hose clamps,*** **Figure 7-15,** were widely used on new vehicles. Hose clamps are no longer used on factory installed air conditioning hoses. Screw type hose clamps are often used by repair shops or aftermarket installers. This type of hose clamp is easy to install, and can be useful when an air conditioning hose has to be made up in the shop.

However, the sealing action of screw type hose clamps is not uniform. The side of the clamp closest to the screw must rub against the hose as it is tightened. This side of the hose will be tightened less than the other side, resulting in a deformed fitting. Refrigerant can escape from the undertightened part of the hose. If the hose is overtightened to compensate, the overtightened side may distort and leak. To compensate, some technicians install two hose clamps and place the screws on opposite sides of the hose.

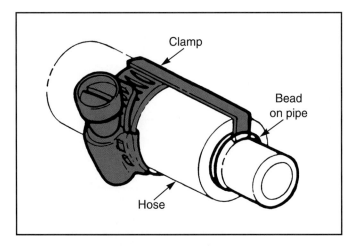

Figure 7-15. *A screw clamp used on many older refrigeration hoses. Screw clamps work in the same manner as those on cooling system hoses. Screw clamps should never be used on R-134a systems. (General Motors)*

equipment. Service fittings are sometimes called service valves or service ports. There are three main types of service fittings.

Push-on Service Fittings

Push-on service fittings are used on R-134a systems. They work like air hose connections. Pushing the charge valve over the fitting locks it into place. The valve is removed by pulling up on the outer sleeve of the valve. Once the valve is installed, the knob at the top of the charge valve is turned clockwise to depress a positive seal inside the fitting. This connects the refrigeration system to the service hose. The high and low side fittings are different sizes to prevent incorrect connections. **Figure 7-16** shows typical push-on service fittings.

Figure 7-16. *Push-on service fitting used on R-134a systems. To prevent misconnection, the low and high sides have different size couplings.*

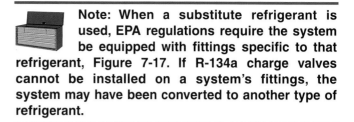

Note: When a substitute refrigerant is used, EPA regulations require the system be equipped with fittings specific to that refrigerant, Figure 7-17. If R-134a charge valves cannot be installed on a system's fittings, the system may have been converted to another type of refrigerant.

Screw-on Service Fittings

Screw-on service fittings use a Schrader valve and are found only on R-12 systems. A *Schrader valve* is a spring-loaded valve similar in operation to a tire valve. The hose connection is threaded onto the valve, causing the pin to be depressed. Removing the hose causes the valve to be closed by spring pressure. A Schrader valve assembly is shown in **Figure 7-18.** Schrader valves on vehicles built after 1986 have different size fittings on the high and low pressure sides of the system. The different size threads ensure the hose connections cannot be crossed.

Manual Service Valves

Manual service valves, **Figure 7-19,** were found on older vehicles. A manual service valve was opened and closed by turning it with a wrench. These valves were usually installed on the compressor. Position A was used during normal system operation. The valve closes off the hose connector port from the system. Position B was the position used for service operations. After the hose was connected, the valve was turned to connect the system to the hose port. Position C was used to isolate the compressor from the rest of the refrigeration system. Position C allowed the compressor to be removed without discharging the system. Manual valves are no longer used on modern refrigeration systems.

Refrigerant	High Side Service Port			Low Side Service Port		
	Diameter (inches)	Thread pitch	Thread direction	Diameter	Thread pitch	Thread direction
HFC-134a	Quick-connect			Quick-connect		
CFC-12 (Pre 1987)	7/16	20	Right	7/16	20	Right
CFC-12 (Post 1987)	6/16	24	Right	7/16	20	Right
Free Zone/RB-276	8/16	13	Right	9/16	18	Right
Hot Shot	10/16	18	Left	10/16	18	Right
GHG-X4/Autofrost	.305	32	Right	.368	26	Right
GHG-X5	1/2	20	Left	9/16	18	Left
R-406A	.305	32	Left	.368	26	Left
Freeze 12	7/16	14	Left	1/2	18	Right
FRIGC FR-12	Quick-connect, different from HFC-134a			Quick-connect, different from HFC-134a		

Figure 7-17. *Fitting sizes are different for each refrigerant. (EPA)*

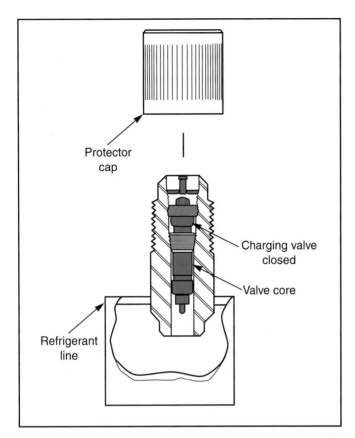

Figure 7-18. *The Schrader valve used on R-12 refrigeration service fittings operates in a manner similar to a tire valve. Seals on the valve core prevent refrigerant loss. (Ford)*

Making Hoses, Lines, and Fittings

Sometimes, a shop may decide to have a replacement hose made locally, rather than purchasing a new hose. This may be done to save time or because a new hose would be expensive or is no longer available. The shop may be able to make the hose, or may have it made by a local company specializing in hose fabrication. Shops making their own hoses will keep a stock of common hoses and fittings. When a hose is needed, the technician can cut the needed length and size hose, then attach the fittings. Sometimes the original fittings or headers are used with new hoses, as in a retrofit where barrier hoses are needed.

If possible, the hose should be attached to the fitting using a hose crimping machine. Types of hose crimping machines were discussed in Chapter 3. Do not use screw type hose clamps unless there is no other way to connect the hose and fitting.

Refrigeration Sealing Devices

Modern refrigeration systems use several types of sealing devices. Where each type of sealing device is used at a particular location depends on the temperature, relative movement of parts, and whether the refrigerant is a liquid or a gas at that location.

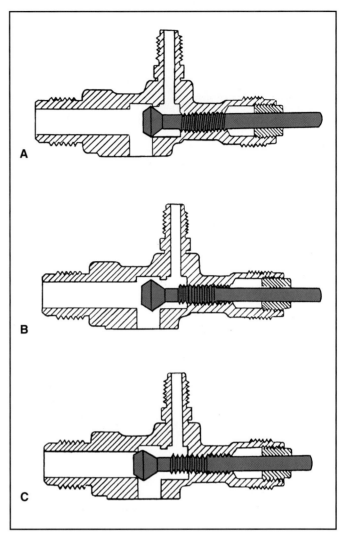

Figure 7-19. *Manual service valves were used on older cars and trucks. Manual valves are occasionally found on some vehicles into the 1980s.*

Compressor Shaft Seals

Lip seals are commonly used to seal the rotating compressor shaft. A cross-section of a lip type shaft seal is shown in **Figure 7-20.** The lip of the seal is held against the shaft by a garter spring. Refrigerant pressure places additional pressure on the lip. These two forces keep refrigerant from leaking at the shaft. A slight amount of refrigerant oil leaks into the space between the seal and shaft. This oil film reduces seal and shaft wear and provides additional sealing. Some compressor manufacturers use a double lip seal for extra sealing.

Some older compressors used a *carbon ring seal*, **Figure 7-21.** The flat surface of the carbon ring rotated against a matching flat area on the pulley assembly. A thin film of oil provided the actual seal. The carbon ring seal is no longer used.

O-ring Seals

O-rings are found throughout the refrigeration system. As its name implies, an O-ring is a rubber ring shaped like the letter *O*. O-rings are always used to seal stationary joints and fittings. As discussed earlier in this chapter, O-rings are compressed between two parts. The O-ring is always installed between machined sealing areas of the mating parts. When the parts are tightened, the O-ring compresses and conforms to any imperfections in the sealing area metal. This prevents refrigerant leaks. O-rings are made of many types of natural and artificial rubber. Often, O-rings are made of the same materials used in refrigeration hoses. Common O-ring materials are neoprene, nitirile, and Buna rubber. Various plastics and polymers are also used. Connections in newer system often have two O-rings per joint.

O-ring Sizes and Shapes

The right size O-ring must be used or the fitting will leak. Three dimensions are used to determine O-ring sizes These three dimensions cover all possible O-ring size measurements. Dimensions may be given in standard (English) or metric units:

❑ Inside diameter, or ID.
❑ Outside diameter, or OD.
❑ Thickness.

There are two general shapes of O-rings, round and square cut. As a rule, round O-rings are used when the O-ring must be slipped over a fitting, while square cut

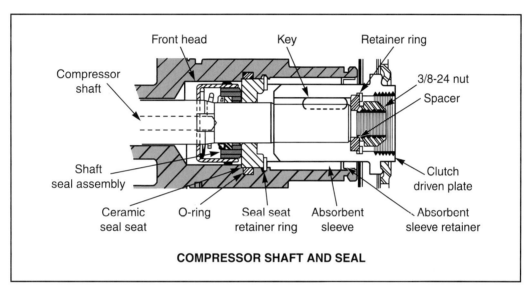

Figure 7-20. *This illustration shows a cutaway view of a lip seal used to seal the compressor shaft of modern compressors. Lip seals were also used on many older compressors. (Ford)*

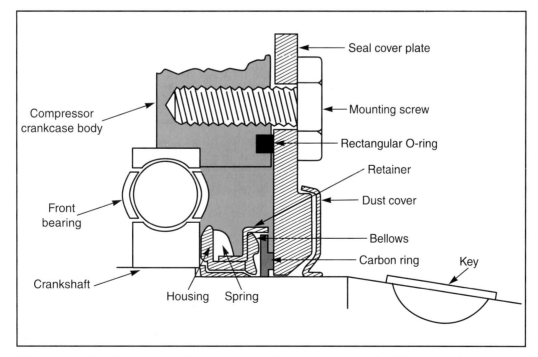

Figure 7-21. *A carbon seal used to seal some older compressor shafts. This seal design has been discontinued. (Ford)*

O-rings are used to seal two flat surfaces. However, there are exceptions to this general rule. The service literature should always be consulted when the original O-ring type cannot be determined.

O-ring Color

In the past, all O-rings were made of black neoprene, and could be used in any location. Modern refrigerant systems use varying refrigerants and compressor oils. They also operate at higher temperatures. This means different O-ring materials must be used. Many modern seals are colored red, green, yellow, blue, or purple, and may have stripes or other identifying markings. These colors and markings determine the proper seal to be used with a particular fitting or type of refrigerant. You should always make sure to use the proper O-ring.

Replacement O-rings are usually colored black (natural rubber or neoprene), blue (R-134a impervious neoprene), or green (various types of Buna rubber). Blue or green O-rings are found on R-134a systems and can be used to replace black O-rings on R-12 systems.

 Note: All manufacturers intend to discontinue making black O-rings. All replacement O-rings will be blue or green, however they will be available in all standard O-ring sizes.

Sealing Washers

A few systems use **sealing washers.** These washers resemble the washers used on oil drain plugs, but are made of special plastics that seal against refrigerant. A sealing washer is shown in **Figure 7-22.**

Gaskets

Gaskets are often used to seal the compressor. Compressor gaskets will be discussed in more detail in

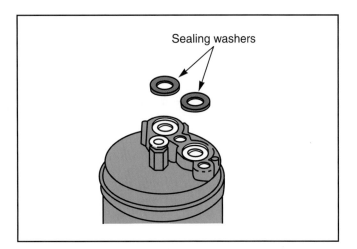

Figure 7-22. *A sealing washer as used to seal a compressor to refrigerant line connection. (General Motors)*

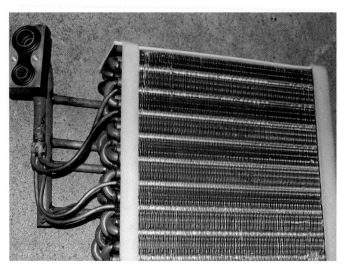

Figure 7-23. *This figure shows a typical use for gaskets in a refrigeration system. Gaskets are used in only a few spots on some refrigeration systems.*

Chapter 17. Gaskets are sometimes used in other places in the refrigeration system. Refrigeration gaskets are made of rubber impregnated materials that resist refrigerant leakage. A typical gasket is shown in **Figure 7-23.**

Summary

The refrigeration system lines and hoses connect the other parts of the refrigeration system. Flexible hoses are used to connect parts that move in relation to each other. Hoses are made of synthetic rubber with an internal braided lining for strength. The inside of the hose is made to be as smooth as possible.

Some high pressure hoses are coated with hard plastic or metal mesh for added protection. To prevent refrigerant leaks, barrier hoses are used on R-134a systems. A barrier hose has an inner lining, usually made of nylon. R-12 hoses in good condition do not have to be replaced as part of a retrofit. Special hoses must be used when retrofitting to a blended refrigerant containing R-22. Some new hoses will seal R-134a, R-12, and R-22.

Metal lines connect parts that do not move in relation to each other. Metal vapor lines are always larger than liquid lines. Most metal lines have fittings formed into the end of the line. Most refrigeration systems have lines and hoses formed into single assemblies.

Several types of fittings are used on modern refrigeration systems. They include compression fittings, crimp fittings, push-on fittings, and screw type hose clamps. Compression fittings seal by compressing an O-ring between two metal parts. Crimped fittings are metal sleeves or rings collapsed around a hose and metal line. Push-on fittings use a cage and garter spring to hold the fitting together. Two O-rings seal in refrigerant. Hose clamps are no longer used on factory installed hoses because refrigerant, especially R-134a, can escape from under the fitting.

Service fittings connect the refrigeration system to the service hoses and through them, the service equipment. Push-on service fittings are used on R-134a systems, while screw-on service valves are used on R-12 systems. Some older vehicles used manual service valves.

Aftermarket hoses, lines, and fittings can be made in the shop or by a company specializing in hoses. If possible, a made-up hose should be crimped to the fitting using a hose crimping machine. Screw type hose clamps should not be used unless there is no alternative.

Modern refrigeration systems use several sealing devices. Lip seals are commonly used to seal compressor shafts. O-rings are used to seal stationary fittings throughout the refrigeration system. O-rings vary by size, shape, and type of material. A color or other identification mark often identifies the type of O-ring material. A few places in the refrigeration system use sealing washers and gaskets. Most gaskets are used on the compressor.

Review Questions—Chapter 7

Please do not write in this text. Write your answers on a separate sheet of paper.

1. What is the purpose of refrigeration system hoses and lines?

2. Hoses are made of a type of _____ rubber.

3. Hoses that carry refrigerant _____ must have a large internal diameter.

4. Compression fittings always use _____ as a seal.

5. Which of the following fittings requires a special machine to install?
   (A) Compression.
   (B) Crimp.
   (C) Spring lock.
   (D) Screw clamp.

6. _____ are used to connect the refrigeration system to the service hoses.

7. Schrader valves on R-12 systems built after _____ have different size high and low side fittings.

8. List the three ways to measure an O-ring.

9. Sealing washers resemble the washers used on _____.

10. Most refrigeration system gaskets are found on the _____.

ASE Certification-Type Questions

1. Technician A says barrier hoses are used on all refrigeration systems originally equipped with R-134a. Technician B says barrier hoses must be installed on all systems retrofitted from R-12 to R-134a. Who is right?
   (A) A only.
   (B) B only.
   (C) Both A & B.
   (D) Neither A nor B.

2. All of the statements about metal refrigerant lines are true, *except:*
   (A) metal lines connect parts that do not move in relation to each other.
   (B) metal lines are more expensive than hoses.
   (C) metal lines may be held by clips.
   (D) metal lines can make tighter turns than hoses.

3. The spring lock coupling contains all of the following parts, *except:*
   (A) compression nut
   (B) garter spring
   (C) female fitting
   (D) two O-rings

4. Technician A says screw type hose clamps are easy to install. Technician B says screw type hose clamps provide uniform pressure all the way around the hose and fitting. Who is right?
   (A) A only.
   (B) B only.
   (C) A and B.
   (D) Neither A nor B.

5. Technician A says Schrader valves are used on R-134a system service valves. Technician B says every type of refrigerant is supposed to have a different service connection. Who is right?
   (A) A only.
   (B) B only.
   (C) Both A & B.
   (D) Neither A nor B.

6. Manual service valves are found on _____ vehicles.
   (A) R-134a
   (B) R-406a
   (C) R-22
   (D) Older

7. All of the following statements about making replacement hoses are true, *except:*

   (A) making a replacement hose saves time.
   (B) the exact replacement hose may not be available.
   (C) screw clamps should be used to attach the hose to metal lines.
   (D) sometimes the original fittings or headers are used with new hoses.

8. Compressor shaft lip seals are being discussed. Technician A says the seal lip is held against the shaft by a garter spring. Technician B says refrigerant pressure holds the lip against the shaft. Who is right?

   (A) A only.
   (B) B only.
   (C) Both A & B.
   (D) Neither A nor B.

9. Which O-rings are used on R-134a systems?

   (A) Green.
   (B) Blue.
   (C) Black.
   (D) Both A & B.

10. All of the following are reasons for using different O-ring materials, *except:*

   (A) modern refrigerant systems use new design compressors.
   (B) modern refrigerant systems use different refrigerants.
   (C) modern refrigerant systems use different compressor oils.
   (D) modern refrigerant systems operate at higher temperatures.

Chapter 8

Compressors, Clutches, and Drives

After studying this chapter, you will be able to:
- ❏ Explain the purpose of a refrigeration compressor.
- ❏ Identify the major parts of radial and axial piston compressors.
- ❏ Explain the operation of radial and axial piston compressors.
- ❏ Identify the major parts and explain the operation of rotary vane compressors.
- ❏ Identify the major parts and explain the operation of scroll compressors.
- ❏ Explain the operation of a capacity control valve.
- ❏ Explain how the engine drives the compressor.
- ❏ Explain the purpose of a compressor clutch.
- ❏ Identify the major parts of a compressor clutch.
- ❏ Explain the operation of a compressor clutch.

Technical Terms

Compressor	Radial compressor	Scroll	Driven pulley
Body	Radial piston compressor	Moving scroll	Idler pulley
Ears	Axial compressor	Orbital path	Pulley ratio
Endplate	Single-ended pistons	Electric compressors	Compressor clutch
Crankshaft	Double-ended piston	Capacity control valve	Armature plate
Crankcase	Wobble plate	Belt	Electromagnet
Lip seal	Swash plate	V-belts	Magnetic field
Piston compressor	Rotary vane compressors	Serpentine belts	Engaged
Reed valves	Rotor	Belt tensioner	Disengaged
Exhaust valve plate	Vanes	Pulley	
Intake valve plate	Scroll compressors	Driving pulley	

The compressor is a device that pressurizes the refrigerant and pumps it through the refrigeration system. While there are many types of compressors, they all do the same job, and most of them are similar in construction and operation. This chapter identifies common compressor designs and explains how they work. This chapter also covers capacity control valves used on many newer compressors. It also explains how compressor clutches work, and why they are used.

Compressor Purposes

As discussed in Chapter 6, the refrigerant must be pressurized to absorb and release heat properly. This is the job of the *compressor*, **Figure 8-1.** The compressor must keep the pressure in the evaporator low enough so the refrigerant can vaporize. It must also keep the pressure high enough so refrigerant can condense. The compressor draws refrigerant from the evaporator and pushes it into the condenser and through the entire system.

Since the vehicle's engine drives the compressor in most cases, compressor speed varies with engine speed. At idle and low speeds, the compressor must turn fast enough to operate the refrigeration system. At high engine speeds, the compressor turns at high speed and produces much more pumping capacity than the refrigeration system needs. Various control devices are used to reduce compressor output as necessary.

Basic Compressor Parts

All compressors have some common parts. Every compressor has a body or housing. The *body* is the mounting place for the other compressor parts. The body also holds the refrigerant and lubricating oil. Bodies are made of aluminum on newer vehicles. Older compressors were

Figure 8-1. *The compressor is the heart of the air conditioning system. (Delphi)*

made of cast iron. The body has built in mounting points or *ears* for attachment to the engine. Attached to the compressor body is at least one cylinder head or *endplate.* Low and high side hose connections are installed on the body or endplate. In some cases, the service fittings may also be installed on the compressor. Other components installed in the compressor include pressure relief valves and electrical high pressure or superheat shutoff switches.

Every compressor has a *crankshaft* that rides on bearings in the compressor body. The part holding the crankshaft is called the *crankcase.* The crankshaft drive end extends out from the body. A pulley and clutch assembly is installed on the crankshaft drive end. **Figure 8-2** shows a typical compressor with pulley and clutch assembly.

A *lip seal* is always installed where the crankshaft drive end extends from the compressor body. Gaskets are used where the end plate or cylinder head connects to the body. Many of the refrigerant line connections use O-rings for sealing.

The compressor is attached to the engine with various brackets. Some compressor mounting points are attached directly to the engine. A slotted bracket may be provided to allow for compressor belt adjustment.

Types of Compressors

Three types of compressors are used on modern vehicles. Most compressors are piston types, with the only differences being the arrangement of the pistons. In the past, a few vehicles used rotary vane compressors. Some new vehicles are equipped with scroll compressors.

Piston Compressors

The most common compressor is the *piston compressor.* Piston compressors resemble small engines, with pistons moving inside cylinders, **Figure 8-2.** Instead of the crankshaft delivering power from the pistons to the drive line, it receives power from the engine and delivers it to the pistons. The moving parts of the compressor are lubricated by oil in the crankcase and by oil carried through the system with the refrigerant. Some compressors have small oil pumps with pickup tubes that draw oil from the bottom of the crankcase.

Older compressors used only two cylinders. For increased smoothness, modern compressors usually have at least four or five cylinders. Some compressors may have ten or more cylinders. In most compressors, both pistons and cylinders are aluminum to reduce weight. Some compressor bodies have iron cylinder liners or cast iron pistons.

The operation of a compressor piston in the cylinder is identical to an engine piston. Compressor pistons have at least one piston ring to seal pressure between the cylinder and piston. The piston's design and crankshaft connection vary with the type of compressor. Piston compres-

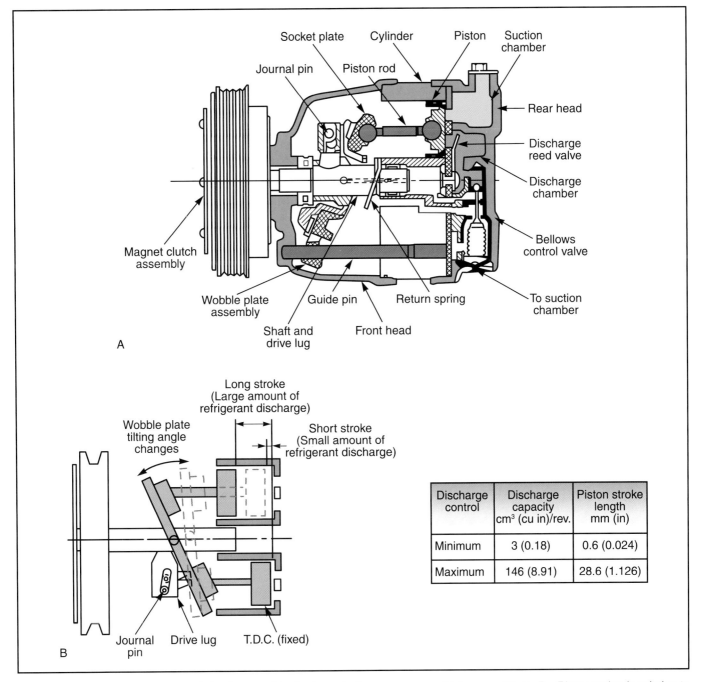

Figure 8-2. *Compressor pulley and clutch assembly. A—Layout of a compressor which uses pistons. B—Piston action in relation to the pulley and clutch. (Nissan)*

Discharge control	Discharge capacity cm³ (cu in)/rev.	Piston stroke length mm (in)
Minimum	3 (0.18)	0.6 (0.024)
Maximum	146 (8.91)	28.6 (1.126)

sor displacements vary from 5 cubic inches (82 cc) on smaller compressors, to 12 cubic inches (200 cc) for the largest compressors.

Reed Valve Operation

All piston compressors use *reed valves* to allow refrigerant to enter and exit in only one direction. A reed valve is a movable flap that closes a passage in the compressor, **Figure 8-3.** When pressure is greater on one side of the flap, it is pushed open. When the pressure is greater on the other side, the flap is pushed closed. The valve seats on the metal passage. The valve usually does not make a perfect seal and slight refrigerant leakage past the valve is acceptable.

Each compressor cylinder has two reed valves, one for intake and one for exhaust. The intake reed valve seals the passage between the cylinder and the low pressure (evaporator) side of the system. The exhaust reed valve seals the passage between the cylinder and the high pressure (condenser) side of the system.

The intake reed valve is positioned so the flap opens when the pressure in the cylinder is less than the low side pressure. The exhaust reed valve flap opens when the pressure in the cylinder is more than the high side pressure. As shown in **Figure 8-4,** this reed valve arrangement ensures the refrigerant moves through the cylinder in only one direction. When the piston is moving down in the cylinder, the intake reed valve is pulled open by the suction in the cylinder. The exhaust reed valve is kept closed by high side pressure. This allows the piston to draw refrigerant from the low pressure side of the system only.

When the piston moves up, the increasing cylinder pressure pushes the intake reed valve shut. When the pressure in the cylinder becomes greater than the high side pressure, the exhaust reed valve is pushed open and refrigerant exits into the high pressure side of the system. This cycle repeats each time the piston travels up and down in the cylinder.

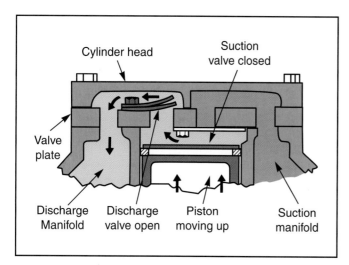

Figure 8-3. *This simplified illustration of reed valve action shows the discharge reed valve being opened by refrigerant pressure as the compressor piston moves up in the cylinder. The intake reed valve is held shut by the pressure in the cylinder. Pressurized refrigerant exits to the high pressure side of the system.*

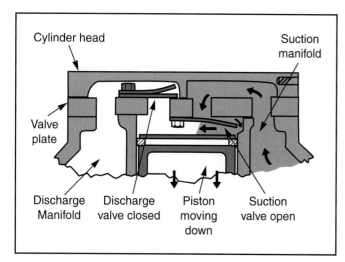

Figure 8-4. *The intake reed is opened by low side refrigerant pressure as the compressor piston moves down in the cylinder. The discharge reed valve is held shut by the pressure in the high side of the system. Refrigerant is drawn into the cylinder by the pressure on the low side.*

Since the reed valves must open and close millions of times during the life of the compressor, they are made of spring steel that can flex without becoming fatigued. The reed valve and seat are lubricated by oil carried with the refrigerant.

All of the discharge reed valves on one side of a compressor are installed in an *exhaust valve plate,* **Figure 8-5.** One side of the reed valve is bolted or riveted to the plate. There may be a stationary cover over the valve. The cover keeps the valve from moving too far and becoming sprung when the compressor is turning at high speeds. On a few compressors, each exhaust reed valve is installed individually at the top of the cylinder.

The intake reed valves are often part of a stamped sheet metal plate. The sheet metal plate may be called the *intake valve plate.* A typical intake valve plate is shown in **Figure 8-6.** The intake valve plate fits against the exhaust valve plate. The intake reed valves seat against holes in the exhaust valve plate. A few intake reed valves are separate units, with an individual valve for each cylinder.

Radial and Axial Piston Compressors

There are two kinds of piston compressors, the radial compressor and the axial compressor. Both types use pistons, cylinders, and reed valves. The major difference is the placement of the pistons in relation to the crankshaft.

Radial Piston Compressors

In the *radial compressor,* the pistons extend at right angles to the crankshaft. There are three general types of radial compressors. The radial compressors used on older vehicles had two pistons placed side by side, **Figure 8-7.** A feature of these compressors is they may be installed with the pistons pointing upward, or to the side. Operation is the same in either position. Other older vehicles have compressors made in the form of a V. A typical V-type compressor is shown in **Figure 8-8.** Each side of the V contains one piston. These types of compressors have been replaced by axial designs.

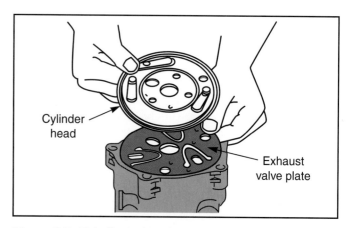

Figure 8-5. *This illustration shows a plate containing the discharge valves. The discharge valves are separate assemblies. Each valve uses a retainer to keep it from being moved too far by the exiting gases and being bent. (Subaru)*

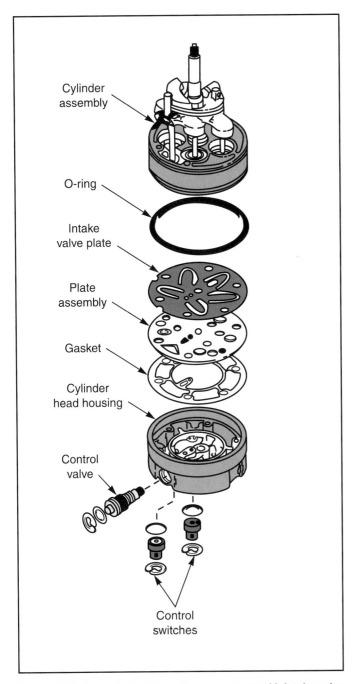

Figure 8-6. *An exploded view of a compressor with intake valve plate. Punching the correct pattern of holes creates the valve plate from a single plate of sheet metal. (General Motors)*

Radial Piston Compressor

Modern radial compressors have four pistons radiating out from the crankshaft at 90° intervals, **Figure 8-9.** In service literature, this kind of compressor is sometimes called a *radial piston compressor.* A reed valve assembly is used at the top of each piston, **Figure 8-10.** This compressor is used on some late-model vehicles as well as older vehicles. Another version of the radial compressor uses an eccentric cam to move the pistons. This compressor had durability problems and was hard to rebuild. Most defective compressors of this type are replaced with conventional radial compressors.

Figure 8-7. *A common two cylinder radial compressor. These compressors were original equipment on many older vehicles, and were often installed as part of aftermarket air conditioners. They are rare today.*

Figure 8-8. *A V-type radial compressor. Each bank of the compressor has one cylinder. This compressor is no longer used.*

Figure 8-9. *The radial four cylinder compressor has been widely used during the last 20 years. It is still installed on a few new vehicles.*

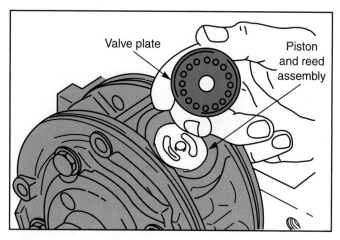

Figure 8-10. *A unique feature of the four cylinder compressor is the individual reed valves used at the top of each piston. Each reed valve assembly can be replaced individually. (General Motors)*

Axial Compressors

In an *axial compressor,* the pistons and cylinders are parallel to the crankshaft. Axial compressors are smoother and quieter than radial compressor designs. **Figure 8-11** shows a typical axial compressor layout.

Note: Axial piston compressors are sometimes called rotary compressors. However, they are not rotary designs.

Some axial compressors have pistons mounted on only one side of the compressor, **Figure 8-12.** These pistons are called *single-ended pistons.* The easiest way to tell the two types apart is by whether there is an odd or even number of pistons. Compressors using double-ended pistons have an even number of cylinders. Compressors with single-ended pistons will have an odd number of cylinders.

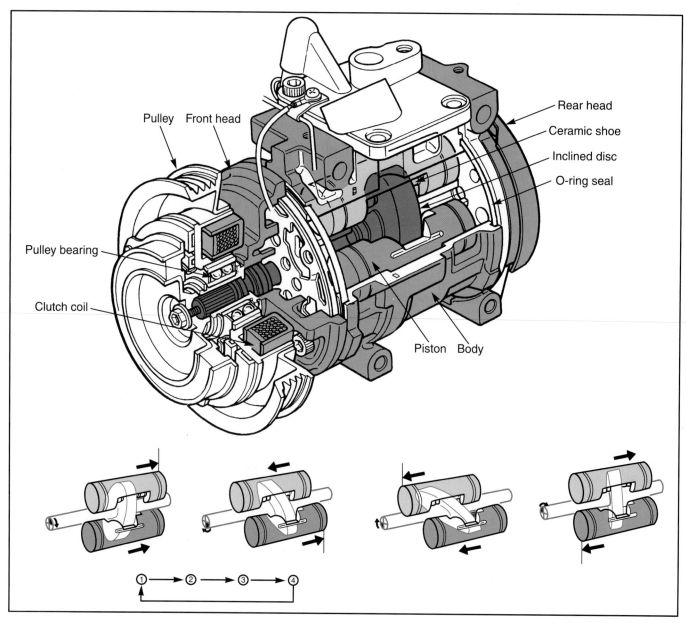

Figure 8-11. *Cutaway view of an axial compressor. Note the pistons are parallel with the crankshaft. Action of the wobble plate and pistons is shown at the bottom of the figure. (Honda)*

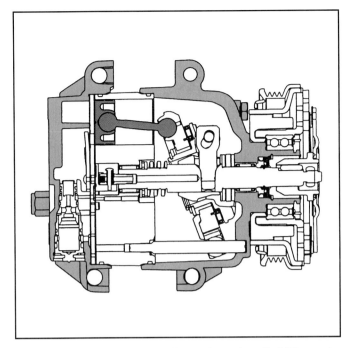

Figure 8-12. *A single-ended piston shown installed. These pistons are often used on the compressors installed on smaller vehicles. Single-ended pistons were also used on older radial compressors. (Delphi)*

Many axial compressors have two pistons combined into one assembly. This assembly is called a **double-ended piston.** The assembly is installed in cylinders opposite from each other, **Figure 8-13.** Therefore, one of the pistons is always drawing refrigerant in as the other piston is pumping refrigerant out.

A **wobble plate,** sometimes called a **swash plate,** drives the pistons. The plate turns with the crankshaft. The wobble plate is slanted, so rotation of the crankshaft causes it to move back and forth in relation to any stationary point. A typical wobble plate is shown in relation to the pistons and cylinders in **Figure 8-13.** On a single-ended piston, the side of the piston away from the head rests against the wobble plate. To reduce friction, a large single ball bearing separates the piston from the wobble plate. A double-ended piston is notched in the center. The wobble plate fits inside of this notch. Ball bearings on each side reduce friction between the double-ended piston assembly and the plate.

Since the cylinders keep the pistons from rotating, the rotating wobble plate slides in relation to the piston assemblies. As the plate slides along the piston assemblies, it moves the pistons back and forth in the cylinders. Reed valve assemblies control refrigerant movement through the cylinders.

Rotary Vane Compressors

A few older vehicles used **rotary vane compressors.** The rotary vane compressor had a series of sliding vanes that drew in and compressed refrigerant. A rotary vane compressor consisted of a **rotor** (turning unit) mounted

Figure 8-13. *A—The main parts of an axial compressor. The wobble plate is in the center of the photograph. Two pistons are shown on each side of the plate, and one-half of the compressor cylinder assembly is shown at the left. B—An assembled view of an axial compressor, with one-half of the cylinder assembly removed. Notice how the position of the wobble plate determines the movement of the pistons.*

off-center inside a large round casing. The compressor body formed the casing. The compressor crankshaft drove the rotor. Lengthwise slots in the rotor held the **vanes.** **Figure 8-14** shows a cross-section of a rotary vane assembly. When the compressor is turning, centrifugal force threw the vanes outward to contact the inside of the casing. The vanes divided the space between the rotor and the casing into a series of separate chambers. Since the turning rotor was mounted off-center in the casing, the chambers grew larger on one side of the casing, and became smaller on the other side.

An opening (usually a slot) in the casing allowed low pressure refrigerant to be drawn into the chambers as they became larger. The chambers reached their largest volume on the side opposite from the rotor. As the chambers continued moving around the casing, they begin to become smaller. This compressed the refrigerant. At its smallest point, the chamber reached a discharge port. The pressurized refrigerant was pushed out of the discharge port to the system high side. **Figure 8-15** shows the suction and compression process.

One advantage of the rotary vane compressor was no reed valves were needed. Without the back and forth motion of pistons, the rotary vane compressor was also smoother. Rotary vane compressors were used on smaller

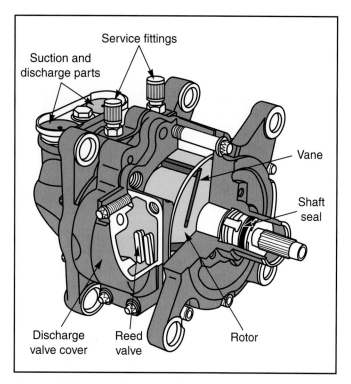

Figure 8-14. *A phantom view of a vane type compressor. The position of the vanes in the rotor causes them to be thrown outward as the rotor turns. (Sun)*

vehicles where smoothness is a major consideration. Most rotary vane compressors have a pumping capacity similar to a piston compressor of about 5 or 6 cubic inches (82 to 98 cc). Rotary vane compressors are no longer used because of service problems.

Scroll Compressors

Some vehicles have *scroll compressors,* **Figure 8-16.** Scroll compressors are sometimes called *spiral* or *orbital* compressors. Scroll compressors are smoother than piston compressors and more durable than rotary vane compressors.

The main components of a scroll compressor are the scrolls. A *scroll* is a length of flat metal, formed into a spiral shape. The scrolls are placed together as shown in **Figure 8-17.** One scroll is fixed to the compressor body and does not move. The other scroll is attached to the crankshaft in an eccentric (off center) position. This is the *moving scroll.* The moving scroll does not rotate with the crankshaft, but circles around the centerline as the crankshaft turns. This circular motion is called an *orbital path.* The scroll compressor crankshaft has a counterweight that compensates for vibrations caused by the orbiting scroll.

The fixed and moving scrolls are shaped and carefully machined so they do not touch during operation. A film of refrigeration oil forms during operation and seals the contact points. The compressor passages are designed so low pressure refrigerant enters the outside of the scroll assembly. Compressed refrigerant exits at the center of the scroll assembly. Scroll compressors do not use intake reed valves, but do have a discharge reed valve to prevent back flow from the high side of the system.

The orbital path of the moving scroll in relation to the fixed scroll creates pockets that trap incoming refrigerant. As the moving scroll continues to orbit, the pocket is moved along between it and the fixed scroll. As the pocket moves, it also becomes smaller, compressing the refrigerant. Finally the pocket of compressed refrigerant exits at the center of the scroll assembly. **Figure 8-18** shows the operation of the scroll compressor as the crankshaft revolves approximately 2.6 times. Notice that new pockets are formed as the existing pockets are moved. Most scroll compressors are used on vans or large pickup trucks. Scroll compressor pumping capacity is roughly equivalent to a piston compressor of about 8 cubic inches (131 cc).

Electric Compressors

Manufacturers are beginning to provide *electric compressors.* These compressors may be found as original equipment in the near future on super ultra low or ultra low emission vehicles (SULEV/ULEV) or electric cars and trucks, **Figure 8-19.**

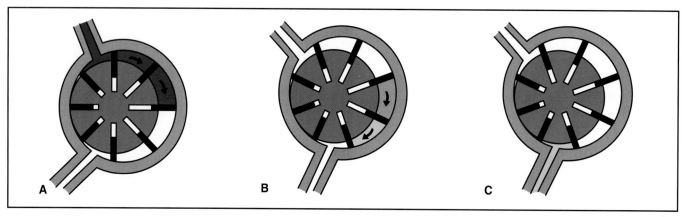

Figure 8-15. *The compressing action of a vane compressor. A—Refrigerant is drawn in from the intake opening as the space between the vanes increases. B—The refrigerant is carried around the rotor housing and compressed, as the space becomes smaller. C—The compressed refrigerant exits through the discharge opening.*

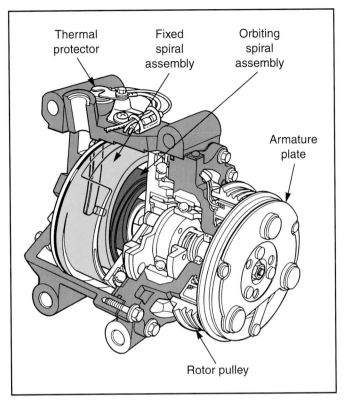

Figure 8-16. *Cutaway view of a scroll compressor. It mates with a second scroll that is similar in appearance. Scroll compressors always have two scrolls. (Honda)*

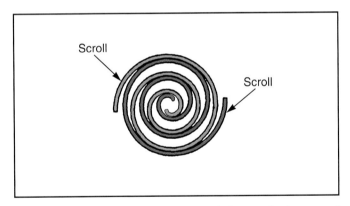

Figure 8-17. *The relative positions of the scrolls in a scroll compressor. (Sanden)*

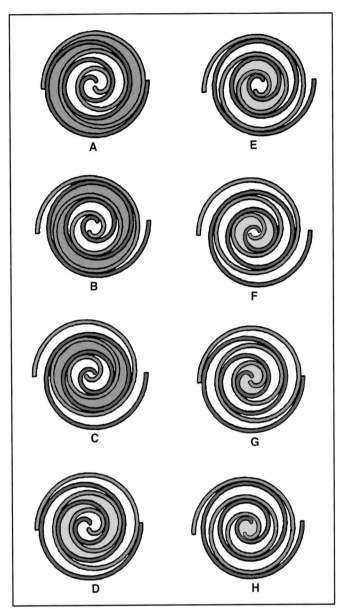

Figure 8-18. *This figure shows the compression process of a scroll compressor. Air is drawn in and compressed in a continuous process that is extremely smooth.*

The compressor is operated by a high voltage motor. This motor is either direct current or three-phase ac current. A separate power box is used to convert vehicle primary voltage for the compressor's needs. The major advantage of an electric compressor is it does not require engine horsepower for operation. On an electric vehicle, the electric compressor system can be programmed using the onboard vehicle computer to precondition the interior while the vehicle is still connected to its charging station. The electric compressor can also operate at higher speeds, providing better cooling performance when the vehicle is sitting in traffic.

The electric compressor system operates similar to a residential heat pump system. In the cooling mode, heat is

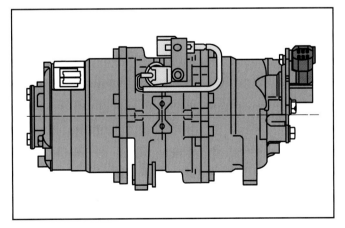

Figure 8-19. *Electric compressors are available for some applications. You are likely to find these on low and zero emissions vehicles. (Sanden)*

removed from the passenger compartment. In the heating mode, heat is taken from the atmosphere and delivered to the passenger compartment.

Compressor Capacity Control Valves

As discussed in earlier chapters, it is important to control the evaporator pressure for maximum cooling. Some piston compressors have a *capacity control valve* for this purpose. The capacity control valve is installed in the compressor body, **Figure 8-20,** and varies the pumping capacity (pumping ability) of the compressor. This design eliminates the need to cycle the compressor clutch on and off. The capacity control valve is operated by pressure signals from the low side of the refrigeration system.

On a compressor with a capacity control valve, the wobble plate is designed so its angle can change. The wobble plate can slide on the crankshaft. The wobble plate is also a two-piece assembly. A guide pin keeps the side that drives the pistons from rotating. See **Figure 8-21.** The stationary side can move up and down on the guide pin.

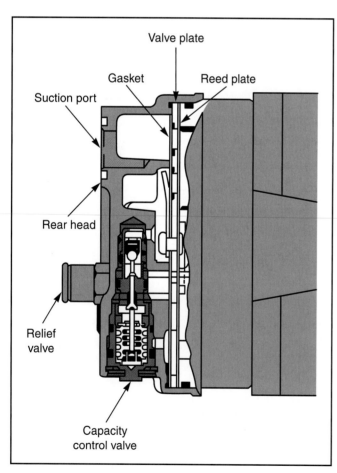

Figure 8-20. *This figure shows the location of the capacity control valve on a common compressor. Most capacity control valves are mounted at the rear of the compressor. (General Motors)*

When the wobble plate moves up or down on the guide pin, its angle changes. This changes the stroke of the pistons and therefore the pumping capacity of the compressor. Crankcase pressure determines piston stroke. The capacity control valve regulates crankcase pressure.

When low side pressure is above the control valve set point, the valve is positioned to open a passage to the low side of the system. This keeps the crankcase pressure low, and the pistons are at their maximum stroke. When low side pressure drops below the valve set point, the valve moves. When the valve moves, it closes off the crankcase passage to the low side of the system and opens a passage to the high side. High side pressure enters the crankcase. High crankcase pressure at the back of the pistons keeps them from moving down the cylinder. This reduces piston stroke and therefore compressor capacity. The wobble plate compensates for reduced piston movement by moving downward on the guide pin, reducing its angle.

When low side pressure rises past the control valve set point, the valve moves to its original position. This closes off the crankcase passage to the high side of the system and opens the passage to the low side. Crankcase pressure is lowered and the pistons resume pumping to their full capacity. This process repeats as necessary.

Some compressors have a provision for adjusting the capacity control valve. However, capacity control valve adjustment should only be made to correct specific problems, and only according to specific manufacturer's instructions.

Compressor Drives and Clutches

The most common method of operating the compressor is drive it with the vehicle engine. Whenever the engine is running, power is available to operate the compressor. For long compressor life and increased fuel mileage, the compressor should not be driven when it is not needed. The following sections explain how engine power reaches the compressor and how it can be disconnected.

Compressor Drive

This section discusses the components that connect the engine to the compressor; the belts and pulleys. Belts and pulleys are always used for simplicity and low cost. An advantage of the belt and pulley arrangement is the belt will slip if the compressor seizes. Although the belt is destroyed, there is no damage to the engine or other vehicle parts.

Types of Belts

A *belt* is a continuous band of flexible material, **Figure 8-22.** Its flexibility makes a belt the ideal method of transmitting engine power to the compressor. Belts are also used to power the engine coolant pump, alternator,

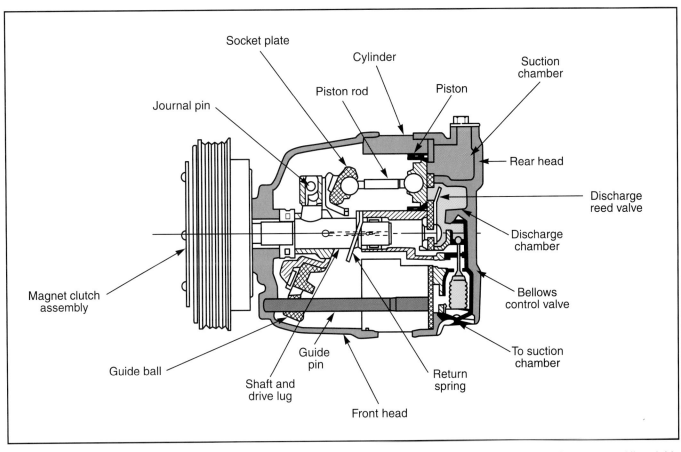

Figure 8-21. *A cross-section view of a variable capacity compressor. Note the relationship of the internal passages. All variable capacity compressors operate by varying the pressure in the compressor crankcase. (Subaru)*

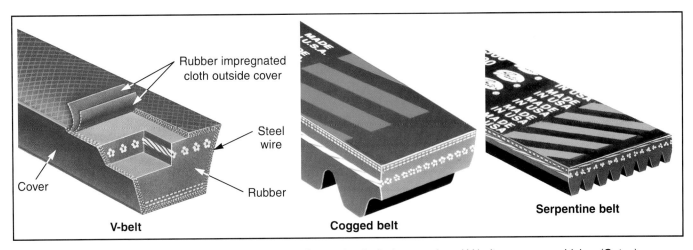

Figure 8-22. *Common types of compressor drive belts. Serpentine belts have replaced V-belts on many vehicles. (Gates)*

power steering pump, and air pump. ***V-belts*** were used exclusively in the past, and are still found on some vehicles. The term V-belt is taken from its cross-section, which resembles the letter V. Some compressors were driven by two V-belts, **Figure 8-23.** ***Serpentine belts*** are flat belts with internal ridges. Serpentine belts are used on almost all new vehicles. They are called serpentine belts because when installed, their bends and direction changes resemble a snake. **Figure 8-24** shows a typical serpentine belt installation.

Belt Adjusters

The adjustment of almost all serpentine belts is controlled by a ***belt tensioner.*** The belt tensioner is a separate pulley that is held against the belt by spring tension, **Figure 8-24.** No adjustment is needed. If the tensioner cannot hold the belt tightly, the tensioner or belt should be replaced.

Most V-belts use a manual adjusting device of some sort. Most V-belt adjusting devices consist of slotted holes in one of the driven accessories. To adjust the belt, loosen the attaching bolts and pull the driven accessory toward or away from the engine as necessary. To tighten the belt, the accessory is usually moved away from the engine. While holding the accessory in the adjusted position, tighten the attaching bolts.

Pulleys

A *pulley* can be thought of as a wheel with a grooved rim. The groove or grooves are designed to match the drive belt. There are two types of pulleys, driving and driven. The engine pulley is always the *driving pulley;* the compressor pulley is always the *driven pulley.* The engine pulley is driven directly by the engine crankshaft. Some engine pulleys are made of two halves of stamped steel, welded together to create a pulley. Other pulleys are made of cast aluminum. Some pulleys consist of grooves cut into the cast iron vibration damper.

Figure 8-23. *This photograph shows a dual V-belt arrangement on an older vehicle. Dual V-belts have been eliminated to reduce size and complexity.*

Figure 8-24. *This late-model engine has one serpentine belt. The belt is arranged so it drives all of the accessories. Serpentine belts are almost always used with self-tensioning devices similar to the one shown here. The tensioner is spring loaded to place a constant tension on the belt.*

Many engines have one or more idler pulleys on their belt drive systems. An *idler pulley* is an extra pulley that does not connect to any belt driven accessory. It usually rides on a separate bracket through a sealed bearing. Idler pulleys are used as guides to route the belt around an obstruction, or to keep a very long belt from vibrating. Some idler pulleys are part of the drive belt tensioning device.

Compressor pulleys may be made of steel or cast aluminum. Some compressor pulleys are vulcanized to a rubber core for shock absorption. A large cast iron vibration damper may be installed at the rear of the pulley on some older cars. The front of the pulley is machined to form one face of the compressor clutch.

The compressor pulley assembly is attached to the stationary compressor through a large bearing. The bearing is usually a double-row ball bearing. Most bearings are held to the compressor nose with a large snap ring. The pulley may be pressed on the bearing, or held by a retainer plate or snap ring. Compressor pulley bearings are permanently greased and do not require repacking. **Figure 8-25** illustrates a bearing in position on the compressor.

Pulley Ratios

Compressor performance depends to a large extent on the speed at which it is driven. The compressor must turn fast enough to cool the vehicle at low speeds. However, it must not turn so fast the belt comes off the pulley due to centrifugal force. The major factor in compressor speed is the speed of the engine. Another factor in compressor speed is the difference in the size of the driving and driven pulleys. This is called the *pulley ratio.*

Pulley ratio can be roughly determined by comparing the diameter of the engine pulley with the diameter of the compressor pulley. If the engine and compressor pulleys are the same size, the compressor will turn at the same speed as the engine. The pulley ratio is 1:1. If the engine pulley is twice as large as the compressor pulley, the compressor will turn twice as fast as the engine. The pulley ratio is 2:1. If the engine pulley is smaller than the compressor pulley, the compressor will turn more slowly than the engine. **Figure 8-26** illustrates this principle.

Pulley size will vary according to the type of engine and the vehicle's intended use. Some vehicles with high performance engines, or small engines with manual transmissions, will have a large compressor pulley to prevent overspeeding. Luxury cars normally driven at low speeds may have a compressor pulley the same size or slightly smaller than the engine pulley.

You will usually not be called on to calculate the pulley ratio. In some cases, however, it will be necessary to substitute a smaller compressor pulley to improve cooling, or install a larger pulley on a compressor used on a high speed engine. In this situation, you can use the following formula:

$$\text{Pulley ratio} = \frac{\text{driving (crankshaft) pulley diameter}}{\text{driven (compressor) pulley diameter}}$$

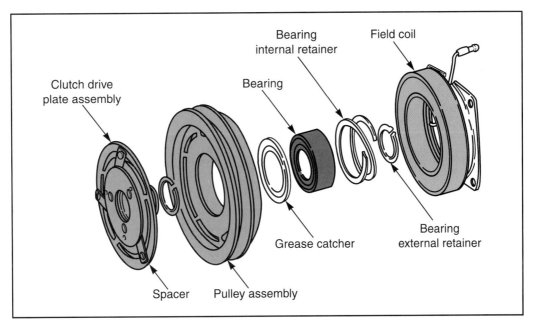

Figure 8-25. *This exploded view of a pulley assembly shows the internal parts. Major components are the pulley and drive plate, driven plate, electromagnet, and bearing. (Ford)*

For instance, if the diameter of the crankshaft pulley is 9″ (23 mm) and the diameter of the compressor pulley is 6″ (15.3 mm):

$$\frac{9''}{6''} = 1.5$$

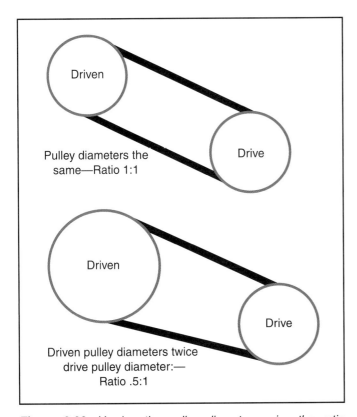

Figure 8-26. *Varying the pulley diameter varies the ratio between the drive and driven pulleys. The larger the driven pulley in relation to the drive pulley, the slower the engine drives the compressor.*

The pulley ratio is 1.5:1. Actual crankshaft and compressor pulley ratios vary greatly, depending on the engine, vehicle, and expected use. Pulley ratios on most vehicles, however, are between 1:1 and 5:1.

Compressor Clutches

Since the compressor pulley turns whenever the engine is running, there must be some way of disconnecting the compressor from the pulley when air conditioning is not needed. The **compressor clutch** is the means of engaging and disengaging the compressor from the engine.

The main parts of the compressor clutch are the machined face on the pulley, the matching **armature plate,** and the **electromagnet.** These parts are discussed in the following paragraphs. **Figure 8-27** shows a cutaway view of a compressor clutch assembly.

The armature plate is made of iron, iron alloy, or steel. It is pressed onto the crankshaft drive end, or may be held by a nut installed on the compressor crankshaft. A key and matching keyway are usually used to ensure the armature plate does not spin on the crankshaft. The armature plate is installed so its machined face almost touches the machined face on the pulley, **Figure 8-28.** The gap between the two faces is usually between .020-.060″ (.05-.15 mm). The connection between the center of the armature plate and the machined face is flexible. The connection may be rubber vulcanized to the metal parts, or may be a series of metal straps that transmit power while still allowing movement.

The electromagnet is installed under the pulley. An input wire connects the electromagnet to the vehicle electrical system through one or more switches. On most newer vehicles, the compressor is grounded through a wire connected to chassis ground. A diode is installed in

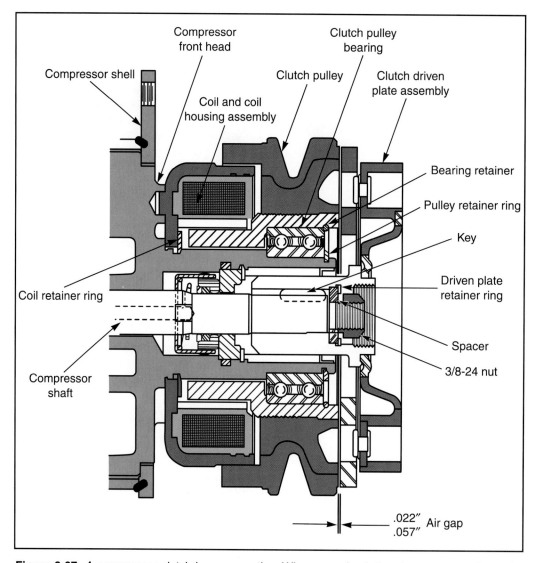

Figure 8-27. *A compressor clutch in cross section. When energized, the electromagnet draws the driven pulley into contact with the drive pulley. (Ford)*

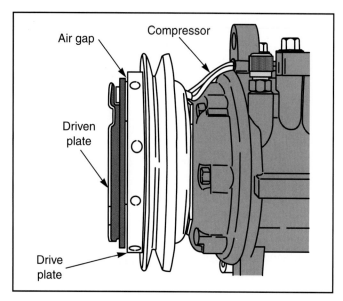

Figure 8-28. *The air gap between the drive and driven pulleys determines both clutch slippage and whether the pulley will drag when disengaged. Proper air gap is critical. (DaimlerChrysler)*

the wire to prevent current from returning through the circuit. In some cases, another wire allows the electromagnet to ground through the compressor body. On some older vehicles a set of brushes similar to those used in an alternator made the electrical connection between the electromagnet and the power source and ground.

If the switch or switches are in the off position, no current flows through the electromagnet. When all switches are in the on position, current begins flowing in the electromagnet. Current flow through the electromagnet creates a ***magnetic field.*** This field attracts the armature plate, which moves toward the pulley. When the machined faces on the armature plate and pulley contact, the pulley begins driving the armature plate and compressor crankshaft. The compressor clutch is ***engaged,*** and the compressor is operating.

When the current flow through the electromagnet is stopped, the magnetic field collapses, and the armature plate moves away from the pulley. The clutch is now ***disengaged,*** and the compressor stops operating.

Compressor Clutch Control Systems

Compressors can draw up to 5 horsepower from the engine. This can reduce fuel mileage by as much as 2 miles per gallon (about 8 liters per kilometer). It is therefore important to ensure the compressor is operating only when it is needed.

To control the flow of current to the electromagnet in the compressor clutch, several switches may be used. These switches are sometimes called relays. The most obvious switch is the on-off switch located on the HVAC control panel. This is one of the switches that are directly controlled by the driver. In most cases, the dashboard switch not only turns on the compressor, but also operates the blower motor. It may also control the movement of various airflow control doors. Other compressor clutch control switches include the following:

❑ Evaporator temperature switch. Opens the circuit to the clutch if the evaporator temperature becomes too low.

❑ Low pressure switch. Cuts off current to the clutch if the low side pressure becomes less than a specified amount.

❑ High pressure switch. Cuts off current when the high side pressure becomes excessive.

❑ Ambient air switch. Will not allow the compressor clutch to engage when the ambient (outside) air temperature is below roughly 32-40°F (0-4°C).

❑ Power steering switch. Turns off the compressor clutch when the power steering pressure rises. This reduces the load on the engine.

❑ Wide open throttle switch. Used on carbureted and some fuel injected engines. Breaks the circuit to the electromagnet when the throttle is opened completely. This eliminates compressor drag when maximum acceleration is needed. On newer vehicles, this function has been taken over by the throttle position sensor.

❑ Time delay switch. Delays compressor clutch engagement for a few seconds when the vehicle is first started. This allows the engine idle to stabilize before the compressor load is applied. The delay switch has been eliminated on newer vehicles and is now a direct computer control function.

❑ Anti-dieseling switch. Turns the compressor clutch on when the engine is turned off. The extra drag of the compressor reduces engine dieseling (run-on when the key is turned off). Computer control has eliminated the need for this switch.

Some older cars were equipped with a fuse that burned out and broke the compressor circuit if the low side pressure closed a switch on the rear of the compressor. Older aftermarket systems sometimes used a thermostatic switch that operated somewhat like an expansion valve. These controls will be discussed in more detail in Chapters 10 and 13. **Figure 8-29** shows some of the compressor clutch control switches used on an older HVAC system.

Some of these switches are now monitored and in some cases, have been eliminated through the use of computer control. Newer vehicles control the compressor through the powertrain control module (PCM) or body control module (BCM). The control module receives inputs from sensors and HVAC system components, then uses these inputs to control the compressor clutch. **Figure 8-30** is a schematic of a typical modern compressor clutch control.

Constantly Engaged Compressor Clutch

Some of the latest vehicles have a **constantly engaged compressor clutch.** Constantly engaged clutches are used on capacity-controlled compressors, and the compressor pumping capacity can be reduced to 10% of maximum when the system is not being used. If the compressor fails and locks up, a rubber disc in the clutch deforms and causes the compressor to disengage from the drive belt.

Compressor Makes and Models

Compressors can be identified by their manufacturer and by a plate or sticker on the compressor body. Some manufacturers such as Sanden, make compressors that are interchangeable. You should use a compressor that is as

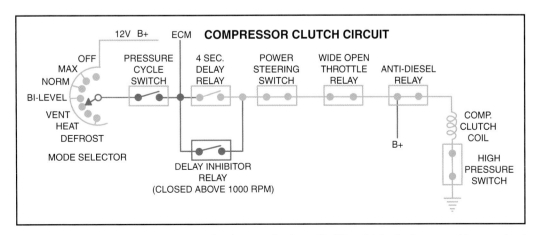

Figure 8-29. *A common compressor clutch control schematic. The clutch electromagnet is energized only after passing through several switches. Most of the switches shown here cut off the compressor as a safety precaution, or to reduce compressor load on the engine. (General Motors)*

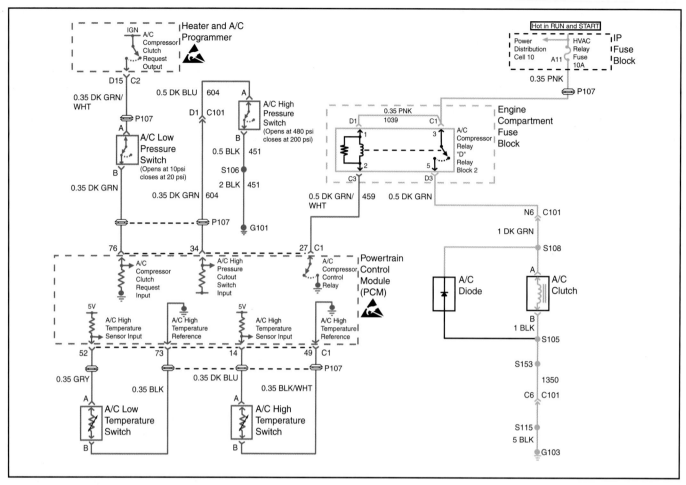

Figure 8-30. *A simplified view of a computer compressor control. The computer receives inputs and energizes the compressor. The computer may also control blower speeds or air door motors. (General Motors)*

close to the original design and performance of the old compressor as possible.

 Note: Not all compressors will operate satisfactorily when an R-12 system is retrofitted to use R-134a or another refrigerant. Always consult the vehicle or compressor manufacturer before retrofitting any vehicle air conditioning system.

Bosch

The Bosch axial piston compressor was used on some BMW models between 1979 and 1992. There are three versions of this compressor, with external bracket and hose connection differences only. Bosch compressors were not used on any other vehicles. This compressor is always equipped with a single V-belt type pulley.

Calsonic

Calsonic compressors were used on Nissan Z-cars and pickup trucks in 1987-1989, and have been used in Subaru vehicles since 1992. These compressors are alu-minum case axial piston types. Calsonic compressors are usually interchangeable.

Chrysler

Since 1984, all axial piston Chrysler compressors have been made by Nippondenso. This includes the A590 and C171 fixed displacement models, and the 6C17 variable displacement model, **Figure 8-31.** These compressors may have Chrysler or Nippondenso labels and model numbers. Chrysler compressors use single V-belt, double V-belt, and serpentine pulleys, depending on the engine and vehicle.

Frigidaire/Harrison/Delco Air/Delphi

 Note: General Motors manufactured some of the following compressors under the names Frigidaire, Harrison, and Delco Air. An independent company named Delphi now manufactures all versions of these compressors under the Delphi and Harrison brand names.

Frigidaire compressors are classed according to whether they are fixed or variable displacement types. Frigidaire compressors have been equipped with single V-belt, double V-belt, and serpentine pulleys. Often the only differences between the same model Frigidaire compressors are the type and location of various pressure and temperature switches at the rear of the compressor body. Some compressors have no switch ports. Frigidaire compressors have also been labeled as Harrison, Delco Air, or Delphi units. In some cases, the mounting ears are slightly different between the same model installed on different engines. Determine the compressor model number, vehicle model and year, and engine displacement before obtaining a replacement compressor.

Fixed Displacement Compressors

There have been three major fixed displacement Frigidaire compressor designs. The A6 was a 6-cylinder axial piston type used on General Motors vehicles from 1962 until 1980, Ford vehicles from 1970 to 1979, and some European vehicles from the 1960s to the 1980s. The A6 was used on Jaguars until 1992. There are three different models of the A6, each one can be identified by the position of a hole in a port at the back of the compressor, **Figure 8-32.** This port takes either a high pressure or superheat switch. The A6 used on Ford vehicles does not mount a switch on the compressor.

The R4 is a four cylinder radial compressor which replaced the A6. It was used on General Motors vehicles and some Mercedes-Benz, Volvo, and Isuzu models between 1973 and 1993. It is used on rear-wheel drive and a few front-wheel drive vehicles.

There are two different styles of R4 compressors. The standard R4 uses a larger clutch design while the clutch on the lightweight R4 is much smaller, **Figure 8-33.** Depending on the application, the compressor could have a port for a superheat switch, high pressure switch, or no port. The two types of R4 are not interchangeable. Later model R4s are improved and produce less noise and vibration. These compressors can be identified by an orange label on the compressor body.

The H6 series (sometimes called the DA6 or HR6) has been used from the early 1980s to present on General Motors and European vehicles. The downsized axial DA6 was used until 1987, when it was replaced by the HR6. The Harrison redesigned HR6 is interchangeable on vehicles equipped with the DA6, **Figure 8-34A.** Two different styles of lip seals are used on these compressors; they are not interchangeable. All DA6 and HR6 compressors come with plugged switch ports, available for use as needed by application. Both compressor models are similar to the A6, but are lighter and have a different mounting system. They will not interchange with the A6.

Standard Variable Displacement Compressors

Standard variable displacement compressors have been installed on some General Motors vehicles since 1985. The 5-cylinder V-5, shown in **Figure 8-34B,** was the earliest version. Later versions are the V-6 and V-7. The V-7 has been used on GM light trucks since 1997.

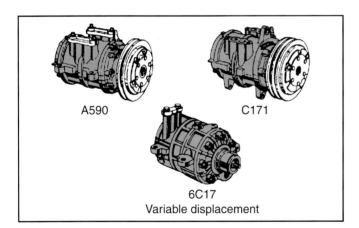

A590   C171

6C17
Variable displacement

Figure 8-31. Vehicles built by DaimlerChrysler usually have compressors manufactured by Nippondenso. (Four Seasons)

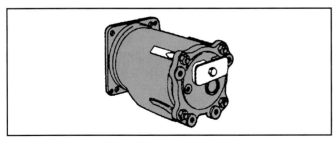

Figure 8-32. The A6 was used by General Motors and other manufacturers for many years. (Four Seasons)

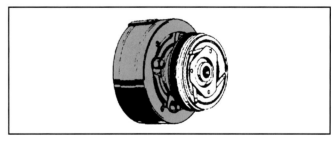

Figure 8-33. R4 compressors are still used by General Motors on a few vehicles. (Four Seasons)

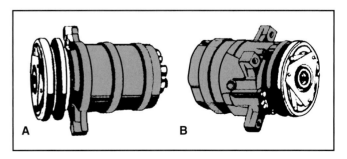

A   B

Figure 8-34. These compressors appear to be similar, but they are very different. A—This is the DA6 compressor. The HR6 is almost identical. B—This is a V-5 compressor. It is a variable displacement unit. (Four Seasons)

Compact Variable Displacement Compressors

Delphi manufactures a series of **compact variable displacement compressors,** or **CVCs.** The CVC series has the same basic design and operating principles as the V-6 and V-7 series. However, CVCs are smaller and lighter than V-6 and V-7 compressors, weighing between 11 and 14 pounds (5.2 to 6.3 kg) each. CVCs are generally found on small, late-model GM vehicles, but they may appear on other makes in the future. The current CVC designs include the CVC125 and CVC135 six-piston models and the CVC165 and CVC185 seven-piston versions.

Ford

Newer Ford compressors are aluminum bodied axial piston designs, **Figure 8-35.** There are three major Ford models; the FS6, the FS10, and the FX15. The FS6 was used on Ford cars and trucks from 1981 to 1994 and on Lincoln models from 1981 to 1987. The FS10 has been used on Ford and Mercury vehicles from 1991 to 1995 and Lincoln from 1992 to the present. The FX15 was used on various Ford Motor Company vehicles between 1989 and 1994. Ford serpentine belt pulleys have either 4, 6, 7, or 8 grooves, and the technician must count carefully to determine the correct compressor and clutch model. Most compressors on late-model Ford Motor Company vehicles are made by Nippondenso and carry Ford part numbers.

Hitachi

Hitachi axial compressors were used on some Nissan vehicles between 1976 and 1989. Hitachi no longer makes automotive air conditioning compressors. Other compressors (usually Sanden or Nippondenso) can be substituted if a similar displacement compressor is used and hose connections are the same.

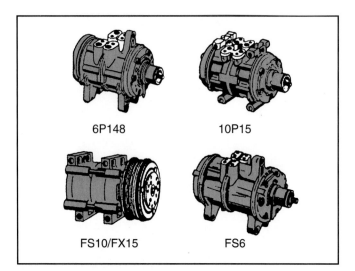

6P148   10P15

FS10/FX15   FS6

Figure 8-35. *Compressors used by Ford on its late-model vehicles. The 6P and 10P are Nippondenso compressors.*

Keihin

Keihin compressors were axial piston types used on the Honda Accord and Prelude between 1983 and 1987. These compressors are no longer manufactured. If a Keihin compressor is defective, conversion kits allow the use of another type of compressor.

Matsushita/Panasonic

This line of axial piston compressors has been called either Matsushita or Panasonic at various times. They were used on Honda vehicles in 1988-1989, Subaru from in 1985-1989, and on Mazda vehicles throughout the 1990s. Matsushita/Panasonic compressors can be identified by their model numbers, which always begin with NL 130 or NL 1300.

Mitsubishi

Mitsubishi axial piston compressors are used on Mitsubishi vehicles, 1994-present. Mitsubishi vehicles may have other manufacturers' compressors, and the technician must check a parts catalogs to obtain the proper replacement compressor.

Nihon/Nihon Radiator

Various kinds of Nihon or Nihon Radiator compressors were used on some Nissan and Infiniti vehicles between 1983 and 1993. Rebuilt versions of these compressors are still available.

Nippondenso

Almost every domestic and imported vehicle manufacturer has used one or more Nippondenso compressor models since they were first manufactured in 1974. These compressors are currently being installed on many vehicles. Almost all Nippondenso compressors are axial piston types, but a few inline radial compressors were made during the 1970s and used by Toyota. Due to the vast number of Nippondenso models and their complex numbering system, the technician must determine the compressor model number, vehicle model and year, number of ears on the compressor body, and engine displacement before obtaining a replacement compressor.

Sanden/Sankyo

Sanden has been manufacturing compressors since 1971. Almost all vehicle manufacturers have used Sanden compressors. Compressors made between 1971 and 1982 were called Sankyo compressors. All Sanden compressors with the SD model number are axial piston types. The number immediately following the SD prefix indicates the number of cylinders and the last number indicates the displacement in cubic centimeters divided by ten. Later models have the letters B or H included; this indicates the location of the port. The letters HD on any Sanden compressor stand for heavy duty. All compressors with the TR model number are scroll compressors. Although most Sanden and Sankyo compressors interchange, the replacement compressor must have the same displacement as the original, **Figure 8-36.**

Seiko-Seki

Seiko-Seki axial compressors have been used on various BMW and Saab models since 1979. Only Saab is currently using the Seiko-Seki compressor.

Seltec

Seltec compressors are used on off-road vehicles and large trucks. These compressors will be found only on these types of vehicles and are well marked.

Tecumseh/York

The Tecumseh model HR 980 was used on many Ford vehicles between 1983 and 1988. The HR 980 was a four piston radial model. These compressors are no longer manufactured. If a model HR 980 compressor is defective, a conversion kit can be used to allow the installation of another type of compressor.

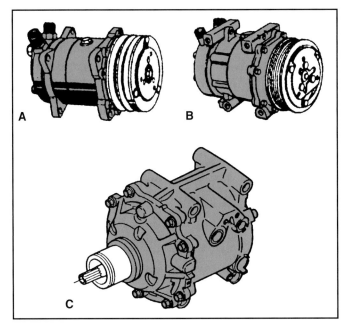

Figure 8-36. *Almost every manufacturer uses Sanden compressors. Sanden compressors are often used in aftermarket air conditioning systems. A—SD-5. B—SD-7. C—TR variable displacement compressor. (Four Seasons, Sanden)*

Unicla

Unicla compressors are not offered as original equipment on any vehicles sold in the United States. However, some parts stores in the United States sell Unicla compressors as replacements for original equipment compressors.

Zexel/Diesel Kiki

Zexel compressors have been used on many domestic and imported vehicles from 1963 to the present. Diesel Kiki was the original name of the Zexel company and many compressors were sold under the Diesel Kiki name. The company became Zexel in 1990. Some Diesel Kiki compressors were sold as York compressors. Some current Zexel compressors are sold under the Seltec brand name. Most Diesel Kiki and Zexel compressors are axial piston types. However, a few rotary vane Zexel compressors have been produced.

Older Compressors

While no longer used, these compressors were installed on many domestic and some imported cars, as well as large trucks and farm equipment. Many are still in service.

Chrysler V-type Radial

This compressor was the only compressor used on Chrysler products from 1961 until the early 1980s. It had a large V-shaped cast iron case with two cylinder heads, **Figure 8-37.** This compressor was sometimes called an Airtemp or RV2 compressor. Some Chrysler V-types used on larger vehicles had a pressure control valve called an EPR valve installed in the inlet port.

York and Tecumseh Radial

York and Tecumseh radials are inline two cylinder compressors that strongly resemble each other, **Figure 8-38.** The York compressor case is aluminum and the Tecumseh case is cast iron. These compressors were widely used on Ford

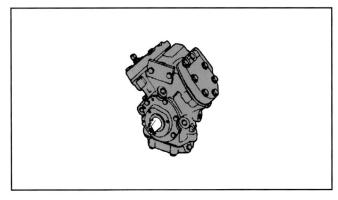

Figure 8-37. *Chrysler used the RV2 compressor for many years. (Four Seasons)*

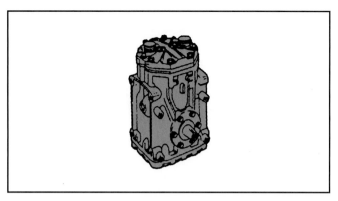

Figure 8-38. *General Motors and Ford both used the York/Tecumseh compressor on many vehicles. (Four Seasons)*

Motor Company and American Motors products, and some imports. York compressors were part of most older aftermarket air conditioner packages. When used on Ford vehicles, these compressors were original equipment into the mid-1980s. Major differences between the various models of these two compressors are the types of hose connection fittings. York and Tecumseh compressors can be interchanged as long as the hose fittings are correct. Some vehicles used a small York compressor called a York Mini. A defective York Mini must be replaced with another York Mini.

General Motors Five Cylinder Axial

The General Motors five cylinder axial was used on older (pre-1962) General Motors vehicles. It was a cast iron compressor that resembled a General Motors A6. This compressor was not rebuildable and replacements will be difficult to locate.

Summary

All refrigerant compressors do the same job, pressurizing the refrigerant and pumping it through the refrigeration system. The compressor and its related parts must be designed to operate at low and high engine speeds.

All compressors have a body or housing made of aluminum or cast iron. The compressor body is the mounting point for the other components. The cylinder head(s) or end plate is installed on the compressor body. Other components installed in the compressor are refrigerant lines, service fittings, pressure relief valves, and electrical high pressure shutoff switches. The crankshaft rides on bearings in the compressor body. The drive end of the crankshaft extends from the body and attaches to the clutch and pulley assembly.

Modern vehicles use three types of compressors. The piston compressor is the most common. Pistons can be arranged in many ways. A few past vehicles used rotary vane compressors. Some new vehicles have scroll compressors.

Piston compressors look like small engines. Older compressors had only two cylinders, but modern compressors may have as many as ten or more. The compressor body and cylinders on older vehicles were made of cast iron. Some compressors have iron cylinder liners or cast iron pistons. On other compressors, both pistons and cylinders are aluminum to reduce weight.

The operation of a compressor piston in the cylinder is identical to an engine piston. Compressor pistons have at least one piston ring to seal pressure between the cylinder and piston. All piston compressors use reed valves. There are two reed valves for each compressor cylinder, one intake and one exhaust. Reed valves may be installed in a valve plate, or be separate for each cylinder.

There are two kinds of piston compressors, the radial compressor with the pistons extending at right angles to the crankshaft, and the axial compressor with the pistons and cylinders parallel to the crankshaft. Both types have similar pistons, cylinders, and reed valves. A few older vehicles used rotary vane compressors. The crankshaft drives the rotor, which is offset in the chamber. Vanes create separate chambers that carry the refrigerant around and compress it. An advantage of the rotary vane compressor is it needs no reed valves.

Scroll compressors are sometimes used. The scroll compressor consists of two flat pieces of metal made into spiral shapes. The movable scroll orbits within the fixed scroll. This creates pockets that move and compress the refrigerant. Scroll compressors have a discharge reed valve to prevent backflow.

Some piston compressors have a capacity control valve installed in the compressor body. The capacity control valve varies the pumping capacity of the compressor by regulating crankcase pressure.

A belt and pulleys connect the engine to the compressor. V-belts and serpentine belts may be used. The engine pulley is the driving pulley. The compressor pulley is the driven pulley. The compressor pulley is attached to the compressor through a bearing. Engine speed and pulley ratio determine compressor speed.

A compressor clutch is used to engage and disengage the compressor and the engine. The electromagnet is energized, and pulls the armature plate into contact with the pulley. The pulley drives the armature plate and compressor crankshaft. The dashboard control switch controls the compressor clutch electromagnet. Other compressor clutch control switches include evaporator temperature switches, low and high pressure switches, and ambient air switches.

Review Questions—Chapter 8

Please do not write in this text. Write your answers on a separate sheet of paper.

1. Compressor speed varies with _____ speed.

2. What is the most common of the three types of compressors?

3. A movable flap that closes an opening in the compressor is a _____.

4. In an axial compressor, the cylinders are _____ to the crankshaft.

5. A rotary vane compressor consists of a _____, mounted off-center inside of a large _____ casing.

6. What are two names for a scroll compressor?

7. What type of valve does a scroll compressor have?

8. To control compressor output, the capacity control valve modifies the pressure in the compressor _____.

9. Which two compressor clutch parts have machined faces that contact each other?

10. When the compressor clutch is _____, the compressor is operating.

ASE Certification-Type Questions

1. All of the following statements about compressors are true, *except:*
   (A) every compressor has a crankshaft.
   (B) every compressor has pistons.
   (C) all compressors use a front lip seal.
   (D) all compressors have an electromagnetic clutch.

2. A reed valve is opened and closed by:
   (A) oil pressure.
   (B) refrigerant pressure.
   (C) a guide rod.
   (D) an electromagnet.

3. Technician A says double-ended pistons are used on radial compressors. Technician B says reed valves are used on all piston compressors. Who is right?
   (A) A only.
   (B) B only.
   (C) Both A and B.
   (D) Neither A nor B.

4. All of the following statements about axial compressors are true, *except:*
   (A) the wobble plate can turn on the crankshaft.
   (B) compressors with double-ended pistons have an even number of cylinders.
   (C) compressors with single ended pistons have an odd number of cylinders.
   (D) rotation of the wobble plate causes it to move back and forth.

5. On a rotary vane compressor, which of the following parts does the crankshaft drive?
   (A) Vanes.
   (B) Casing.
   (C) Discharge.
   (D) Rotor.

6. All of the following statements about scroll compressors are true, *except:*
   (A) both scrolls can move.
   (B) one scroll is attached to the crankshaft in an off-center position.
   (C) the scroll circles around the crankshaft centerline in an orbital motion.
   (D) a discharge reed valve is installed in the compressor.

7. Technician A says the capacity control valve is operated by pressure signals from the crankcase. Technician B says the capacity control valve is operated by pressure signals from the low side of the refrigeration system. Who is right?
   (A) A only.
   (B) B only.
   (C) Both A and B.
   (D) Neither A nor B.

8. All of the following are types of belt drives, *except:*
   (A) serpentine.
   (B) electromagnetic.
   (C) V-belt.
   (D) double V-belt.

9. Based on the information in this chapter, which of the following would be an incorrect air gap between the faces of the compressor clutch armature and pulley?
   (A) .010″ (.025 mm).
   (B) .030″ (.075 mm).
   (C) .045″ (.115 mm).
   (D) .060″ (.150 mm).

10. What is used to ensure the armature plate does not spin on the crankshaft?
    (A) Electromagnet.
    (B) Flexible connection.
    (C) Key and keyway.
    (D) Double V-belt.

Evaporators transfer the heat from the incoming air to the refrigerant. They are located inside the blower case, near the point where air enters the passenger compartment. (Delphi)

Chapter 9

Evaporators, Condensers, Accumulators, and Receiver-Driers

After studying this chapter, you will be able to:
- ❏ Explain the purpose of the evaporator.
- ❏ Identify the major types of evaporators.
- ❏ Explain the purpose of the condenser.
- ❏ Explain the purpose of the accumulator.
- ❏ Explain the purpose of the receiver-drier.
- ❏ Describe the operation of pressure relief devices.

Technical Terms

Evaporator	Condenser	Adsorption
Heat exchanger	Tube and fin	Silica gel
Single-pass evaporator	Serpentine	Molecular sieve
Multiple-pass evaporator	Multiflow condensers	XH-7
Tube and fin evaporator	Header	XH-5
Plate and fin evaporator	Accumulator	XH-9
Serpentine evaporator	Oil return hole	Pressure relief valve
Orifice tube	Receiver-drier	Fusible plug
Expansion valve	Desiccant	High pressure switch
Evaporator temperature control	Absorption	Refrigerant containment switch
Interior air filter		

The evaporator and condenser are used on every refrigeration system. Almost all refrigeration systems have either an accumulator or a receiver-drier. The evaporator and condenser work with the compressor and flow control device to cause the refrigerant to change state. The accumulator and receiver-drier are used to store, separate, and remove any moisture from the refrigerant. This chapter will cover the design and operation of these parts. It also explains how they can affect the operation of other refrigeration system parts. Studying this chapter will enable you to better diagnose and service these components when you encounter them in the service chapters.

Evaporators

Chapter 5 discussed how refrigerant changes state from liquid to vapor by absorbing heat from the *evaporator.* Refrigerant flow through the evaporator can be downward, upward, or sideways, depending on the design. Air flows through the entire evaporator, usually in as straight a line as possible. To keep air from flowing around the evaporator, it is usually a close fit in the evaporator case, with a rubber or foam seal at the top. The following sections cover the types of evaporators, how evaporators are constructed, and other parts that may be attached to them.

 Note: Some vans and trucks may have two evaporators. Each evaporator on these systems is made and operates in the same way as a single evaporator.

Evaporator Design and Construction

The basic design of an evaporator allows the air entering the passenger compartment to flow through it with little resistance. The contact between the evaporator metal and the air transfers (exchanges) heat in the air to the evaporator, which is then absorbed by the refrigerant. For this reason, the evaporator is often called a *heat exchanger.* The evaporator must allow maximum airflow while being as compact as possible. Moisture collected on the surface of the evaporator must be channeled to the bottom of the evaporator case. The lowest point at the bottom of the evaporator case has a drain, usually with a rubber flap or neck that allows water to drip out. The flap seals against the entry of dirt, insects, and outside air.

Evaporator tubes and fins are usually constructed from aluminum or aluminum alloy. Aluminum transfers heat very well and makes an excellent heat exchanger. Almost all original equipment tubing is made of aluminum. Some replacement evaporators may use copper tubing. Aluminum tubing provides better heat transfer than copper. However, leaks in copper tubing can be more easily repaired.

Fins are attached to the coils to aid heat transfer into the tubing. Fins may be soldered in place, but most are simply pressed tightly over the tubing. The fins are made of aluminum sheets so they are as thin and flat as possible. The use of thin flat fins allows maximum airflow and heat transfer. The fins may be bent into V or accordion shapes to maximize the contact between the fin metal and the moving air. Bending the fins also makes them stronger, and they will be less likely to be bent out of shape by debris.

Types of Evaporators

There are three refrigerant evaporator designs. The use of each depends on the manufacturer and the space and airflow limitations of the system. Manufacturers often use more than one type of evaporator. All evaporators can be classified as either single-pass or multiple-pass.

Single- and Multiple-pass Evaporators

Many evaporators use one tube, in multiple rows. Others are long thin designs to fit in the available space of a small car. Some evaporators have headers that direct the refrigerant through several sets of tubes at the same time.

Evaporators may use one tube throughout to move the refrigerant. This type is known as a *single-pass evaporator.* Single-pass evaporators are made from a single piece of tubing folded into a series of coils. Single-pass evaporators are often used in smaller vehicles, where underdash space is at a premium.

Some evaporators are designed so refrigerant enters through the top and passes through three or four of the plates before being redirected upward through the next set of plates. This design is called a *multiple-pass evaporator.* The multiple-pass design keeps refrigerant velocity high, preventing oil from dropping out and collecting in the bottom of the evaporator.

Tube and Fin Evaporator

The *tube and fin evaporator* is usually a single-pass evaporator. The coils are formed to reduce evaporator size as much as possible without causing restriction to the refrigerant flow.

Figure 9-1 illustrates a tube and fin evaporator. Tube and fin evaporators are made in many shapes and sizes. Some tube and fin evaporators are double or even triple row designs with sets of coils arranged in several rows.

Refrigerant flowing through the coils absorbs the heat transferred from the air by way of the fins and tubing. If the evaporator and pressure controls are working properly, very little liquid refrigerant will be left to exit the evaporator. During hot weather, the refrigerant will be superheated. As was explained in Chapter 5, superheat is any extra heat added after the refrigerant changes to a vapor.

Figure 9-1. *This figure shows a simple tube and fin evaporator. The fins conduct heat from the air passing through the case to the tubes, where it is absorbed by the refrigerant.*

Plate and Fin Evaporator

Instead of a tube surrounded by fins, the **plate and fin evaporator** is a set of flat aluminum plates between two aluminum end tanks. Fins are installed between the plates. The flat tube design reduces resistance to airflow while allowing more contact between refrigerant and air.

On early plate and fin evaporators, the refrigerant entered at the top of the evaporator and passed through a header that distributed equal amounts of refrigerant through each fin. This design caused oil to drop out of the refrigerant (along with some liquid refrigerant) and collect at the bottom of the evaporator. A separate oil return line, sometimes called a bleed line or a liquid return line, was

needed. **Figure 9-2** is an illustration of a plate and fin evaporator with an oil return line.

Later model plate and fin evaporators are multiple-pass types. They are similar in appearance to the earlier models but can be identified by the lack of an external oil return line, **Figure 9-3.** As a general rule, multiple-pass plate and fin evaporators are used with cycling clutch systems, while older plate and fin evaporators are used on systems with expansion valves. There are, however, exceptions to this rule depending on the manufacturer.

Serpentine Evaporator

The **serpentine evaporator** is a single-pass evaporator. It consists of a flat tube coiled to allow refrigerant to pass through the entire tube. Fins similar to the ones used on other evaporator designs are attached to the tube. The shape of the serpentine evaporator tube resembles a serpent, **Figure 9-4.**

Parts Attached to the Evaporator

The **orifice tube** or **expansion valve** is installed at the inlet of the evaporator. The fitting is either a compression type using an O-ring or a spring lock fitting. When an expansion valve is used, the sensing bulb will be located at the outlet tube of the evaporator. The bulb will be placed in a pocket made into the tube or clamped to the tube exterior and covered with insulation. **Figure 9-5** illustrates the two sensing bulb attachment methods.

Some type of **evaporator temperature control** is often used to prevent icing. They are either installed on the evaporator outlet or in the evaporator fins. This may be an

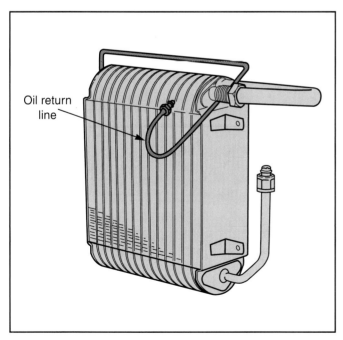

Figure 9-2. *This plate and fin evaporator distributes equal amounts of refrigerant through each fin. This type can be recognized by its use of an oil return line. (Ford)*

Figure 9-3. *This plate and fin evaporator is designed so refrigerant makes several passes through the fins. It has no oil return line.*

Figure 9-4. *A serpentine evaporator is a single flat tube that passes snakelike through the cooling fins. (Subaru)*

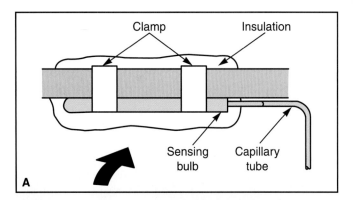

Figure 9-5. *A—Most sensing bulbs are attached to the evaporator with clamps and covered with insulating material, often called dum-dum or presstape. B—A few expansion valve bulbs are installed in a cavity formed in the evaporator outlet tube.*

electrical switch that opens and closes at certain temperatures, or a temperature sensor connected to a control module. A typical evaporator temperature control switch is shown in **Figure 9-6.** The design and function of various temperature sensors will be discussed in Chapter 10.

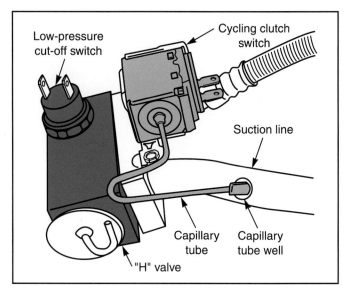

Figure 9-6. *On some refrigeration systems an evaporator temperature control device is installed at the evaporator outlet or in the evaporator fins. This sensor is used to control clutch cycling. (DaimlerChrysler)*

Interior Air Filter

The air intake side of some blower case assemblies contain an *interior air filter.* This filter is usually a conventional filter. In some cases, it may be electrostatically charged. The filter removes dust, pollen, and in some cases, odors from the incoming air. This keeps the evaporator from becoming restricted, and the inside of the vehicle cleaner, **Figure 9-7.**

Condensers

The *condenser* is the heat exchanger that removes heat from the refrigerant, allowing it to condense from a vapor to a liquid. On vehicles with water-cooled engines, the condenser is either located ahead or mounted beside the vehicle radiator. On an air-cooled engine, the condenser is placed where the cooling system blower can draw air through it. Since the condenser operates under higher pressures than the evaporator, it is made from a heavier gauge of aluminum or copper tubing, usually painted black to aid in heat transfer.

Condenser Design

Most modern condensers are cross-flow types, which means the tubes are placed so refrigerant flows across the condenser. There are three primary types of condensers in use. The first condenser is the *tube and fin* design. Tube and fin condensers have been in use for many years. The one most often used is referred to as a 3/8 inch tube design, **Figure 9-8A.**

A variation on the tube and fin design has a smaller diameter tube (6 mm or 1/4 inch), **Figure 9-8B.** In some

Figure 9-7. *This evaporator case is equipped with an air filter that removes dust, pollen, and odors from the incoming air. This filter is electrostatically charged to assist in filtration. (Lexus)*

cases, the 6 mm condenser may have an internal baffle which gives the condenser similar flow characteristics as the multiflow design, explained later in this section.

The second type is the **serpentine** type, or *single-pass condenser.* Serpentine condensers use a single length of tubing throughout its length, **Figure 9-8C.** The tube has multiple internal passages. These are slightly more efficient than tube and fin designs.

Many R-134a systems use large **multiflow condensers.** The multiflow condenser that directs refrigerant through a small number of tubes laid in rows across the condenser, **Figure 9-8D.** The refrigerant usually passes through the condenser assembly three or four times before exiting.

Both serpentine and multiflow condensers use headers. A **header** is simply a pipe which, in the case of a condenser, distributes and collects refrigerant. The inlet header distributes refrigerant to each tube while the outlet header collects refrigerant. As with the evaporator, metal fins are pressed over the tubing. The fins on a condenser are

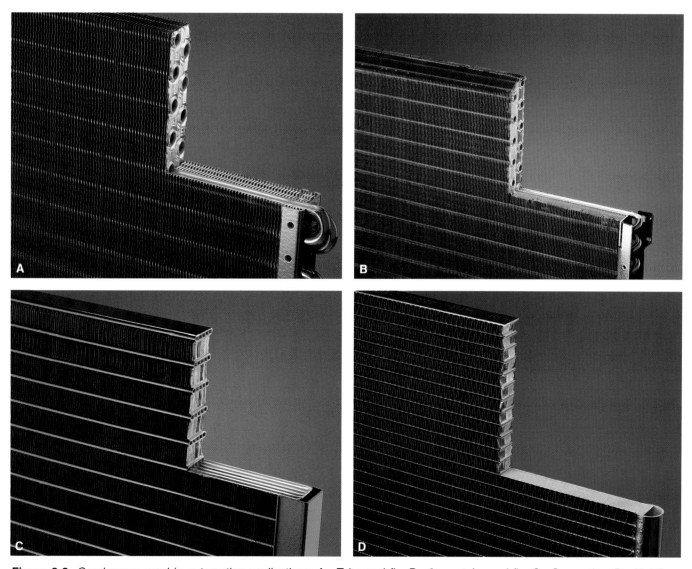

Figure 9-8. *Condensers used in automotive applications. A—Tube and fin. B—6 mm tube and fin. C—Serpentine. D—Multiflow. (Modine)*

similar to those on the engine radiator. The majority of condensers have high pressure compression fittings using an O-ring. A few condensers use spring lock fittings. The sides of the condenser are usually equipped with seals to keep air from bypassing the condenser. One or more electric fans or an engine-driven fan ensures airflow at low speeds.

Accumulators

Accumulators are always located between the evaporator and the compressor inlet, **Figure 9-9.** Accumulators store vaporized refrigerant from the evaporator. The accumulator absorbs some of the pressure and flow fluctuations that occur as the compressor cycles on and off. It also holds enough extra refrigerant to allow for slight variations in the system charge. The accumulator is a large container, usually made of aluminum. If underhood space permits, the accumulator is located close to the evaporator outlet.

 Note: Some vehicles with two evaporators will have more than one inlet to the accumulator. Operation is the same as with an accumulator with one inlet.

Accumulator Operation

During cool weather operation, or on startup, some liquid refrigerant may exit the evaporator. To keep liquid refrigerant from getting to the compressor where it could cause damage, the accumulator separates liquid and vaporized refrigerant. The inlet pipe opens directly into the large chamber of the accumulator. The refrigerant slows down as it enters the chamber. When the refrigerant slows down, any liquid refrigerant or oil will go to the bottom of

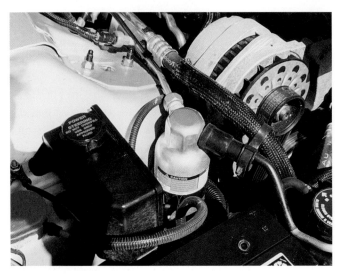

Figure 9-9. *The accumulator is always placed between the evaporator and the condenser.*

the accumulator. Dirt or debris will also go to the bottom of the accumulator. An internal pipe is connected to the accumulator outlet. This internal pipe runs to the top of the accumulator. Since the liquid is at the bottom, only vapor can enter the pipe. The pipe is made in a loop that takes it almost to the bottom of the accumulator.

At the lowest point of the pipe, a small *oil return hole* is open to the bottom of the accumulator. This hole allows any refrigeration oil at the bottom of the accumulator to be returned to the compressor. The hole is small enough so the compressor does not receive too much liquid oil or refrigerant. A screen surrounds the oil return hole. The screen keeps any system debris from clogging the hole or entering the compressor. See **Figure 9-10.**

Figure 9-10. *The accumulator screen catches debris that would clog the oil return hole. The screen and oil return hole are always placed at the bottom of the internal separator tube.*

Receiver-Driers

The *receiver-drier* is a tank that stores condensed refrigerant and a small amount of vaporized refrigerant. The receiver-drier is always located at the condenser outlet. Connections are made using compression fittings and O-rings. Receiver-driers are manufactured in many sizes and shapes. However, they all perform the same function.

The main purposes of the receiver-drier are to store refrigerant to make up for slight losses and to remove dirt and moisture. The pickup tube leading to the expansion valve is at the bottom of the receiver-drier. Since any vaporized refrigerant rises to the top of the receiver-drier, only liquid refrigerant goes to the expansion valve. This allows the expansion valve and evaporator to operate with peak efficiency.

A small screen filter at the pickup tube catches any particles of dirt or metal that do not settle to the bottom of the tank. A receiver-drier in an older system usually had a sight glass located at the top of the tank, **Figure 9-11.** The sight glass gave the technician a way to make a rough determination of the amount of refrigerant in the system. The technician could check for bubbles in the sight glass after the system had been operating for a few minutes. A slightly low system would cause a few bubbles in the sight glass. A system that was very low would have many bubbles or foam in the glass. No bubbles meant the system was either full or completely empty. Sometimes the sight glass is located at another place in the liquid line between the receiver-drier and the expansion valve.

The receiver-drier may have a pressure switch. The function of this switch is to cut off the compressor when the high side pressure becomes excessive. Once the pressure returns to normal, the switch resets and the compressor begins operating again. A dual function switch is used on some receiver-driers. The dual function switch will cut off the compressor if the pressure is too high or too low.

Desiccants

Receiver-driers and accumulators use a **desiccant** to remove moisture and a filter to catch debris. The desiccant bag is located at the bottom of the tank. In this position, almost all of the liquid refrigerant will contact it before entering the pickup tube.

Desiccants hold water by one of two methods, absorption and adsorption. **Absorption** means moisture enters the desiccant and is held there. **Adsorption** means moisture sticks to the surface of the desiccant. Most modern desiccant used with R-134a refrigerant works by absorption and adsorption.

Desiccant bags are made of nylon mesh or some other fabric that will not break down under normal operating conditions. The desiccant bag is usually placed in the tube loop

Figure 9-11. A sight glass is sometimes installed at the top the receiver drier. Other refrigeration systems have the sight glass installed elsewhere, or do not use a sight glass.

directly above the oil return hole. **Figure 9-12** shows the relationship of the desiccant inside an accumulator.

In most systems, only part of the refrigerant passes through the desiccant as it circulates through the system. Allowing the refrigerant to bypass the desiccant bag keeps moisture from being pulled out of the desiccant if it is completely saturated (filled with water). Bypassing the desiccant also allows the refrigerant to continue flowing if the desiccant bag becomes restricted.

Figure 9-12. This illustration shows the internal parts of a typical accumulator. The desiccant removed any moisture from the refrigerant. (Jack Klasey)

On many vehicles, a pressure switch is installed on the accumulator. See **Figure 9-13.** This switch may be used to operate the cycling clutch or may send a signal to an onboard computer.

Desiccant Types

In the past, all desiccants were made of **silica gel.** Silica gel is a combination of silicon, carbon, and sulfur that absorbs moisture. Silica gel granules look like large sand particles. In addition to silica gel, a modern desiccant may contain calcium compounds and activated carbon to absorb moisture. It also contains various polymers (plastic compounds) that hold moisture by adsorption. Some polymers

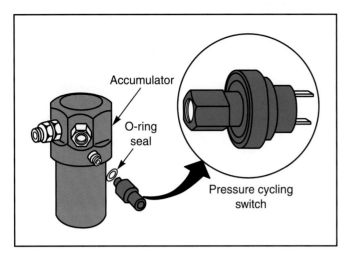

Figure 9-13. *The pressure cycling switch is often installed on the accumulator. When low side pressure becomes lower than the set point, the switch contacts open, de-energizing the compressor clutch. (General Motors)*

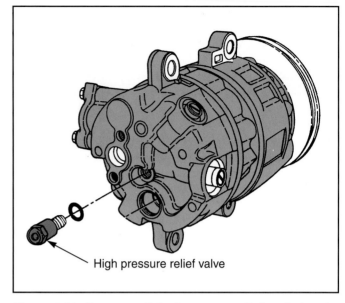

Figure 9-14. *Pressure relief valves are usually installed on the rear of the compressor body. Relief valves are not used on all systems. (DaimlerChrysler)*

have microscopic openings that allow R-134a to pass but trap the larger water molecules. This polymer design is called a *molecular sieve.* The desiccant mixture is pressed into small pellets. The pellets resist disintegration and allow maximum refrigerant flow through the desiccant bag.

The desiccant used in R-134a systems is called *XH-7* desiccant. Desiccants used with R-12 are sometimes called *XH-5* desiccants. The technician must use the right type of desiccant. R-134a and R-12 desiccants are not interchangeable.

XH-9 desiccant can be used with R-134a or R-12 refrigerants. This desiccant is usually used in aftermarket receiver-driers and accumulators; especially those designed as universal replacements for original equipment units. One difference between XH-9 and XH-7 desiccant is XH-9 contains additional silica gel. Some original equipment manufacturers do not recommend the use of an XH-9 desiccant in R-134a systems.

Pressure Relief Devices

Some R-12 systems contained one of two devices to prevent excessive pressure on the high side of the system. The *pressure relief valve* is a spring loaded valve that remains closed during normal operation. If the system pressure becomes too high, the spring pressure will be overcome and the valve will open to relieve pressure. After enough refrigerant has been released, the lowered pressure will not be enough to overcome spring pressure and the valve will close. **Figure 9-14** shows the location of a pressure relief valve on a late-model compressor.

The *fusible plug* resembles a pipe plug with a center section made of a material softer and more heat sensitive than any other part of the refrigeration system. This plug may also be referred to as a *melting bolt.* During normal system operation, the fusible plug has no effect. If the temperature and pressure become too high, the fusible

plug will soften and blow out. This removes system pressure and prevents damage to the compressor and other system components. Once the fusible plug blows, it must be replaced and the system evacuated and recharged. Both pressure relief valves and fusible plugs are no longer used because of the potential to discharge refrigerant to the atmosphere.

Some R-12 systems with pressure relief valves may also contain a *high pressure switch.* This switch is also referred to as a *refrigerant containment switch,* and cuts off the compressor before the pressure reaches the relief valve opening point. If the system is not equipped with one, a high pressure switch must be installed if the system is retrofitted to accept a different refrigerant.

 Note: Early R-12 systems did not use a high pressure switch.

Summary

Every refrigeration system has an evaporator and condenser and every refrigeration system has either an accumulator or a receiver-drier. The evaporator and condenser work with the compressor and flow control device to cause the refrigerant to change state. The accumulator and receiver-drier store refrigerant, separate liquid and vapor refrigerant, and remove moisture.

Refrigerant boils in the evaporator and absorbs heat from the air. There are two refrigerant evaporator designs, the tube and fin and the plate and fin. The tube and fin evaporator is made from a single length of aluminum or copper tubing, folded into a series of serpentine coils. Tube and fin evaporators can be made in any shape and size.

The plate and fin evaporator has less resistance to airflow while allowing more contact between refrigerant and air. Plate and fin evaporators are roughly the same size and shape. The first plate and fin evaporators had the refrigerant passing in equal amounts through each fin. This design caused oil to drop out of the refrigerant and an external oil return line was needed. The multiple-pass evaporator eliminated this problem and does not have an external oil line. Parts that may be attached to the evaporator include the expansion valve and sensing bulb, orifice tube, and temperature probe.

All condensers are tube and fin types, and most are cross-flow types with the refrigerant flowing across the condenser. Metal fins are pressed over the tubing. Other condensers are header types with an inlet header that distributes refrigerant to each tube. R-134a systems use multiflow condensers.

Accumulators are located between the evaporator and the compressor inlet and store vaporized refrigerant from the evaporator. To keep liquid refrigerant from getting to the compressor, the accumulator separates liquid and vaporized refrigerant. A small hole at the bottom of the refrigerant pickup tube allows oil to return to the compressor. The accumulator also contains the desiccant bag.

The receiver-drier is a tank that stores condensed refrigerant and removes moisture and debris. Vaporized refrigerant rises to the top of the receiver-drier, and only liquid refrigerant exits through the bottom mounted pickup tube. Some receiver-driers have pressure switches. The switch turns off the compressor if the pressure goes too high. Older receiver-driers contained a pressure relief valve or a fusible plug. These devices remove system pressure and prevent compressor and system damage.

Pressure relief valves were installed on the compressor body of older vehicles. Pressure relief valves are no longer used. They consisted of a spring-loaded valve that opened when system pressures became too high. When pressure was reduced, the valve closed. The fusible plug was designed to melt and blow out when system pressure and temperature became too high. Fusible plugs are no longer used since they cause the entire refrigerant charge to be lost.

Review Questions—Chapter 9

Please do not write in this text. Write your answers on a separate sheet of paper.

1. Why is an evaporator often called a heat exchanger?

2. What is the purpose of the rubber flap at the evaporator case drain?

3. Describe a multiple-pass evaporator.

4. The expansion valve sensing bulb is located on the _____ side of the evaporator.

5. Which of the following desiccants can be used with R-134a only?
   (A) XH-5.
   (B) XH-7.
   (C) XH-9.

6. _____ keep the air from bypassing the condenser.

7. Accumulators are located at what point in the refrigeration system?

8. No bubbles in a receiver-drier sight glass meant the system was _____.

9. Liquid refrigerant will fall to the _____ of the receiver-drier.

10. A high pressure switch cuts off the _____ if the pressure becomes too high.

ASE Certification-Type Questions

1. Technician A says some vans and trucks have two evaporators. Technician B says some evaporators have more than one row of tubes. Who is right?
   (A) A only.
   (B) B only.
   (C) Both A and B.
   (D) Neither A nor B.

2. All of the following statements about tube and fin coils are true, *except:*
   (A) fins are usually pressed tightly over the tubing.
   (B) some tube and fin evaporators are double or triple row designs.
   (C) bending the fins makes them transfer more heat
   (D) bending the fins makes them weaker.

3. Technician A says a multiple-pass plate and fin evaporator can be spotted by the use of an oil return line. Technician B says the multiple-pass plate and fin evaporator has been replaced by a later design. Who is right?
   (A) A only.
   (B) B only.
   (C) Both A and B.
   (D) Neither A nor B.

4. Evaporators with expansion valves are being discussed. Technician A says the sensing bulb is located at the outlet tube of the evaporator. Technician B says the bulb can be clamped to the tube exterior and covered with insulation. Who is right?
   (A) A only.
   (B) B only.
   (C) Both A and B.
   (D) Neither A nor B.

5. All of the following statements about condensers are true, *except:*

   (A) the condenser tubing is a heavier gauge than the evaporator tubing.

   (B) modern condensers are cross-flow types.

   (C) all condensers have a single serpentine tube.

   (D) R-134a systems are likely to use multiflow condensers.

6. The accumulator is located close to the:

   (A) evaporator inlet.

   (B) evaporator outlet.

   (C) condenser inlet.

   (D) condenser outlet.

7. Which of the following is the *best* description of the purpose of the accumulator oil return hole?

   (A) Allows refrigerant to return to the compressor.

   (B) Allows oil to return to the compressor.

   (C) Traps debris in the accumulator.

   (D) Separates liquid and vaporized refrigerant.

8. All of the following are jobs of the receiver-drier, *except:*

   (A) hold condensed oil.

   (B) hold condensed refrigerant.

   (C) remove moisture from the refrigerant.

   (D) trap loose metal or dirt particles.

9. The sight glass is used to determine the refrigerant:

   (A) cleanliness.

   (B) charge level.

   (C) temperature.

   (D) flow direction.

10. A fusible plug on the receiver-drier prevents all of the following, *except:*

   (A) compressor damage.

   (B) loss of refrigerant.

   (C) hose rupture.

   (D) system overheating.

Chapter 10

Control Valves and Switches

After studying this chapter, you will be able to:
- ❏ Explain the importance of refrigerant flow and pressure control in the evaporator.
- ❏ Explain the construction and purpose of thermostatic expansion valves.
- ❏ Explain the construction and purpose of evaporator pressure control devices.
- ❏ Identify the major types of evaporator pressure control devices.
- ❏ Explain the construction and purpose of compressor clutch control devices.
- ❏ Identify the major types of compressor clutch control devices.
- ❏ Explain the construction and purpose of high and low pressure cutoff switches.
- ❏ Explain the purpose of pressure relief valves.

Technical Terms

Thermostatic expansion valve (TXV)
Flexible diaphragm
Sensing bulb
Capillary tube
Equalizer
Internal equalizer
External equalizer
Superheat spring
H-block
Fixed orifice tube

Evaporator pressure control
Pilot operated absolute (POA) valve
Control piston
Evacuated bellows
Evaporator pressure regulator (EPR) valve
Valves-in-receiver (VIR)
Suction throttling valve (STV)
Hot gas valve

Clutch cycling
Pressure cycling switches
Thermostatic temperature cycling switch
Thermistors
Pressure transducers
Pressure cutoff switches
Cutout switches
High pressure cutoff switches
Low pressure cutoff switches

Trinary cutoff switch
Temperature cutoff switches
Temperature cutout
High temperature cutout switch
Ambient air switch
Ambient temperature switch
Thermal limiter
Superheat switch

This chapter addresses the ways evaporator pressure and refrigerant flow are monitored and controlled. Included in this chapter are the switches and valves that safeguard the refrigeration system against excessively high or low pressures. After studying this chapter, you will understand how pressure controls operate. This will be the foundation for diagnosing and servicing these controls in later chapters.

Two Ways to Control Evaporator Temperature

For maximum cooling, the amount of refrigerant flowing through the evaporator must be carefully regulated. Pressure in the evaporator also must be carefully regulated. If not enough refrigerant flows through the evaporator, it will cool poorly. If too much refrigerant enters the evaporator, it will remain liquid and again, poor cooling will result. Excess liquid refrigerant entering the compressor can damage the reed valves. If pressure is too high, the refrigerant will not boil in the evaporator and there will be no cooling. If pressure is too low, ice will form on the evaporator surface, restricting airflow and cooling.

Many methods have been used to control evaporator flow and pressure. The evaporator temperature must be kept within a narrow range for maximum cooling without freezing the moisture on the evaporator. The lowest part of this range is usually 32°F (0°C), or the freezing point of water. The upper part of this range is about 47°F (8°C). Therefore, the range is approximately 15°F (8°C). There are two ways to keep the evaporator in this range:

❑ Control the flow of refrigerant through the evaporator.
❑ Control refrigerant pressure in the evaporator.

Some valves and switches control refrigerant flow, while others control pressure. Older systems had a combination system with a valve to control flow and a valve to control pressure. Newer systems use a cycling clutch or capacity control valve to control both flow and pressure. Some older systems controlled flow only.

The following sections identify and describe various evaporator temperature control devices. As you read, notice which devices are flow controls and which are pressure controls.

Thermostatic Expansion Valves

The *thermostatic expansion valve (TXV)* is a refrigerant flow control device. It is sometimes called an *expansion valve* or an *X valve* for short. While fixed orifice tubes are used on many vehicles, expansion valves are also installed on some new cars and trucks from the factory or as part of an aftermarket air conditioning system.

All expansion valves have a passage or opening between the high side line and the evaporator. The passage is partially blocked by a small valve. The valve is a pin

with an enlarged end, **Figure 10-1.** The end can seal the opening when it is properly positioned. The pin is attached to a *flexible diaphragm.* On the opposite side of the diaphragm is a chamber connected to a *sensing bulb* by a tube. The tube is often called a *capillary tube.* A refrigerant gas is sealed in the sensing bulb, tube, and chamber. The gas used is usually R-12 or R-134a to match the refrigerant in the system. Some valves use a different refrigerant. The gas used will react more readily to temperature

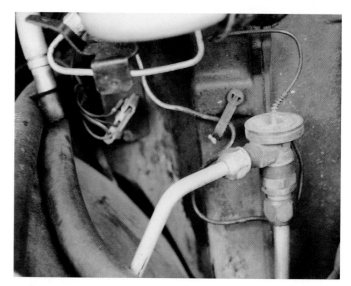

Figure 10-1. *A thermostatic expansion valve installed at the front of the evaporator case. The high pressure line is the horizontal line at the left; the inlet to the evaporator is the vertical line leaving the bottom of the valve. The capillary tube and equalizer line can also be seen.*

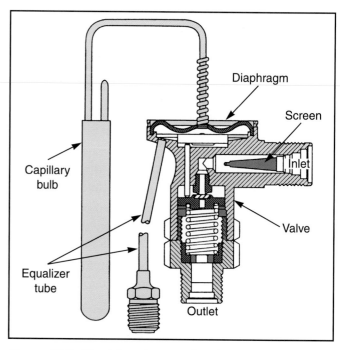

Figure 10-2. *The internal parts of a thermostatic expansion valve. Note the connection between the sensing bulb and the valve diaphragm. The screen keeps debris out of the valve. (General Motors)*

changes than the system refrigerant. Use of this gas allows the expansion valve to react quickly to system temperature changes. The sensing bulb is placed in a cavity at the outlet end of the evaporator, or attached to the outlet tube and covered with insulating tape. **Figure 10-2** is a cutaway view of an expansion valve.

When the air conditioner is first started, the evaporator outlet is warm. This warmth causes the gas in the sensing bulb to expand. The expanded gas produces pressure that travels through the tube and presses on the diaphragm. The diaphragm expands and opens the expansion valve as

fully as possible, **Figure 10-3.** The maximum amount of refrigerant enters the evaporator.

After the system has been running for a few minutes, the entire evaporator becomes cold, including the outlet tube. The cool outlet tube causes the gas in the sensing bulb to cool and contract. The contracting gas in the bulb pulls gas through the tube, reducing the pressure on the diaphragm. The diaphragm moves in the opposite direction and begins to close the expansion valve. See **Figure 10-4.** The amount of refrigerant entering the evaporator is reduced.

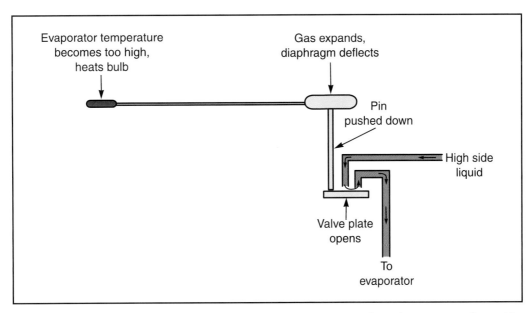

Figure 10-3. *This drawing is a simplified illustration of how an expansion valve opens to allow refrigerant to flow into the evaporator. Higher temperatures at the sensing bulb always cause the expansion valve to open. Increased heat at the sensing bulb causes the expansion valve diaphragm to pull the valve open.*

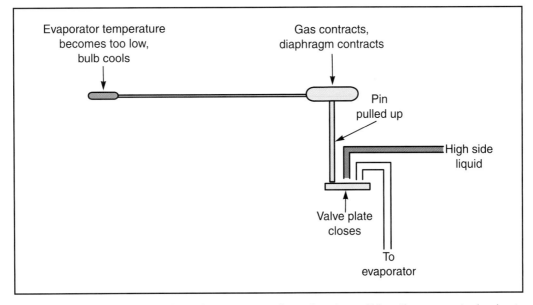

Figure 10-4. *This illustration shows how an expansion valve closes. When the evaporator begins to cool too much, the temperature at the sensing bulb becomes low and the gas inside contracts. Contraction of the gas in the sensing bulb causes the diaphragm in the expansion valve to pull the valve closed.*

When less refrigerant enters the evaporator, it begins to warm up. This causes the gas in the sensing bulb to expand and slightly open the expansion valve. The process repeats over and over as long as the refrigeration system is operating.

Equalizers and Springs

To further control the operation of the expansion valves, two devices are installed in the expansion valve. Both are used to keep the valve from opening too much when extra flow is not necessary.

An *equalizer* is used on all but very old systems. The equalizer is a passage that connects evaporator pressure to the underside of the expansion valve diaphragm. The *internal equalizer* is a passage through the expansion valve body. The *external equalizer* is a separate tube from the evaporator outlet. Both types of equalizers operate by pressurizing the underside of the diaphragm with the exact amount of pressure that is present at the evaporator outlet. This pressure opposes the pressure created by the sensing bulb. The equalizer causes less refrigerant flow when pressure rises in hot weather and compressor speed is low. When the compressor speeds up and lowers the evaporator pressure, the expansion valve can open to allow more refrigerant to enter the evaporator. The equalizer keeps the evaporator from being flooded with liquid refrigerant when the vehicle is operating at low speeds on a hot day, and when the system is first started.

A *superheat spring* is used to oppose the opening of the valve. The spring is placed between the diaphragm and the valve. Spring tension is calibrated into the opening pressure of the expansion valve diaphragm. The spring ensures the valve returns to the closed position when the diaphragm contracts. The spring also helps to keep the valve from opening too much when the refrigerant contains superheat. If the refrigerant is superheated, the sensing bulb will receive too much heat from the evaporator outlet. Pressure from the sensing bulb will open the valve too much if the spring does not place opposing tension on the valve.

Expansion valve operation is similar for all refrigeration systems. However, the shape and location of the expansion valve does vary. The valve and diaphragm are installed on the evaporator inlet, and the sensing bulb is installed on the evaporator outlet. A tube connects the diaphragm and sensing bulb. The valve may be internally or externally equalized.

H-blocks

The *H-block*, or *block expansion valve* combines the valve and diaphragm into one assembly, **Figure 10-5.** There is no separate sensing bulb and capillary tube. The evaporator inlet and outlet tubing are shaped so all refrigerant enters and exits through the H-block. Internal passages connect all parts. The evaporator outlet passes through the part of the H-block containing the diaphragm.

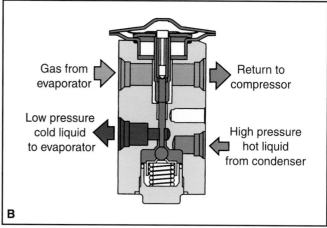

Figure 10-5. *A—Block type expansion valve. It operates in the same manner as other expansion valves, but the sensing bulb, diaphragm, valve, and equalizer are all located in a single housing. B—A cutaway view showing the flow of refrigerant through the block expansion valve. (DaimlerChrysler)*

Temperature changes in the outlet gases work directly on the diaphragm. The diaphragm is connected to the valve by a pin that passes through the evaporator outlet. When increasing evaporator outlet temperature expands the gas in the diaphragm, it opens the valve. When evaporator outlet temperature decreases, the gas contracts and the diaphragm starts to close the valve. H-block expansion valves were commonly used on factory air conditioners well into the 1980s. A few refrigeration systems still use H-block valves.

Fixed Orifice Tubes

Some vehicles use a fixed opening to do the job of the expansion valve. This opening is a *fixed orifice tube.* The plastic or metal tube houses a fine screen and a small fixed opening through which a metered amount of refrigerant can pass, **Figure 10-6.** The orifice tube is usually installed at the evaporator inlet in the same place as the expansion valve. A few orifice tubes are installed at the outlet of the condenser.

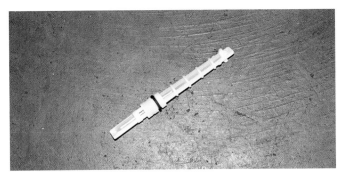

Figure 10-6. *Fixed orifice tube. Most orifice tubes are made of plastic, some are made of metal. Both have screens to filter debris.*

To control the flow of refrigerant through an orifice tube system, a cycling clutch compressor is used. Compressor pressure controls are discussed later in this chapter.

Evaporator Pressure Controls

The expansion valve does a good job of controlling the flow of refrigerant through the evaporator. However, pressure must also be controlled to keep the evaporator temperature from becoming too low. Many older vehicles with an expansion valve also had an ***evaporator pressure control.*** The evaporator pressure control was always installed in the suction line between the evaporator and the compressor. The primary purpose of evaporator pressure control was to keep pressure above the point that would allow the temperature to drop below 32°F (0°C). Pressure controls can be thought of as devices that keep condensed moisture from freezing on the evaporator.

> **Note: Some parts catalogs and service manuals refer to all evaporator pressure controls as suction throttling valves. The** term *suction throttling valve* applies to only one kind of pressure control device, as explained later in this chapter.

Evaporator pressure controls are pressure regulating valves. They operate when the evaporator pressure drops below a set point. Since most evaporator pressure controls were designed to work with R-12, this point is set at about 28-30 psi (193-206 kPa). The few systems that use R-134a with an evaporator pressure control will have a set point of about 21 psi (145 kPa). With any type of refrigerant, the evaporator pressure control valve is open when evaporator pressure is above the set point. When the valve is open, the compressor can draw refrigerant from the evaporator.

On cool days, or during sustained highway driving, the evaporator pressure will drop below the set point. The valve then closes, shutting off the connection between the compressor and the evaporator. The expansion valve is not directly affected by the operation of the pressure control

valve. It continues to allow refrigerant to enter the evaporator, based on outlet temperature. More refrigerant entering the evaporator causes its pressure to rise. When the evaporator pressure goes above the set point, the evaporator pressure control valve reopens. This cycle continues as long as evaporator pressure is at or near the set point.

If the temperature and humidity are very high, or the vehicle is idling, the pressure remains above the set point. The evaporator pressure control valve has no effect on system operation above the set point. With the evaporator pressure above the set point, the expansion valve alone controls evaporator temperature.

You can see when evaporator pressure is at the set point, both the pressure control valve and expansion valve work together to control evaporator temperature. When evaporator pressure is above the set point, the expansion valve alone controls temperature.

An advantage of pressure controls was they allowed the compressor to run at all times. This resulted in smoother engine operation. However, when fuel economy became a prime concern of vehicle manufacturers, running the compressor at all times became a disadvantage.

Types of Evaporator Pressure Controls

Many types of evaporator pressure controls have been used in the past. Most of these are no longer used. However, some evaporator pressure controls were installed on new vehicles as recently as the late 1980s. You may encounter one of these devices when working on an older car. The various types of pressure controls are discussed in the following paragraphs, starting with the most common.

Pilot Operated Absolute (POA) Valve

The ***pilot operated absolute (POA) valve*** is the most common type of pressure control device, **Figure 10-7**. It has no external controls. The main POA valve is a ***control piston*** that seals the passage between the evaporator and compressor. The control piston is operated by two springs, and a chamber called an ***evacuated bellows.*** The evacuated bellows is a sealed chamber made of flexible metal. The body of the chamber is pleated to be able to expand and contract with changes in pressure. The bellows is evacuated (pumped down) so its interior is close to a perfect vacuum. The springs are placed on either side of the bellows, and perform different jobs. The spring on the evaporator side of the bellows is called spring A. The spring on the compressor side is called spring B.

The job of the evacuated bellows is to place the POA valve in operation. When the air conditioner is first started, evaporator refrigerant pressures are relatively high. This pressure compresses the bellows, and the needle valve is held open by spring B. See **Figure 10-8A.** With the needle valve open, the compressor can draw refrigerant from the underside of the control piston. Since the compressor is trying to draw refrigerant from the evaporator, pressure on

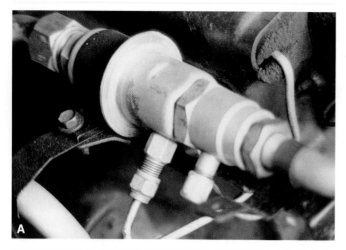

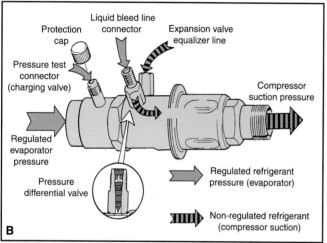

Figure 10-7. *The pilot operated absolute (POA) valve is the most common type of evaporator pressure control valve. Versions of this valve were used until the early 1990s. Studying the connections to this POA valve will make it easier to understand its operation. (Ford)*

Spring A functions to control most of the evaporator pressure. The evacuated bellows and the needle valve may move to open the control piston when the compressor suction pressure is high, such as when the vehicle is idling on a very hot day. However, during almost all other operating conditions, the needle valve will remain closed.

Other Types of POA Valves

While all POA valves operate in roughly the same manner, there are some variations in the placement and shape of the valve. These variations are discussed in the following paragraphs.

Evaporator Pressure Regulator (EPR) Valve

The *evaporator pressure regulator (EPR) valve* is a version of the POA valve. It operates in the same way as other POA valves, but is smaller and is located in the inlet to the compressor. **Figure 10-9** shows an EPR valve.

Valves-in-receiver (VIR)

A variation of the POA valve is the *valves-in-receiver (VIR)*. The VIR is a single housing containing the expansion valve, POA valve, and receiver-drier. See **Figure 10-10.** The expansion valves and POA operate in the same manner as conventional units. One difference between the VIR system and a system where the parts are separated is the location of the receiver-drier. The receiver-drier is on the low side of the system instead of the high side.

Older Pressure Controls

Two pressure controls used on older air conditioners are discussed in the following paragraphs. You can find these pressure control devices on domestic vehicles made in the 1960s or earlier.

Suction Throttling Valve (STV)

The *suction throttling valve (STV)* controls evaporator pressure with a spring-loaded shutoff valve. **Figure 10-11** shows an STV. The spring tension is set so the valve will not allow evaporator pressure to go below about 28 psi (193 kPa).

A vacuum diaphragm modifies spring tension. When vacuum is present, the diaphragm pulls against spring tension, and allows slightly lower evaporator temperatures. A vacuum control valve in the HVAC control panel modifies engine manifold vacuum before it goes to the diaphragm. The temperature control lever is connected to the vacuum control valve. When the driver moves the temperature control lever, the vacuum control valve moves. This changes the amount of vacuum reaching the diaphragm, varying the effect it has on the spring. An advantage of using engine manifold vacuum is it compensates for high altitudes. When the vehicle is operated at higher altitudes, manifold vacuum is less than it would be at sea level. This causes the diaphragm to have less effect on the spring, resulting in higher evaporator pressure at higher altitudes. This reduces the chance of evaporator icing at higher altitudes.

the underside of the control piston is less than the pressure in the evaporator. The difference in pressure causes the control piston to be pushed away from its seat and to compress spring A. To ensure the control piston can push against the spring, the bellows side of the piston has a greater area than the evaporator side. With the control valve off its seat, refrigerant can flow freely through the POA valve to the compressor. A bleed hole allows refrigerant to be drawn into the bellows area from the evaporator inlet.

As the compressor continues to remove refrigerant from the evaporator, pressure begins to drop. When pressure gets to roughly 28.5 psi (196 kPa), the evacuated bellows expands and pushes against spring B to close the needle valve. This is shown in **Figure 10-8B.** Once the needle valve closes, evaporator pressure enters through the bleed hole. Pressure on both sides of the control piston is now equal, and spring A pushes the control piston closed. If pressure in the evaporator increases, the control piston will be pushed open against spring pressure, **Figure 10-8C.** Refrigerant will be drawn into the compressor to lower evaporator pressure.

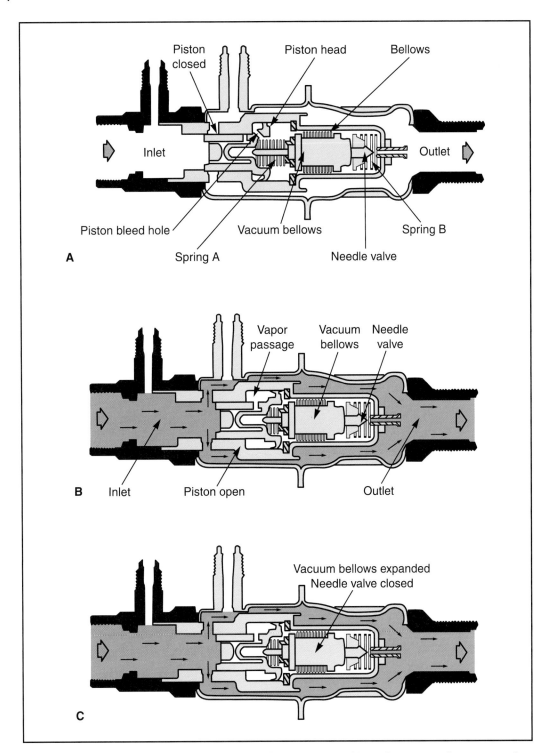

Figure 10-8. *A—In this cutaway view of a POA valve the refrigeration system is not operating. Spring A keeps the piston closed. B—The refrigeration system begins operating and compressor suction compresses spring A and opens the piston. C—When evaporator pressure becomes too low, the bellows will expand and cause the needle valve to close. This allows spring A to close the piston, sealing the passages between the evaporator and compressor. (General Motors)*

Hot Gas Valve

The **hot gas valve** controlled evaporator pressure by bleeding compressed gas into the evaporator from the compressor discharge. This valve was sometimes called a *bypass valve* or *bypass solenoid.* The valve contained a spring-loaded valve that opened a passage between the suction and discharge sides of the compressor when evaporator pressure went below about 30 psi (207 kPa). A cable attached to the HVAC control panel varied the spring tension of the hot gas valve to modify the evaporator pressure. However, the pressure could not go below 30 psi. **Figure 10-12** shows a hot gas valve circuit. Most hot gas valve designs will be similar to the one shown in **Figure 10-12.**

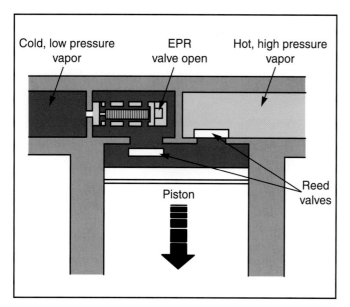

Figure 10-9. *The evaporator pressure regulator (EPR) valve is a POA valve that is installed in the compressor inlet port. An EPR valve operates in the same manner as a standard POA valve.*

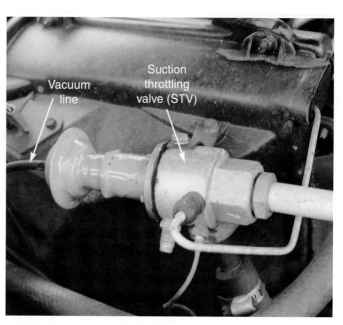

Figure 10-11. *Suction throttling valve (STV) installed on an older car. The vacuum line at the left is attached to the intake manifold through the HVAC control panel.*

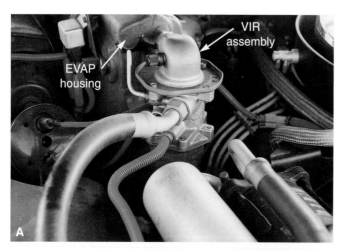

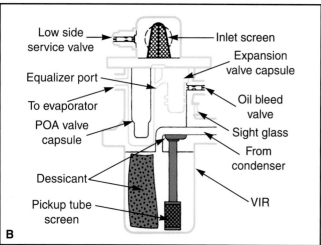

Figure 10-10. *A—A typical valves-in-receiver (VIR) assembly shown installed on a vehicle. B—This cutaway view of a VIR shows the relative positions of the expansion valve, POA valve, and desiccant bag. (Four Seasons)*

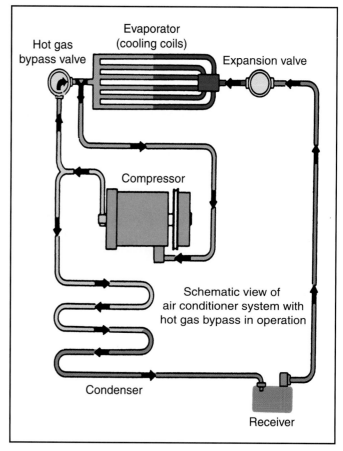

Figure 10-12. *A system equipped with a hot gas bypass valve. The hot gas bypass valve operates by forming a connection between the high and low sides of the refrigeration system. High pressure refrigerant from the high side is bled into the low side raising low side pressures. A lever operated from the dashboard was used to change the set point of the hot gas valve. (General Motors)*

Compressor Clutch Controls

The most common way to control evaporator pressure and flow on modern vehicles is to control the operation of the compressor. Turning off the compressor reduces both pressure and refrigerant flow.

Some vehicles have variable capacity compressors. Other HVAC systems control compressor operation by applying (turning on) or releasing (turning off) the compressor clutch. This is called **clutch cycling.** The compressor clutch contains an electromagnet energized by battery power. Energizing the electromagnet applies the clutch. De-energizing the electromagnet releases the clutch. Clutch cycling causes a slight change in engine speed as the clutch turns on or off. This is considered a minor problem compared with the savings in fuel resulting from eliminating the drag of the compressor when it is not needed.

To control the clutch, switches must be installed in the circuit between the battery and the electromagnet. In the simplest system, battery power passes through the ignition switch and HVAC control panel to reach the electromagnet. In a cycling clutch system, one or more pressure or temperature switches can break the circuit to the compressor clutch electromagnet. These switches are discussed in the next sections.

Pressure Cycling Switches

Pressure cycling switches are the most common clutch control switches. This switch opens and closes the electrical circuit based on low side pressure. Pressure works on a flexible diaphragm to open and close the contacts. Most pressure switches are similar to the one shown

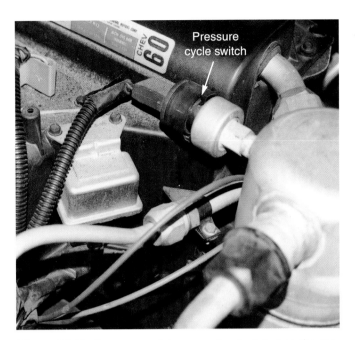

Figure 10-13. *Pressure switches are often installed on the system accumulator. Some are installed on the low pressure line leading to the accumulator or compressor.*

in **Figure 10-13,** although they may have different shapes and connectors. Some pressure switches can be adjusted. Many are sealed units that must be replaced if they do not operate correctly.

The pressure cycling switch is installed on the low pressure side of the system. It is often installed on the accumulator. The compressor clutch electromagnet is wired to the HVAC control panel through this switch, as shown in **Figure 10-14.** When low side pressure is above a certain value, the diaphragm expands to close the switch

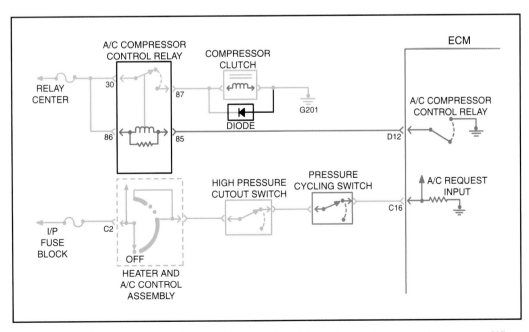

Figure 10-14. *The pressure cycling switch reacts to refrigeration system low side pressure. When pressure drops too low, the switch opens, and the clutch electromagnet disengages. (General Motors)*

contacts and the clutch is engaged. If pressure falls below this value, the diaphragm contracts, the switch opens, and the clutch is disengaged.

On refrigeration systems using R-134a, the pressure switch opens at about 21 psi (145 kPa). On systems using R-12, the pressure switch is set at about 28 psi (193 kPa). When the compressor stops operating, evaporator pressure rises, the switch closes and the clutch is re-engaged. The range between the opening and closing pressures is about 20 psi (138 kPa).

Thermostatic Temperature Cycling Switches

The *thermostatic temperature cycling switch* is not as widely used as the low pressure cycling switch. However, thermostatic temperature switches are used on a few imported vehicles, many aftermarket air conditioners, and some older domestic cars, **Figure 10-15.**

The thermostatic temperature cycling switch uses a gas sealed in a sensing bulb. The bulb is connected to a diaphragm that operates electrical contacts. The bulb and diaphragm are connected through a capillary tube. The thermostatic temperature switch looks like an expansion valve with an electrical connector. Instead of operating a valve, the diaphragm operates a set of contact points.

The sensing bulb is sometimes located on the evaporator outlet, or may be located between the evaporator coils. Some thermostatic temperature cycling switches do not have a bulb on the end of the tubing. They rely on the gas inside the tube to provide a temperature signal.

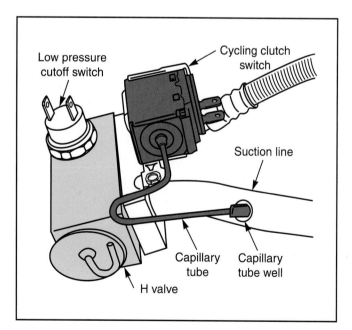

Figure 10-15. *A thermostatic clutch control. When the evaporator becomes too cold, the gas in the capillary tube causes the bellows to contact. When the bellows contracts, it pulls the points open. This breaks the circuit to the clutch electromagnet. Rising evaporator temperature causes the bellows to expand, and the return spring can close the points. (DaimlerChrysler)*

When the evaporator temperature is above 32°F (0°C), the gas in the tube is warm enough to expand and cause the diaphragm to move outward. In the expanded position, the diaphragm pushes on the contact points, keeping them closed. When the temperature approaches 32°F (0°C), the gas in the sensing bulb and capillary tube cools off and contracts. This causes the gas in the diaphragm to contract, and the diaphragm moves enough to open the points. The compressor clutch is de-energized, and the compressor stops. When the temperature rises several degrees above freezing, the gas in the bulb, tube, and diaphragm expands, causing the diaphragm to push the contact points closed. When the points close, the clutch is reapplied. The minimum range between point opening and closing is 7°F (3.8°C). The temperature range is higher on some systems.

Adjustable Thermostatic Temperature Cycling Switch

Some older factory air conditioners used an adjustable thermostatic temperature cycling switch. Some current aftermarket units use these switches. The adjustable thermostatic temperature switch allows the driver to vary the evaporator temperature by varying when the clutch is engaged. The adjustment range is usually about 20°F (11°C).

The simplest adjustable thermostatic switch consisted of a sealed diaphragm and contact point set. In addition to these parts, the adjustable switch had a control knob that turned a threaded shaft. Turning the knob caused the shaft to move the pivot point of the contact points. This changed the position of the points in relation to the diaphragm. Moving the points away from the diaphragm caused them to close at a higher temperature. The gas had to become hotter and expand more to close the points. Moving the points closer to the diaphragm allowed the points to close with less gas pressure, and the clutch would reengage at a lower temperature.

Pressure and Temperature Sensors

Some vehicles have computer-controlled air conditioning systems. Pressure and temperature sensors are used to communicate with the on-board computer. Temperature sensors are sometimes called *thermistors.* A thermistor is a kind of resistor. Thermistor resistance changes when its temperature changes. The computer uses these sensor inputs to control the operation of the compressor clutch, other parts of the refrigeration system, and airflow.

Other computer control systems use pressure sensitive crystals called *pressure transducers* to control system operation. A pressure transducer performs the same function as mechanical thermostatic switches, but have no moving parts. **Figure 10-16** is a circuit for a pressure transducer input to the vehicle computer.

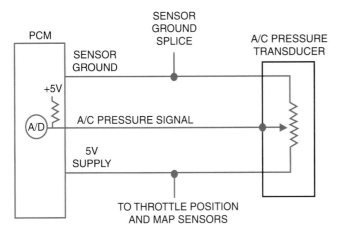

Figure 10-16. *A pressure transducer uses a pressure sensitive crystal that communicates with the computer. Many modern vehicles use the computer to operate the compressor and other HVAC system parts. (DaimlerChrysler)*

System Protection Switches and Valves

To protect the refrigeration system some pressure switches and valves are used. The switches turn off the compressor clutch when operation with very high or low pressures could cause damage. Pressure relief valves are designed to open and release pressure before another refrigeration system part fails. These protection devices are discussed in the following paragraphs.

Pressure Cutoff Switches

Pressure cutoff switches are designed to interrupt, or cut off, power to the compressor clutch electromagnet when certain pressures are present in the refrigeration system. Pressure cutoff switches are sometimes called *cutout switches.*

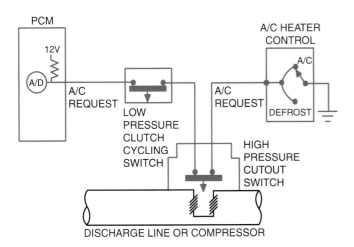

Figure 10-17. *A high pressure cutoff switch disengages the clutch when the high side pressure becomes excessive. Most high pressure cutoff switches are located in the compressor outlet line. The cutoff pressure varies with manufacturers. (General Motors)*

High Pressure Cutoff Switches

High pressure cutoff switches are always installed on the high side of the refrigeration system. See **Figure 10-17.** A high pressure cutoff switch is similar to a cycling clutch pressure switch. The switch is wired in series with the compressor clutch electromagnet. The switch contacts remain closed when system pressures are normal. When pressures become too high, the switch contacts will open, cutting off current to the compressor clutch. When the compressor clutch disengages, the refrigeration system stops operating but airflow continues through the condenser. Airflow through the condenser lowers the high side pressure to a point below the opening point of the pressure cutoff switch. The switch closes and re-energizes the compressor clutch. Once the compressor clutch is re-energized, the refrigeration system begins working normally. Normal opening pressure is about 450 psi (3110 kPa). The cutoff switch will usually reclose at about 250 psi (1723 kPa).

Low Pressure Cutoff Switches

Low pressure cutoff switches can be installed on the low or high side of the refrigeration system, **Figure 10-18.** No matter where they are installed, they perform the same job. If a leak develops and the system is low on refrigerant, system pressures will drop. Oil also leaks out with the refrigerant. If the system runs with a low charge of refrigerant or oil, the compressor will be damaged. To prevent damage, the low pressure cutoff switch will open when refrigerant pressure falls below a certain point. The compressor clutch is disengaged and the system shuts down.

When the system is recharged, the cutoff switch will close and the refrigeration system will begin operating normally. In newer vehicles, a low pressure reading causes the ECM to shut off and prevent compressor operation until the

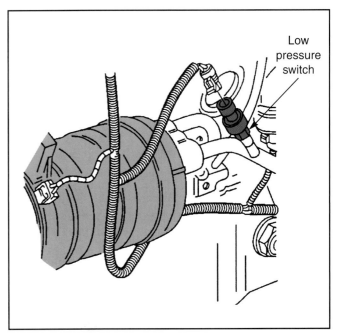

Figure 10-18. *The low pressure switch shuts off compressor operation if the refrigerant level is too low. (General Motors)*

system is serviced. Low pressure cutoff switch opening and closing pressures vary, depending on which side of the system it is installed.

Trinary Cutoff Switch

The **trinary cutoff switch** is a combination of low and high pressure switches. It will cut off current to the compressor if refrigerant pressure becomes too high or too low. Some trinary switches also control the operation of the radiator fan.

Temperature Cutoff Switches

Temperature cutoff switches are used in two situations: when the refrigerant temperature is too hot and when the outside air is too cold. These switches are sometimes called **temperature cutout** switches.

High Temperature Cutout Switch

The **high temperature cutout switch** was used on a few vehicles. It is clamped on the condenser outlet tube. If the condenser outlet temperature rises above a certain value, the switch breaks the circuit to the compressor clutch.

Ambient Air Switch

At temperatures near or below freezing, the refrigeration system will not work and the compressor could be damaged. When the refrigeration system operates at low temperatures, the refrigerant flow through the system will be very sluggish. Oil will drop out of the refrigerant stream and collect in the condenser and evaporator. The compressor will run without enough oil and be damaged. The **ambient air switch,** or **ambient temperature switch,** prevents compressor clutch engagement when the ambient air temperature is too low, **Figure 10-19.** Ambient air is simply the air surrounding the vehicle. The usual ambient air switch setting is at or below freezing (32°F or 0°C). A few ambient air switches are set at slightly above freezing at roughly 35-40°F (1-4°C).

Some ambient air switches use a bimetallic strip to sense temperatures. Bonding together two metals with different rates of expansion creates a bimetallic strip. The different expansion rates cause the strip to bend with changes in temperature. A set of contact points are connected to the strip. When the temperature drops to freezing, the strip bends enough to pull the contacts apart. This breaks the circuit to the compressor clutch. When the temperature rises above freezing, the strip straightens out and allows the contact points to close. The system begins operating normally.

Some later ambient air switches have temperature sensitive materials that can change from conductors to insulators when the temperature changes. Other ambient air switches use thermistors to sense temperature, which then shuts off the compressor through other electronic circuits. They perform the same job as the bimetallic strip but with no moving parts.

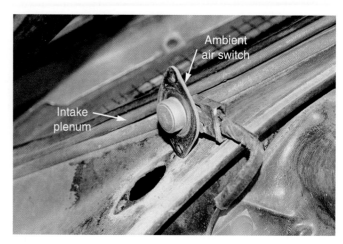

Figure 10-19. *An ambient air temperature switch may be installed in the air intake assembly as shown here. Ambient air temperature switches are also placed in the vehicle grille ahead of the radiator and condenser.*

Thermal Limiters

Some older vehicles had a **thermal limiter.** The thermal limiter was a fuse with three electrical terminals, **Figure 10-20.** One terminal was connected to the HVAC control system. This terminal received battery current whenever the air conditioning was operating. Another terminal was connected to the compressor clutch electromagnet. The third terminal was connected to a pressure switch on the rear of the compressor. This switch was sometimes called a **superheat switch.** The superheat

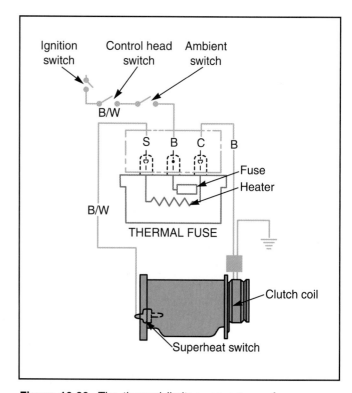

Figure 10-20. *The thermal limiter was a type of compressor fuse used on some older cars. If the superheat (low pressure) switch contacts closed, the limiter burned out and the clutch coil was de-energized.*

switch was designed to close if the low side pressure went below a preset value, or if the compressor temperature became higher than a certain value. Closing the valve completed a circuit between the fuse and ground.

During normal operation, current flowed from the 12-volt terminal, through the fuse, and to the clutch electromagnet. If the refrigeration charge became too low, the switch on the back of the compressor closed. This caused current to flow through the fuse to ground. The amount of current flowing in the fuse created enough heat to melt, or blow, the fuse. When the fuse melted, no current could flow to the clutch electromagnet and the system was shut down. The melted fuse had to be replaced and the cause of the problem located and corrected.

Summary

The modern refrigeration system controls evaporator temperature in two ways. Valves and switches can be used to control the flow of refrigerant through the evaporator, or can be used to control the pressure of the refrigerant in the evaporator. The most common refrigerant flow control device is the thermostatic expansion valve. Thermostatic expansion valves consist of a pressurized diaphragm that operates an inlet valve. The diaphragm is connected to a sensing bulb through a capillary tube. A gas is sealed in the unit. The gas expands and contracts with temperature changes. Cooler evaporator temperatures tend to close the valve. Warmer temperatures tend to open the valve. To more precisely control temperature, an equalizer passage and a superheat spring are installed in the expansion valve. Both of these devices oppose diaphragm pressure under different conditions. Most expansion valves have a separate diaphragm and bulb, connected by the capillary tubing. A separate equalizer line may also be used. H-block expansion valves combine all parts into one unit.

Older vehicles with expansion valves may also use some kind of evaporator pressure control. The evaporator pressure control is installed in the suction line between the evaporator and the compressor. The valves are open when evaporator pressure is above a set point, usually around 28 to 30 psi (193 to 207 kPa). When the valve is open the compressor can draw refrigerant from the evaporator. Above the set point the evaporator pressure control valve has no effect on system operation. Evaporator pressure controls begin to work when evaporator pressure drops below the set point.

The most common later evaporator pressure control is the pilot operated absolute (POA) valve. Versions of the POA valve include the EPR and VIR valve. Older pressure controls include the suction throttling valve (STV) and the hot gas valve.

Cycling the compressor controls evaporator temperature on most modern vehicles. Cycling the compressor means turning it on and off. Turning off the compressor reduces pressure and refrigerant flow. In a cycling clutch system, one or more pressure or temperature switches is used to break the circuit to the compressor clutch electromagnet. Pressure cycling switches are the most common clutch control switches. Some pressure switches can be adjusted. Thermostatic temperature cycling switches are also used. Some thermostatic temperature cycling switches are adjustable. Some vehicles have computer-controlled air conditioning systems. These systems use pressure and temperature sensors.

Pressure switches are often installed to protect the refrigeration system from excessively high or low pressures. Pressure switches respond to pressure problems by breaking the electrical circuit to the compressor clutch electromagnet. High and low pressure switches are used. High pressure switches are always used on the high side of the system, while low pressure switches can be found on either side.

Ambient air switches prevent compressor clutch engagement when the surrounding air temperature is too low. This prevents damage to the compressor.

Review Questions—Chapter 10

Please do not write in this text. Write your answers on a separate sheet of paper.

1. The thermostatic expansion valve is a refrigerant _____ control device.

2. An expansion valve must have an opening between the high side line and the _____.

3. When the outlet of the evaporator is warm, the gas in the sensing bulb causes the flexible diaphragm to _____.

4. What does the expansion valve equalizer passage prevent?

5. The superheat spring tries to _____ the expansion valve.

6. What is the primary job of any pressure control?

7. Define clutch cycling.

8. The cycling clutch pressure switch is located on the _____ pressure side of the system.

9. The thermostatic temperature switch looks like an _____ with an _____.

10. Which of the following system pressure switches and valves will cause loss of refrigerant from the system?
    (A) High pressure cutoff switch.
    (B) Low pressure cutoff switch.
    (C) Ambient air switch.
    (D) Pressure relief valve.

ASE Certification-Type Questions

1. Technician A says evaporator temperature controls control refrigerant flow through the evaporator. Technician B says evaporator temperature controls control refrigerant pressure. Who is right?
   (A) A only.
   (B) B only.
   (C) Both A and B.
   (D) Neither A nor B.

2. All of the following statements about expansion valves are true, *except:*
   (A) the passage between the high and low sides is partially blocked by a small valve.
   (B) the valve is attached to a flexible diaphragm.
   (C) the sensing bulb is installed on the inlet end of the evaporator.
   (D) a capillary tube connects the sensing bulb and diaphragm.

3. Which of the following expansion valve parts is used to prevent evaporator flooding?
   (A) Superheat spring.
   (B) Equalizer passage.
   (C) Sensing bulb.
   (D) Capillary tube.

4. Which of the following parts is *not* used on a POA valve?
   (A) Control piston.
   (B) Evacuated bellows.
   (C) Vacuum diaphragm.
   (D) Springs.

5. Technician A says the VIR valve contains the POA valve. Technician B says the VIR valve contains the expansion valve. Who is right?
   (A) A only.
   (B) B only.
   (C) Both A and B.
   (D) Neither A nor B.

6. All of the following statements about pressure cycling switches are true, *except:*
   (A) pressure cycling switches open and close the electrical circuit to the HVAC control panel.
   (B) pressure cycling switches are the most common clutch control switches.
   (C) pressure cycling switches are installed on the low side of the system.
   (D) some pressure cycling switches can be adjusted.

7. Thermostatic temperature cycling switches are being discussed. Technician A says the switch uses a gas sealed in a sensing bulb and diaphragm. Technician B says the diaphragm operates the expansion valve. Who is right?
   (A) A only.
   (B) B only.
   (C) Both A and B.
   (D) Neither A nor B.

8. The high pressure cutoff switch contacts will open when system pressures are:
   (A) too low.
   (B) too high.
   (C) normal.
   (D) varies with refrigerant temperature.

9. The *most likely* condition under which the low pressure cutoff switch would cut off current to the compressor clutch is:
   (A) a refrigerant leak develops.
   (B) very high pressures develop on the low side of the system.
   (C) very high pressures develop on the high side of the system.
   (D) the compressor clutch begins slipping.

10. The ambient air switch is designed to disengage the compressor clutch when the temperature is at all of the following except:
    (A) 32°F.
    (B) 0°C.
    (C) 47°F.
    (D) 1.6°C.

Chapter 11

Engine Cooling Systems and Vehicle Heaters

After studying this chapter, you will be able to:

- ❏ Explain how an engine liquid cooling system operates.
- ❏ Identify the major parts of liquid cooling systems.
- ❏ Explain how an engine air cooling system operates.
- ❏ Identify the major parts of air cooling systems.
- ❏ Explain how a heating system operates on a vehicle with a liquid cooling system.
- ❏ Identify the major parts of heating systems on vehicles with liquid cooling system.
- ❏ Explain how a heating system operates on a vehicle with an air cooling system.
- ❏ Identify the major parts of heating systems on vehicles with air cooling system.

Technical Terms

Liquid-cooled engine	Downflow	Thermostat monitor	Recovery tank
Coolant	Crossflow	Hoses	Overflow tank
Antifreeze	Transmission oil cooler	Neck	Drain
Ethylene glycol	Ram air	Radiator hoses	Air bleed
Propylene glycol	Radiator fan	Heater hoses	Air-cooled engines
Long-life coolants	Bite	Molded radiator hoses	Cooling fins
Organic acid technology	Pitch	Hose clamps	Cylinder barrels
(OAT)	Belt-driven fan	Quick disconnect hoses	Blower
Coolant passages	Flex fan	Bypass	Shrouds
Water jackets	Fan clutch	Electrochemical	Engine oil cooler
Coolant pump	Centrifugal fan clutch	degradation (ECD)	Heater core
Water pump	Thermostatic fan clutch	Pressure cap	Heater hoses
V-belts	Electric fans	Coolant recovery system	Heater shutoff valve
Serpentine belt	Two-speed fans	Reservoir tank	Heater door
Radiator	Thermostat	Reservoir bottle	Heater ducts

The engine cooling system and the heater play a major part in the operation of the overall HVAC system. The design and operation of the cooling system and heater is much simpler than the air conditioner. However, it must be thoroughly understood before any problems in the system can be diagnosed and corrected. By studying this chapter, you will become familiar with the operation of the engine cooling system and heater parts.

Cooling System

While the cooling system's primary job is to maintain engine temperature, it is considered part of the heating and air conditioning system. The cooling system has an extra heat load placed on it when the air conditioning compressor is operating. During air conditioner operation, the compressor consumes a small portion of the engine's horsepower. As the condenser removes heat from the refrigerant, it delivers this heat to the air entering the radiator. The cooling system must be in good condition to handle these extra loads.

If the heater does not deliver enough heat, driving in cold weather will be uncomfortable. If the heater puts out too much heat or heat at the wrong time, air conditioner efficiency will be reduced. Coolant leaks can cause a haze on the windshield and odors in the passenger compartment.

Engine Cooling System Operation

Because engines produce much more heat than they can use, they are the source of heat for the heater. Burning fuel in the combustion chamber can produce temperatures of over 5000°F (2760°C). Only about one-third of this heat is actually used to push on the pistons and produce power. Another third of the heat leaves the engine with the exhaust gases. The cooling system must remove the remaining heat.

To remove heat from the engine before it causes damage, the cooling system uses a liquid or direct air contact to absorb it. The heat is then transferred to the outside air. Even on very hot days, the outside air is always cooler than an engine at operating temperature. Since heat always moves from a warm object to a cooler one, a properly working cooling system will always transfer heat from the engine to the outside air.

Engine Cooling System Components

The cooling system is usually liquid-cooled, but a few air-cooled vehicles are still being manufactured. As you learned in Chapter 5, heat tends to flow from a warmer to a cooler substance. Both air and liquid cooling system involve circulating a cooler material around a warmer material to cause heat to be transferred from the warmer to the cooler material.

Liquid-cooled Engines

In a *liquid-cooled engine,* the liquid coolant is the medium of heat transfer. Coolant must circulate through the engine, transfer heat easily, and not cause damage to any cooling system components.

Coolants

Coolant is a mixture of water and other chemicals to create a compound called antifreeze, **Figure 11-1.** *Antifreeze* is made primarily from one of two compounds. The first is called *ethylene glycol,* with small amounts of anti-corrosion ingredients and lubricating oils. This antifreeze is sometimes called EG coolant. It has a semi-sweet taste. Newer vehicles use antifreeze containing *propylene glycol* or PG. It has a more bitter taste than EG coolants. Propylene glycol coolants are non-toxic and are considered to be "environmentally safe." However, since all coolants contain some levels of heavy metals from the cooling system, used antifreeze must be properly disposed. EG and PG coolants should not be mixed.

> ⚠ **Warning: While some coolants are considered to be biodegradable in their pure, unused form, ingesting coolant is dangerous and can make you sick or even kill you.**

Ethylene glycol will freeze at roughly 9°F (-13°C). When ethylene glycol is mixed with water, which has a freezing point of 32°F (0°C), the resulting mixture has a

Figure 11-1. *There are two types of coolants available for vehicles, standard and long-life. Keep in mind the color of the antifreeze should not be used as an indicator of whether it is standard or long-life coolant. (Jack Klasey)*

freezing point lower than either liquid by itself. A half-and-half (50-50) mixture of antifreeze and water has a freezing point of about –35°F (-37°C). A mixture of 70% antifreeze and 30% water will not freeze until the temperature reaches –67°F (-55°C).

A 50-50 mixture of propylene glycol antifreeze and water will freeze at -26°F (-32°C). A 60-40 mixture will freeze at -54°F (-48°C). Increasing the antifreeze concentration above 70% will cause the mixture's freezing point to rise. Most manufacturers recommend a 50-50 mixture of water and antifreeze.

Another reason to use antifreeze is it raises the boiling point of the coolant. Water boils at 212°F (100°C). A 50-50 mixture of water and any antifreeze will not boil until the temperature reaches 223°F (106°C). Therefore, using antifreeze reduces the chance of cooling system boil over in hot weather.

 Note: The color of coolants can vary (green, blue, red, yellow, orange, or pink) even between manufacturers' models and has no bearing on composition or service life. Dyes are used to give antifreeze its color.

Long-life Coolants

Most antifreeze is designed to last for one or two years, or about 30,000 miles (48 000 km). Newer vehicle have coolant which can be used for a longer period of time. These are referred to as *long-life coolants.* Long-life coolants can be left in the cooling system for up to 100,000 miles (160 000 km).

Long-life coolants are usually ethylene glycol based. Some long-life coolants are based on *organic acid technology (OAT).* OAT coolants are based on a blend of two or more organic acids which have long-life properties. All are formulated without silicates, phosphates, and other minerals and substances that can form deposits in the cooling system.

Long-life coolants have approximately the same freeze points as conventional ethylene glycol coolant. A 50-50 mixture will protect the cooling system down to -34°F (-36°C). The boiling point is raised about the same amount as with conventional coolant. Also, like regular coolants, mixtures greater than 67% coolant are not recommended. Some long-life coolants come prediluted 50-50 from the manufacturer, eliminating the need to add water.

Today, there are many types of long-life coolant on the market. The type of coolant required varies from manufacturer to manufacturer. Some long-life coolants may damage cooling systems that are not designed for them. Always check the proper service literature before adding coolant to any system.

Engine Water Jackets

The source of heat for the HVAC system is the engine itself. Heat generated inside the combustion chamber enters the metal of the cylinder walls and cylinder head.

This heat is then transferred to the coolant as it circulates through internal **coolant passages,** sometimes called **water jackets.** Coolant passages are cast into the engine block, cylinder heads, and sometimes the intake manifold. **Figure 11-2** illustrates coolant passages on a common engine. Coolant must circulate equally through every part of the coolant passages to remove heat from the engine. If the coolant does not circulate equally, cooling will be uneven, and some parts of the engine will become too hot. It is especially important to cool the cylinder heads to prevent valve burning.

Coolant Pump

The **coolant pump,** often called the **water pump,** is used to circulate coolant through the cooling system. The coolant pump is almost always a centrifugal pump. Centrifugal pumps are designed to pump large amounts of liquid without developing much pressure. The pump draws in coolant at its center, and uses centrifugal force to spin it outward. The rotating force causes the coolant to circulate through the cooling system. **Figure 11-3** is a cutaway view of a typical water pump. As coolant enters the suction area at the center of the blade assembly, it is thrown outward by the rotating blades. The clearance between the pump blades and the housing is relatively large, and some coolant can leak back to the suction area. Since the pump is not designed to produce pressure, this clearance does not affect efficiency.

The engine turns the coolant pump through a drive belt and pulleys. The pump shaft rides on one or two bearings pressed into the pump housing. A seal keeps coolant from leaking out through the pump shaft. A weep hole is usually drilled into the pump housing to allow slight coolant leakage to drip out instead of damaging the bearings. Many pump housings have internal baffles or sheet metal spacers to direct the pump output. Coolant always enters the pump through the bottom radiator hose. The pump pushes the coolant through the engine. The coolant exits through the top radiator hose, then passes through the radiator before returning to the pump through the lower radiator hose.

Coolant Pump Drive Belts

The engine turns the coolant pump. The most common drive method is an arrangement connecting the pump to the engine crankshaft through pulleys and one or more flexible drive belts. In the past, all drive belts were **V-belts.** The coolant pump pulley is often driven by two V-belts. This is usually done to ensure the pump continues to operate if one of the belts fails, and to provide a less complex belt path for other engine driven components.

Today, many vehicle engines have **serpentine belts.** Usually one serpentine belt drives the coolant pump, compressor, alternator, power steering pump, and air pump, when used. A gear attached to the engine timing chain or belt is used to drive a few coolant pumps. Gear driven pumps are rare, however.

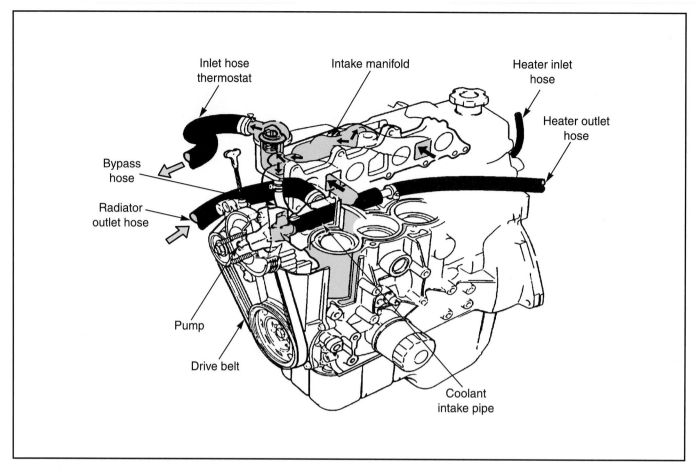

Figure 11-2. *The coolant passages shown in this illustration allow engine heat to transfer into the radiator. Coolant passages must be carefully designed to carry heat away from the hottest engine parts without overcooling other engine areas. (General Motors)*

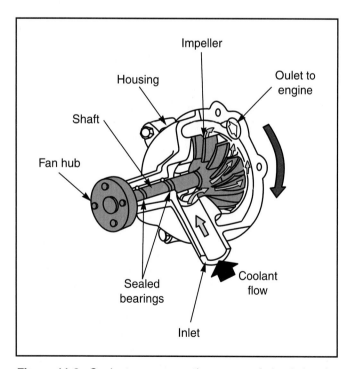

Figure 11-3. *Coolant pumps are the means of circulating the engine coolant through the coolant passages. This cutaway view shows how the pump uses centrifugal force to draw in coolant at its center and push it outward into the coolant passages.*

Radiator

The **radiator** is a large heat exchanger installed in front of the engine compartment. The radiator is a set of flattened tubes through which coolant can flow. Attached to the tubes are metal fins that transfer heat from the coolant to the air. The design of the tubes and fins allows air to flow with as little restriction as possible. Tanks at each end of the tube and fin assembly channel coolant through the tubes. Radiator tanks are usually open designs that allow coolant to flow through the tubes. This design allows the radiator to continue removing heat if some of the tubes become plugged. The top of the tank contains a filler neck used to add coolant to the system. The pressure cap is installed on the filler neck.

Almost all radiators receive heated coolant from the upper radiator hose and return the coolant to the lower radiator hose. Older radiators were **downflow** types, **Figure 11-4A**. Modern radiators are **crossflow** types, **Figure 11-4B**. Many older and many newer radiators have copper tubes and tanks. Some newer radiators are made of aluminum. Other newer radiators have a metal tube and fin assembly with plastic tanks.

Automatic transmission vehicles will have a radiator with a **transmission oil cooler** in the return tank. Transmission fluid is pumped through the cooler by transmission pressures and gives up its heat to the coolant.

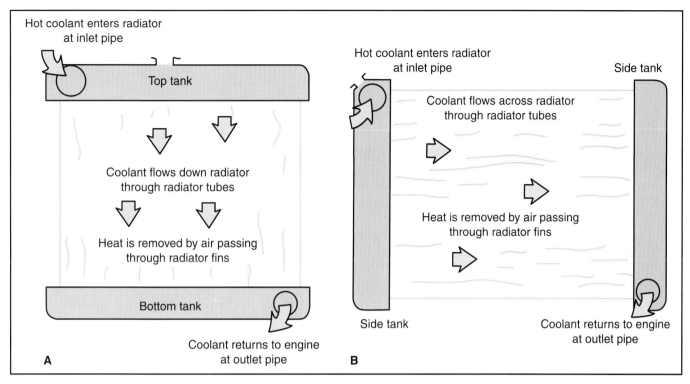

Figure 11-4. *A—Downflow radiator. Coolant enters at the top tank and flows down to the bottom tank. These radiators are usually used in trucks. B—A crossflow radiator. Coolant enters at one of the side tanks and flows across to the tank on the other side. Most modern radiators are crossflow types.*

Radiator Fans

To remove heat from the coolant in the radiator, air must be forced through the radiator fins. At higher speeds, air is forced through the radiator by vehicle movement. This is called **ram air** and removes heat easily. At lower speeds, a **radiator fan** must be used to draw enough air through the radiator to remove heat. Every fan consists of a set of blades curved to intercept, or **bite** the air. The amount of fan blade curve is called the **pitch.** More pitch causes the fan to bite more air, but uses more engine power. Modern fan pitch is a compromise between maximum airflow and minimum power consumption. Fan blade assemblies are carefully balanced to eliminate vibration. There are two types of fan drive systems.

Engine Driven Fans and Fan Clutches

Older rear-wheel drive vehicles use a **belt-driven fan.** The fan hub is attached to a pulley, which is driven by one of the engine drive belts, **Figure 11-5.** Even numbers of blades are used on fans for small engines. Fans used on larger engines may have an odd number of blades. The blades are often unevenly spaced around the hub. This uneven spacing allows the blade to draw in more air at low speeds.

Some fans have flexible metal or fiberglass blades. These blades flatten out at higher engine speeds. This reduces the blade pitch and the fan will use less horsepower and fuel. This type is referred to as a **flex fan.**

Figure 11-5. *Radiator fans used on rear wheel drive vehicles are usually belt driven from the engine crankshaft. (Ford)*

A **fan clutch** is often used with a belt-driven fan. The fan clutch is installed between the drive pulley and the fan assembly, **Figure 11-6.** There are two types of fan clutch; centrifugal and thermostatic. The **centrifugal fan clutch** is composed of two internal sets of ridges that can move in relation to each other. One set of ridges is attached to the drive pulley, while the other set is attached to the fan assembly. The ridges are close together but do not touch. The space between the sets of ridges is filled with a thick silicone fluid compound. At low speeds, the silicone acts as a clutch and causes pulley rotation to be transmitted to the fan. At higher speeds, when ram air is providing enough airflow, the silicone allows the ridges to slip in

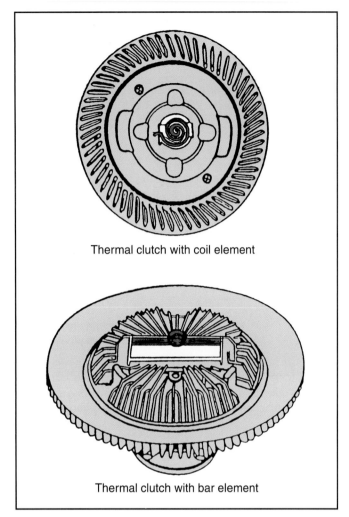

Thermal clutch with coil element

Thermal clutch with bar element

Figure 11-6. *A fan clutch is used to reduce noise and power consumption. The clutch drives the fan at full power when the engine is running slowly. At higher speeds or when the engine is cold the clutch slips to reduce fan action. (General Motors)*

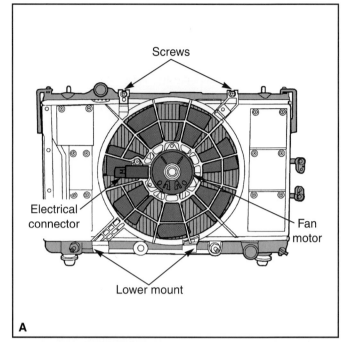

Figure 11-7. *A—This illustration shows a typical electric radiator fan assembly. Note the drive motor is directly connected to the fan blade assemblies. Many vehicles have only one fan. B—Some vehicles have two or more fans. This fan assembly is operated by separate control systems related to engine temperature and AC operation. (DaimlerChrysler)*

relation to each other. This reduces the horsepower needed to drive the fan at high speeds.

The ***thermostatic fan clutch*** is a centrifugal clutch with a thermostatic flow control added. The thermostatic control is a bimetallic strip or coil that closes a passage to the space between the two ridges. Silicone cannot enter the spaces between the ridges until the radiator warms up enough to allow the bimetallic strip to open the passage. When the passage opens, the silicone can circulate between the ridges and cause the fan to begin turning at the same speed as the pulley.

Electric Fans and Controls

Electric fans are used on all front-wheel and many rear-wheel drive vehicles. The fan is driven by an electric motor. Fan motors are small direct current motors designed to run for long periods without overheating. The fan and motor are usually installed in a frame assembly that fastens to the rear of the radiator core. The two most common types of electric fan arrangements are shown in **Figure 11-7.**

One or more temperature switches and relays are used to control the motor. Temperature sensors are installed so they contact the coolant inside an engine water jacket. The temperature switch may be mounted on the radiator, but is usually installed on the engine. Most temperature switches operate through a relay to reduce current flow through the sensor. On some vehicles, turning on the air conditioner also energizes a relay that operates the fan motor. Many modern vehicles use the on-board computer to control the fan. The fan is usually operated by a relay controlled by the computer, but a few fans are driven directly from the computer driver circuits.

The fan motor is usually off when a cold engine is first started. This allows the engine to reach operating temperature quickly. The fan motor is turned on in one of two ways:

❑ The engine reaches a certain temperature.
❑ The air conditioner is turned on.

If the air conditioner is turned on after the engine is first started, the fan will come on almost immediately. These conditions vary with each manufacturer and among vehicles and engines made by a single manufacturer. Many modern systems use input from the engine temperature sensor to signal the engine control computer when the engine is at operating temperature. The computer then energizes the radiator fan motor.

Many vehicles are equipped with two cooling fans. Sometimes this is done to obtain maximum airflow through a long narrow radiator, or to obtain more precise control of engine temperature. In applications that use a side-by-side radiator/condenser arrangement, two fans are used to pull air through each heat exchanger.

The fans may come on together, or may be designed so one fan comes on at a lower temperature than the other fan. Some systems are designed so one fan is controlled by engine temperature. The other fan runs only when the air conditioner is turned on. Some of these fan controls are pressure switches installed on the refrigeration system.

When high side pressure reaches a certain point, the fan is energized.

Two-speed Fans

Some vehicles use *two-speed fans.* These systems will always have two fans and motors. Two-speed fans work through a set of two or three relays, **Figure 11-8.** The input to fan relays 1 and 2 has current available (hot) at all times. When no fan operation is needed, all relays are deenergized. When low speed fan operation is needed, the control module energizes cooling fan relay 1. Current flows through the relay, the left side fan motor, the fan mode relay, the right side fan motor, and on to ground. Since both fans receive current through the same series circuit, the resistance causes low current flow, and the fans run slowly.

When outside temperatures are very high, the coolant temperature or refrigerant pressure may become high enough that high speed fan operation is necessary. When high speed operation is needed, fan relay 2 and the fan mode relay are energized. Each fan then operates through a separate circuit. With only one fan motor winding in each circuit, the resistance is halved, and twice as much current can flow. Fan speed increases to about twice low speed.

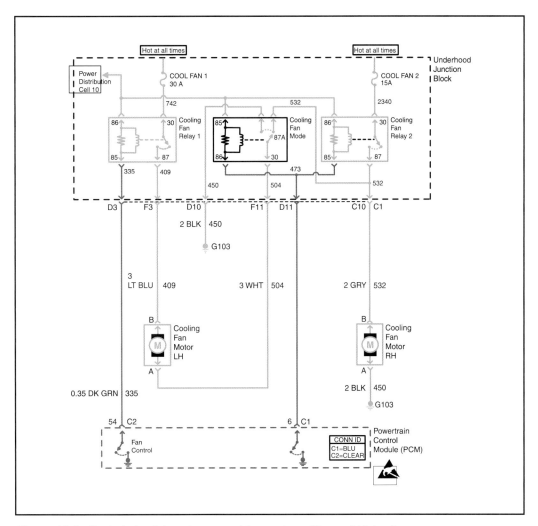

Figure 11-8. *Control circuit for a two-speed fan system. (General Motors)*

Thermostat

The *thermostat* is a temperature control device that allows a cold engine to warm up quickly. When the engine is cold, the thermostat is closed. Coolant circulates inside the engine and picks up heat from the combustion chambers. As the engine warms up, the thermostat begins to open. Older thermostats began to open at 160-180°F (71- 82°C). Since a hotter running engine produces fewer exhaust emissions, thermostats used on newer vehicles do not start to open until the temperature is over 200°F (93°C). Some thermostats remain closed until the temperature is as high as 220°F (104°C). In **Figure 11-9,** the engine is warmed up before the thermostat opens. When the engine is at its normal operating temperature, the thermostat will be completely open.

Once the engine reaches normal operating temperatures, the thermostat has no effect on cooling system operation. When the vehicle is operated in extremely cold weather, the engine may be overcooled when the thermostat opens. In this case, the cold coolant reaching the thermostat will cause it to re-close.

Thermostats consist of a valve (usually a flat metal flap or disc) that closes the passageway from the engine to the upper radiator hose. Connected to the valve is a sealed chamber filled with a substance that expands and contracts with temperature changes. The sealed chamber is sometimes called a *pellet*. See **Figure 11-10.**

When the engine is cold, the chamber contracts and holds the valve closed. As the engine warms up, the chamber expands and pushes the valve open. The gradual expansion of the chamber allows the valve to open slowly. This reduces thermal shock on the radiator and cooling system.

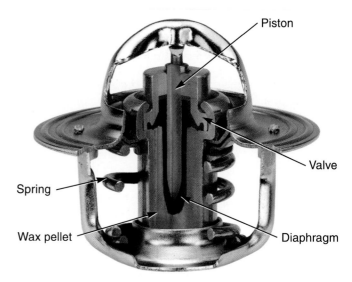

Figure 11-10. *This cross-section of a thermostat shows the heat sensitive element that controls the opening and closing of the thermostat. The wax in the element expands with increasing heat and closes when the heat is reduced. (Gates)*

OBD II Thermostat Monitor

For the vehicle emission controls to operate properly, the engine must reach its normal operating temperature as quickly as possible. If the thermostat sticks open, the engine will warm up slowly. During cold weather, the engine may never reach operating temperature. Cold engine operation keeps the emission control system from operating properly, resulting in high emissions and poor fuel mileage.

All 2000 and newer vehicles will have a **thermostat monitor** to alert the vehicle driver when the thermostat sticks open. The thermostat monitor is part of the OBD II emissions control system. The monitor consists of the coolant temperature sensor and an internal timer in the vehicle computer. When the engine is first started, the timer begins counting. A target time is set, based on ambient air temperature at engine start-up. If the coolant has not warmed to approximately 80% of the thermostat opening temperature within 5-14 minutes, the computer will set a trouble code and illuminate the malfunction indicator lamp (MIL). The monitor runs only after a minimum two-hour engine off soak time. This permits the engine coolant to drop below the warm-up temperature on a hot engine.

> **Note: If the thermostat sticks closed, the engine will quickly overheat. The temperature gauge or warning light will warn the driver of overheating, even if steam is not visible from the engine compartment.**

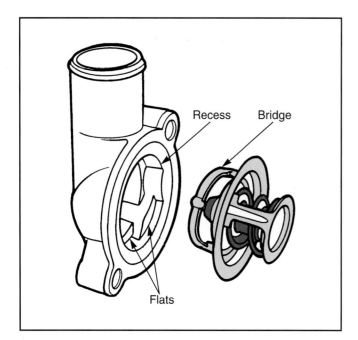

Figure 11-9. *This figure shows a typical thermostat and housing. The thermostat element always faces the engine cooling passage. (Ford)*

Coolant Hoses

As it operates, the engine moves slightly in relation to the vehicle body. The radiator is solidly bolted to the vehicle body, as is the heater core. To connect the engine to the other parts of the cooling system, and to allow for expansion as engine temperature changes, flexible coolant *hoses* are needed, **Figure 11-11.** These hoses consist of plies of fabric or nylon cords covered with synthetic rubber. The rubber covers the inside and outside of the hose. Hoses fit over the opening or *neck,* of the part to which they are attached.

The two main types of flexible hoses are *radiator hoses* and *heater hoses.* Hose size depends on its use. Large hoses are used to connect the engine, and radiator and smaller hoses are used to connect the engine to the heater core. Larger engines usually have larger radiator hoses than smaller engines. *Molded radiator hoses* are specially formed, or molded to a specific shape, **Figure 11-12.** The hose's specific shape allows it to make sharp bends and clear obstructions without kinking. Molded hoses are used to connect the engine and radiator of most vehicles. Heater hoses are sometimes molded.

Hoses are attached to the engine, radiator, and heater core by *hose clamps.* The hose clamp holds the hose to the neck of the mating part and prevents leaks. Some common hose clamps are shown in **Figure 11-13.** Some vehicles have *quick disconnect hoses.* Quick disconnect hoses are heater hoses with male and female connectors, **Figure 11-14.** The male and female ends are pushed together and snap to form a tight connection.

Figure 11-12. *Molded hoses are formed to a specific shape to fit a particular engine/vehicle application. The shape prevents the hose from kinking. (Gates)*

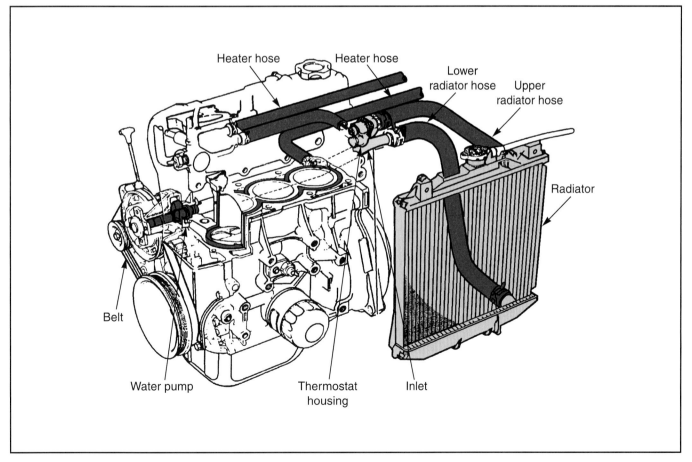

Figure 11-11. *Hoses connect the engine coolant passages, radiator, and heater core. Hose size and construction varies between different vehicles and engines. (General Motors)*

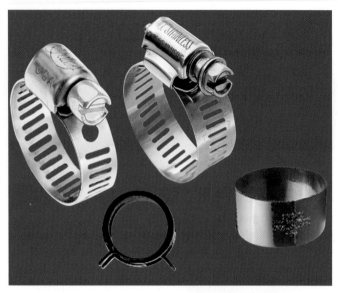

Figure 11-13. *The main types of clamps are screw clamps, spring clamps, and polymeric band clamps. Manufacturers use all three as original equipment. (Gates)*

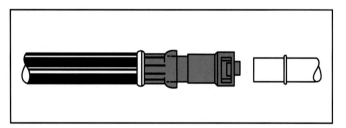

Figure 11-14. *Quick-connect coolant hoses snap into place over special fittings. These hoses are often used where tight clearances make conventional hose clamps impractical. (Gates)*

Bypass Hose or Passage

The overall design of the cooling system allows coolant to circulate when the thermostat is open. When the thermostat is closed, the coolant may be unable to circulate within the engine. With no circulation, the hottest parts of the engine, such as the cylinder head and valves, may overheat before the thermostat can open. To allow coolant to circulate with the thermostat closed, a **bypass** is built into the engine. The bypass allows the pump to move coolant through the water jackets, although it cannot exit the engine. On some engines, a molded bypass hose, **Figure 11-15**, is used. Other engines have a bypass built into the engine block. In some cases, the heater core and hoses are the bypass. With the thermostat closed, coolant is pushed through the heater core by the pump. After passing through the heater, the coolant reenters the engine and circulates through the water jackets before returning to the pump.

Hose Failures

All hoses have a limited useful life. It was thought for years that hose failure was caused by cracking due to heat and cold or material failure. Research may have shown another reason hoses fail is due to **electrochemical degradation (ECD).**

Figure 11-15. *When used, a bypass hose allows coolant to circulate within the engine. This prevents localized hot spots in the engine. (General Motors)*

ECD is an electrochemical assault on the rubber compounds in the hoses. This occurs when dissimilar metals in the engine, radiator, and heater core combine with the salts and compounds within the liquid coolant to form a galvanic cell (battery). This causes current flow through the hose, which causes microcracks to form inside the hose reinforcement. When combined with normal flexing and high temperatures, the hoses will eventually leak or rupture. More information on ECD is located in Chapter 20.

An increasing cause of electrolysis is defective or missing ground wires as well as poorly grounded aftermarket accessory systems. Electrolysis due to poor grounds can destroy a radiator, heater core, or an entire engine.

Pressure Cap

The **pressure cap** is installed on the radiator filler neck, or the coolant reservoir opening. The pressure cap may be called the radiator cap. It prevents coolant loss and allows pressure to build to a certain point as the engine warms up. On some engines, the thermostat does not open until the coolant temperature is higher than the boiling point of water. To prevent boil over, the system must be pressurized. Pressurizing the system raises the boiling point by 3°F (1.7°C) for every pound (6.89 kPa) of pressure. A 15 psi (103 kPa) pressure cap will raise the boiling point of a 50-50 coolant mixture from 223°F (106°C) to 268°F (131°C).

 Note: Some newer vehicles have pressure caps rated up to 25 psi (172.36 kPa).

The primary part of the pressure cap, **Figure 11-16,** is a seal held against the filler neck by a spring. As pressure builds up in the quickly warming system, the spring continues to hold the seal against the neck. Eventually, system pressure becomes high enough to overcome spring tension and open the valve, relieving excess pressure. Spring tension determines the exact system pressure. Pressure and any extra coolant exit through an overflow hose.

When the engine cools off, a separate check valve opens to allow air to enter the system. Without this valve, the hoses would collapse as system pressure became less than outside air pressure. Caps used with coolant recovery systems have a seal between the underside of the cap and the filler neck, **Figure 11-17.** The seal keeps air from entering the system as the engine cools off.

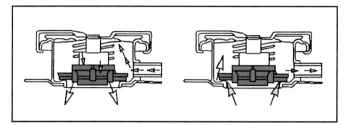

Figure 11-16. *A common modern radiator cap. The double action of the cap valves allows coolant to enter and leave the cooling system while maintaining pressure. (General Motors)*

Figure 11-17. *The dual seals allow coolant to be drawn into the cooling system from the coolant recovery tank while keeping air out.*

Coolant Recovery Systems

The **coolant recovery system** keeps air out of the cooling system. Less air means less oxygen to cause corrosion, and more coolant available to transfer heat. As the engine heats up enough to open the pressure cap, pressurized coolant travels through a hose to a separate **reservoir tank,** sometimes called **reservoir bottle,** a **recovery tank,** or an **overflow tank.** See **Figure 11-18.**

As the engine cools off, pressure drops off and eventually becomes less than outside air pressure. The cooling system then draws coolant from the reservoir. In fact, coolant is pushed back into the system by outside air pressure. On some systems, the reservoir tank and hoses are also pressurized. On these systems, the pressure cap is installed on the reservoir tank, **Figure 11-19.**

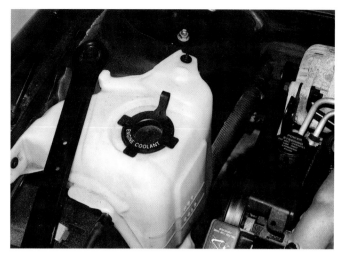

Figure 11-18. *The coolant recovery tank is usually installed in the inner fender or on the radiator. It stores coolant expelled from the cooling system as it heats up. When the engine cools off, the coolant is drawn back into the system. (General Motors)*

Figure 11-19. *This type of coolant recovery tank is pressurized. The filler cap is actually the radiator cap. This type of tank is often called a surge tank. (Jack Klasey)*

Drains and Air Bleeds

To remove old coolant from the system, a **drain** must be provided. The drain is usually located on the radiator. Some vehicles have additional drains on the engine block. Drains may be old style petcocks or plastic valves. Some manufacturers provide a pipe plug that must be removed to drain the coolant. **Figure 11-20** shows a typical drain.

On older vehicles, air could be removed from the cooling system by operating the engine until the thermostat opened. Since the radiator filler neck was higher than the engine, air would rise to the filler neck and be displaced as more coolant was added. On many newer vehicles, the radiator is lower than the engine. Air cannot be easily removed from the engine coolant passages on these vehicles unless an **air bleed** is installed on the engine, **Figure 11-21.** The air bleed is opened with the engine off. Then coolant is added through the radiator filler neck. Air exits the engine passages as the coolant level rises. Some

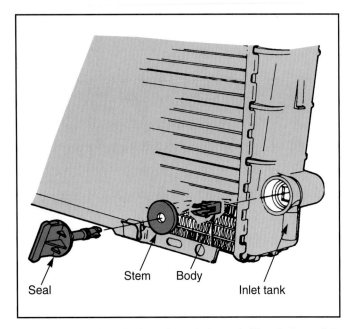

Figure 11-20. *A typical modern drain cock. The drain cock is always located at the lowest point in the radiator. (Ford)*

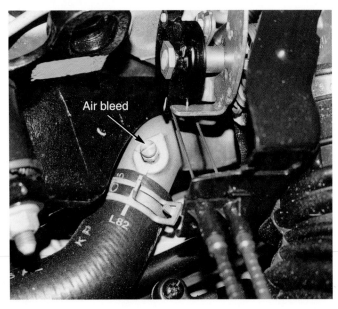

Figure 11-21. *Air bleeds are installed on engines in front-wheel drive vehicles. They are usually located close to the highest point on the engine.*

vehicle manufacturers recommend the air bleed be opened after the engine has been running a while to remove any additional air. Other methods can be used to bleed air on vehicles without air bleeders.

Air-cooled Engines

Air-cooled engines are much simpler than liquid-cooled engines. Air is blown over the engine and heat is transferred by convection. Most air-cooled engines, sometimes called *pancake engines,* are four or six cylinders.

Placing two banks of cylinders across from each other makes it easier to design the cooling system. The various parts of an air cooling system are explained in the following paragraphs.

Engine Cooling Fins

Instead of coolant passages, the hottest parts of an air-cooled engine are equipped with *cooling fins.* The cylinder heads are cast with cooling fins that surround the intake and exhaust valves, **Figure 11-22.** The cylinders are individual units called *cylinder barrels.* Fins are cast into the cylinder barrels to remove heat from the cylinder walls. For the fins to effectively remove heat, outside air must be blown over them. This is done by the next two components.

Blower

An engine driven *blower* pushes air over the cooling fins. The blower is turned by a belt and pulley arrangement similar to the drive system for a coolant pump, **Figure 11-23.**

Figure 11-22. *Air-cooled engines have fins on the engine and head, close to the valves.*

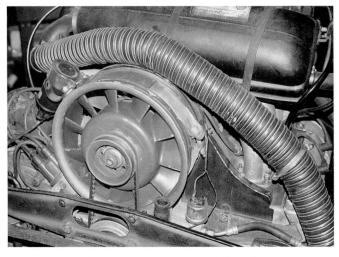

Figure 11-23. *Blower fan on an air-cooled engine.*

The blower turns at about twice engine speed. There is usually no fan clutch or other speed adjusting device. On some older vehicles, the drive belt also operated the vehicle generator. Blowers are centrifugal designs that draw air in at the center and blow it out at the edges.

Shrouds

To accurately direct the blower output over the fins, sheet metal **shrouds** surround the engine. The shrouds force the moving air to pass between the fins, picking up heat and discharging it to the atmosphere. The shrouds are designed so air from the blower is channeled over each bank of cylinders and exits through an opening at the rear of each bank. **Figure 11-24** illustrates a shroud on an air-cooled engine. Note how the shroud directs air from the blower across the fins.

Oil Cooler

Almost all air-cooled and some liquid-cooled engines used in automobiles have an **engine oil cooler**. Oil coolers are heat exchangers that remove heat from the oil by conduction and convection. Coolers look like small radiators and are constructed the same way, **Figure 11-25.** The cooler is placed in the shrouds so air passes through it whenever the engine is running. The engine oil pump pushes oil through the cooler. The oil gives up its heat to the passing air and then enters the engine lubrication system. On most liquid-cooled engines, the oil cooler uses coolant to carry heat from the oil to the radiator.

Thermostat

To quickly warm a cold engine, some air cooling systems use a thermostat. The thermostat is a liquid-filled bellows attached to a damper door at the air exit point. **Figure 11-26** illustrates a thermostat. On a cold engine, the bellows contracts and holds the damper door closed. As the engine heats up, the

Figure 11-25. *Oil coolers are used to dissipate the heat from air-cooled engines.*

bellows expands and pushes the damper door open. Most air-cooled engines have a thermostat and damper on each bank of cylinders.

Heating System

The heater system uses heat from the engine cooling system. The cooling system supplies heat that would be transferred to the outside air and wasted. The heater takes this heat and sends it to the passenger compartment. The cooling system is usually liquid-cooled, but a few air-cooled vehicles are still being manufactured. How cooling system type affects heater design is explained in the following paragraphs.

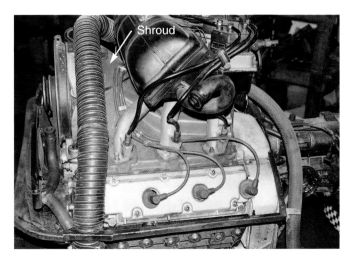

Figure 11-24. *Air-cooled engines have shrouds that route air over the engine surface.*

Figure 11-26. *Bellows type thermostat used on the air flap of an air-cooled engine.*

Heater—Liquid-Cooled Engine

In a liquid-cooled system, heat is transferred from the engine to the liquid coolant. The heat then passes from the coolant to air entering the passenger compartment. In operation, the coolant pump circulates heated engine coolant through the heater core. Convection causes air passing through the heater core to pick up heat from the coolant. This air enters the passenger compartment. The coolant returns to the engine to pick up more heat.

Heater—Air-Cooled Engine

In an air-cooled heating system, heat is transferred from the engine directly to air passing over the engine. The heated air then enters the passenger compartment. In operation a blower drives air over the hottest parts of the engine, then into the passenger compartment. The heated air exits the passenger compartment as fresh air is drawn in to be heated.

Heating System Components

The following section covers the heater and associated components. The section is divided into two parts; heaters on vehicles with liquid-cooled engines and heaters on vehicles with air-cooled engines.

Liquid-Cooled System

If the engine is liquid-cooled, the parts of the heater must contain and direct liquid coolant with maximum flow and no leaks. The parts of the liquid operated heater are covered in the following paragraphs.

Heater Core

The *heater core* is an example of a tube and fin heat exchanger. The heater core is a chamber containing flattened tubes through which coolant can flow. Attached to the tubes are metal fins that transfer heat from the coolant to the air. Sometimes the tubes are slightly bent to increase coolant-air contact. The design of the tubes and fins allows air to flow with as little restriction as possible. Tanks at each end of the core channel the coolant through the tubes.

The inlet tank is usually designed with two isolated sections. Coolant entering the inlet tank flows through half of the tubes, turns at the end tank, and returns through the other tubes. The coolant then exits through the outlet tank. The heater can function even if some of the tubes become plugged. For maximum heat transfer, heater cores must be made of metals that transfer heat easily. Most older and some current heater cores have copper tubes and tanks. Most newer heater cores are made of aluminum, or have an aluminum tube and fin assembly with plastic end tanks. **Figure 11-27A** shows a typical modern heater core. Note

the aluminum fins and plastic tanks. **Figure 11-27B** illustrates an older copper heater core.

Like the condenser and evaporator, the heater core is a heat exchanger. It exchanges heat between the engine coolant and the outside air. Unlike the refrigerant in the evaporator or condenser, the coolant does not change state inside of the heater core. After the engine has been operating for a few minutes, the coolant is hot enough to transfer significant amounts of heat to the air as it passes through the core.

Heater Hoses

Heater hoses are flexible reinforced rubber hoses that connect the heater core to the engine. Heater hoses must be flexible to allow for movement between the engine and body. Heater hoses are attached to the heater core with clamps. **Figure 11-28** illustrates a typical heater hose arrangement.

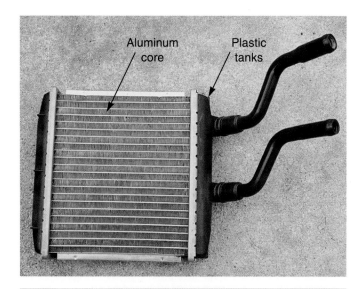

Figure 11-27. *A—An aluminum heater core, commonly used on late-model vehicles. The header tanks can be made of aluminum or plastic, as shown here. B—An all copper heater core as used on many older vehicles.*

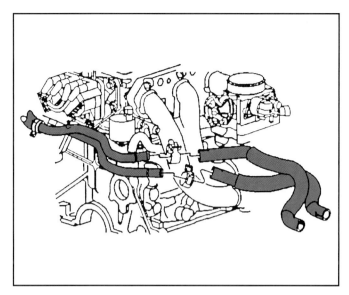

Figure 11-28. *Radiator hoses connect the heater core to the engine cooling passages. They may be molded to fit around various underhood components. (General Motors)*

Heater Shutoff Valve

The **heater shutoff valve** is installed in one of the heater hoses and stops the flow of coolant when necessary. The shutoff valve is usually a small flap that pivots to open or close the coolant passage to the heater core, **Figure 11-29.** Some shutoff valves have plungers that move up and down to close off the passage, **Figure 11-30.**

Heater shutoff valves are controlled using one of two methods. A cable operates the manual shutoff valve. The cable is attached to a temperature control lever on the heater control panel. Moving the lever moves the cable, which opens or closes the valve.

A vacuum diaphragm operates the vacuum shutoff valve. Applying or removing vacuum causes the diaphragm

Figure 11-29. *A flap type of heater shutoff valve. It is usually installed in the hoses leading to the heater core. This shutoff valve is vacuum operated.*

Figure 11-30. *The plunger heater control valve is often installed by threading one end into the engine block. The other end connects to the heater inlet hose. The vacuum line leads to the control panel.*

to open or close the valve. A vacuum control valve in the heater control panel controls the application of engine manifold vacuum to the diaphragm. The vacuum control valve is usually connected to the HVAC mode control lever. When the driver moves the HVAC control lever, the vacuum control valve is moved. This applies or releases vacuum from the diaphragm.

Air-Cooled System

The heater used on an air-cooled system is extremely simple. The components of the air-cooled heater are explained in the following paragraphs.

Heater Door

A **heater door** is installed on the engine shroud on one bank of cylinders. The purpose of the heater door is to redirect heated air into the passenger compartment. A dashboard cable operates the heater door. When the heater is not needed, the heater door seals off the passage to the heater system. Air exits from the rear of the engine.

When the heater lever is operated, the heater door opens the heater passages while closing off the passage to the rear of the engine. The blower forces heated air into the passenger compartment. On some vehicles, the heater door does not seal off the exhaust passage. The HVAC system blower draws heated air into the passenger compartment while allowing some air to exit through the exhaust passage.

Heater Ducts

Since most air-cooled engines are mounted in the rear of the vehicle, a system of **heater ducts** is needed to carry the heated air to the front of the vehicle. Sometimes two

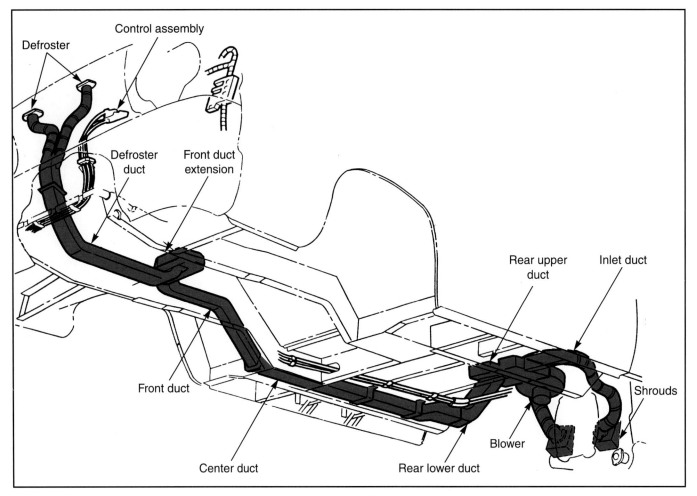

Figure 11-31. *The heater ducts on an air-cooled engine are usually directed under the vehicle. The fan and air doors are usually located at the engine. (General Motors)*

ducts are placed around the sides of the passenger compartment. Other systems have a large central duct at the center of the passenger compartment floor. See **Figure 11-31.**

Exhaust Sealing

It is vitally important no exhaust gases reach the passenger compartment. For this reason, the engine cooling shrouds and heater ducts are carefully sealed off from the exhaust system.

Summary

The heater has a major part in the operation of the HVAC system. The engine is the source of heat for the heater system. Engines produce much more heat than they can use, and this heat must be removed to prevent engine damage. Most cooling systems use liquid cooling. A few vehicles are air-cooled.

In a liquid cooling system, liquid coolant circulates through the engine. Coolant must transfer heat easily and not cause damage to the cooling system. The liquid cooling system consists of engine coolant passages, a belt driven coolant pump, a radiator, an engine-driven or electric fan, hoses and clamps, thermostat, pressure cap, coolant recovery system, drains, and air bleeds.

Air-cooled engines are simpler than liquid cooling systems. Blowing air over the engine cools it. Heat is transferred by convection. Most air-cooled engines are flat fours or sixes. The hottest parts of an air-cooled engine are equipped with cooling fins instead of coolant passages. A belt driven blower pushes air over the cooling fins. Shrouds direct the air through the fins. Almost all automobile air-cooled engines have an oil cooler. Some air cooling systems use one or more thermostats consisting of liquid filled bellows and an air damper.

Both types of heater systems use heat from the cooling system. In a liquid-cooled system, heat is transferred from the engine to the coolant. The coolant pump circulates heated engine coolant through the heater core. The heat then passes from the coolant to air entering the passenger compartment. Coolant returns to the engine to absorb more heat.

In an air-cooled heating system, heat is transferred from the engine directly to air passing over the engine. This air then enters the passenger compartment. The heated air exits the passenger compartment as fresh air is drawn into the engine to be heated.

Liquid-cooled system components include the heater core, heater hoses, and heater shutoff valve. Air-cooled system components include the heater door, heater ducts, and sometimes a blower and motor. The shrouds and ducts must be carefully sealed to keep exhaust gases out of the passenger compartment.

Review Questions—Chapter 11

Please do not write in this text. Write your answers on a separate sheet of paper.

1. What percentage of the heat produced by the engine is actually used to produce power?
   (A) None.
   (B) 1/3.
   (C) 2/3.
   (D) All of it.

2. The outside air is always _____ than an engine at operating temperature.

3. Most vehicle manufacturers recommend a _____ mixture of water and antifreeze.
   (A) 10-90
   (B) 20-80
   (C) 30-70
   (D) 50-50

4. Coolant passages are cast into the _____.

5. Centrifugal pumps are designed to pump large amounts of liquid while developing little _____.

6. All of the following materials are used in radiators, *except:*
   (A) copper.
   (B) steel.
   (C) aluminum.
   (D) plastic.

7. Radiators can be _____ flow types or _____ flow types.

8. Under what circumstances would an electric fan use a fan clutch?

9. Under what three conditions would the radiator fan motor be turned on?

10. The two main types of flexible hoses are _____ and _____ hoses.

11. On an air-cooled engine, fins are made into the cylinder _____ and cylinder _____.

12. The air-cooled engine blower is driven by the engine through a _____.

13. What is the purpose of having coolant flow through several tubes of the heater core at the same time?

14. Cable and vacuum are the two ways the _____ can be controlled.

15. The heater door used on an air-cooled engine heater is installed on one of the cooling system _____.

ASE Certification-Type Questions

1. Technician A says using antifreeze in the cooling system reduces the chance of freezing in cold weather. Technician B says using antifreeze in the cooling system reduces the chance of boil over in hot weather. Who is right?
   (A) A only.
   (B) B only.
   (C) Both A and B.
   (D) Neither A nor B.

2. The cylinder heads must be especially well cooled to prevent:
   (A) valve burning.
   (B) system corrosion.
   (C) excessive heater output.
   (D) water pump damage.

3. All of the following statements about the coolant pump are true, *except:*
   (A) the engine turns the coolant pump through a belt and pulleys.
   (B) a weep hole is usually drilled into the pump housing.
   (C) many pump housings have internal baffles.
   (D) coolant always enters the pump through the bottom radiator hose.

4. The fan clutch _____ at high speeds to reduce power loss.
   (A) disengages
   (B) overheats
   (C) slips
   (D) locks up

5. Technician A says modern thermostats remain closed until coolant temperature is as high as 180°F (82°C). Technician B says once the engine is completely warmed up, the thermostat has no effect on cooling system operation. Who is right?
   (A) A only.
   (B) B only.
   (C) Both A and B.
   (D) Neither A nor B.

6. What determines the pressure in the cooling system?
   (A) Engine temperature.
   (B) Thermostat opening temperature.
   (C) The pressure cap seal.
   (D) The pressure cap spring.

7. All of the following statements about the coolant recovery system are true, *except:*
   (A) keeping air out of the cooling system reduces corrosion.
   (B) coolant passes through the radiator pressure cap to reach the reservoir.
   (C) cooling system pressure is never less than outside air pressure.
   (D) the pressure cap may be installed on the reservoir tank.

8. Which of the following is *not* used on an air-cooled engine?
   (A) Oil cooler.
   (B) Bypass hose.
   (C) Blower.
   (D) Thermostat.

9. How does the heater core resemble a condenser or evaporator?
   (A) A change of state takes place inside of the core.
   (B) It is a heat exchanger.
   (C) It is under very high or low pressure.
   (D) It is made of copper.

10. The heater shutoff valve is usually installed in the heater:
    (A) hose.
    (B) core.
    (C) ductwork.
    (D) control panel.

Chapter 12

Air Delivery Systems

After studying this chapter, you will be able to:

❑ Explain how an HVAC system blower operates.
❑ Identify the major parts of an HVAC system blower and motor.
❑ Explain the operation of HVAC blower speed controls.
❑ Identify the purposes of HVAC system air ducts.
❑ Explain the design and placement of HVAC system air ducts.
❑ Identify the types and purposes of HVAC system air doors.
❑ Explain how HVAC system air doors are operated.
❑ Identify and explain HVAC system air door operating devices.

Technical Terms

Air handling system	Plenums	Independent vents	Dual Zone
Upstream	Intake ductwork	Air doors	Air bypass door
Downstream	Evaporator cases	Blend door	Upper and lower mode doors
Blower wheel	Heater cases	Temperature door	Cables
Squirrel cage blowers	HVAC modules	Diverter doors	Lever
Blower motor	Split case	Mode doors	Vacuum diaphragms
Resistor pack	Modules	Air conditioning-heater door	Vacuum motors
Blower relay	Independent cases	Heater-defroster door	Stepper motors
Afterblow module	Output ductwork	Recirculation door	Cabin filters
Air ducts	Vents	Air inlet door	Evaporator water drain
Ductwork	Registers	Heater restrictor door	

You cannot understand the refrigeration and heating systems without knowledge of the air delivery system. Some of the most common problems affecting the HVAC system are caused by defects in the blower, ductwork, and air doors. The parts discussed in this chapter are used when the HVAC system is set to ventilation, heating, air conditioning, and window defrosting.

Air Handling Systems

The delivery system that sends the cooled or heated air to the passenger compartment is as important as the operation of the refrigeration system and heater. These parts are sometimes called the *air handling system.* This system generates airflow, ensures the air is heated or cooled as necessary, and delivers the air to the proper place in the passenger compartment.

Note: The terms *upstream* and *downstream* are used to locate HVAC air handling components in relation to other components. A component receiving incoming air before another component is *upstream* of that component. A component receiving incoming air after another component is *downstream* of that component.

Blowers and Motors

The HVAC system could not function properly without a blower fan and motor. The blower fan is usually placed upstream of the evaporator and heater core. On a few vehicles, the blower is downstream from the heater core or between the heater core and evaporator. Some vehicles with rear heating/cooling systems may have a second blower and motor located at the rear of the passenger compartment.

Figure 12-1. *The blower wheel, or squirrel cage, delivers air to the other HVAC passages.*

Blower Wheels

The blower fan consists of a *blower wheel* resembling a hamster exercise wheel. These blowers are referred to as *squirrel cage blowers.* **Figure 12-1** shows the shape and blade design of a typical squirrel cage blower wheel. The blower wheel is actually a centrifugal air pump. Blower wheel blades are shaped so the rotating wheel pulls air into its center and drives it outward. The shape of the blower housing directs the air into the center of the blower wheel and causes it to exit in the proper direction. The blower wheel may be bolted, press-fit, or held to the motor by a clip.

Blower Motors

A direct current motor turns the blower wheel. The *blower motor* is connected to the electrical system through the ignition switch and HVAC controls. When stationary windings in the motor are energized, current is delivered to the movable armature. The interaction between the magnetic fields set by the armature and stationary windings causes the armature to rotate. The armature shaft is directly attached to the blower wheel. See **Figure 12-2.** Some blower motors have a shaft at each end, and turn two blower wheels. Since the motor operates for long periods of time, it is usually vented to cool the windings. The motor must turn in the proper direction, and the fan blades must be pointed in the proper direction. If the motor or blades turn backward, air will be pulled out of the passenger compartment.

Motor Speed Control

For good system operation, it is necessary to be able to control the speed of the blower. In many cases, three or four electrical resistors are used to control blower motor speed. This assembly is usually referred to as a *resistor*

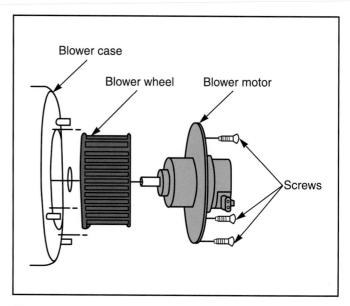

Figure 12-2. *The blower motor drives the blower wheel. (General Motors)*

pack. Selecting the blower speed causes the current to flow through one of these resistors before reaching the blower. Each resistor reduces the amount of current flowing into the blower by a different amount, causing blower speeds to be different in every speed selection. On a few HVAC systems, energizing different windings in the motor controls motor speed. Most HVAC systems have four blower speeds.

The highest blower speed is usually a direct connection that uses no resistors. High speed is usually selected through a *blower relay.* A simplified electrical schematic of the blower control system is shown in **Figure 12-3.**

On the three lower speeds, power travels through the switch and resistor assembly to reach the blower motor. When high blower is selected, a relay is energized, sending current directly to the blower, and bypassing the resistor assembly. On systems that do not use a blower relay, current is sent directly through the high blower switch contacts to the blower motor.

Automatic Motor Speed Control

Some modern vehicles use an on-board computer to energize the blower. Blower speeds are selected according to computer inputs. Instead of using mechanical contacts, the computer uses internal electronic circuitry with power transistors and resistors. To reduce current flow through the computer, the control system may use a high blower relay to control the blower motor.

Afterblow Module

The evaporator core on some vehicles may have a tendency to retain condensed moisture. This moisture causes these evaporators to develop mold on their surfaces. Mold spores are always present in the air and may form large colonies on the evaporator surface if moisture does not dry between system uses. Mold can cause a musty odor in the evaporator case. These odors will enter the passenger compartment when the HVAC system is used. To prevent this, the evaporator must be allowed to dry after each use.

To dry out the evaporator, an *afterblow module* energizes the blower motor for a set period of time after the vehicle's engine has been shut off. Some afterblow modules allow the blower to run as soon as the engine is turned off. Others energize the blower motor after the vehicle has been parked for awhile, usually anywhere from 30-60 minutes. Air movement over the warmed evaporator core helps to remove any remaining moisture, reducing the chance for mold formation.

Exhaust Gas Purging

On all vehicles made since 1977, the blower motor runs when the engine is running, whether or not the HVAC system is being used. Running the blower at all times provides a constant flow of air through the passenger compartment. This purges any carbon monoxide that may enter from a leaking exhaust system. The ignition switch energizes the blower through the resistor assembly, causing the motor to turn at its slowest speed.

Most modern vehicles have a temperature switch installed between the ignition switch and the blower motor. The switch opens to de-energize the blower motor when coolant temperatures are low. This reduces driver discomfort when the vehicle is first started in cold weather. Modern engines heat up quickly and the blower is energized before significant amounts of carbon monoxide can form.

Air Ducts and Heater/Evaporator Cases

Air ducts are tubes or passageways the HVAC system air passes through. The sets of ducts in a vehicle are called the *ductwork.* Air also passes through the heater and evaporator cases on its way to the passenger compartment. Note many of the components are combined into a single module. For purposes of explanation, this discussion will treat the various components as though they were separate units.

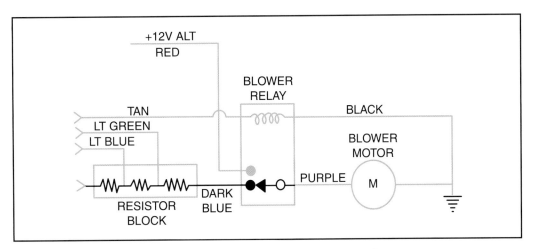

Figure 12-3. *Simple schematic for the blower circuit. (General Motors)*

Note: Sometimes the cases and ducts are called *plenums*. A plenum is a pressurized chamber. Pressure differences occur when air is flowing through the ducts and cases. Therefore, these components can be called plenums.

Intake Ductwork

Most of the ***intake ductwork*** is formed by the body sheet metal in the cowl and firewall areas and is not a separate duct assembly. The blower draws air into the HVAC system through a grill or screen just ahead of the windshield, **Figure 12-4.** A seal keeps exhaust gases and engine vapors from entering the intake area. Air flows through the evaporator and heater even when they are not in operation. Some vehicles have a filter in the intake area, **Figure 12-5.** The intake area leads to the intake of the blower. The blower directs the air to the heater and evaporator cases.

Heater and Evaporator Cases

The blower output is directed through the heater and evaporator cases. These cases are called ***evaporator cases, heater cases,*** or ***HVAC modules.*** The heater case contains the heater core. The evaporator case contains the evaporator. In most vehicles, both the evaporator and heater will be housed in the same case. These cases may also house the blower motor. The case is usually installed under the dashboard on the passenger side of the firewall or cowl. Placing the case away from the driver side of the dashboard allows room for the steering column and control pedals.

Most cases extend into the right side of the engine compartment. The case may be a two-piece unit bolted to either side of the firewall, **Figure 12-6.** When assembled, the case halves and firewall form the case assembly. This is called a ***split case.*** Other cases are single piece units that fit under the dashboard. These are called ***modules*** or ***independent cases.*** The module may also contain the blower motor and other HVAC parts. Usually some components extend under the hood. **Figure 12-7** is an illustration of a module.

Output Ductwork

From the heater and evaporator cases, air is directed into the passenger compartment by the ***output ductwork.*** There are three main outlets for the air entering the passenger compartment:

❑ Heater outlets.
❑ Defroster outlets.
❑ Air conditioner outlets.

Some vehicles also have side defroster vents, rear heater vents, and other outlets. The output ducts connect the case with the heater, air conditioner, and defroster outlet vents. Ducts can be hard plastic tubes, **Figure 12-8A,** flexible hoses, **Figure 12-8B,** or both. The ducts are

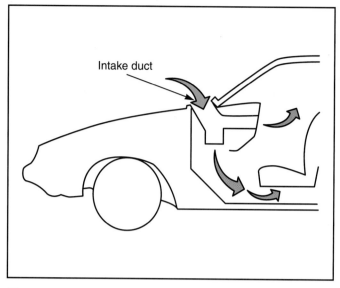

Figure 12-4. *To ventilate the vehicle interior, air is drawn in at the cowl, directly under the windshield. Air circulates through the vehicle interior and through vents in the door pillars. These vents have flaps that open when the interior air pressure becomes higher than outside pressure. (General Motors)*

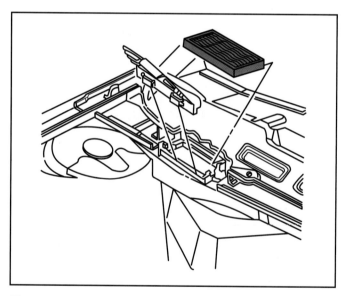

Figure 12-5. *Many newer cars and vans have air filters in the intake system. The filter may be located in the cowl, as shown here, or in the HVAC case under the dashboard. (General Motors)*

clipped or bolted into place or may be a light press fit onto the mating parts. Some plastic ducts have internal baffles, **Figure 12-8C.** These baffles direct airflow inside of the ductwork.

Vents

Vents are outlets from the ductwork into the passenger compartment. Ventilation and air conditioning vents are installed in the dashboard. Heater vents are

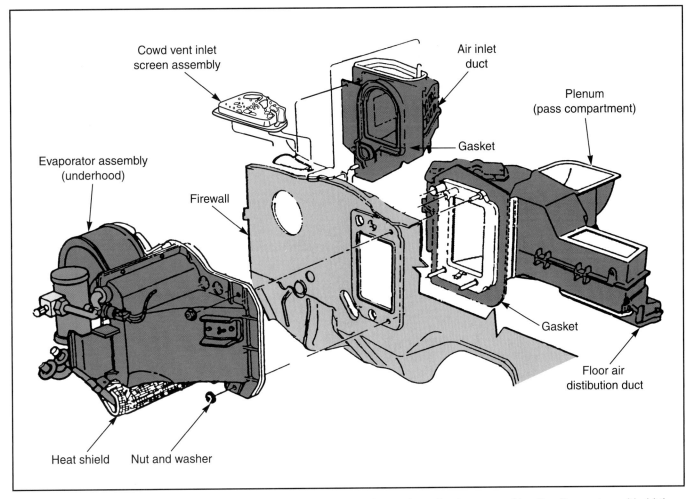

Figure 12-6. *A typical split case used on larger vehicles. Note how the studs and nuts are used to align the parts and hold them tightly to the firewall. (Ford)*

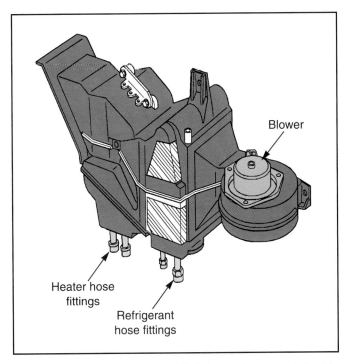

Figure 12-7. *A modular case. All of the parts are contained in a single assembly. Modular cases are usually installed under the vehicle dashboard. (General Motors)*

placed under the dashboard at the rear of the firewall, **Figure 12-9.** Defroster vents are placed at the forward part of the upper dashboard so air can strike the windshield. To defog the side windows, many vehicles have small demister vents at the side pillars to direct air onto the side glass, **Figure 12-10.** Air conditioner vents are usually adjustable and may have shutoff valves. Other air conditioner vents can be closed off by turning the fins to the fully closed position. Vents are sometimes referred to as ***registers.***

Independent Vents

Some vehicles have air inlets independent of the HVAC system. These ***independent vents*** are usually installed on the kick panels, **Figure 12-11.** Ram air flows through these ducts when the vehicle is moving. A separate control is used to operate these vents. HVAC system operation has no effect on the operation of these vents. Independent vents are usually found on older cars, trucks, and vans.

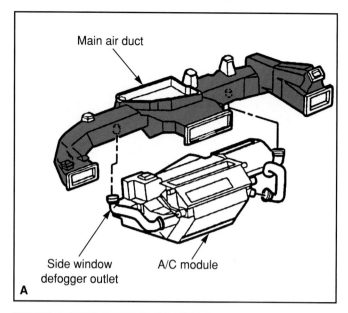

A

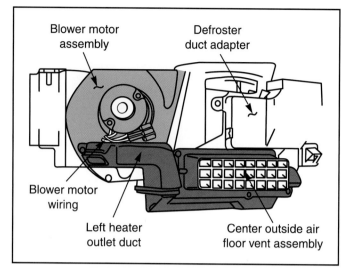

Figure 12-9. *Heater vents are placed under the dashboard. Heated air tends to rise, and flows upward to heat the entire passenger compartment. (DaimlerChrysler)*

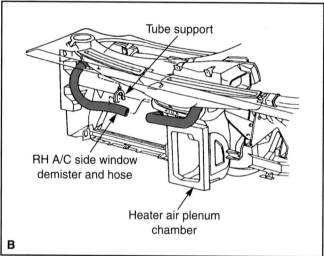

B

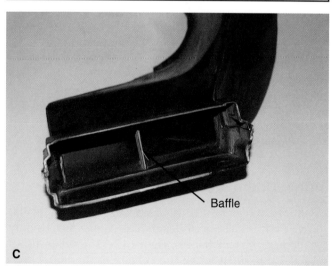

C

Figure 12-8. *A—This hard plastic ductwork connects the case outlets to the dashboard vents. B—Flexible ducts like these are often used to attach the case outlets to the defroster vents. Flexible ducts are also used to connect short sections of hard ductwork. C—Baffles are often installed near the end of a duct to direct air more accurately into the passenger compartment. (General Motors, DaimlerChrysler, Ford)*

Air Doors and Operating Devices

Airflow through the ducts is controlled by *air doors.* Air doors are made of plastic or stamped sheet metal and may have foam seals on their outer edges. Doors pivot on a hinge and are moved by an offset crank arm. The crank arm is operated by linkage attached to one of the operating devices discussed later in this chapter. **Figure 12-12** is an illustration of a typical air door. Doors are installed in the intake ducts, cases, and output ducts. Air doors are divided into blend doors and diverter doors. Every HVAC system has one blend door and several diverter doors.

Blend Door

The *blend door* is positioned to combine cooled air from the evaporator with warm air from the heater core. This door is sometimes called the *temperature door.* Positioning the blend door allows the driver to obtain the desired HVAC system outlet temperature. In **Figure 12-13,** the door is positioned to open or close off the entrance to the heater core.

Positioning the blend door allows some air from the evaporator (or outside air on vehicles without air conditioning) to pass through the heater core. The amount of air warmed by the heater core governs the temperature of the air entering the passenger compartment. If the blend door is completely closed, no air passes through the heater core, **Figure 12-14A.** This position would be selected when no air heating is wanted, such as when the air conditioner is being used on a hot day. If the blend door is opened, air passes through the heater core, **Figure 12-14B.** This position would be used when warm air is needed.

Notice the blend door is adjustable to any position between fully open and fully closed. The blend door position can be varied no matter what mode is being used and

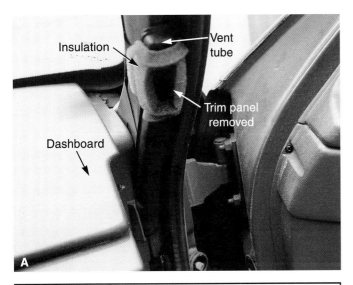

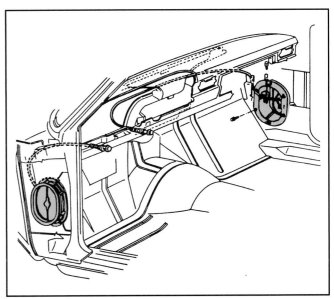

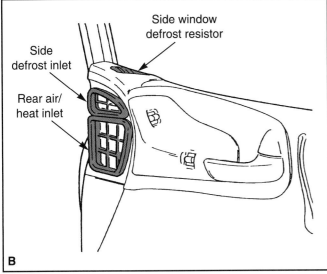

Figure 12-10. *A—Side defroster vents direct heated and dried air against the side windows. This improves side vision in cold, wet weather. B—The outlets of the defroster vents direct air against the windshield. Heated and dried air from the HVAC system remove fog from the inside of the windshield, and helps melt ice on the outside of the windshield. (DaimlerChrysler)*

Figure 12-11. *An independent vent installed on the right kick panel. A knob works through a cable to open and close the vent. (General Motors)*

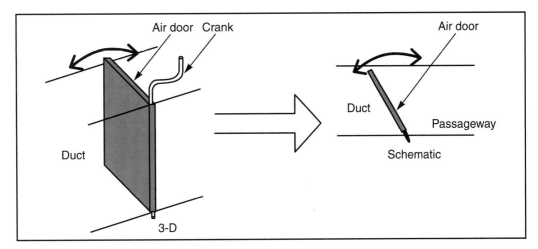

Figure 12-12. *This drawing of an air door shows how the door opens and closes to control airflow. The schematic view shows how the door is usually drawn in a diagram or schematic showing HVAC system airflow.*

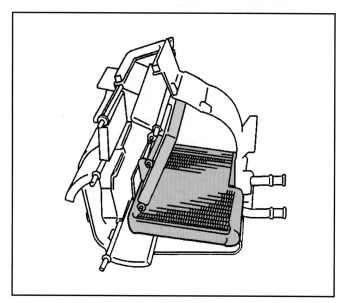

Figure 12-13. *This cutaway view of a blend door shows the relationship of the blend door and heater core. In this position, the door allows air to flow through the heater core, warming it. When the door is positioned against the core, air flows around the core and is not heated. Any position between fully closed and fully open provides a mixture of heated and unheated air. (General Motors)*

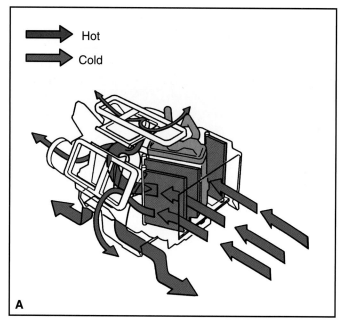

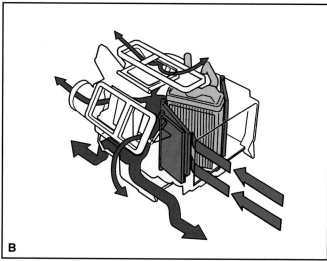

Figure 12-14. *A—Blend door completely closed. No air passes through the heater core. B—Blend door fully open. All air passes through the heater core and is warmed. (Honda)*

independently of whether the heater or air conditioner is operating. For instance, the blend door can be positioned so incoming air is cooled by the evaporator, then reheated by the heater core. This position can be used when warm dry air is needed, such as for defrosting the windows on a cold damp day

Diverter Doors

Diverter doors are used to change the direction of air-flow during various operating modes, **Figure 12-15.** This is referred to as diverting the airflow. Some manufacturers refer to diverter doors as *mode doors.* Diverter doors change airflow direction so it flows out of the floor heater, air conditioner outlet vents, windshield defroster vents, or all vents at the same time. To accomplish these functions, all HVAC systems have a minimum of three diverter doors. The three basic diverter doors are described in the following paragraphs.

Some systems use additional diverter doors. These doors are used to further control the airflow, and are usually used on the HVAC systems of more expensive vehicles.

Air Conditioning-heater Door

The *air conditioning-heater door* or *ac-heater door* directs airflow to either the floor heater outlet or the dashboard vents. This door is sometimes called a *panel-heater door.* The "panel" referred to in the name panel-heater is the dashboard panel. The air conditioning-heater door is an open-close door with no in-between positions. Refer to **Figure 12-15.** If the system has a bi-level mode, the air

conditioning-heater door can be positioned between open and closed. This allows air to go to the air conditioning and heater vents in the bi-level setting.

Heater-defroster Door

The *heater-defroster door* is used to direct air between the heater and the defroster outlets. This diverter door is always downstream from the air conditioning-heater door. In **Figure 12-15,** the relative position of the heater-defroster door is shown. On most systems, this door is either fully open or fully closed. When the system features a bi-level mode, the door may be positioned midway between open and closed to allow flow to the heater and defroster outlets. On a few systems, the doors are arranged so one door diverts air from the air conditioning passages to the heater outlets and the other door diverts air from air the conditioning passages to the defroster vents.

Recirculation Door

The **recirculation door,** sometimes called the **air inlet door,** closes off the outside air supply to recirculate the air in the passenger compartment, **Figure 12-16.** The recirculation door has two positions. One door position allows outside air to enter the HVAC system. When the door is in the other position, air is drawn from the passenger compartment and reused, or recirculated. To purge carbon monoxide, some outside air enters the passenger compartment even when the door is in the recirculation position.

Sometimes two recirculation doors are used. One door allows outside air in, while the other allows passenger compartment air to recirculate. **Figure 12-16** is a schematic of a system with two recirculation doors. On modern vehicles, vacuum diaphragms operate these doors. This door is always upstream of the blower, the ducting, and doors. Most air recirculation doors are open-close doors with no intermediate positions.

Heater Restrictor Door

The **heater restrictor door** completely blocks off the heater core to ensure no heat gets into the passenger compartment during maximum air conditioner operation. **Figure 12-16** illustrates a system with a heater restrictor door.

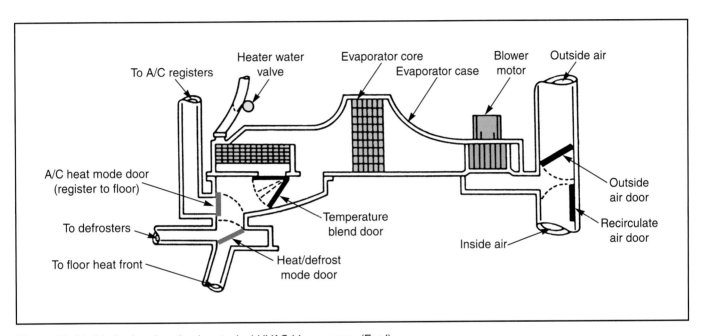

Figure 12-15. *Mode door location in a typical HVAC blower case. (Ford)*

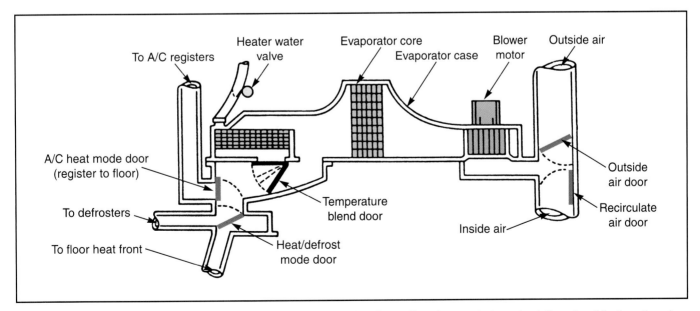

Figure 12-16. *This schematic shows a system with two recirculation doors. One door controls recirculating air, while the other door controls the entry of outside air. The air conditioning-heater door directs air to either the upper (air conditioning) outlets or the floor (heater) outlets. This door has only two positions. The heater-defroster door directs air to either the heater or the defroster outlets. This door can be positioned between the full heater and full defrost positions. (Ford)*

Dual Zone Blend Doors

Some late model HVAC systems have a feature called **Dual Zone,** with separate temperature controls for the driver and passenger. The Dual Zone system has separate driver and passenger blend doors, **Figure 12-17.** Separate driver and passenger controls allow the front seat occupants to move each blend door independently of the other. Moving the passenger blend door raises or lowers the temperature of the air exiting the passenger side vents. Moving the driver blend door raises or lowers air temperature at the vents on the driver's side. Some Dual Zone systems also have a main blend door.

Air Bypass Door

The **air bypass door** is used during maximum air conditioning operation to direct additional cooled air around the blend door. This aids cooling under extremely hot and humid conditions.

Upper and Lower Mode (Bi-Level) Doors

Two doors are sometimes used in place of a single air conditioning-heater door. These are called **upper and lower mode doors.** One door is opened to direct air to the air conditioner outlets. The other door opens to allow air into the heater and defroster outlets. When both doors are open, air flows to the air conditioner, heater, and defroster outlets. Upper and lower mode doors are shown in **Figure 12-18.**

Door Operating Devices

To operate the blend and diverter doors, various operating devices are used. The most common types are covered in the following paragraphs.

Cables and Levers

Cables were widely used on older vehicles. A cable operates the blend door on many modern vehicles. A cable consists of a wire inside a covering or sheath, **Figure 12-19.** The wire can slide back and forth inside the sheath. The ends of the sheath are attached solidly to non-moving parts of the vehicle, usually with brackets formed into the sheath. The wire is attached to the dashboard **lever** and the door crank. Moving the lever causes the cable to move the door crank.

Vacuum Diaphragms

Vacuum diaphragms, Figure 12-20, are sometimes used to operate various diverter doors. They are sometimes called **vacuum motors.** The diaphragm consists of a sealed chamber connected to the engine manifold vacuum

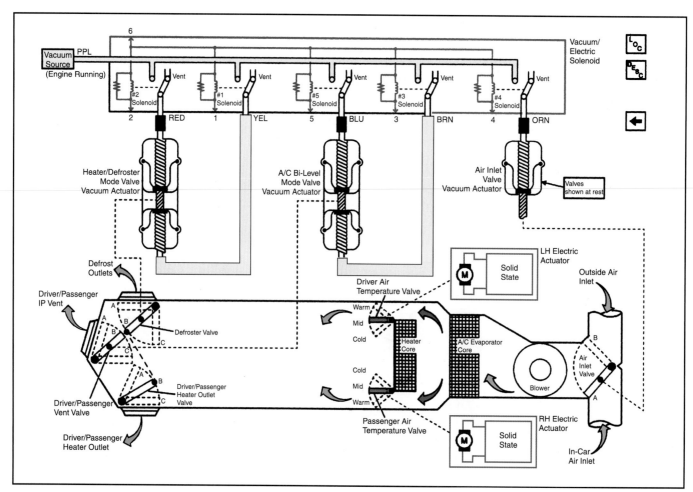

Figure 12-17. An HVAC Dual Zone air delivery system. The purpose of the extra doors is to better control the comfort levels between the driver and front seat passenger. (General Motors)

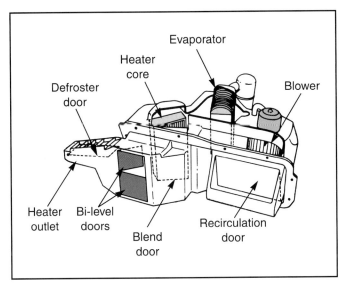

Figure 12-18. *A system with upper and lower mode doors. The upper and lower door concept is widely used on General Motors vehicles. (General Motors)*

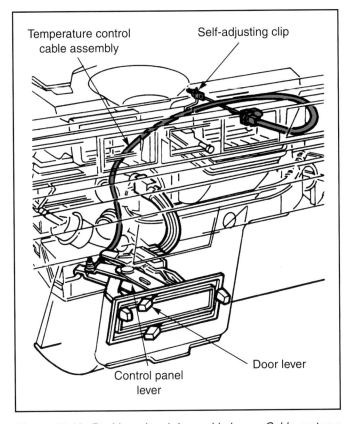

Figure 12-19. *Dashboard and door cable levers. Cable systems have been largely replaced with electronic and vacuum controls. (Ford)*

through hoses. The hoses are directed through the HVAC control system, which applies or releases vacuum according to driver input.

The chamber is connected to the door through linkage. A spring inside of the diaphragm opposes movement of the diaphragm. Applying vacuum causes the diaphragm to move against spring pressure to move the door. When

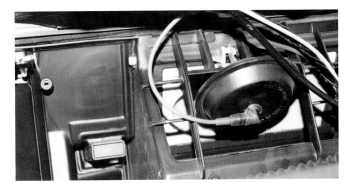

Figure 12-20. *This illustration shows a typical vacuum diaphragm used to operate a diverter door. Notice how the straight line movement of the diaphragm rod is converted to rotary motion to move the door.*

vacuum is removed, the spring returns the door to its original position. Some diaphragms have more than one spring or a variable rate spring. This spring arrangement allows the diverter door to be partially opened when a certain amount of vacuum is applied, and to open completely when more vacuum is applied.

Some diaphragms are double diaphragm types. Using vacuum on both sides of the diaphragm allows it to be positioned exactly, or to perform more than one function. Vacuum diaphragms are sometimes controlled by electric solenoids. Solenoids will be covered in more detail in Chapter 13.

Stepper Motors

On some modern vehicles the blend door is operated by an electric motor. These motors are called **stepper motors.** Unlike most electric motors, stepper motors move only a small amount, or step, each time they are energized. This small and precise movement allows them to accurately position the blend door. **Figure 12-21** shows the schematic of a stepper motor. The motor is controlled from the temperature lever on HVAC control panel.

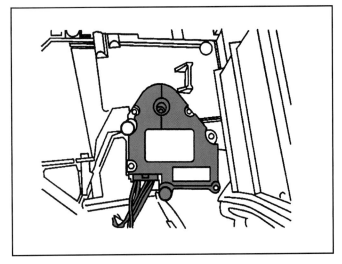

Figure 12-21. *A small electric stepper motor used to operate an air door. These motors are sometimes called actuators or drivers. (General Motors)*

Cabin Filters

Some newer vehicles have **cabin filters,** sometimes called *intake air filters*. Cabin filters are always installed ahead (upstream) of the evaporator. The position of the cabin filter means all incoming air must pass through the filter. The filter traps dust, pollen, and in some cases, odors, before they can reach the evaporator and passenger compartment. Clean, incoming air reduces the chance of the evaporator clogging and increases the comfort level in the passenger compartment.

Cabin filters have paper filtering elements and somewhat resemble engine air filters. They may be installed in the passenger compartment, at the bottom of the HVAC case near the firewall, behind or under the glove compartment, or at the air intake duct under the cowl. Like any filter, cabin filters can become plugged and must be replaced every year or 12,000 miles (19 000 km). The filter usually slides into place and can be reached after a cover panel has been removed. A few light trucks have cabin filters made of plastic mesh. Washing the filter with water removes most of the dirt, and the filter can then be dried and reinstalled.

Evaporator Water Drains

When the evaporator is operating, large amounts of condensed water accumulate on its surface. This water drips to the bottom of the evaporator case and is removed through a water drain. The **evaporator water drain** is always located as close to the exact bottom of the case as possible. Some water drains consist of a small hole in the bottom of the case, but most drains have a flexible rubber or plastic flap. The flap remains closed until water begins to collect at the bottom of the case. When enough water collects, its weight pushes the flap open and the water drains out. The flap closes, keeping dust and other debris from entering the evaporator case.

On modern vehicles, the evaporator is often installed in the passenger compartment. A small hose will be installed on the drain opening to direct water outside. Some vehicles use two drains and two hoses.

Summary

The blower consists of a blower wheel and motor. It is usually ahead of the evaporator and heater core. The blower wheel is called a squirrel cage blower. It acts as a centrifugal air pump. The direct current blower motor is connected to the vehicle electrical system through the ignition switch and HVAC electrical controls. The blower wheel is attached to a shaft at the end of the motor.

Resistors are used to control blower motor speed. Energizing different windings in the motor controls motor speed on a few HVAC systems. Modern HVAC systems usually have four blower speeds.

Air ducts are tubes or passageways for the HVAC system air. Air enters behind the windshield and is directed by the blower to the heater and evaporator cases.

The heater case contains the heater core and the evaporator case contains the evaporator. From the heater and evaporator case, ducts direct air to the vents. Some vehicles have independent vents that allow ram air to enter when the vehicle is moving.

Air doors control airflow through the ducts. Doors pivot on hinges and are installed in the intake ducts, cases, and output ducts. Every HVAC system has one blend door and several diverter doors. The blend door combines cooled air from the evaporator with warm air from the heater core. The driver positions the blend door to obtain the desired HVAC system outlet temperature.

Diverter doors are sometimes called mode doors. The three main diverter doors are the recirculation door, the air conditioning-heater door, and the heater-defroster door. Some systems use additional doors. Door operating devices include cables, vacuum diaphragms, and electric motors.

Review Questions—Chapter 12

Please do not write in this text. Write your answers on a separate sheet of paper.

1. If the incoming air passes through the heater core before reaching the blower, the heater core is _____ of the blower.

2. The rotating blower wheel pulls air into its _____.

3. The blower relay is used to select which blower speed?
   (A) First speed.
   (B) Second speed.
   (C) Third speed.
   (D) High speed.

4. What is the purpose of having the blower run at all times, even when the HVAC system is not on?

5. Most of the intake ductwork is formed by the _____.

6. Output ducts can be _____ tubes or _____ hoses.

7. How many blend doors are used on the average HVAC system?

Matching

Match the air door with the description that best fits.

_____ 8. Most likely to be operated by a cable.

_____ 9. Used to block off incoming air

_____10. Used to select the defroster
   (A) Recirculation door
   (B) Heater-defroster door
   (C) Heater-AC door
   (D) Blend door

ASE Certification-Type Questions

1. The squirrel cage blower wheel is attached to the:
   (A) blower case.
   (B) blower motor armature.
   (C) resistor assembly.
   (D) blower motor stationary windings.

2. Technician A says the fan is usually upstream from the evaporator. Technician B says the blower is usually downstream from the heater core. Who is right?
   (A) A only.
   (B) B only.
   (C) Both A and B.
   (D) Neither A nor B.

3. On modern vehicles the blower motor runs whenever the _____ is running.
   (A) engine
   (B) compressor
   (C) stepper motor
   (D) water valve

4. All of the following statements about HVAC blower speed control are true, *except:*
   (A) electrical resistors are used to control blower motor speed.
   (B) selecting high blower speed causes the current to flow through a large resistor.
   (C) on a few HVAC systems, energizing different windings in the motor controls motor speed.
   (D) some modern HVAC blowers are controlled through a computer.

5. Technician A says the intake ductwork is not a separate duct assembly. Technician B says the intake ductwork is downstream from the blower. Who is right?
   (A) A only.
   (B) B only.
   (C) Both A and B.
   (D) Neither A nor B.

6. All of the following parts may be included in the HVAC case assembly, *except:*
   (A) heater core.
   (B) blower.
   (C) vents.
   (D) evaporator.

7. All of the following statements about air doors are true, *except:*
   (A) air doors control airflow leaving the vents.
   (B) air doors pivot on a hinge.
   (C) every HVAC system has one blend door.
   (D) every HVAC system has several diverter doors.

8. Technician A says not every HVAC system has a air-heater conditioner door. Technician B says some HVAC systems use two recirculation doors. Who is right?
   (A) A only.
   (B) B only.
   (C) Both A and B.
   (D) Neither A nor B.

9. Which of the following air doors is used to select airflow from the heater outlet or air conditioner vents?
   (A) Blend door.
   (B) Upper and lower mode doors.
   (C) Heater-defroster door.
   (D) Recirculation door.

10. Which of the following air doors is *least* likely to be operated by a vacuum diaphragm?
    (A) Blend door.
    (B) Recirculation door.
    (C) Air-heater conditioner door.
    (D) Heater-defroster door.

Most manual control heads on modern vehicles use some electronics. The switches on this manual control head are potentiometers.

Chapter 13

Manual HVAC Controls

After studying this chapter, you will be able to:
- ❏ Identify the purposes of HVAC control systems.
- ❏ List and describe common HVAC control system modes.
- ❏ Identify the major parts of a manual HVAC control system.
- ❏ Explain how a manual HVAC control system operates.

Technical Terms

Temperature control	Blower control device	Vacuum restrictor
Humidity	Mode control device	Vacuum source
Airflow control	Temperature control device	Reserve tank
HVAC control system modes	Lever	Check valve
Vent	Turnbuckle	Auxiliary vacuum pump
Bi-level	Vacuum control valves	Blower speed switches
Defrost	Rotary vacuum valves	High blower relay
Control panels	Sliding valves	Vacuum solenoids
Control head	Pushbutton vacuum controls	Stepper motor controls

This chapter explains the operation and modes of common manual HVAC systems. Without knowing how control systems work, you cannot begin to diagnose or repair it or other parts of the HVAC system. This chapter will also prepare you for Chapter 14, which covers automatic temperature controls. Since this chapter is about controlling the various devices in the HVAC system, you must be familiar with the information covered in the earlier chapters. Review any information as necessary, especially Chapters 5, 8, 11, and 12.

Control System Purposes

While not a common source of problems, HVAC control systems do fail. Since they tend to fail less frequently than other HVAC components, it is sometimes difficult for technicians to become familiar with their operation and common problems. As you will learn over the next two chapters, HVAC controls can be simple manual devices or complex systems that make decisions based on sensor inputs.

Temperature Control

The HVAC system controls the passenger compartment heating and cooling components to maintain the temperature desired by the occupant(s). *Temperature control* is accomplished by operating the following HVAC system components:

❑ Compressor clutch.
❑ Heater shutoff valve.
❑ Blend door.
❑ Recirculation door.
❑ Blower motor.

Humidity Control

One of the primary jobs of any HVAC system is to control the amount of humidity in the passenger compartment air. Since the human body produces a great deal of water vapor, in most driving situations the humidity must be reduced. *Humidity* is reduced by controlling the operation of the following HVAC components:

❑ Compressor clutch.
❑ Blend door.
❑ Blower motor.

Airflow Control

Airflow control is the modulation and directing of air as it exits from the HVAC system. The control of airflow allows the vehicle occupant(s) to dictate where and at what speed the air comes into the passenger compartment. Airflow control is performed by controlling the following:

❑ Blower motor.
❑ Recirculation door.

❑ Air conditioner-heater door.
❑ Heater-defroster door.
❑ Other doors when used.

There is almost no situation when the HVAC system controls only one of the listed conditions. When the refrigeration system is operating, the incoming air is both cooled and dehumidified. If the air must be dehumidified in cold weather, the incoming air is cooled and then reheated.

Some control operations are performed independently of the HVAC control system. Examples are maintaining evaporator core temperature, which is handled internally by the refrigeration system; and heater core temperature, which is determined by the engine cooling system. The HVAC control system is designed to work with independent controls as needed.

> **Note: On a few vehicles, the driver operates the compressor clutch and in the case of older cars, the heater shutoff valve through dashboard controls. Driver-operated heater shutoff valve controls are not used on newer vehicles.**

HVAC Control System Modes

All the control operations are combined to obtain various HVAC operating states or modes. *HVAC control system modes* can be defined as the way the HVAC system does something, in this case, delivering the proper kind of air for a specific need. This need may be for warm air, cold air, dehumidified air, outside air, or a combination. The driver selects the modes in a manual system, and the control system selects the modes on an automatic system. Modes are referred to as *functions* in some service literature. **Figure 13-1** shows a typical manual system and its doors in the *off* mode.

This section covers common modes on a manual control HVAC system. On actual vehicles, the modes may be called other names and additional modes may be available. These, however, are the basic modes found on all HVAC systems. Vehicles without air conditioning will usually have only the heat, defrost, and vent modes.

> **Note: The illustrations in this section show a common arrangement of air passages and diverter doors. These illustrations also assume the HVAC system has a heater shutoff valve. As noted in Chapter 12, not all systems will have this exact arrangement. Always consult the appropriate service manual for the exact airflow diagrams of a particular system.**

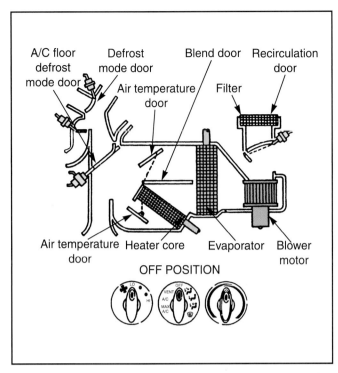

Figure 13-1. *Manual control air conditioning blower case. The controls are in the off position. (Ford)*

Air Conditioning (Inside Air Recirculating)

This mode is sometimes called MAX AIR or RECIRC. It does not use outside air, **Figure 13-2.** Air reenters the evaporator from inside the passenger compartment. The air is cooled and dehumidified as it passes through the evaporator core. Moisture condenses on the cold evaporator core and drains out the bottom of the evaporator case. The cool, dry air is directed past the blend door and through the air conditioning vents. The driver can regulate temperature by sending some air through the heater core using the blend door. The heater shutoff valve is closed and the compressor clutch is engaged.

To reduce the chance of carbon monoxide poisoning, the closed recirculation door allows about 5% outside air to enter the passenger compartment. This is provided by installing a stop that keeps the door from closing fully or by drilling a hole in the door. Some intake systems have a separate intake air bypass opening.

Air Conditioning (Outside Air Entering)

This mode is sometimes called AC, NORMAL, or NORM. It is similar to MAX AC except outside air is cooled. In this mode, incoming air is blown through the evaporator core. The refrigeration system is operating and the incoming air is cooled. As in MAX AC mode, moisture condenses and drains out the evaporator case. The blend door has closed and isolated the heater core. The cool, dry air passes by the closed blend door and exits at the dashboard vents. The vehicle occupant(s) can also redirect the cooled air to the floor, defrost, or a combination of all three vents.

On some vehicles, the heater shutoff valve is open to allow coolant flow through the heater core. The driver can move the blend door to allow some air through the heater core if the temperature becomes too cold. The compressor clutch is engaged and may cycle according to evaporator pressure or temperature. In this mode, the recirculation door is open to allow outside air to enter the passenger compartment through the evaporator. **Figure 13-3** shows the system airflow in this mode.

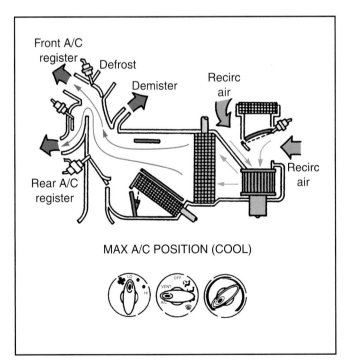

Figure 13-2. *MAX AC operation. The doors are positioned to allow airflow from the interior to be recooled. Some air blows from the demister at all times. (Ford)*

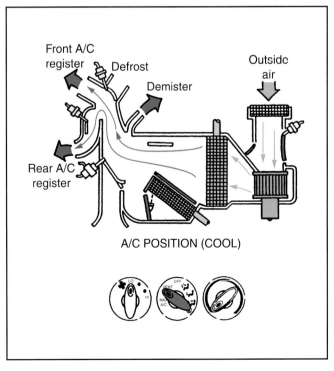

Figure 13-3. *Normal air operation. This mode operates the same as MAX AC except outside air is pulled in and cooled. (Ford)*

Vent

The major difference between the *vent* mode and the air conditioner modes is the refrigeration system is not operating. Incoming air passes through the evaporator core. Since the compressor clutch is disengaged, the evaporator has no effect on the air. From the evaporator, the air passes by the blend door and exits at the dashboard air conditioning vents, **Figure 13-4.** The driver can warm the air by moving the blend door to allow some air to pass through the heater core. In this mode, the heater shutoff valve and recirculation doors are open.

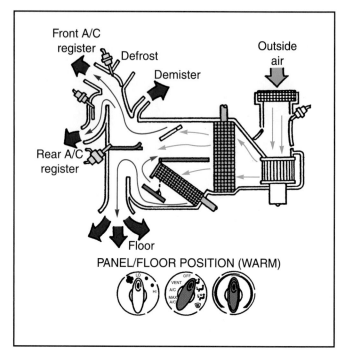

Figure 13-5. *In this position, air is cooled and heated and exits through the heater, air conditioner, and defroster vents. (Ford)*

Bi-Level Heat and Defrost—Cold Weather

In **Figure 13-6,** the system is delivering warm dry air to the side window defrost and floor vents. The diverter door(s) can be repositioned to allow all air to exit at the defrost, dash, or floor. The blend door is moved by the driver to increase or reduce the amount of incoming air passing through the heater core.

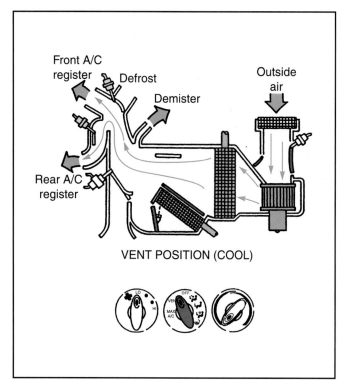

Figure 13-4. *In the Vent mode, door positions allow air to flow through the evaporator core. However, the refrigeration system is off, and no cooling occurs. (Ford)*

Bi-Level

Bi-level mode is similar to the other air conditioning modes, except for the position of the diverter door(s). Although this mode is called bi-level, the air exits at all three major outlet points. Incoming air passes through the evaporator core. The compressor clutch is engaged and the refrigeration system is operational. The air is cooled and dehumidified. The heater shutoff valve is open and hot coolant flows through the heater core. The driver can send some or all the air through the heater core by moving the blend door. The diverter doors are positioned so the air exits at the dash, floor, and defrost vents. Diverter door positions during bi-level operation are shown in **Figure 13-5.** Most HVAC systems keep the recirculation door open in this mode.

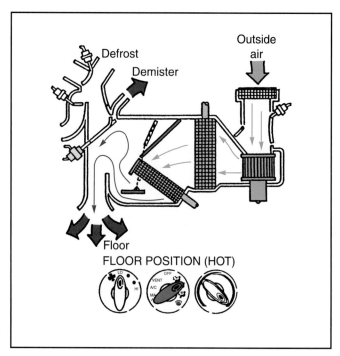

Figure 13-6. *In this figure the control system has positioned the air doors to provide bi-level heating and defrosting. (Ford)*

In this mode, the control system has positioned the doors to provide heating and window defrosting. The recirculation door is positioned to draw in outside air and the heater shutoff valve is open. The incoming air is then directed through the evaporator if the vehicle has air conditioning. If the temperature is several degrees above freezing, the refrigeration system will be operating and moisture is removed from the air. If the temperature is below freezing, the compressor clutch is not engaged. The air passes through the heater core and picks up heat from the coolant. The diverter door(s) is positioned so that warmed air exits at the floor outlets under the dashboard.

The passenger(s) can adjust the temperature control, which moves the blend door to increase or reduce the amount of incoming air passing through the heater core. This varies the temperature of the air entering the passenger compartment. As the cooling system warms up, the occupant(s) will usually want to reduce the percentage of air passing through the heater core. The direction controls can also be adjusted to direct the air to the defrost vents. The position of the diverter door(s) determines whether the dry heated air exits at the defrost, dash, or floor ducts.

Bi-Level Heat and Defrost—Mild Weather

In this mode, the control system adjusts the various doors to reduce the amount of heated and dehumidified air, **Figure 13-7.** If the vehicle has air conditioning, the refrigeration system is operating. The heater shutoff valve is open and engine coolant flows through the heater core. Incoming air passes through the cold evaporator core. The air is cooled and moisture condenses on the core. The

cool, dry air then passes through the heater core and is warmed by the coolant. The compressor clutch may cycle according to evaporator pressure or temperature. If the vehicle does not have air conditioning, the incoming air is heated by the heater core.

Defrost Mode

In the **defrost** mode, the doors are adjusted to channel all the air to defog the windshield and side windows. Unless the temperature is below freezing, the air conditioning compressor is operating. If the vehicle does not have air conditioning, the incoming air is heated by the heater core. The heater shutoff valve is open and coolant flows through the heater core. Outside air passes through the evaporator. The dry air may pass through the heater core if the driver requests heated air. From the heater, the air is channeled to the defrost registers, **Figure 13-8.**

Manual Control Systems

If a vehicle has only a heating and ventilating system, it always has a manual control system. Many vehicles with air conditioning systems have manual control systems. Modern versions of these systems often have electromechanical or electronic parts. Simple heater-only systems often use cables alone. Most manual HVAC controls select mode and temperature by a combination of cable, vacuum, and electrical or electronic parts.

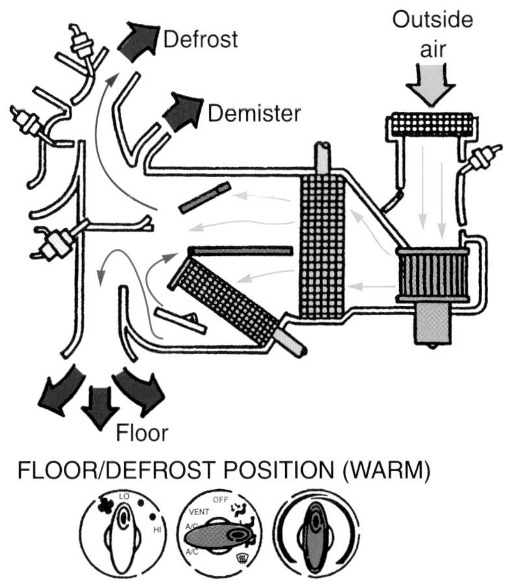

Figure 13-7. *This mode is similar to Figure 13-6, except one of the secondary doors allows warm air to recirculate in the case, which provides warmed air, but not hot air. (Ford)*

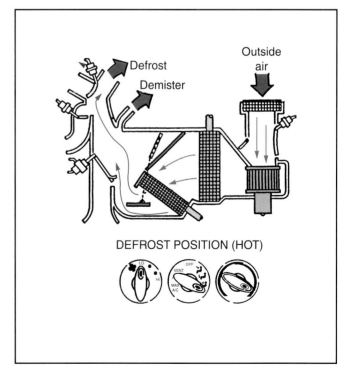

Figure 13-8. *The defrost position sends air to the defrost and demister vents. If the ambient air temperature is above freezing, the refrigeration system is operating. (Ford)*

Control System Parts

The components of a manual control system are relatively simple, but require some study to see how they fit together to produce the various operating modes. These parts can be mechanical, vacuum, or electrical, as explained in the following paragraphs.

Control Panels

HVAC **control panels** are mounted in the dashboard and provide all of the driver and passenger control of the HVAC system. Some automobiles have separate control panels for the driver and front seat passenger. Vans with rear-mounted heaters or air conditioning evaporators often have separate rear seat controls. Control panels may be referred to as a **control head** or *AC head*. Manual control panels consist of at least three control devices:

- ❑ **Blower control device.** The blower control can be a lever, rotary switch, or a set of pushbuttons. It is an electrical connection to the blower motor resistors and high blower relay.
- ❑ **Mode control device.** This can be a lever, a series of pushbuttons, or a rotary switch. This device operates the vacuum valves to correctly position the air doors. It also completes the circuit to the compressor clutch and blower switch.
- ❑ **Temperature control device.** The temperature control device is usually a lever or sliding switch. On some vehicles a rotary switch is used. This device operates the cable or stepper motor that moves the blend door. It may also operate the heater shutoff valve.

The three controls used in a typical electromechanical control panel are shown in **Figure 13-9**. The figure shows the rear of the panel. Vacuum and electrical switches are visible.

Other HVAC control panels are electronic, as shown in **Figure 13-10**. In this figure, there are no vacuum or cable connections. All outputs are electrical signals.

Multiple Control Panel Systems

Many modern vehicles have separate control panels that can be operated by the rear seat passengers, as shown in **Figure 13-11**. These devices are usually electronic to eliminate problems with cables or vacuum lines. Basic control operations are the same as for the main control panel. Multiple control panels are used on vehicles with Dual Zone climate control and vans with rear heating and air conditioning.

The control panels are attached to the HVAC control devices explained in the following paragraphs. The control panel usually contains one or more bulbs for nighttime illumination. These bulbs are usually wired to the other instrument panel lights and are dimmed or brightened at the headlight switch. The HVAC control panels of many newer vehicles use light emitting diodes (LEDs) or vacuum fluorescent displays to indicate modes and other information.

Cable Levers

Many manual control systems have at least one **lever** to operate the blend door cable. The lever, **Figure 13-12**, pivots on a central bearing. The lever extends out of a slot

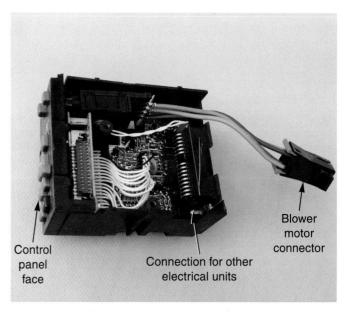

Figure 13-10. *This figure shows the internal portions of a fully electronic control panel. There are no vacuum or cable connections. The heavy wires power the blower motor.*

Figure 13-11. *Many modern vehicles have separate control panels for the front and rear seat passengers. (General Motors)*

Figure 13-9. *A typical electromechanical control panel. (Jack Klasey)*

in the panel. Moving the lever in one direction causes the cable to move in the opposite direction. Since the cable sheath is solidly anchored, the cable moves and operates the blend door.

Some levers are combination units that move the blend door and operate a vacuum control valve or electrical switch. Some heater shutoff valves on older cars are cable operated. Controlling the opening of the shutoff valve varies the coolant flowing through the heater core and therefore, the amount of heat produced by the core.

Most cables have an adjusting device to compensate for manufacturing tolerances and cable wear, **Figure 13-13.** This adjusting device may be located on the lever, or on the door crank end of the cable. Some older cables have a **turnbuckle** to adjust the cable. The turnbuckle is usually located about halfway along the cable.

Vacuum Control Valves

As was discussed in Chapter 12, vacuum diaphragms are often used to move the HVAC system's various doors. **Vacuum control valves** operate the vacuum diaphragms by controlling the connection between the diaphragms and the vacuum source. Older systems used **rotary vacuum valves, Figure 13-14A.** **Sliding valves** are often used as part of an automatic temperature control system. Rotary or sliding vacuum valves are attached by linkage to levers or switches on the control panel. Other valves are pushbutton activated, **Figure 13-14B.** **Pushbutton vacuum controls** may contain internal passages that direct vacuum as

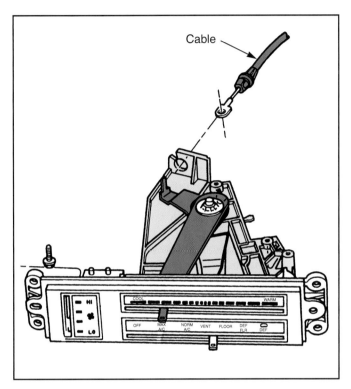

Figure 13-12. *A typical cable and lever setup. Dashboard lever movement is transferred through the cable and moves the lever at the air door. (Ford)*

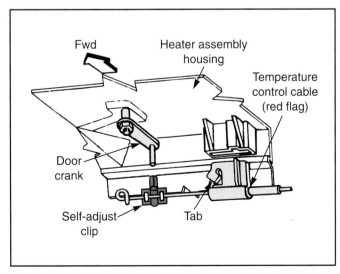

Figure 13-13. *An adjusting clip. The clip can slide along the cable to adjust the amount of blend door movement. (DaimlerChrysler)*

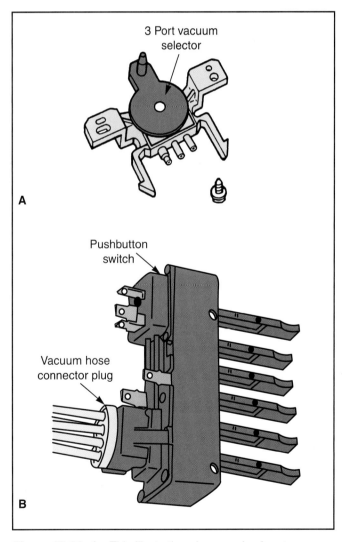

Figure 13-14. *A—This illustration shows a simple rotary vacuum valve. Internal passages in the valve redirect vacuum. Vacuum control valves such as this one were widely used until recently. B—A pushbutton vacuum switch. The pushbutton assembly also operates the electrical components. Pushbutton vacuum controls are relatively rare. (Ford, DaimlerChrysler)*

needed. Other pushbutton vacuum controls move linkage that operates a sliding or rotary valve.

All connections between the vacuum diaphragms, control system, and vacuum source are made through flexible neoprene rubber hoses, **Figure 13-15.** The hoses are color-coded to make tracing easier. T-fittings are used to direct vacuum to two or more vacuum devices.

Often, a **vacuum restrictor** is placed in the vacuum line leading to a particular vacuum door, **Figure 13-16.** This is done to reduce the noise of the door closing and discomfort caused by abrupt changes of airflow direction. The restrictor is a fitting with a small air passage, or orifice. The orifice permits only a small amount of air to pass through the fitting. This causes the vacuum to build up slowly in the diaphragm, and the door will close gently. Some systems use a porous material as the restrictor. Another purpose of the restrictor is to reduce vacuum loss to the rest of the HVAC control system when a large vacuum diaphragm is activated.

Figure 13-15. *HVAC vacuum hoses are color-coded for easier tracing, much like vehicle wiring. Service manuals have hose schematics to assist in diagnosis.*

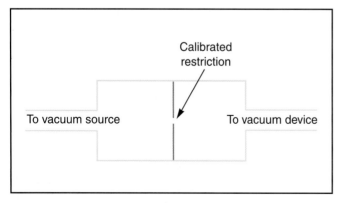

Figure 13-16. *A vacuum restrictor slows down the flow of air through the vacuum hose. Reducing airflow causes the diaphragm to apply slowly and gently.*

Vacuum Sources

The primary **vacuum source** for the HVAC vacuum controls is the engine intake manifold. As the engine operates at idle, some pistons are always moving downward in their cylinders and attempting to draw in air through the open intake valves. With the throttle plate closed, the engine cannot draw in enough air to replace what is taken into the cylinders. This creates a vacuum in the intake manifold. When the throttle is opened, manifold vacuum drops, but because the engine speeds up in response, there is always some vacuum in the intake manifold. **Figure 13-17** illustrates this process. If the engine is placed under a heavy load, the vacuum may drop to almost zero. However, there is always some vacuum present or air would not enter the engine.

Intake manifold vacuum is put to use in several ways. Most vehicles use engine vacuum to operate the power brake booster, cruise control, emission control valves, and positive crankcase ventilation (PCV) system, as well as the HVAC diaphragms. The amount of vacuum changes with throttle opening and demand for vacuum changes as different vacuum devices are operated.

To keep these variations in vacuum from affecting diverter door operation, almost all HVAC systems use a **reserve tank** and **check valve.** Engine operation creates a vacuum in the reserve tank. The check valve is a one-way valve that permits air to exit toward the intake manifold as high vacuum is produced, but closes if the manifold vacuum drops. The vacuum in the reserve tank continues to hold the various diaphragms in position. **Figure 13-18** is a photograph of a reserve tank.

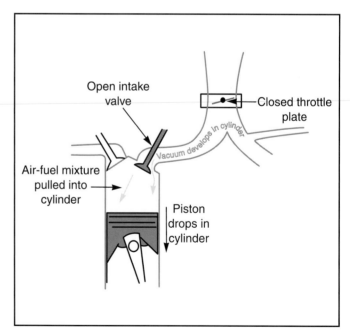

Figure 13-17. *This figure shows how an internal combustion engine develops vacuum in the intake manifold. All gasoline and some diesel engines develop vacuum this way. Opening and closing the throttle plate (accelerating and decelerating) cause variations in vacuum. These variations must not reach the HVAC control system.*

Figure 13-18. *Vacuum tanks often resemble the one shown here. However, vacuum tank size, shape, and placement vary widely. Some vacuum tanks are installed on the HVAC case in the passenger compartment.*

Diesel engines do not have throttle plates and cannot develop vacuum in the intake manifold. On some vehicles with small gasoline engines, the engine load is too great for its intake manifold to be a reliable source of vacuum. These vehicles are equipped with a separate **auxiliary vacuum pump** to provide vacuum. Auxiliary vacuum pumps are usually diaphragm types, similar in operation to mechanical fuel pumps. The pump is operated by an electric motor that moves the diaphragm up and down. One-way check valves in the pump assembly keep the air flowing in the proper direction. A typical pump is shown in **Figure 13-19.** On some cars, the auxiliary vacuum pump is used to supply extra vacuum to the brake booster only and does not assist other vacuum systems.

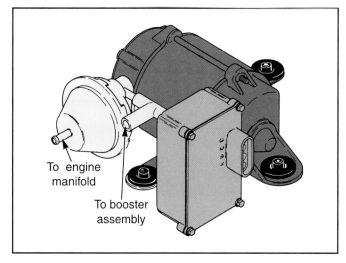

To engine manifold

To booster assembly

Figure 13-19. *If the vehicle has a diesel engine or a relatively small gasoline engine, an auxiliary vacuum pump may be installed. The pump provides additional vacuum to operate the HVAC system, brake booster, and cruise control.*
(General Motors)

Electrical and Electronic Controls

All manual HVAC controls use electrical switches to control blower motor speed. Later model HVAC systems use electrical solenoids to control the operation of the vacuum diaphragms. Some current HVAC systems have a stepper motor that positions the blend door. Controls for these devices are discussed in the following paragraphs.

Blower Speed Switches

Blower speed switches are used on all vehicles with manual control HVAC systems. Most manual blower speed switches have four speeds, and a few late-model systems have five speeds. Some older systems, or heater only systems, may have only three blower speeds.

Several types of blower switches are used. Sliding contact switches were commonly used on older HVAC systems, and are still found on some vehicles. A lever switch is shown in **Figure 13-20.** All of these switches use internal contacts to complete an electrical circuit between a hot wire from the ignition switch and the blower motor. Switch position determines how current flows through the resistor assembly. Resistors with more electrical resistance will reduce the current flowing in the circuit, and reduce blower motor speed.

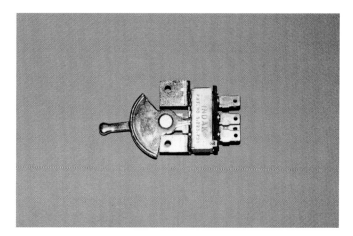

Figure 13-20. *Electrical switches such as this one often have mechanical contacts, sometimes called points. Moving the switch lever selects different sets of contacts, varying current flow.*

Many HVAC blower speed control systems have a **high blower relay.** The purpose of the high blower relay is to allow current flow to bypass the resistor assembly. Bypassing the resistors allows full current to drive the blower motor. The high blower relay is usually mounted near the resistor assembly, **Figure 13-21,** or may be installed in the vehicle fuse and relay box. **Figure 13-22** is a schematic of a typical modern blower control system with four blower speeds. Systems with only three blower speeds usually do not have a high blower relay.

Figure 13-21. *This figure shows a common older type of high blower relay. It is usually located near the resistor assembly at the firewall. On later model vehicles, blower relays may be mounted separately, or in the master fuse and relay box.*

Vacuum Solenoids

Vacuum solenoids are usually on-off devices. The vacuum solenoid consists of a vacuum valve attached to a solenoid plunger. A spring holds the valve against its seat, closing the valve. **Figure 13-23** is a cutaway view of a vacuum solenoid. Energizing the solenoid moves the valve to the open position, allowing vacuum to reach the diaphragm. De-energizing the solenoid closes the valve. Most solenoid and valve assemblies have a bleed hole that allows atmospheric pressure to enter the line to the diaphragm. On-off switches in the control panel operate the solenoids. The solenoids may be installed in the control panel, or mounted in a separate control device. **Figure 13-24** shows a solenoid pack used on a popular automobile.

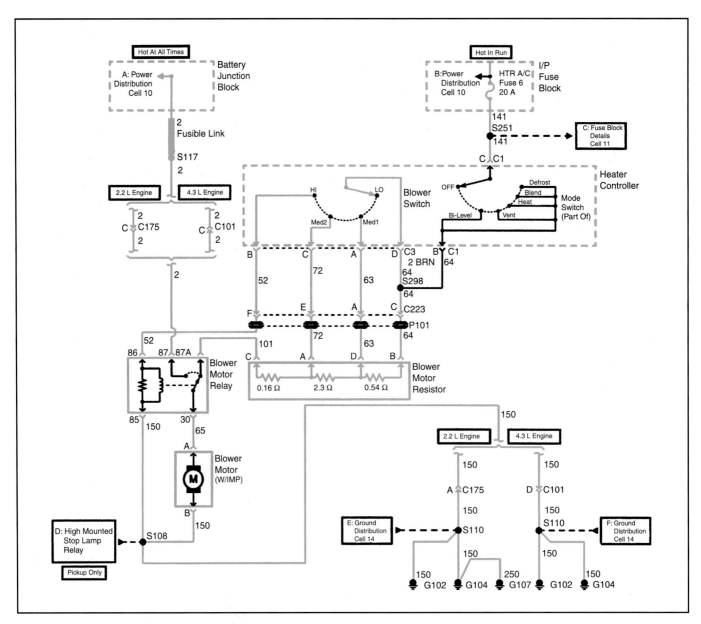

Figure 13-22. *This is a schematic of common blower relay and resistor wiring. Trace the schematic to see how the relay controls current flow into the blower motor. (General Motors)*

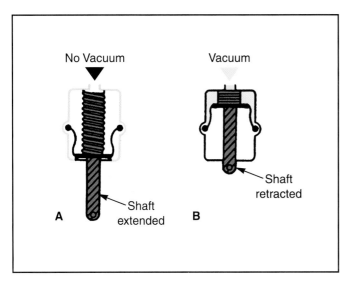

Figure 13-23. *This simple illustration of a vacuum solenoid shows how the solenoid controls vacuum to a diaphragm. A—When the solenoid is de-energized, manifold vacuum is cut off, and a passage is opened to exhaust the vacuum in the diaphragm. B—When the solenoid is energized, manifold vacuum is sent to the diaphragm, causing it to move the proper air door. (General Motors)*

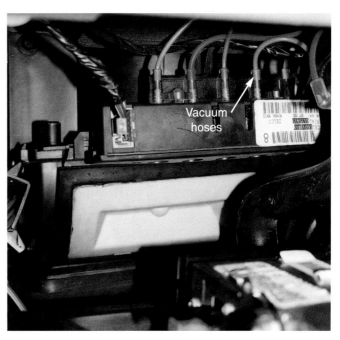

Figure 13-24. *Most modern HVAC systems combine all of the individual vacuum solenoids into a solenoid assembly such as the one shown here. This assembly contains five solenoids.*

Stepper Motor Controls

Stepper motor controls oversee the operation of the stepper motor, discussed in Chapter 12. They are made in two parts: a variable resistor in the control panel and a controller built into the stepper motor. The temperature lever on the control panel is attached to the variable resistor. The sliding contact on the variable resistor changes the amount of voltage sent to the stepper motor. The motor control reads this voltage change as a temperature change signal and energizes the motor as required. The motor is attached to the blend door by linkage. Small changes in lever position cause small voltage changes that cause the motor to move in small steps.

Compressor Controls

As discussed in Chapter 8, many switches are used to control compressor operation. Most of these switches are controlled by changes in system pressure and temperature, and are not directly controlled through the control panel. The major function of the control panel is to energize the compressor clutch when the controls are turned to a mode requiring compressor operation.

Most compressor clutches are energized when the control panel is set to an A/C cooling or defrost mode. A few control panels have a separate A/C switch to energize the compressor clutch. On some newer systems, the compressor clutch is energized only when the fan switch is turned from the *off* position.

HVAC Electrical System Protection Devices

To protect the electrical components of the HVAC system, fuses and circuit breakers are installed in the wiring leading to the control panel. These were discussed in Chapter 4. Some HVAC systems have more than one protection device. For instance, the majority of controls may be directed through one fuse, while the high blower may be energized through another fuse. **Figure 13-25** is a schematic of such a system. The advantage of this system is the high current draw of the blower motor does not have to pass through the wiring and controls of the dashboard control panel.

Control System Operation

Basic control of the system is performed by outside forces. The refrigeration system and engine cooling system operate independently of the HVAC control system. The driver selects the operating modes and blower speeds by moving the dashboard controls. The controls respond only to driver inputs, and do not make any adjustments on their own. In effect the driver is also a part of the HVAC control system.

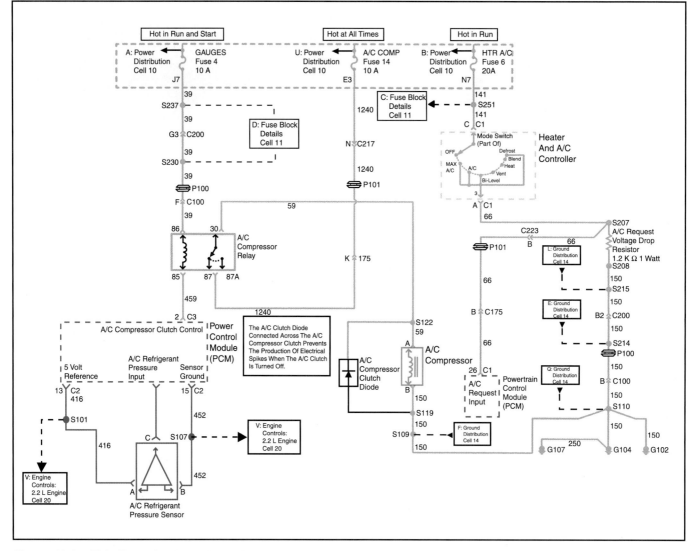

Figure 13-25. *This illustration is a schematic representation of an HVAC electrical control system. Notice how the HVAC system receives electricity through three separate fuses. (General Motors)*

Summary

Control systems are not a common source of HVAC complaints, but they sometimes develop problems. The purpose of any HVAC temperature control system is to control temperature, humidity, and air movement. The HVAC system controls the blower motor, compressor clutch, heater shutoff valve, and air doors to accomplish this. A few control operations such as evaporator and heater core temperature are independent of the HVAC control system. The HVAC control system works with independent controls as needed.

The temperature control system may need to provide air that is warmed, cooled, dehumidified, or a combination. The air may come from outside or be recirculated inside air. This is provided by the use of HVAC modes. Common modes on a manually controlled HVAC system include heating, defrosting, three types of air conditioning, and venting. To obtain these modes the HVAC control system operates the blower motor, compressor clutch, heater shutoff valve, and air doors. This determines not only the type of air exiting to the passenger compartment, but also which outlet(s) it exits from.

The manual control system is used to select mode and temperature by a combination of cable, vacuum, and electrical or electronic parts. The control system includes a dashboard mounted control panel with a mode control, temperature control, and blower control.

Manual control systems have a lever to operate the blend door cable. Some levers move the blend door and also operate a vacuum control valve or the blower switch. Manual control systems also have vacuum control valves to deliver vacuum to the vacuum diaphragms. Rotary, sliding, or pushbutton valves are used. Vacuum valves are attached by linkage to levers or switches on the control. All vacuum system connections are made through hoses and T-fittings. Restrictors slow down vacuum apply. The most common vacuum source for the HVAC vacuum controls is the engine intake manifold. A check valve and reservoir maintain vacuum in the HVAC system under low manifold vacuum conditions. A few vehicles have an auxiliary vacuum pump.

Manual HVAC controls always use electrical devices to control blower motor speed. Some blower control systems use a high blower relay. Newer HVAC systems use electrical solenoids to control the application of the vacuum diaphragms and may have a stepper motor to position the blend door. Fuses, fusible links, or circuit breakers are used to protect the HVAC electrical system.

Review Questions—Chapter 13

Please do not write in this text. Write your answers on a separate sheet of paper.

1. Which of the following is *not* used to control humidity?

    (A) Compressor clutch.

    (B) Blend door.

    (C) Blower motor.

    (D) Heater shutoff valve.

2. In some service literature, modes are referred to as _____.

3. In the heating mode, the heater shutoff valve allows _____ to flow through the heater core.

4. In defrost mode, the air is first _____ and dried, then _____.

5. What is the major difference between the vent and air conditioner modes?

6. To compensate for manufacturing tolerances and cable wear most cables have a _____ device.

7. The vacuum source for the HVAC vacuum controls is the engine _____ manifold.

8. Most manual HVAC blower speed switches have _____ speeds.

9. The vacuum solenoid valve is held closed by a _____.

10. When the HVAC blend door is attached to a stepper motor, the temperature lever on the control panel is attached to a variable _____.

ASE Certification-Type Questions

1. Temperature control is accomplished by operating all of the following HVAC components, *except:*

    (A) compressor clutch.

    (B) diverter door.

    (C) heater shutoff valve.

    (D) blend door.

2. The heater shutoff valve would be open in all of the following modes, *except:*

    (A) defrost.

    (B) heat.

    (C) air conditioning.

    (D) vent.

3. The compressor clutch would be engaged in all of the following modes, *except:*

    (A) vent.

    (B) bi-level.

    (C) defrost.

    (D) heat.

4. Technician A says a cable or vacuum diaphragm operates the heater shutoff valve. Technician B says some vehicles do not use a heater shutoff valve. Who is right?

    (A) A only.

    (B) B only.

    (C) Both A and B.

    (D) Neither A nor B.

5. All of the following statements about HVAC system vacuum restrictors are true, *except:*

    (A) vacuum restrictors are built into the vacuum control valve.

    (B) vacuum restrictors reduce the noise of diverter door closing.

    (C) vacuum restrictors reduce discomfort caused by abrupt changes of airflow.

    (D) vacuum restrictors reduce vacuum loss during apply of a large diaphragm.

6. The vacuum in the reserve tank holds the vacuum diaphragms in position under which of the following conditions?

    (A) High intake manifold vacuum.

    (B) Low intake manifold vacuum.

    (C) One-way check valve open.

    (D) Vacuum valve in off position.

7. On which of the following engines would an auxiliary vacuum pump be *least* likely to be used?

    (A) Diesel engine.

    (B) Large engine in a small car.

    (C) Large engine in a large car.

    (D) Small engine in a large car.

8. All of the following statements about blower speed switches are true, *except:*

    (A) blower switches use internal contacts to complete an electrical circuit.

    (B) switch position determines how current flows through the resistor assembly.

    (C) resistors with more electrical resistance will reduce current flow in the circuit.

    (D) adding resistors to the electrical circuit will increase blower motor speed.

9. Technician A says energizing a vacuum solenoid closes the vacuum valve. Technician B says solenoid and valve assemblies usually have a vacuum bleed hole. Who is right?

    (A) A only.

    (B) B only.

    (C) Both A and B.

    (D) Neither A nor B.

10. The stepper motor controller is built into the:

    (A) stepper motor.

    (B) control panel.

    (C) solenoid block.

    (D) fuse block.

Chapter **14**

Automatic Temperature Control Systems

After studying this chapter, you will be able to:

❑ Explain automatic temperature control systems using control loop theory.
❑ Identify the major parts of electronic automatic temperature control systems.
❑ Explain the operation of electronic automatic temperature control systems.
❑ Identify the components of mechanical automatic temperature control systems.
❑ Explain the operation of mechanical automatic temperature control systems.
❑ Identify the components of electromechanical automatic temperature control systems.
❑ Explain the operation of electromechanical automatic temperature control systems.

Technical Terms

Automatic temperature control systems
Reheat systems
Semiautomatic temperature controls
Temperature control loop
Feedback
Redundant controls
Dual Zone
Inside temperature sensor
Thermistor
Reference voltage
Aspirator
Venturi

Ambient air temperature sensor
Sensor string
Sunload sensor
Photovoltaic cell
Refrigeration temperature sensors
Thermal lockout switch
Set temperature
Temperature control computer
Drivers
Blend door stepper motor
Diverter door actuators

Vacuum solenoids
Slugging
Power module
Purge modes
Snow ingestion mode
Trouble code
Self-diagnosis mode
Mechanical temperature control system
Transducers
Vacuum servo
Programmer

In the last chapter, you covered manual control systems and how they operate the vehicle's HVAC system components. This chapter covers automatic HVAC control systems. This chapter also discusses the specific components and operation of electronic, mechanical, and electromechanical automatic temperature controls. Many of the control devices mentioned in this chapter, such as vacuum valves, motor speed controls, and vacuum solenoids, have already been discussed. If necessary, review the information in Chapters 12 and 13 before proceeding.

Automatic Control Systems

The automatic control system makes the same adjustments the driver would make on a manual control system to maintain passenger compartment temperature. The difference is the automatic temperature control system can do it faster and with greater accuracy.

Early automatic controls were relatively simple mechanical semi-automatic devices. Most modern automatic temperature controls are electronic and control the refrigeration system, heating system, and air distribution system based on sensor inputs. Because of their complexity, it can be difficult at times to properly diagnose and repair these systems. They require a thorough knowledge of HVAC system operation as well as specialized service literature, tools, and test equipment.

Types of Controls

All *automatic temperature control systems* use a vacuum servo or electronic stepper motor to control the position of the blend door. This is similar to the vehicle driver controlling blend door position by moving the temperature lever of a manual system. Whether manual or automatic, moving the blend door reheats the incoming air as much or as little as needed to maintain passenger compartment temperature. Some older automatic temperature control systems are called *reheat systems.*

Other than the blend door, the HVAC components operated by each kind of automatic temperature control varies greatly. Some systems are called *semiautomatic temperature controls* because they allow the driver to directly control some HVAC operations. Typical driver controls in a semiautomatic system are fan speeds or defroster operation. Some semiautomatic systems control temperature only, and allow the driver to select modes and fan speeds.

As a general rule, the fully automatic system controls movement of diverter doors, heater shutoff valve operation, fan speeds, and compressor clutch engagement. The only direct driver controls are of the defrost, and defog modes. A semiautomatic system controls movement of the diverter doors and compressor clutch engagement. It usually does not control fan speed and may allow the driver to select several modes such as heating, air conditioning, defrost, and defog.

 Note: The defog mode is referred to as bi-level (window and floor) on manual systems.

There are exceptions to these general rules, especially in the design of semiautomatic controls. In addition, some fully automatic controls have provisions for driver control of fan speed and mode selection. The simplest way to determine the type of temperature control system is to observe its control panel. If the panel controls are primarily designed to modify temperature rather than selecting specific modes (other than defog/defrost), it is probably a fully automatic system. If the panel has provisions for selecting fan speeds and all HVAC modes, it is a semiautomatic system. Always consult the manufacturers' service literature for exact system components and operation. Since they are similar, fully automatic and semiautomatic systems are discussed together.

 Note: A cycling compressor clutch or variable capacity compressor controls evaporator temperature in the same manner as on manual HVAC systems.

Temperature Control Loops

The *temperature control loop* is a continuous series of operations that keep the passenger compartment temperature at a fixed setting. A temperature control loop is shown in **Figure 14-1.** Inputs to the system are a combination of temperature sensor and driver commands at the control panel. The temperature control computer processes sensor inputs. The temperature control computer sends control commands to the blend door, diverter doors, compressor clutch, blower motor, and heater shutoff valve.

The temperature control loop general operation is as follows:

❑ The input sensors detect changes in passenger compartment temperature.

❑ The input sensors send readings to the temperature control computer.

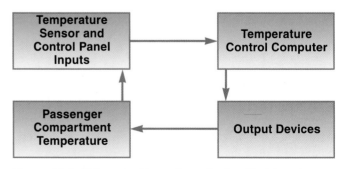

Figure 14-1. *This control loop shows in simplified form how an automatic temperature control unit determines the temperature and adjusts the HVAC system to compensate.*

- The temperature control computer decides what output commands should be issued in relation to the temperature requested by the driver.
- The temperature control computer sends commands to the blend door and other output devices.
- The output devices cause the HVAC system to modify the passenger compartment temperature.
- Changes in passenger compartment temperature affect the input sensors, starting the cycle again.

This basic cycle is used on all automatic and semiautomatic temperature control systems. The process is more complex on an actual system, but the basic cycle is the same.

The term **feedback** is often used to describe control loop components and operation. Feedback means temperature and other information returns, or is fed back to the control components.

Electronic Temperature Control Systems

Some electronic temperature control systems were developed from existing mechanical designs. Most late-model systems are essentially new designs. Electronic temperature control systems, no matter how they were derived, make use of computers and control loop theory. One way to identify an electronic system is by the use of a digital control panel. Older mechanical systems or mechanical systems that have been updated to use electronic components will have some manual switches or levers.

Automatic and Semiautomatic Control Heads

The control head in an automatic system is an electronic device, often with digital readouts indicating temperature and mode, **Figure 14-2.** Some automatic control heads have buttons which remain down when pressed to indicate the mode. Semiautomatic control heads look similar to some manual control heads. One difference is most semi-automatic heads will have a graduated temperature scale instead of a blue and red graduated band.

Redundant Controls

Some vehicles with automatic temperature controls are sometimes equipped with **redundant controls** for air conditioning and heating, **Figure 14-3.** Redundant controls are any control point for the HVAC system other than the dashboard mounted control head. The most common redundant controls are mounted on the steering wheel, usually found on luxury vehicles. This permits the driver to control the air conditioning system without removing his or her hands from the steering wheel. Some are located on the center console. Vans and sport utility vehicles

Figure 14-2. *Modern automatic temperature control heads are fully electronic. A—The control panel shown here has a display screen. B—Semiautomatic control heads look like manual control heads. (Infiniti)*

Redundant controls

Figure 14-3. *Some luxury vehicles have redundant HVAC controls. Most are mounted on the steering wheel or elsewhere if the vehicle is equipped with Dual Zone or rear air conditioning. (Ford)*

equipped with rear air conditioning also have redundant controls. Vehicles with Dual Zone air conditioning and heating also have more than one control head.

Dual Zone Systems

Some late-model automatic temperature control systems have a feature called **Dual Zone.** Systems with Dual Zone have separate controls for the driver and passenger. The driver controls temperature through the main control panel. The passenger control is located on the right side of the vehicle dashboard. Separate drive and passenger controls allow the front seat occupants to raise and lower the temperature of the air exiting on each side of the dashboard. The passenger controls may be a pushbutton design or a rotary knob, **Figure 14-4.** Some Dual Zone systems also have a switch on the main control head that disables the Dual Zone feature.

The driver and passenger Dual Zone Controls operate potentiometers or variable resistors connected to the control knobs or buttons. When the driver or passenger control switch is operated, changes in resistance are sent to the HVAC control computer. The computer then moves separate right and left side blend doors to adjust temperature.

Control System Parts

The major parts of the electronic temperature control system are the input sensor, temperature control computer, and output devices. They are identified in the following sections.

Inside Temperature Sensors

All electronic temperature control systems have an **inside temperature sensor.** The inside temperature sensor is a **thermistor.** Thermistors are made of resistor material that responds to temperature changes. Unlike other kinds of resistors, thermistor resistance goes down as its temperature increases. Changes in temperature quickly affect thermistor resistance, which affects the amount of current that can travel through the thermistor.

A constant **reference voltage** is fed to the sensor from the temperature control computer. Reference voltage is an unchanging voltage source that enters the sensor. Typical reference voltages are around five volts. Changes in sensor temperature cause a change in current flow, and therefore voltage. The computer reads the change in voltage as a change in passenger compartment temperature.

The inside temperature sensor is usually located in the lower part of the dashboard, usually in the kneepad or kickpanel area, **Figure 14-5.** The temperature sensor is part of an **aspirator,** sometimes called a **venturi** assembly. The aspirator provides a continuous flow of passenger compartment air across the sensor. A continuous flow of air provides a more accurate air sample than random air movement inside the passenger compartment.

To operate the aspirator, a small amount of airflow is diverted from the HVAC ductwork. This air travels through a small tube to the aspirator and sensor assembly. The aspirator is constructed with a narrow area called a venturi. The venturi forces the air to flow faster and creates a vacuum at that spot. This is the same principle used in a carburetor venturi.

The venturi vacuum pulls in a small sample of passenger compartment air. The sensor reads this sample and

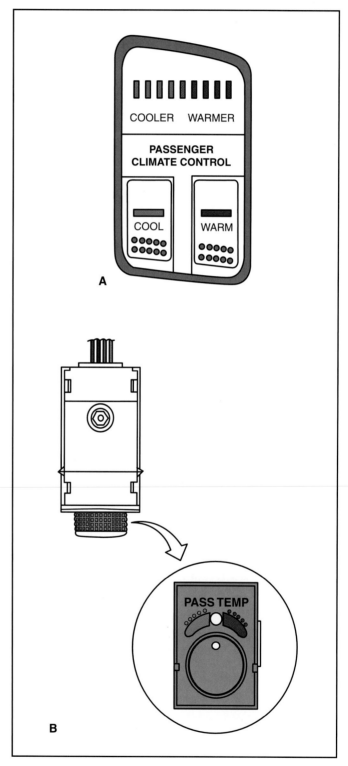

Figure 14-4. *A—Some Dual Zone systems use a pushbutton control for the passenger side. B—Rotary knob control used on Dual Zone air conditioning systems. (General Motors)*

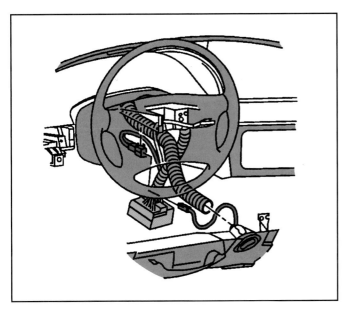

Figure 14-5. *This is a typical location for a sensor used to measure passenger compartment air temperature. These temperature sensors are usually located on the dashboard. Many systems have several passenger compartment temperature sensors. (General Motors)*

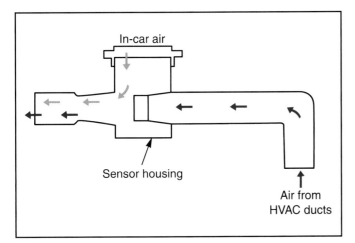

Figure 14-6. *The aspirator uses airflow from the HVAC ducts to pull passenger compartment air into the sensing assembly. (Ford)*

converts it to an electrical signal. **Figure 14-6** is a cross-section of a typical aspirator and sensor assembly. Some aspirators use a small fan to pull air across the sensor and a venturi section is not needed.

Early control systems used only one inside temperature sensor. Most modern systems use other temperature sensors. Other systems use dual sensors for more accurate readings. Some of the more common sensors are discussed in the following paragraphs.

Ambient Air Temperature Sensor

The **ambient air temperature sensor** or *outside temperature sensor* is a thermistor that measures the temperature of the outside air and converts it into an electrical

signal. This sensor is located in the air intake at the vehicle cowl, or in front of the radiator, **Figure 14-7.** Sometimes, the outside temperature sensor is connected in series with the inside temperature sensor. In some cases, the inside temperature sensor is in series with the control panel temperature selecting device. When used, a series sensor circuit is called a **sensor string.**

Sunload Sensor

The **sunload sensor** is usually located on the upper dashboard near the windshield, **Figure 14-8.** Sunlight hitting the sensor tells the temperature control computer when sunlight is creating an additional heat load on the HVAC system. This sensor is a semiconductor device called a **photovoltaic cell** or *photodiode.* These devices produce a small electric current when sunlight falls on them. The sunload sensor produces its own voltage signal without a reference voltage from the temperature control computer.

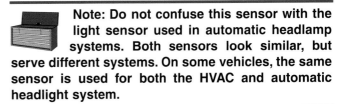

Note: Do not confuse this sensor with the light sensor used in automatic headlamp systems. Both sensors look similar, but serve different systems. On some vehicles, the same sensor is used for both the HVAC and automatic headlight system.

Refrigeration Temperature Sensor

Refrigeration temperature sensors are used on some systems. These sensors measure the temperature of the refrigerant in the system. Some systems measure only the

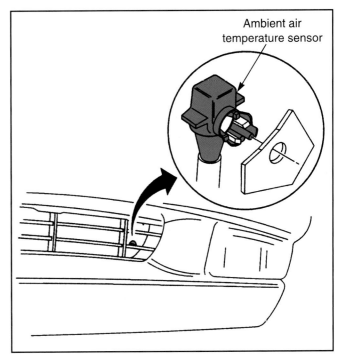

Figure 14-7. *Some outside, or ambient air temperature sensors are located in the front of the vehicle grille, as shown here. (General Motors)*

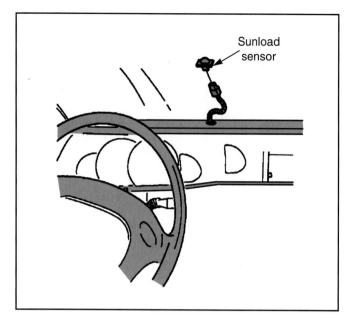

Figure 14-8. *The sunload sensor is often installed at the top of the dashboard. Some automatic climate controls have a sunload sensor at each corner. (General Motors)*

low side temperature, while others measure low and high side temperatures. The control computer uses temperature inputs to determine the pressures in the refrigeration system. This is done as part of malfunction monitoring.

Thermal Lockout Switch

Some vehicles with automatic temperature controls are equipped with a *thermal lockout switch,* located in the heater hose, **Figure 14-9.** A thermal element inside the switch acts as a sensor. The primary function of this switch is to shut off coolant flow to the heater core when the coolant is too cold to heat the passenger compartment. This keeps the system from blowing cold air into the passenger compartment on cold startups. After the engine coolant warms up, the system is allowed to operate normally.

Other Inputs to the Computer

The temperature control computer also receives a desired temperature input from the control panel. This is often referred to as the *set temperature.* The temperature control can be a variable resistor or potentiometer that creates a signal to the computer. On fully electronic systems, the computer is directly attached to the control panel and the temperature signal is created by internal circuitry. Other control panel inputs can override the system to directly select blower speed or defog/defrost modes.

To inform the computer of blend and diverter door positions, potentiometers may be attached to the door linkage. Some door position feedback inputs are made by the stepper motor internal circuitry. A few systems have inputs from the dome light switches on the vehicle doors. These switches tell the system the doors are being opened, and the computer performs a hot air purge cycle before the vehicle is started.

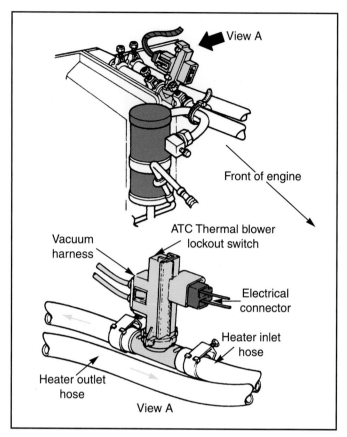

Figure 14-9. *The thermal lockout switch cuts off electricity to the blower when the engine is cold. This keeps cold air from being blown around the passenger compartment. (Ford)*

Input from the control panel is sometimes routed to the powertrain control module (PCM) or body control module (BCM). Typical inputs from the PCM or BCM are engine speed, transmission gear, and vehicle speed. Basic system inputs are the same no matter what the computer design.

Temperature Control Computer

The *temperature control computer* is the decision making part of the temperature control system. It uses internal electronic circuitry to process information and issue output commands. The temperature control computer receives its electrical power from the vehicle electrical system. The computer provides reference voltages for the sensors and battery voltage to the output devices.

The temperature control computer may disregard some sensor inputs when the outside temperature is very low or when placed in maximum cooling. Internal damping circuits to keep the system from overreacting to sudden changes in sensor inputs. For example, a sudden loss of the sunload input when the vehicle enters an underpass will not cause the system to react until several seconds have elapsed.

The temperature control computer is called by many names, including control module, controller, programmer, ECM, ATC box, amplifier, and power module. In some cases, the temperature control computer may be an integral part of the PCM or BCM. **Figure 14-10** shows a typical BCM installation. Some temperature control computers are part of a power head assembly that contains some or all the output devices. The computer in **Figure 14-11** also contains the vacuum and blower controls.

The temperature control computer may monitor refrigeration system pressure and temperature as a safety measure. If the pressures and temperatures indicate a malfunction, the temperature control computer can shut down the refrigeration system.

Output Devices

The most common output devices are solenoids, relays, and motors. On some systems the electrical devices are directly operated through power transistors in the temperature control computer. These power transistors are called *drivers.* Other systems use relays to operate the electrical devices. How these electrical devices are used is discussed in the following paragraphs.

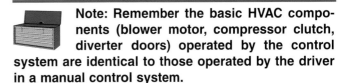

Note: Remember the basic HVAC components (blower motor, compressor clutch, diverter doors) operated by the control system are identical to those operated by the driver in a manual control system.

Blend Door Stepper Motor

Since the common factor in all temperature control systems is control of the blend door, every system has a blend door control as an output device. A common output device is a *blend door stepper motor.* Stepper motors are reversible electric motors that can move in small increments. This motor may be part of an assembly containing other output devices. **Figure 14-12** shows a typical stepper motor used to operate the blend door. Note its resemblance to the door motors discussed in Chapter 13.

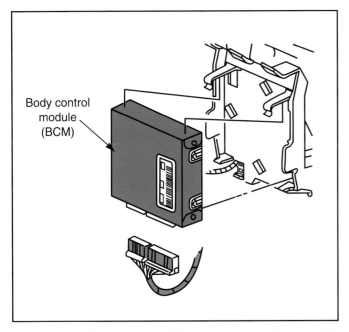

Figure 14-10. *The control module may be a separate assembly, as in this figure, or may be part of the control panel.* (General Motors)

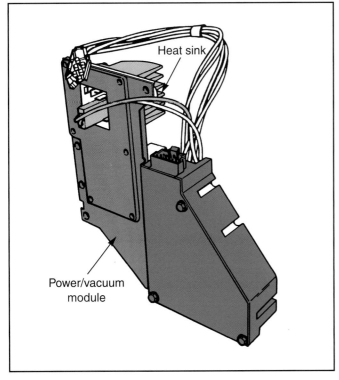

Figure 14-11. *Some control modules also contain the vacuum and electrical controls. These parts may be serviced separately.* (DaimlerChrysler)

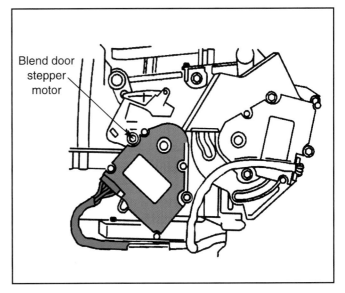

Figure 14-12. *Electronically controlled HVAC systems may use small electric stepper motors called drivers to operate the air doors. The motor also contains the position sensor.* (General Motors)

Diverter Door Actuators

Diverter door actuators are usually vacuum diaphragms. The vacuum diaphragms are controlled by **vacuum solenoids.** Vacuum from the intake manifold is directed through the solenoid and valve assembly, **Figure 14-13.** The computer controls vacuum to the diaphragms by energizing or de-energizing the solenoids. The solenoids operate in the same manner as those in a manual control system. The solenoids are usually installed in a single solenoid assembly.

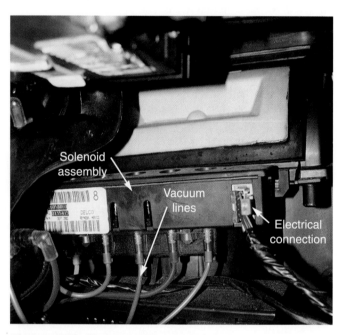

Figure 14-13. *The solenoid assembly shown here uses electrical solenoids to control vacuum flow to the diaphragms. It can be operated manually, or by an automatic control system.*

Blower Motor Control

On some systems, the blower motor is directly energized through drivers. Using drivers allows the system to have many more blower speeds. Some systems have as many as 25 speeds.

On other electronic systems, the computer energizes mechanical relays or electronic modules that operate the blower motor through a resistor assembly. The relays and resistor assemblies closely resemble those used in manual HVAC systems.

Compressor Clutch Control

The computer may directly energize the compressor clutch, but usually operates through a relay because of the high current needed to apply the clutch. See the wiring diagram in **Figure 14-14.** Refrigeration system pressure switches are also used, just as on a manual system. The computer can also control compressor operation to compensate for other vehicle loads and systems. Some of these controls are used on newer vehicles with manual control heads.

Stall Prevention

On some vehicles, the engine ECM will command the compressor clutch off during periods where there is an increase in power steering pressure. These increases take place usually when the steering is turned right or left to its lock point. This is done temporarily and usually for a couple of seconds. The reason for this is to reduce engine load to prevent stalling.

Wide Open Throttle

If the engine ECM detects a wide open throttle (WOT) condition, it will disengage the compressor clutch to reduce engine load. This provides additional engine power for acceleration. If the WOT condition remains, the ECM will reengage the compressor after a set period of time.

Slugging

Slugging is the migration of liquid refrigerant to the inlet side of the compressor. If liquid refrigerant is drawn into the compressor at high speed, it can damage compressor components over time. The ECM can energize the compressor during engine cranking to remove some of the liquid refrigerant. This is not done at every engine start, specific operating conditions must be met before the anti-slugging mode will take place.

Extended and Minimum Compressor On-time

Because a cycling clutch can affect engine idle, the ECM may command the clutch to remain engaged during periods of extended idle. This helps to maintain a smooth idle. The minimum clutch On-time for extended idle is anywhere from 30 seconds to two minutes.

Some vehicles will keep the compressor clutch on for a minimum time before allowing it to be disengaged. This is usually done at speeds above 20 mph.

Power Module

Some systems use a **power module, Figure 14-15.** The power module is a type of relay that uses power transistors. It receives control signals from the temperature control computer and operates the blower motor and compressor clutch directly from battery voltage. Using a power module reduces the amount of current flowing into the computer, increasing its life. The power module is installed in the HVAC system ductwork. Air passing through the ducts cools the power module.

Control System Operation

This section explains how the control system components work together during some typical heating and cooling situations. Input sensors measure passenger compartment temperature, ambient air temperature, evaporator pressure or temperature, and sunload, and convert it to a voltage signal. The driver also inputs desired temperature and operating mode information. The temperature control computer processes the inputs. The computer then makes

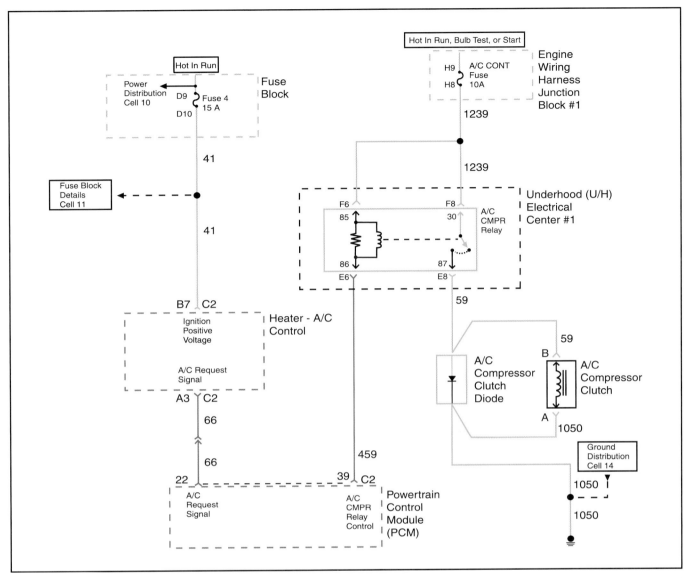

Figure 14-14. *An electrical schematic showing how the compressor and blower are operated on an automatic climate control system. The operation is similar to operation of a manual system. (General Motors)*

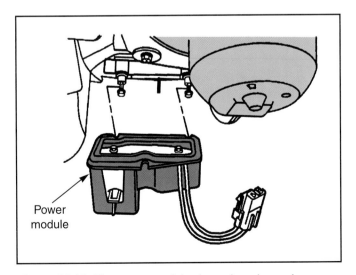

Figure 14-15. *The power module shown here is used on some HVAC systems. It controls blower speeds based on low voltage inputs from the HVAC control system. (General Motors)*

output decisions based on the inputs. The following are examples of how the automatic temperature control system operates in the AUTO (automatic) position during some common cooling and heating conditions. The set temperature at the control panel is 72°F (22°C).

Cool Morning Becoming Warm Afternoon

The vehicle is started on a cool summer morning (50°F or 10°C) and driven until afternoon when the temperature has increased to about 80°F (27°C). At startup, the sensor inputs indicate the passenger compartment and ambient temperature are below 72°F (22°C). The temperature control computer processes this information and opens the heater shutoff valve, allowing engine coolant to flow through the heater core. The module also commands the blend door to open fully and moves the diverter doors to allow heated air to exit at the floor vents. The fan motor is energized as needed. If the system uses input from the coolant temperature sensor, the fan may not operate until the coolant has warmed up, usually to about 130°F (55°C).

When the passenger compartment air temperature reaches 72°F, the module gradually reduces fan speed, partially closes the blend door, and completely closes the heater shutoff valve.

As ambient temperature increases, the sensor inputs indicate the increase in air temperature, passenger compartment temperature, and sunload. The temperature control computer processes the sensor inputs. It then issues output commands that energize the compressor clutch, allowing the air conditioner to begin cooling the passenger compartment. It also commands the blend door partially closed to reduce flow through the heater core. It increases the blower speed as necessary and operates diverter doors to allow the cooled air to exit from the dashboard vents.

If the temperature control computer receives information that interior temperature and sunload are still increasing, it will completely close the blend door to cut off heater core airflow. It will also close the recirculation door to prevent outside air from entering the passenger compartment, and increase blower motor speed.

Rapid Vehicle Cooldown (Rainstorm)

The vehicle is driven on a hot day, and encounters a sudden rainstorm. Ambient air temperatures may be reduced from 90°F (32°C) to 70°F (21°C) or more within a few minutes. At 90°F (32°C), the system is operating at maximum or near maximum cooling.

The temperature and sunload sensors would indicate the rapid drop in temperature. The temperature control computer processes this information and begins to open the blend door to reduce the amount of cooled air entering the passenger compartment. The computer also reduces blower speed and may open the recirculation door. In some cases, the compressor clutch may be turned off for brief periods.

If the temperature drops below 72°F (22°C), the computer will open the blend door an additional amount and reduce blower speed to minimum. The refrigeration system will continue to operate, but its output of cooled and dried air will be reheated as necessary.

Extreme Vehicle Temperature

A vehicle parked in the sun on a hot day may have passenger compartment temperatures over 150°F (66°C). Even with a properly operating air conditioning system, it can take several minutes for the vehicle's passengers to notice a cooling effect. The amount of time needed to cool the interior can be affected by the vehicle's color, as darker colors tend to hold heat.

When the engine is started, the temperature and sunload sensors indicate that temperature is excessively high. The temperature control computer uses this information to place the HVAC system in maximum cooling. The compressor is energized, however, there may be a delay of up to 30 seconds after startup to allow engine operation to

stabilize. The blend door is fully closed to allow maximum cool air to enter the passenger compartment. The recirculation door will be open to draw in cooler air. Fan speed is increased to maximum.

As the temperature begins to drop, the temperature control computer reduces blower speed and may close the recirculation door to allow the partially cooled air to recirculate. When passenger compartment temperature approaches the preset level of 72°F (22°C), the system will begin to open the blend door to maintain this temperature. The system may also reopen the recirculation door.

 Note: Some manufacturers will operate the air conditioning system to prevent instrument panel damage from extreme heat.

If the vehicle is parked outside on a winter day, both the exterior and interior temperatures will be close. When the engine is started, the temperature sensor readings indicate the low temperature. The temperature control computer opens the heater shutoff valve. If the engine coolant is not warm, the computer will limit fan speed until the coolant reaches the proper temperature. Air from both outside and inside the vehicle are circulated and heated. The warmed air comes out of the window and floor vents, but can be directed through the dashboard registers, if requested by the driver. If the temperature is just above the freezing point, the compressor may be engaged to assist in defogging the windows.

Purge Modes

Some automatic temperature controls have one or more **purge modes.** The purpose of the purge mode is to remove excess moisture and moisture laden air from the air distribution system. If this moisture was not removed, it might cause window fogging or the discharge of water from the HVAC system vents. The purge modes operates automatically with no driver input. When the HVAC computer determines excess moisture is present in the distribution system, it overrides the control panel settings and directs air toward the floor. This is referred to as *A/C purge.* The blower motor will not operate (or only operate at the lowest speed) until the engine coolant temperature increases to a set value.

Some purging systems also direct air against the windshield when the vehicle is first started in cold, damp weather. This is referred to as *cold purge.* This reduces windshield fogging caused by the condensed breath of the vehicle occupants. Most purging systems can be bypassed by selecting any of the manual positions on the control panel.

Snow Ingestion Mode

Some automatic control systems will operate the blower motor at a reduced speed in cold weather until the engine coolant warms. This is referred to as the

snow ingestion mode. This mode is usually used with the defrost mode and reduces the possibility of snow being pulled into the vehicle through the blower case. The fan will resume normal operation once the coolant temperature exceeds about 100°F (38°C).

Computer Self-diagnosis

All electronic HVAC control systems use the computer to monitor system operation. When the computer detects a malfunction, it stores a ***trouble code.*** Trouble codes identify either a general HVAC problem or a defective system component. The exact problem identified by the trouble code varies with each system manufacturer. Some systems have a separate warning light or LED on the dashboard control panel. Other systems may use the display itself to indicate a problem, as in **Figure 14-16.** If the

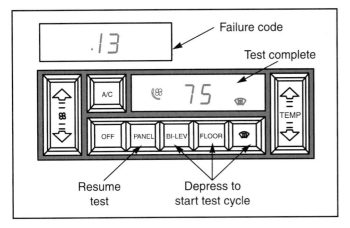

Figure 14-16. *When an automatic control system problem occurs, some panels warn the driver by displaying a failure code instead of the temperature. Other systems have a separate warning light. (DaimlerChrysler)*

problem is a hard (or permanent) failure, the light will remain on. If this light is illuminated, there is a problem in the control system or basic HVAC system.

The technician can sometimes obtain trouble codes by pressing a series of control panel buttons. This causes the system to go into ***self-diagnosis mode*** and display the trouble codes on the control panel. A scan tool can also be used to retrieve trouble codes.

Older Temperature Control Systems

Mechanical and electromechanical temperature control systems were used on many older domestic vehicles. They have now been replaced by electronic systems. A brief discussion of these older systems follows.

Mechanical Temperature Control Systems

The ***mechanical temperature control system*** used a temperature sensor located in the dashboard, **Figure 14-17.** These sensors are usually called ***transducers.*** Transducers are devices that convert a mechanical or electrical signal to a vacuum signal.

Vacuum from the intake manifold was delivered to the transducer by hoses. The transducer used a valve to modify the vacuum before it was distributed to the rest of the control system. The valve was a calibrated opening and a tapered pin. The valve was operated by a bimetal spring, **Figure 14-18.** Changes in temperature caused the spring to move the valve, controlling engine vacuum to the other vacuum components.

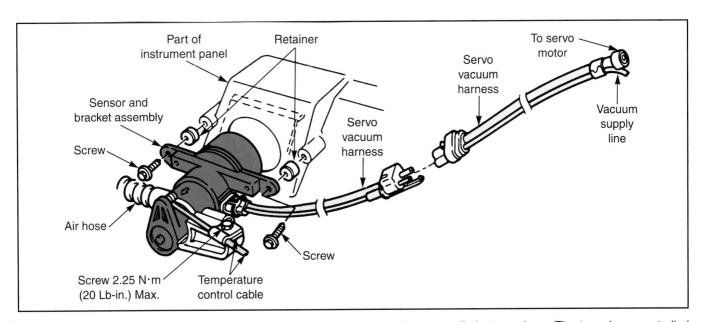

Figure 14-17. *Older automatic temperature controls used a mechanical sensor called a transducer. The transducer controlled vacuum to a vacuum power servo, and was adjusted by a cable. (Ford)*

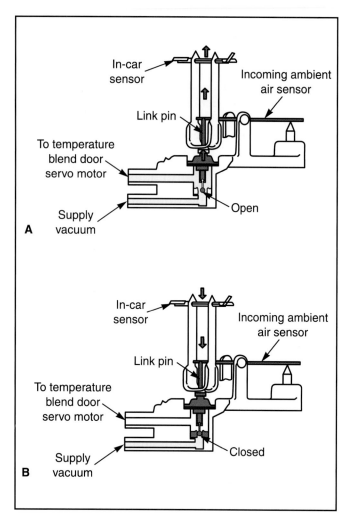

Figure 14-18. *The vacuum supply to the power servo was varied by a bimetallic spring in the transducer. A—When the passenger compartment temperature rose, the spring expanded and allowed vacuum to move the servo to increase the cooling effect. B—When the passenger compartment temperature reached its preset level, the spring contracted to lessen the cooling effect. (Ford)*

The temperature lever on the control panel operated a cable attached to the transducer. Moving the cable changed the amount of tension on the bimetal spring. This affected the amount of vacuum allowed through the transducer, and ultimately, the temperature produced by the control system.

The modified vacuum went to a *vacuum servo.* The vacuum servo positioned the blend door and operated the electrical components of the HVAC system. The main diaphragm moved a rod or lever. The rod operated the blend door and also operated a vacuum valve and electrical switch. The servo was arranged so changes in vacuum from the transducer opened and closed the blend door and also operated the related diverter door diaphragms and electrical devices.

The relation of the vacuum delivered to the servo by the transducer is the same for most systems:

❏ High vacuum from the transducer caused the servo diaphragm to move to the full cooling position.

❏ Low vacuum from the transducer caused the servo diaphragm to move to the full heating position.

The control loop formed by the mechanical temperature system used the transducer as an input sensor and controller. Temperature changes affected vacuum to the servo, which moved the blend door, diverter doors, and the heater control valve. It also operated electrical components such as the blower motor and compressor clutch. These changes in door position and airflow adjusted the temperature. The temperature change was sensed by the transducer, which started the cycle again. **Figure 14-19** shows the control loop of a mechanical temperature control system.

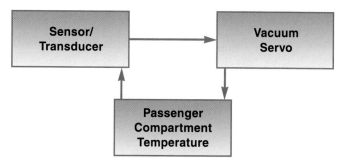

Figure 14-19. *A control loop for a mechanical temperature control system. The transducer sensed the interior temperature and directly operated the power servo to control temperature.*

Electromechanical Temperature Control Systems

Electromechanical temperature control systems are similar to mechanical systems. They use a vacuum servo and transducer, but use electrical or semi-electronic devices to operate the transducer assembly. The sensors shown in the figure modify a reference voltage to produce a signal to the amplifier. The amplifier then opens the vacuum valve in the transducer as necessary. This affects the temperature in the passenger compartment. The sensor string senses the change in temperature and causes the amplifier to change the vacuum output of the transducer. A typical electromechanical temperature control loop is shown in **Figure 14-20.** On some electromechanical systems, the amplifier performs the electrical switching functions instead of the power servo.

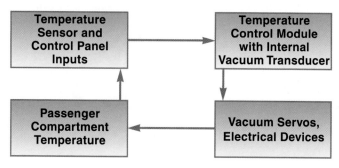

Figure 14-20. *A control loop for a mechanical temperature control system. Operation is similar to a fully electronic system.*

A few older electromechanical systems use a **programmer.** The programmer uses a small stepper electric motor instead of a vacuum diaphragm, **Figure 14-21.** Resistors and potentiometers in the HVAC panel control motor operation. The motor moves the blend door and operates other mechanical and electrical controls. This kind of programmer should not be confused with the electronic temperature control computers that are sometimes called programmers.

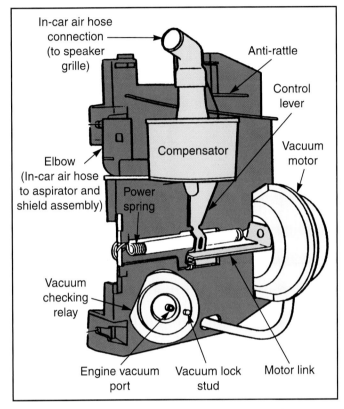

In-car air hose connection (to speaker grille)

Anti-rattle

Control lever

Compensator

Vacuum motor

Elbow (In-car air hose to aspirator and shield assembly)

Power spring

Vacuum checking relay

Engine vacuum port

Vacuum lock stud

Motor link

Figure 14-21. *Some older systems used an electric motor to operate the blend door. The motor was installed in a single housing with its associated controls and relays. The assembly was sometimes called a programmer. (General Motors)*

Summary

The automatic temperature control system duplicates human control of the HVAC system, only more quickly and precisely. Temperature controls can be simple semi-automatic devices, or complex electronic systems. All automatic temperature control systems use a vacuum servo or stepper motor to control the position of the blend door. What other HVAC system devices are controlled varies with manufacturers. Semiautomatic temperature controls allow the driver to directly control some HVAC operating parameters. Fully automatic control systems operate all HVAC modes except defog and defrost modes. The best way to determine the nature of a temperature control system is to observe its control panel.

Temperature control loops are a continuous series of operations that monitor and adjust passenger compartment temperature. Temperature sensors and driver commands are sent to the temperature control computer. The computer reads and processes sensor inputs and operates the blend door, diverter doors, compressor clutch, blower motor, and heater shutoff valve.

The major parts of the electronic temperature control system are the input sensor, temperature control computer, and output devices. All electronic temperature control systems have an inside temperature sensor called a thermistor. Modern systems use other temperature sensors including an outside temperature sensor, sunload sensor, and evaporator temperature sensor. Other inputs come from the driver, selected switches, and other computers when used.

The temperature control computer makes decisions based on inputs and issues output commands. The computer always controls the blend door. A stepper motor or a vacuum diaphragm moves the blend door. Diverter door vacuum diaphragms are controlled by vacuum solenoids. The blower motor is operated by drivers in the computer or by relays. The compressor clutch is usually controlled by the computer through a relay.

Mechanical temperature control systems have been replaced by electronic systems. The mechanical temperature control system uses a temperature sensor in the dashboard. The vacuum servo positions the blend door and operates the electrical components of the HVAC system. The control loop formed by the mechanical temperature system uses the transducer as an input sensor and controller. Temperature changes affect vacuum to the servo, which moves the blend door, diverter doors, and the heater control valve.

Electromechanical temperature control systems are similar to mechanical systems. However, sensors and an amplifier are used to control the transducer.

Review Questions—Chapter 14

Please do not write in this text. Write your answers on a separate sheet of paper.

1. All automatic temperature control systems control the movement of the _____ door.

2. Which of the following would be controlled by the driver on a semi-automatic temperature control system?
   (A) Compressor clutch engagement.
   (B) Blower speeds.
   (C) Recirculation door opening.
   (D) Recirculation door closing.

3. Define the term *feedback.*

4. A sensor reading is a control loop _____.

5. *True or False?* The driver setting the temperature is a control loop output.

6. *True or False?* The computer processes outputs and makes decisions.

7. *True or False?* The output devices may control electrical or vacuum devices.

8. Diverter doors are usually operated by _____, which are usually controlled by _____.

9. The blower motor may be operated through internal computer _____ or separate _____.

10. A mechanical temperature control system contains all of the following components, *except:*

    (A)   vacuum transducer.
    (B)   power servo.
    (C)   temperature control computer.
    (D)   control cable.

ASE Certification-Type Questions

1. Moving the blend door to the open position causes the air to be _____.

    (A)   cooled
    (B)   dehumidified
    (C)   reheated
    (D)   recooled

2. Technician A says a cycling compressor clutch is operated by the temperature control system. Technician B says a variable capacity compressor controls evaporator temperature independently of the temperature control system. Who is right?

    (A)   A only.
    (B)   B only.
    (C)   Both A and B.
    (D)   Neither A nor B.

3. The temperature control computer sends control commands to all of the following, *except:*

    (A)   blend door.
    (B)   sunload sensor.
    (C)   compressor clutch.
    (D)   heater shutoff valve.

4. All of the following statements about thermistor type sensors are true, *except:*

    (A)   thermistor resistor material responds quickly to temperature changes.
    (B)   thermistor resistance goes down as its temperature decreases.
    (C)   thermistor resistance is quickly affected by changes in temperature.
    (D)   changes in temperature affect the amount of current traveling through the thermistor.

5. Technician A says an aspirator pulls air over the temperature sensor using the venturi principle. Technician B says pulling air over the temperature sensor causes it to generate its own current. Who is right?

    (A)   A only.
    (B)   B only.
    (C)   Both A and B.
    (D)   Neither A nor B.

6. The sunload sensor is usually located on the:

    (A)   cowl.
    (B)   upper dashboard.
    (C)   front of the radiator.
    (D)   rear package shelf.

7. Damping circuits in the computer keep the system from overreacting to changes in:

    (A)   sensor inputs.
    (B)   reference voltage changes.
    (C)   output loads.
    (D)   voltage spikes.

8. Technician A says a computer driver is a type of power transistor. Technician B says a computer driver is a type of relay. Who is right?

    (A)   A only.
    (B)   B only.
    (C)   Both A and B.
    (D)   Neither A nor B.

9. All of the following statements about the temperature control loop are true, *except:*

    (A)   input sensors measure passenger compartment and ambient air temperature.
    (B)   input sensors convert temperature into a voltage signal.
    (C)   the driver also inputs desired temperature and operating mode information.
    (D)   the computer always makes output decisions based on inputs.

10. The temperature control computer will display trouble codes in what mode?

    (A)   AUTO
    (B)   ECON
    (C)   DEFOG
    (D)   None of the above.

Chapter 15

Refrigeration System Diagnosis and Leak Detection

After studying this chapter, you will be able to:
- ❑ Explain the seven step troubleshooting process.
- ❑ Make a refrigeration system and HVAC system performance check.
- ❑ Correctly attach gauges to a refrigeration system.
- ❑ Diagnose problems in a refrigeration system.
- ❑ Determine the type of refrigerant in a refrigeration system.
- ❑ Locate refrigeration system leaks.

Technical Terms

Strategy-based diagnostics	Performance test	Soap solution
Logic	System undercharge	Dyes
Seven-step process	System overcharge	Special tools
Intermittent problems	System restrictions	Follow-up
Functional test	Electronic leak detectors	Documentation

All previous chapters concentrated on HVAC components and how various HVAC systems operate. This chapter begins the discussion of HVAC service. In this chapter, you will learn how to diagnose refrigeration system problems and quickly identify defective parts. The seven-step troubleshooting process outlined here will enable you to quickly locate and correct refrigeration system problems. Be sure you know how to perform every diagnosis and service procedure in this chapter. You will need all of the information presented here to successfully complete the remaining chapters in this text.

Strategy-based Diagnostics

In the past, it was fairly easy to find and locate a problem, since most vehicle systems were simple and common to many, if not all vehicles. As vehicles became more and more complex, the methods used to diagnose them became obsolete and in some cases, inapplicable. Technicians who were accustomed to using the older diagnostic routines, or no routine at all, began to simply replace parts hoping to correct the problem, often with little or no success. Unfortunately, this process was very expensive, not only to the customer, but to the shop owner as well.

In response to this problem, a routine involving the use of logical processes to find the solution to a problem was devised for use by technicians. This routine is called *strategy-based diagnostics.* The strategy-based diagnostic routine involves the use of a logical step-by-step process, explained in the next sections. Variations of strategy-based diagnostics are used in many fields outside of automotive repair. A flowchart of this process as recommended by one vehicle manufacturer is shown in **Figure 15-1.**

The Importance of Proceeding Logically

When troubleshooting any refrigeration system or other HVAC system problem, always proceed logically. *Logic* is a form of mental discipline in which you weigh all factors without jumping to conclusions.

To work logically, the first thing you must know is how the refrigeration system works and how it affects and is affected by its components as well as other vehicle systems. This has been covered in earlier chapters. The knowledge you have gained can be put to use when a refrigeration system problem occurs.

The second thing you need is a logical approach. To diagnose a problem, think about the possible causes of the problem, and just as important, the things that cannot cause it. You can then proceed from the simplest things to check, to the most complex. Do not guess at possible solutions, and do not panic if the problem takes a little while

to find. If you remember these points, you will be able to diagnose most refrigeration and HVAC problems with a minimum of trouble.

The Seven-Step Troubleshooting Procedure

Troubleshooting is a process of taking logical steps to reach a solution to a problem. It involves reasoning through a problem in a series of logical steps. The *seven-step process* will, in the majority of cases, be the quickest way to isolate and correct a problem. Refer to **Figure 15-2** as you read the following sections.

Step 1—Determine the Exact Problem

Do not expect the vehicle's driver to tell you what is wrong in a way that will immediately lead you to the problem area. Most drivers will state the problem in layman's terms, such as "It doesn't cool." or "The air conditioner is noisy." The first step is to determine the exact problem. You determine the driver's exact complaint and its symptoms. Many times, the complaint has nothing to do with the HVAC system. This process involves talking to the driver and road testing the vehicle.

Talking to the Driver

Obtaining information from the driver is the first and most important part of troubleshooting. Information from the driver will sometimes allow you to bypass some preliminary testing and go straight to the most likely problem. In one sense, the driver begins the diagnostic process by realizing the vehicle has a problem.

Question the vehicle driver to find out exactly what he or she is unhappy about. Try to get an accurate description of the problem before beginning work on the vehicle. Try to translate the driver comments into commonly accepted automotive diagnostic terms. The easiest way to do this is with a series of basic questions:

❑ When does the problem occur?
❑ How often does the problem occur?
❑ Does the problem only occur in a certain mode or all modes?
❑ Do you hear any unusual noises?
❑ Are there any unusual odors?
❑ Does there seem to be any airflow when the problem occurs?
❑ Does air come out of the wrong vents when the problem occurs?
❑ Did the vehicle have any recent HVAC service, cooling system service, or any other type of repairs?
❑ Did the problem start suddenly, or gradually develop?

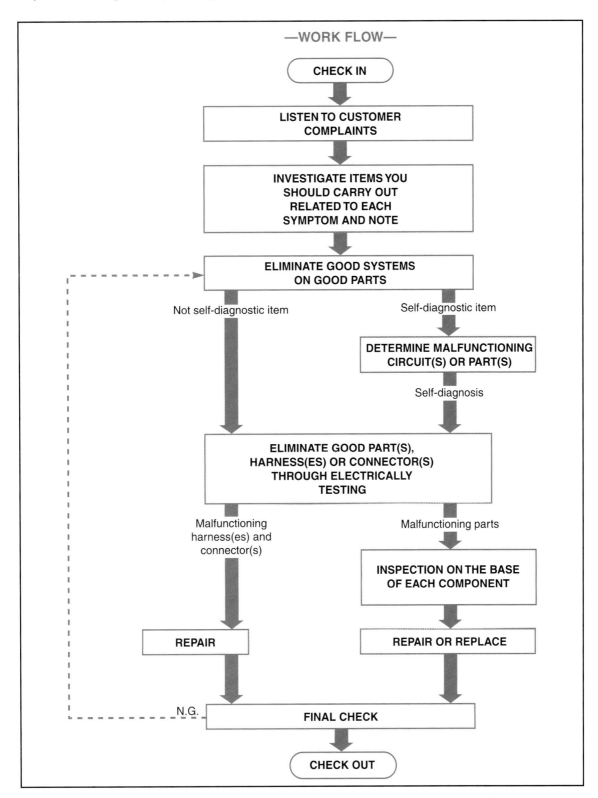

Figure 15-1. *Strategy-based diagnostic flowchart as recommended by one manufacturer. (Nissan)*

You may think of other questions depending on the answers you get to these questions. Write down the driver's comments on an inspection form such as the one shown in **Figure 15-3.** If an inspection form is not available, use the back of the repair order or a sheet of paper. Before going on to Step 2, make sure you have a good idea of the driver's complaint.

Assessing Driver Input

While taking into account what the driver says, try to estimate his or her attitude and level of automotive knowledge. Because drivers are not usually familiar with the operation of automobiles, they often unintentionally mislead technicians when describing symptoms or may have

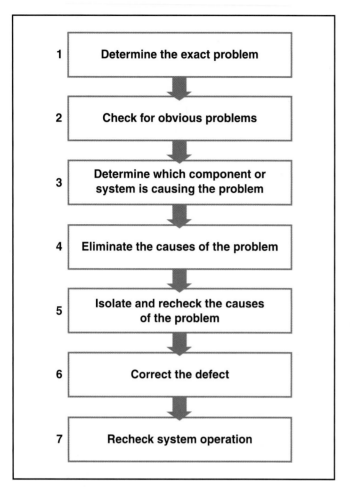

Figure 15-2. *A logical troubleshooting process will enable you to quickly diagnose and repair refrigeration system and other HVAC system problems.*

reached their own conclusion about the problem. In describing vehicle problems, drivers have been known to use hand gestures, body language, and even simulate noises they have heard. While this can sometimes be fun to watch, keep in mind it is a part of the diagnostic process. Many times, important clues can be found simply by observing a driver's physical actions while describing a particular problem.

In many cases, the person bringing in the vehicle has already formed an opinion as to what is wrong. These opinions are a common occurrence, often based on poor or incomplete understanding of vehicle operation, advice from uninformed friends, or other failures to fully comprehend the problem. The best course is to listen closely to the driver's description of the symptoms. Some drivers will be sensitive to even slight changes, and may be overreacting to a normal condition. Never accept a driver's or another shop's diagnosis until you can verify it.

Often, the owner is concerned about the cost of repairs. Some will even downplay the symptoms, hoping for an inexpensive repair. Very few vehicle owners are unconcerned about the cost of vehicle repairs and maintenance. Do not give any type of uninformed estimate, even though you may have a good idea of the problem. Giving

an estimate without diagnosis is a mistake made by many technicians. This practice invites one of two things to occur; either the recommended repair will not correct the problem or it will frighten the driver, who may decide to take his or her vehicle to another shop or not have the repair done at all. Explain that the charge for diagnosing the problem is actually more cost effective than paying for a service, which in many cases may not fix the problem. Before going on to the road test, be sure you have a good idea of the driver's complaint.

Road Testing

In the case of many HVAC problems, it is usually not necessary to perform an extensive road test. However, in some cases, performing a short road test is the fastest way to confirm a problem. Before beginning a road test, make a few quick checks to ensure the vehicle can be safely road tested. Walk around the vehicle's exterior and make a note of any damage that is present. Check each tire to ensure they are inflated properly and in good condition. Also make sure that all safety-related equipment, such as the turn signals and horn are working properly.

 Warning: Do not road test a vehicle that is not safe to drive. Low or no brake pedal, tires with exposed steel or cloth cord, and slipping transmissions are all examples of problems that would render a vehicle unsafe.

Turn the steering wheel and make sure the steering system does not have excessive play. Depress the brake pedal to ensure the brake system has at least a minimal pedal. Also make sure the vehicle has enough fuel to conduct a road test. Do not adjust anything in the passenger compartment, such as mirror, seat, and tilt steering wheel position, other than what is absolutely necessary. If the radio is on, turn it off so that you can listen for unusual noises.

Wear your seat belt at all times during the road test. Try to duplicate the exact conditions under which the driver says the problem occurs. Unfortunately, duplicating some conditions is not always possible. Always try to road test the vehicle with the owner. This will ensure you are both talking about the same problem, and will save valuable diagnostic time.

Drive slowly as you leave the service area to ensure that no obvious mechanical problems exist that could further damage the vehicle or cause personal injury. Make one or two slow speed stops to verify the brakes work properly. Drive the vehicle carefully and do not do anything that could be viewed as abuse. Tire squealing take-offs, speed shifts, fast cornering, and speeding can all be interpreted as misuse of the vehicle.

While road testing, obey all traffic rules, and do not exceed the speed limit. It is especially important to keep in mind that you are under no obligation to break any laws while test driving a customer's vehicle. Also be alert while driving. It is easy to become so involved in diagnosing the

International Mobile Air Conditioning Association

Inspection Report

CODE OF PROFESSIONAL PRACTICE

Customer Name:		License No.:
Automobile Year/Make:	Model:	Engine Size:
Inspection Performed By:		Date:

Procedure | Recommendations | Estimated Cost of Repairs

		Parts	Labor

VISUAL INSPECTION - Engine Compartment

1) Hoses, tubing and connections (Suction, Discharge & Liquid Lines)

2) Compressor

3) Compressor Clutch

4) Service Ports

5) Condenser

6) Expansion Valve/Orifice Tube

7) Evaporator Pressure Regulator (POA, STV or VIR)

8) Cabin Air (Evaporator) Filter (if equipped)

9) Accumulator/Drier

10) Drive Belts, Pulleys and Tensioners

11) O-rings, Gaskets, Seals and Spring Locks

12) Inline Filter

13) Electric Fan, Fan Clutch & Fan Blade

14) Electrical Components

VISUAL INSPECTION - Passenger Compartment

1) Air ducts, louvers, sensors, control knobs and cables

2) Control Head

3) Interior Condition

LEAK CHECK - Engine Compartment (NOTE: Engine must be off during this procedure)

1) Refrigerant Check

2) Results of Leak Check

Subtotal of Estimated Repair Costs

INITIAL PERFORMANCE EVALUATION

Type of Refrigerant: Purity ☐ Yes _____ % ☐ No   High-Side Press.:   Low-Side Press.:

Louver Temperature: Interior Temperature: Ambient Temperature:

Amount of Refrigerant Added to system: Amount of Refrigerant Recovered from system:

FINAL PERFORMANCE EVALUATION

Type of Refrigerant: Purity ☐ Yes _____ % ☐ No   High-Side Press.:   Low-Side Press.:

Louver Temperature: Interior Temperature: Ambient Temperature:

Amount of Refrigerant Added to system:

Total Estimated Cost of Repairs Based On This Inspection:

The above inspection was done in accordance with the IMACA Code of Professional Practice procedures manual. If repairs are recommended, you will be provided an estimate and only the repairs authorized by you (the customer) will be made. If further repairs are necessary, you will be informed of and approve the additional parts and labor costs before the repairs are performed.
Thank you for your business!

Manager:
Date:

Figure 15-3. *This sample inspection report form can be used to accurately diagnose problems. Filling out this form as you check the refrigeration system will enable you to tell exactly what is wrong with the system. (IMACA)*

problem, that you forget to pay attention to the road or the traffic around you. If it is necessary to look for a problem or monitor a scan tool's readout while the vehicle is driven, get someone (not the vehicle's owner) to drive for you. Once you have verified the problem exists, proceed to Step 2.

Diagnosing Intermittent Problems

If the problem does not occur either in the shop or during the road test, it is tempting to dismiss it as the owner's imagination or as normal vehicle operation, but the problem may well be real. *Intermittent problems* are the most difficult to diagnose, because they usually occur only when certain conditions are met. Intermittent malfunctions can be related to temperature, humidity, certain vehicle operations, or in response to certain tests performed by a vehicle computer. While most problems in the HVAC system are usually easily spotted, like other vehicle systems, it can develop problems that occur intermittently.

When dealing with an intermittent malfunction, always try to recreate the *exact* conditions in which the problem occurred. Unfortunately, most drivers do not relate intermittent problems to external conditions. Intermittent problems cannot always be duplicated. If a road test of reasonable duration does not duplicate the problem, it is time to try other types of testing. It is essential the principles of strategy-based diagnostics be followed closely when diagnosing intermittent malfunctions.

Step 2—Check for Obvious Problems

Most of your time in Step 2 will be spent checking for obvious causes of the problem, including possible causes that can be easily tested. Visual checks and simple tests take only a little time, and might save more time later. As a minimum, open the hood and check the following items before you start the engine and HVAC system:

- ❏ Retrofit label. A retrofit label indicates the original refrigerant has been replaced with a substitute. If the retrofit was done properly, the service fittings should be different from the originals. Always use a refrigerant identifier whether a retrofit label is present or not.
- ❏ Service fittings. The type and style of the service fittings are the other indication the system may have been retrofitted to another refrigerant. Note the size, shape, and location of the high and low side fittings. Keep in mind some vehicles have an additional fitting that was used at the factory, and should not be used. Service fittings are also the cause of some refrigerant leaks.
- ❏ Obvious refrigerant leaks. Since refrigerant oil leaks out with the refrigerant, leaks can usually be spotted by the presence of oil at the leak site. **Figure 15-4** shows some typical refrigeration system leak locations.

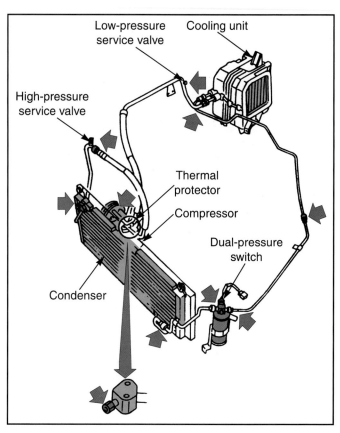

Figure 15-4. *Leaks can occur at many places in the refrigeration system. This diagram shows some of the most likely locations on one kind of vehicle. (Nissan)*

- ❏ Belt condition. Check the belt for tightness and condition. Sometimes you may find the belt is missing. Also check the condition of the belt pulleys. If the belt is missing or badly burned, try to turn the compressor by hand to ensure it is not locked up.
- ❏ Refrigeration lines and hoses. Check for obvious damage such as frayed rubber or cuts. Also look for kinks or improper bends in lines and hoses.
- ❏ Compressor clutch. Check for evidence of slippage, excessive clearance, and overheating, **Figure 15-5.** This check is especially important if the belt is damaged.
- ❏ Radiator fan. Check for bent or missing blades and loose attaching bolts. If the fan is electric, make sure the motor works properly.
- ❏ Fan clutch (when used). If the center of the fan clutch is leaking oil, the front of the clutch will be oily.

Look for other obvious problems such as loose or missing compressor mounting bolts, loose electrical wires, dented or damaged system components, and missing shrouds around the condenser and radiator. Check vacuum hoses to ensure they are not cracked, misrouted, or disconnected.

Check the vehicle dashboard for damaged HVAC controls. Operate the dashboard controls and ensure they are working. Look for levers that are stuck or do not appear to be connected, sticking pushbuttons or knobs, or hissing noises when certain modes are selected. Turn the ignition

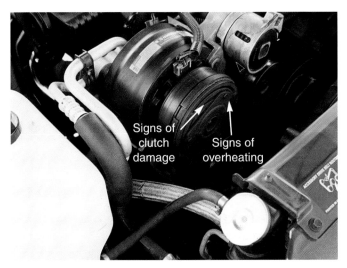

Figure 15-5. *Make a visual check of the compressor. Typical problems are a loose and slipping drive belt and signs of refrigerant leaks at the front seals or hose fittings.*

key to the on position and check if the blower operates on all speeds, and if any indicator screens or other electrical indicators are working.

If the problem appears to be electrical or electronic, you may want to visually check the fuses, related electrical connections, and grounds. In many cases, these simple checks will uncover the problem, or give you a likely place to start in Step 3.

Refrigerant Identification

Check the refrigerant type to determine whether it agrees with the manufacturer's label or retrofit label. Even if the label and fittings indicate the system has not been retrofitted, it is a very good idea to check the refrigerant composition. **Figure 15-6** shows a refrigerant identifier

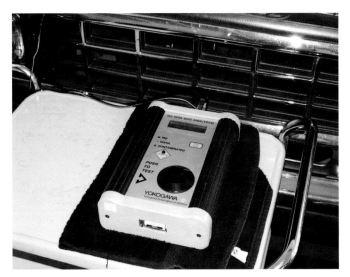

Figure 15-6. *Before performing any service to the refrigeration system, always identify the refrigerant. This may save you a lot of trouble later. This refrigerant identifier will identify the refrigerant as R-134a, R-12, or as unknown. It will also give the percentages of R-134a, R-12, and unknown components.*

being used to check an air conditioner. The type of refrigerant should match the retrofit label (if present) and the type of service ports. A good refrigerant identifier will also check for unknown refrigerants, R-22 blends, and for contamination by unknown gases.

Caution: Do not attach a refrigerant service center to a system until you can verify the composition of the refrigerant. If the refrigerant is OK, you can attach a refrigerant recovery/recycling service center. If the refrigerant is contaminated or you cannot verify its composition, use a set of manifold gauges to initially check the system charge.

Diagnosing Odor Complaint

Because of the dampness and cool conditions, the evaporator and blower case create an environment for the growth of mold and mildew. This problem is usually seen in areas with hot humid climates. Mold and mildew will cause the air coming out of the vents to have a musty smell. In most cases, the problem will disappear over time as climate conditions change. However, in some cases, the problem may persist due to leaves or other debris in the evaporator case or microbial growth on the evaporator core face. In these cases, the evaporator and case needs to be disinfected. If debris is present in the case, it must be removed or else the problem will return in a short period of time.

If the customer complains the windows frequently fog up coupled with the smell of coolant, a leaking heater core may be the cause. A massive refrigerant leak from the evaporator could cause refrigerant oil to be sprayed in the blower case, giving the air an oily smell. There are many other causes of blower case odors, ranging from malfunctioning electrical components to dead vermin in the case.

Step 3—Determine Which System Is Causing the Problem

The third step is to determine which HVAC system components or vehicle systems could cause the problem. The first reaction to what appears to be a refrigeration problem is to decide whether or not the refrigeration system is defective. However, the refrigeration system is composed of mechanical and electrical parts and interacts with other vehicle systems. To determine the source of the problem, you must combine the information you obtained in Steps 1 and 2 with the knowledge you obtain by making a system performance test as part of this step.

Instead of looking for something obviously wrong, as you did in Step two, you are using the performance test to check for something that could cause the specific problem. This will also help you to eliminate things that could not cause the problem, so in Step 4 you can concentrate on any suspected components.

Functional and Performance Tests

The following is a general system function and performance test. A *functional test* checks for proper system operation at different settings. The *performance test* checks the refrigeration and heating system components for proper pressures and temperatures. Some of the test procedures do not apply to every system. Always make sure you obtain and use the manufacturer's procedures and specifications for function and performance tests.

The functional test can be performed without gauges or a refrigerant service center. Start the engine and allow it to run for five minutes. Then, perform the functional test steps outlined in the service manual. A typical functional test procedure is shown in **Figure 15-7**.

To make the performance test, shut off the engine, make sure the transmission is in Park or Neutral, and set the parking brake. Attach gauges or a refrigerant service center as shown in **Figure 15-8**. Ensure the high and low side hoses are attached properly.

 Caution: If there is no refrigerant in the system, do not attempt the performance test. Instead go to Step 4.

Once the gauges are attached, check static pressure. A normally charged system will have 70-125 PSI (482-861 kPa) when it has been inactive for about one hour. If the

Figure 15-8. *This illustration shows a typical refrigerant service center used for servicing the refrigeration system.*

gauges show low or no pressure in the system, you can be sure there is a leak somewhere. Be sure the hoses do not contact any moving parts.

Install a temperature gauge in the vent nearest the evaporator, **Figure 15-9**. Then start the engine and set it to run at approximately 1500 to 2000 RPM (this will vary by manufacturer). Turn the HVAC control panel settings to the maximum cooling position and set the temperature switch to the maximum cold position. Turn the blower speed switch to the high position and open the front windows.

	Control Settings			System Response					
Step	Mode	Temp Set	Blower Switch	Blower Speed	Heater Outlet	A/C Outlet	Def. Outlet	Side Wind Def	A/C Comp
1	Defr	Full hot	Off	Off	No air flow	No air flow	No air flow	No air flow	Off
2	Defr	Full hot	Hi	Hi	Some hot air flow	No air flow	Hot air flow	Some hot air flow	On
3	Defog	Full hot	Hi	Hi	Hot air flow	No air flow	Hot air flow	Some hot air flow	Off
4	Heat	Full hot	Hi	Hi	Hot air flow	No air flow	Some hot air flow	Some hot air flow	Off
5	Vent	Full hot	Hi	Hi	No air flow	Hot air flow	No air flow	No air flow	Off
6	Bi-lev	Full cold	Hi	Hi	Air flow	Cold air flow	Some cold airflow	Some cold air flow	On
7	A/C	Full cold	Hi	Hi	No air flow	Cold air flow	No air flow	No air flow	On
8	A/C	Full cold	Lo	Lo	No air flow	Cold air flow	No air flow	No air flow	On
9	Max	Full cold	Hi	Hi	No air flow	Cold air flow	No air flow	No air flow	On
10	Max	Full cold	Lo	Lo	No air flow	Cold air flow	No air flow	No air flow	On

Figure 15-7. *Functional test chart as used by one manufacturer. Each vehicle typically has its own test chart sequence. (General Motors)*

Figure 15-9. *A mechanical temperature gauge installed in the outlet nearest the evaporator provides an accurate reading of evaporator temperature.*

Check the compressor clutch to make sure it is engaged. If the clutch does not engage, shut off the HVAC system and engine and check the clutch, relay, switches, and wiring. Basic electrical system checks were outlined in Chapter 4.

If the compressor clutch engages, allow the refrigeration system to operate for about five minutes to stabilize the gauge readings. Monitor the cooling system gauge or light to ensure the engine does not overheat. Observe the fan clutch or fan motor(s) and ensure they are operating and moving air through the condenser and radiator.

 Caution: If the cooling system fans are not operating, or if the high side pressure exceeds 325 psi (2467 kPa), stop the performance test immediately and determine the cause.

Go through all the steps outlined in the service manual for testing the system. A typical performance test chart is shown in **Figure 15-10.**

Relative Humidity (%)	Ambient Air Temp		Maximum Low Side Pressure		Engine Speed (rpm)	Maximum Right Center Air Outlet Temperature		Maximum High Side Pressure	
	°F	°C	PSIG	kPaG		°F	°C	PSIG	kPaG
20	70	21	37	255	2000	46	8	225	1551
	80	27	37	255		47	8	275	1896
	90	32	37	255		53	12	325	2241
	100	38	38	262		54	12	325	2241
30	70	21	37	255	2000	48	9	240	1655
	80	27	37	255		50	10	285	1965
	90	32	39	269		57	14	340	2344
	100	38	43	296		60	16	360	2482
40	70	21	37	255	2000	49	9	260	1793
	80	27	37	255		53	12	305	2103
	90	32	42	290		60	16	355	3137
	100	38	49	338		66	19	395	2724
50	70	21	37	255	2000	51	11	275	1896
	80	27	39	269		56	13	320	2206
	90	32	46	317		63	17	375	2586
	100	38	55	379		72	22	430	2965
60	70	21	37	255	2000	53	12	290	2000
	80	27	42	290		59	15	340	2344
	90	32	49	338		66	19	390	2689
	100	38	60	414		78	26	445	3068
70	70	21	37	255	2000	55	13	305	2103
	80	27	45	310		62	17	355	2448
	90	32	53	365		70	21	405	2792
80	70	21	41	283	2000	56	13	320	2206
	80	27	48	331		65	18	370	2551
	90	32	57	393		73	23	420	2896
90	70	21	45	310	2000	58	14	335	2310
	80	27	52	359		68	20	385	2655

Figure 15-10. *After the ambient temperatures and system pressures have been determined, the technician can refer to a chart showing the relationship of pressures and temperatures of a given refrigerant. He or she can use the chart to determine whether the system has a problem. (General Motors)*

Note: Gauge readings will vary with temperature, humidity, system design, and type of refrigerant. For this reason, you should always refer to the manufacturer's specifications before deciding the refrigeration system is defective.

Normal Operation

If the system is operating normally, high side pressure should be between 150 and 300 PSI (1034 to 2067 kPa) on a R-134a system. R-12 high pressures are usually somewhat lower, about 250 PSI (1723 kPa). Low side pressures should be between 30 and 40 PSI (208 and 276 kPa). Normal system pressures for each system type are shown in **Figure 15-11.** Based on the pressure gauge readings, you may want to proceed to Step 4 and consult available diagnostic and troubleshooting charts.

If the pressure readings are within specifications, observe the evaporator outlet line. It should be covered with condensed moisture, possibly frozen. This is a visual sign the system is working properly. If the system uses an accumulator, touch the inlet and outlet tubes, **Figure 15-12.** If the system is fully charged and working properly, the temperature should be roughly equal at both pipes.

> ⚠ **Warning: High pressure lines can become extremely hot. Touching a line for an extended period of time can result in a burn.**

Briefly touch the low and high pressure lines. All high side lines should be hot, while the low pressure lines should be cold. If the system uses a sight glass, check the

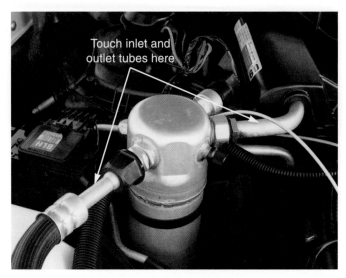

Figure 15-12. *Touching the lines going to and from the accumulator is a quick, but good test of system performance. Do not attempt this on receiver-drier systems.*

glass for foaming. After the system has been running for five minutes on a reasonably warm day (70°F or 20°C), the glass should be clear. See the additional sight glass information included in **Figure 15-13.**

Check the temperature gauge in the outlet vent. Outlet temperature will vary with outside air temperature and humidity. As a general rule, the outlet temperature should be about 30°F (17°C) lower than the outside air temperature after the system has been operating for 5-10 minutes. If the gauge is not showing a reasonable drop in temperature, something is wrong.

Next, unplug the blower motor connection. This is usually easiest to do at the resistor assembly. With the

Normal System Operation		
Low Side (Suction)	**High Side (Discharge)**	
(Varies with ambient temperature)	(Varies with ambient temperature)	
Orifice Tube	25-35 PSI (172-241 kPa)	150-285 PSI (1034-1965 kPa)
Exp. Valve	15-35 PSI (103-241 kPa)	150-285 PSI (1034-1965 kPa)
VDOT	26-32 PSI (179-221 kPa)	150-285 PSI (1034-1965 kPa)
Other	Sight glass: Clear	Max A/C air temp: 40-50°F (4-10°C)

Figure 15-11. *Gauge readings during normal system operation. The color areas indicate low- and high-side pressures, as well as the typical numerical values indicated on the bottom. The normal color region will be used in other gauge examples in this chapter.*

Item to check	Adequate	Insufficient	Almost no refrigerant	Too much refrigerant
State in sight glass	CLEAR Vapor bubbles sometimes appear when engine speed is increased or decreased. 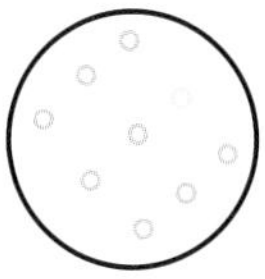	FOAMY or BUBBLY Vapor bubbles always appear. 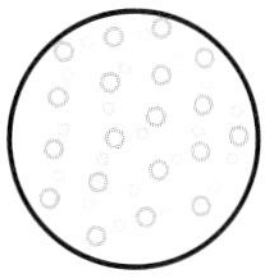	FROSTY Frost appears. 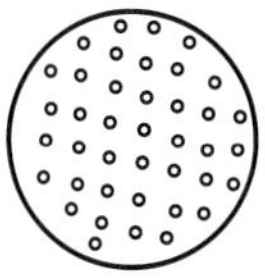	NO FOAM No vapor bubbles appear. 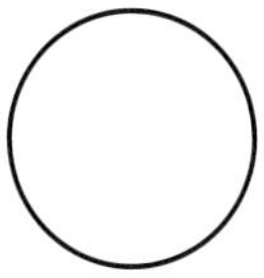
Temperature of high- and low-pressure lines	High-pressure side is hot while low-pressure side is cold. (A big temperature difference between high- and low-pressure side.)	High-pressure side is warm and low-pressure side is slightly cold. (Not so large a temperature difference between high- and low-pressure side.)	There is almost no temperature difference between high- and low-pressure side.	High-pressure side is hot and low-pressure side is slightly warm. (Slight temperature difference between high- and low-pressure side.)
Pressure of system	Both pressures on high- and low-pressure sides are normal.	Both pressures on high- and low-pressure sides are slightly normal.	High-pressure side is abnormally low.	Both pressure on high- and low-pressure sides are abnormally high.

Note: The condition of the bubbles in the sight glass, temperatures, and pressure are affected by ambient temperature and relative humidity.

Figure 15-13. *Many older refrigeration systems have a sight glass. Observing the sight glass after the refrigeration system has been running for a few minutes will tell the technician approximately how much refrigerant is in the system. (Nissan)*

blower not turning, there is no heat load on the evaporator, and the compressor clutch should cycle off within 30 seconds. If the system uses an evaporator pressure control valve (STV, POA, VIR, EPR) pressure should drop to 28-30 psi (193-207 kPa).

System Undercharge

A *system undercharge* is the most frequent problem found in a refrigeration system. Depending on how much refrigerant is left in the system, gauge pressures will read much lower than normal, even when factoring in air temperature. Low pressure tubing and hoses will feel warmer while high pressure tubing will feel cooler. The outlet temperature will be higher than normal. If the system has a sight glass, bubbles or foam will be present. **Figure 15-14** shows typical readings from an undercharged system.

The cause of a system undercharge is usually leaks, but could be caused by failing to fill the system with the proper charge. An undercharged system will not only provide inadequate cooling, but will fail to carry the necessary lubricants through the system. This can lead to reduced compressor life and eventual failure.

System Overcharge

System overcharge occurs quite frequently, in fact, almost as frequently as system undercharges. The first sign of a system overcharge is much higher than normal system

pressures, **Figure 15-15.** Cooling will be affected as the evaporator, accumulator/receiver-drier, and other system components are flooded with refrigerant.

In some cases, a system appearing to be overcharged contains air (sometimes called *non-condensable gas* or *NCG*). An overcharged system should be checked for leaks as the extra refrigerant may have been added because the system was undercharged. If no leaks are found, recover the refrigerant charge, evacuate, and recharge the system.

Restriction in Lines, Orifice Tubes, and Expansion Valves

System restrictions can easily be found by feeling the system's lines, hoses, and components. If the high side becomes cold at any point before the orifice tube or expansion valve, that spot is restricted. A restriction in the high side is usually located at the orifice tube or expansion valve, depending on the system. However, restrictions in lines and components, such as evaporators and condensers, can occur. **Figure 15-16** shows typical system pressures and symptoms when a line, orifice tube, or expansion valve is restricted.

Gauge pressures may be affected by the presence of a restriction. However, as mentioned earlier, variable displacement compressor systems may show little or no change. If a restriction is present, gauge pressures will usually be lower than normal and there will be no cooling. Orifice tube and expansion valve restrictions can be

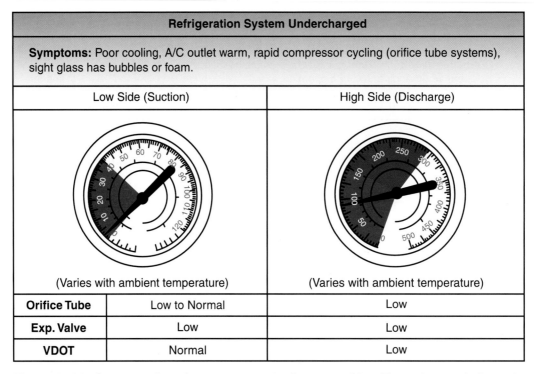

Figure 15-14. *Gauge readings for a system undercharge condition. The red areas indicate the regions where the gauge needles will typically be. They may cycle (increase and decrease in pressure) if the compressor clutch is cycling.*

Refrigeration System Undercharged	
Symptoms: Poor cooling, A/C outlet warm, rapid compressor cycling (orifice tube systems), sight glass has bubbles or foam.	
Low Side (Suction)	High Side (Discharge)
(Varies with ambient temperature)	(Varies with ambient temperature)
Orifice Tube — Low to Normal	Low
Exp. Valve — Low	Low
VDOT — Normal	Low

Refrigeration System Overcharged	
Symptoms: Fair to poor cooling, continuous compressor operation (orifice tube systems), sight glass clear or foamy.	
Low Side (Suction)	High Side (Discharge)
(Varies with ambient temperature)	(Varies with ambient temperature)
Orifice Tube — High	High
Exp. Valve — Normal to High	High
VDOT — Normal to High	High

Figure 15-15. *Gauge pressures for a system overcharge condition.*

caused by a defective compressor, a ruptured desiccant bag, or contaminants such as dirt or corrosion. If the system uses a thermostatic expansion valve, the sensing bulb should be tested for proper operation. A defective expansion valve can give readings similar to a plugged orifice.

Defective Accumulator/Receiver-Drier

Usually, accumulators and receiver-driers are not the source of air conditioning system problems. However, they can cause other system problems, such as restrictions in the orifice tube or expansion valve should the desiccant bag rupture. On accumulators and driers with an oil bleed

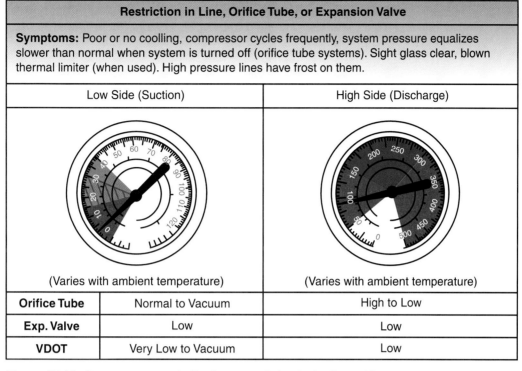

Restriction in Line, Orifice Tube, or Expansion Valve	
Symptoms: Poor or no coolling, compressor cycles frequently, system pressure equalizes slower than normal when system is turned off (orifice tube systems). Sight glass clear, blown thermal limiter (when used). High pressure lines have frost on them.	
Low Side (Suction)	High Side (Discharge)
(Varies with ambient temperature)	(Varies with ambient temperature)

	Low Side (Suction)	High Side (Discharge)
Orifice Tube	Normal to Vacuum	High to Low
Exp. Valve	Low	Low
VDOT	Very Low to Vacuum	Low

Figure 15-16. *Gauge pressures indicating a restriction in the line, orifice tube, or expansion valve. Do not allow a restricted system to run for an extended period of time as high pressures may cause a hose or line to burst.*

hole, compressor failure can be caused if this hole is restricted. A problem in an accumulator will usually show up as another problem in the system, and cannot be detected by gauge readings.

 Note: On receiver-drier systems, if the receiver-drier is hot to the touch, the expansion valve is defective or plugged. If the receiver-drier is cool, the receiver-drier is defective.

Defective Compressor

After the engine and refrigeration system have been operating for five minutes, observe and listen to the compressor. Watch the clutch operation carefully. When the outside temperature is low, (60°F or 16°C) the clutch may cycle every 20 seconds. When the air temperature is high (90°F or 30°C), the clutch may cycle every one or two minutes or more. On very hot and humid days, the clutch may not cycle. On a vehicle with an evaporator control valve, the clutch should remain engaged. If the clutch cycles excessively, the system may have a low charge.

Some compressor problems are easy to diagnose. A noisy compressor has usually failed or is about to fail. If the compressor will not turn or makes an extremely loud noise when engaged, it may be seized. Adding oil to the refrigeration system can sometimes quiet older compressors.

Other compressor problems are more difficult to detect. Diagnosing internal compressor problems requires skill at reading gauge pressures, **Figure 15-17.** Variable

displacement compressors are sometimes difficult to diagnose as some of them are able to adjust pressure so that even a system restriction would cause only a very minor pressure change. Usually, a good indicator of possible internal compressor problems is slightly lower than normal high side pressure with a confirmed full system charge. However, before the compressor is suspected, the system should be checked for restrictions and proper refrigerant charge.

Defective or Restricted Condenser

A defective condenser will usually show up as a leak, allowing the refrigerant charge to escape. Because the condenser handles high refrigerant pressures, a leak will usually be evident, even without the use of a leak detector. However, a slow leak from the condenser will allow refrigerant oil to escape, possibly leading to compressor failure.

Restrictions in the condenser can be either internal or external. An internal restriction will create higher than normal gauge pressure readings, **Figure 15-18.** In some cases, a restriction may cause ice or frost to form on the condenser. An external restriction will cause higher than normal gauge readings due to the lack of air passing through the condenser.

Defective or Restricted Evaporator

A defective evaporator will usually show up as inadequate cooling caused by a leak in the core. A restriction in the evaporator may cause ice to form on the high pressure tube leading to the evaporator. Both problems will cause lower than normal system pressures.

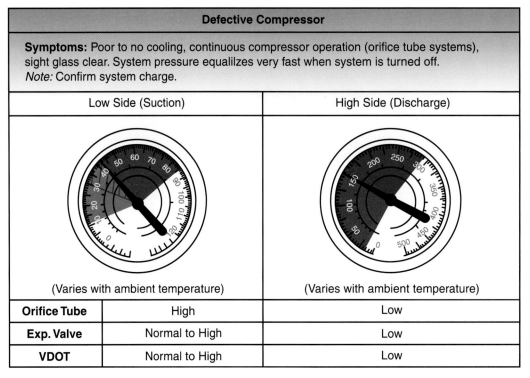

Defective Compressor	
Symptoms: Poor to no cooling, continuous compressor operation (orifice tube systems), sight glass clear. System pressure equalilzes very fast when system is turned off. *Note:* Confirm system charge.	
Low Side (Suction)	High Side (Discharge)
(Varies with ambient temperature)	(Varies with ambient temperature)

	Low Side (Suction)	High Side (Discharge)
Orifice Tube	High	Low
Exp. Valve	Normal to High	Low
VDOT	Normal to High	Low

Figure 15-17. *Gauge readings for a defective compressor.*

Condenser Restriction	
Symptoms: Poor cooling at low speeds, engine overheats, sight glass clear. Ice or frost forms on the condenser. Verify proper fan and cooling system operation.	
Low Side (Suction)	High Side (Discharge)
(Varies with ambient temperature)	(Varies with ambient temperature)

	Low Side (Suction)	High Side (Discharge)
Orifice Tube	Normal to High	High
Exp. Valve	Normal to High	High
VDOT	Normal to High	High

Figure 15-18. *Gauge readings for a restricted condenser.*

Defective or Misadjusted Switch

A defective pressure or thermostatic switch can also cause problems. Usually, problems caused by one of these switches can be diagnosed by looking for a frozen condensation at the evaporator inlet. The compressor on an orifice tube system will operate continuously. Typical system pressures are shown in **Figure 15-19.**

After all checks have been made, return the engine to idle, shut off the HVAC system and engine and go to Step 4.

Leak Detection

One of the most common refrigeration system diagnostic jobs you will do is locating leaks. It has been estimated that over 50% of all refrigeration problems are caused

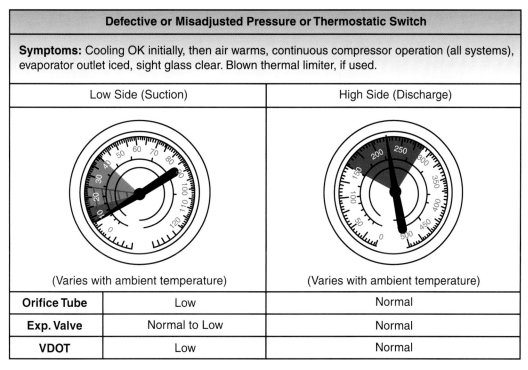

Defective or Misadjusted Pressure or Thermostatic Switch

Symptoms: Cooling OK initially, then air warms, continuous compressor operation (all systems), evaporator outlet iced, sight glass clear. Blown thermal limiter, if used.

	Low Side (Suction)	High Side (Discharge)
	(Varies with ambient temperature)	(Varies with ambient temperature)
Orifice Tube	Low	Normal
Exp. Valve	Normal to Low	Normal
VDOT	Low	Normal

Figure 15-19. *Gauge readings for a misadjusted or defective pressure switch.*

by system leaks. Leaks either cause performance problems or lead to failure of a system part, usually the compressor. Leak detection can be done by one of several methods. Visible evidence of oil on the refrigeration system fittings, compressor shaft, or evaporator drain hole means there is a leak. Oil, swelling, or a torn spot on the rubber covering of a hose usually mean that refrigerant is leaking from the hose.

If an obvious leak cannot be found, test for leaks using one of the methods explained in the following paragraphs.

Note: Due to the expense and potential environmental damage of refrigerants, any leak detected, no matter how insignificant, must be fixed. Do not simply add refrigerant because the leak does not seem to be excessive.

Ensure the System Is Charged

If there is no refrigerant in the system, none can leak out to be detected. To make a leak check, there should be a minimum low side refrigerant charge of 50 PSI (345 kPa) with the engine off. Some leaks, especially those on the high side of the system, may require a higher charge. If an obvious leak is so severe the system will not hold any pressure, repair that leak first, then pressurize the system.

Note: The compressors on some vehicles are disabled by the engine control computer if the refrigeration system loses its charge. When this occurs, a trouble code is usually set. A scan tool is usually required to clear this code in order for the compressor to operate.

Some technicians prefer to pressurize completely empty systems with nitrogen. If the system has only recently begun leaking, there may be enough refrigerant left to be detected by a sensitive leak detector. Pressurizing with nitrogen will also allow the technician to find a relatively large leak by using the *soap solution* method. Refrigeration systems can be pressurized up to about 150 PSI (1033 kPa) without damaging any of the low side components.

Remove Stray Refrigerant Vapors

Any leak detection device will produce a false leak signal if it contacts refrigerant vapors built up under the hood or in the shop. Before starting the leak checking procedure, run the engine briefly to remove any vapors from the engine compartment. If you suspect refrigerant vapor has built up in the shop, clear the vapor using fans or the shop ventilation system.

Using Leak Detection Devices

Modern HVAC shops use several leak testing devices. At one time, the flame type halide leak detector was widely used. Today, however, it has been largely replaced by electronic and dye detection devices. The following sections explain how to use various types of leak testing devices. Leak testing device construction was explained in Chapter 3.

Electronic Leak Detectors

Electronic leak detectors are more refrigerant sensitive than the other leak detection methods. Modern electronic detectors are extremely sensitive and can locate a leak as small as 1/2 ounce (15 ml) of refrigerant per year.

Begin the leak detection process by turning the detector on and allowing it to warm up for about one minute away from the refrigeration system components. Most leak detectors will make a ticking noise that increases when the probe encounters refrigerant. Large leaks raise the ticking to a high pitched squeal. Many leak detectors have a display which indicates the leak rate.

In some cases, the electronic detector's sensitivity must be reduced when a large leak is present or when other engine fumes trigger the detector. A satisfactory initial detector sensitivity setting would be to detect a leak rate of about 1 1/2 ounce (45 ml) per year. The sensitivity

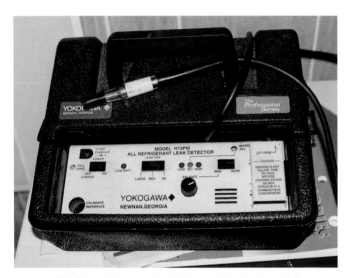

Figure 15-20. *An electronic detector will quickly locate the smallest R-12 and R-134a leaks. Electronic detectors can be adjusted to detect any size leak.*

adjustment knob is usually located on the detector face, **Figure 15-20.**

After setting sensitivity, slowly pass the sensing tip closely around possible leak areas and check for an increase in the ticking noise. Also remember to pass the tip under suspected leak areas. See **Figure 15-21.**

Dyes

Another leak detection method involves using **dyes.** A dye is injected into the refrigeration system and allowed to circulate. The dye will leak out along with any refrigerant and stain the components at the site of the leak.

The first refrigerant dyes were colored orange and were contained in a small can resembling a one-pound refrigerant can. The can was connected to the system low side through the gauge manifold. With the system operating, the dye was drawn into the system. After the dye circulated for a few minutes, the technician could look for orange dye at the site of leaks.

Modern dye injectors are designed to inject a fluorescent dye directly into the refrigeration system. The injector is attached to one of the system service ports and the handle is turned to force the dye into the system. The engine and HVAC system are started and the dye allowed to circulate for a few minutes. Then the technician shines a black light, **Figure 15-22,** onto the suspected leak areas. The black light makes the dye fluoresce, or shine, identifying the leak.

The technician should make sure the propellant and dye are compatible with the type of refrigerant and oil on the system. The dye must be soluble (dissolve) in the oil, and not affect the oil's lubricating properties.

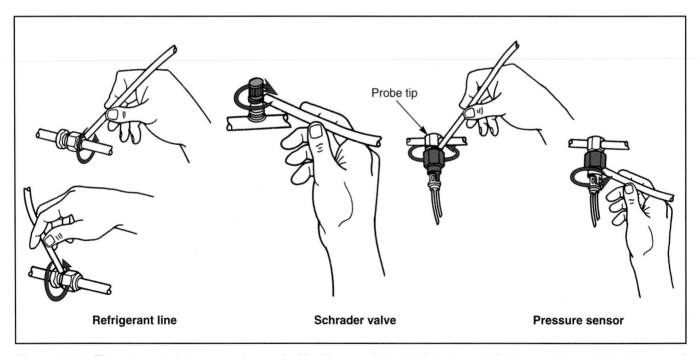

Probe tip

Refrigerant line   Schrader valve   Pressure sensor

Figure 15-21. *These trace techniques can be used with either an electronic detector or halide torch. Always check the area all around the suspected leak location. (General Motors)*

Figure 15-22. *Dye and black light are often used to check for small leaks. The dye is injected into the system and allowed to circulate. The black light will illuminate the dye as it leaks out. (Tracerline)*

 Note: Some R-134a refrigerant manufacturers add fluorescent dye to their refrigerant. They are marketed under several names.

Soap Solution

The **soap solution** method will find large leaks only, and should not be relied on to locate small leaks or leaks in inaccessible locations. Soap solutions are often used with nitrogen pressurizing to check for leaks. It is also an easy way to confirm what appears to be an obvious leak. To make a soap solution test, ensure the refrigeration system has pressure. Then mix a small amount of soap with water. Dishwashing liquid is best, but almost any kind of soap will work.

 Note: Commercial leak checking solutions are available.

Spray or pour the soap solution on the area of the suspected leak. Leaking refrigerant will form bubbles. The size of the bubbles and how rapidly they form will increase with the size of the leak, **Figure 15-23.** Slight foaming indicates a small leak, while large bubbles are a sign of a serious leak. If bubbles form at a rate faster than one per second, the leak is severe.

Figure 15-23. *A soap solution can be used to locate large leaks. The rate and size of the bubbles indicate the size of the leak. (Saturn)*

Flame Leak Detector

The flame leak detector, sometimes called the *halide torch,* was used for many years and still does a good job of finding moderate to large leaks on systems. The principle of the flame type leak detector is simple: a flame from a propane cylinder changes color when refrigerant enters through a sensing hose. The color and intensity of the flame can be used to determine the size of the leak.

⚠ **Warning: If the system is filled with an unidentified refrigerant blend, *do not* use a halide torch to check for leaks. Some blends may contain propane or butane, and the leak detector flame may cause a fire or explosion.**

Before lighting the flame detector, always make sure the vehicle has no fuel leaks, and no flammable fumes are present in the shop. Be sure the flame detector is used in a well ventilated shop. If at all possible, try to make the test outdoors.

⚠ **Warning: The halide leak detector flame breaks down R-12 refrigerant, creating phosgene, a poisonous gas. Always make sure the work area is well ventilated before using a flame type detector.**

To use the flame leak detector, light the torch and allow the flame to heat the copper reaction plate, **Figure 15-24.** Adjust the burner until it gives a yellow flame. Hold the detector upright as you slowly pass the hose around all joints, hoses, sealing flanges, and other potential leak spots. Also

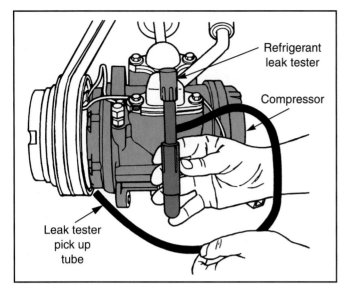

Figure 15-24. *Using a torch to check for refrigerant leaks. Remember a torch is less efficient than an electronic detector. Also keep in mind that breathing the torch fumes will expose you to poison gas. (DaimlerChrysler)*

pass the hose under suspected leak areas. Since refrigerant is heavier than air, it should flow downward from a leak. Do not try to check for leaks with the engine running, since refrigerant will be blown away from the leak.

Observe the flame as the hose moves under each potential leak area. If the flame turns blue or green, the hose has passed near a leak. A small leak will give the flame a greenish tint, while a bright blue flame indicates a large leak, Recheck the suspect area until the leak has been pinpointed. Do not breathe the fumes from the leak detector. When you are through using the flame detector, make sure the propane valve is closed tightly.

Step 4—Eliminate Other Causes of the Problem

In the fourth step, you think about the observations you made in Step 3 and begin eliminating the causes of the problem, one by one. Always begin by checking the components or systems that are the most likely sources of the problem. For instance, you may need to search for a hidden or slight refrigerant leak, as will be explained later in this chapter. In many cases, you may need to raise the vehicle, or remove shrouds or parts of the blower case.

If there are no obvious problems, go on to make more involved checks. During this step, you should check for problems in related systems such as the blower motor, cabin filter (when used), and diverter doors. Checking related systems is very important if the refrigeration system pressures are good but the discharge air is not cold. You can spend a lot of time working on the refrigeration system if you do not realize the blend door cable is broken. Also check for unusual problems such as a condenser or evaporator core clogged with dirt, leaves, or lint.

Troubleshooting charts are useful for determining what is wrong with a refrigeration system. Always obtain the correct manufacturer's chart when troubleshooting an actual HVAC system.

Step 5—Recheck the Cause of the Problem

In this step, the cause of the problem determined in Step 4 is rechecked. This step requires reviewing the various test procedures performed in the last step, and determining whether the suspect component is likely to be the source of the problem. It is often helpful to take a short break to consider all possible causes and determine if what you have found is the most likely cause of the problem or the only thing that could be defective. Review how the particular system works, and how the defect could cause the system problem.

Before going to the next step, recheck the condition of the suspect part as much as possible. Also recheck all other related parts. This will ensure you have not condemned the wrong part or overlooked another defect. For instance, if one O-ring is leaking, do not assume it is the only defective seal. Thoroughly check the rest of the refrigeration system for leaks.

Always Perform Additional Tests

Additional testing is especially important when the suspected part is a solid-state or an otherwise untestable device, such as an automatic temperature control assembly. Such parts are too expensive to simply replace without knowing for sure whether they are good or bad. Making further checks to confirm the problem is always a good idea, if only to increase your confidence about finding the defective part. Not many technicians are sorry they made further checks, but a lot of them are sorry they did not.

Deciding on Needed Work

Deciding on needed work is a process of interpreting the results of all diagnostic tests. It is simply a matter of taking all test readings and deciding what they mean. As discussed earlier, the test results can be simple observations of visible defects, detailed readings from elaborate test equipment, or any procedure in between.

Before condemning any part based on test results, mentally review its interaction within the system and with the various engine and vehicle systems. Then decide whether the part in question can cause the particular test reading or symptom. For instance, if the HVAC system is losing refrigerant and you have located a leak at the evaporator, do not assume it is the only source of leaks. Check the entire system thoroughly before giving an estimate.

Troubleshooting charts and other diagnostic information can be a great asset to this process, **Figure 15-25.** If

International Mobile Air Conditioning Association

CODE OF PROFESSIONAL PRACTICE
Inspection Checklist

This Inspection Checklist is a summary of the steps detailed in the IMACA Code of Professional Practice manual. For detailed information, consult the manual.

VISUAL INSPECTION - Engine Compartment

1) Hoses, tubing and connections (Suction, Discharge & Liquid lines)
A. Examine exterior for deterioration, blistering, bubbling refrigerant, oil stains and battery acid damage or burns. Oil stains could indicate leakage.
B. Check for incorrect routing, rubbing, missing hardware or loose hoses, bent or collapsed tubing.

2) Compressor
A. Examine exterior for damaged or missing bolts / hardware, broken housing, oil stains.
B. Internal - rotate compressor 2 complete turns by hand to determine if seized or locked up.

3) Clutch
A. Examine for broken springs, burnt face, damaged grooves, oil stains from seal leak.

4) Service Ports
A. Check size and thread to determine type of refrigerant - confirm with engine compartment label.
B. Examine ports for missing caps, damaged threads and leaking Schrader valves.

5) Condenser
A. Check for loose or damaged connections, loose or missing hardware or air dams.
B. Examine coil for bent or damaged fins, restriction due to debris or dirt, oil stains.

6) Expansion Device (if possible)
A. Examine for physical damage or oil stains.

7) Evaporator Pressure Regulator (early models)
A. Check POA, EPR, STV (if equipped) for physical damage or oil stains.

8) Cabin Air (Evaporator) Filter (if equipped)
A. Check for physical damage, oil stains and for proper installation.

9) Accumulator or Receiver/Drier
A. Check for physical damage, loose or missing hardware, loose connections or oil stains
B. Examine sight glass (Receiver/Drier only) for stains.

10) Drive Belts
A. Check for missing or damaged pulleys and tensioners; routing, tension and alignment.
B. Examine condition of belts for cracking, checking and excessive wear.

11) O-Rings, Gaskets, Spring Locks (if equipped)
A. Examine all connections not previously inspected for loose or missing parts and oil stains.

12) Inline Filter
A. Check for physical damage or oil stains.

13) Fan Clutch & Blade
A. Examine for fluid leakage or excessive bearing wear.
B. Check for damaged or bent blades on fan.

14) Electrical Components
A. Examine connectors for loose, burnt, broken parts or corrosion.
B. Examine wiring harness for burns, cracks or rubbing on insulation.

VISUAL INSPECTION - Passenger Compartment

1) Air Flow
A. Check all louvers for directional movement and air flow.

2) Control Head
A. Check all blower speeds.
B. Check controls for proper movement and function.
C. Check operation for heater, defrost, and A/C.

3) Interior Condition
A. Check evaporator / heater case for water leakage.
B. Check carpet for water damage.

LEAK CHECK

1) Preparation (Engine Off)
A. A refrigerant identification check is recommended to verify the refrigerant in use or to detect flammables, unknown or contaminated refrigerant.
B. Connect manifold gauge - refrigerant pressure must read 50 psig or more. (Add refrigerant to increase the pressure if necessary).
C. Clean all connections using a clean dry rag.

2) Leak Check (Engine Off)
A. Start at the compressor discharge port and follow the flow of refrigerant through the system. Move the detector sensor completely around each connection.
B. Refrigerant is heavier than air; leak-check the underside of the hoses, clean and leak-check the condensate drain tube(s).
C. Leak check the compressor shaft seal.
D. Leak check evaporator through dash vents.

REPAIR SYSTEM PER CUSTOMER APPROVAL

FINAL PERFORMANCE EVALUATION

1) Functional Inspection (Engine On)
A. Check the compressor clutch for proper operation.
B. Check evaporator blower at all speeds
C. Check operation of function control doors for fresh air/recirculate (A/C - Max A/C), Dash louvers, Floor outlets, Defroster outlets
D. Check operation of heater flow control (if equipped).
E. Check fan clutch (if equipped). Once engine has reached normal operating temperature, turn off engine and "soak" fan clutch for 2 minutes. Restart engine, fan clutch should be engaged. With engine OFF, spin fan, should rotate maximum of 2 turns.
F. Check electrically driven condenser/radiator fan(s) (if equipped).

2) System Checkout (1,200 rpm's, condenser air flow = 35 mph)
A. Measure and record (on Inspection Report) temperature 2" in front of center of radiator. (Ambient temperature.)
B. Set A/C controls to OEM specs. Allow system to stabilize 5-10 minutes.
C. Record the high- and low-side pressure.
D. Record the center louver temperature and interior temperature.
E. Check operation of temperature controls.

3) Post Service Inspection
A. A final refrigerant check is optional to verify purity of refrigerant and absence of air.
B. Install service port caps and perform final leak check.
C. Perform final visual inspection. Check for tools and loose components.
D. Record results on Inspection Report.

CAUTION: Safety Glasses must be worn during any A/C diagnosis or repair procedure. Refrigerant may cause blindness if it comes in contact with eyes.

Figure 15-25. *The troubleshooting inspection checklist covers most modern refrigeration systems. For detailed testing, always use the correct troubleshooting chart. (IMACA)*

researched and prepared correctly, the troubleshooting chart will list all the possible causes of the problem, allowing you to check everything in a logical sequence. Properly used, such information will speed up the checking and isolating process.

Deciding on the Proper Repair Steps to Take

The amount and type of corrective action must also be determined. In some cases, the repair is as simple as reattaching a vacuum hose, removing grease, dirt, or debris from a sensor connection, or tightening a belt. In other cases, major unserviceable parts, such as the evaporator or condenser, must be replaced to correct the problem. To reduce the possibility of future problems, you should also service parts that interact with the defective part. An example is replacing a fixed orifice tube when the compressor is replaced. In all cases, the technician must thoroughly determine the extent of the repairs before proceeding.

Factors that must be considered when deciding to adjust, rebuild, or replace a part are ease of adjustment, the need for special tools, cost of the replacement part, and the possibility the old part will fail again.

If a part is easily adjustable, you can try the adjustment procedure before rebuilding or replacing. Generally, most HVAC parts cannot be adjusted. If adjusting the part does not restore its original performance, the part can still be rebuilt or replaced with little time lost. If there is any doubt about whether an adjustment has corrected a problem, replace the part.

Rebuild or Replace?

In cases when a defective component can be rebuilt, the investment in materials and time must be weighed against the possibility that rebuilding the part may not fix the problem. It is often cheaper to install a new part than to spend time rebuilding the old one. Many repair shops, and even some new vehicle manufacturers, are going increasingly to a policy of replacing complete assemblies. You must determine if rebuilding is cost effective. However, keep in mind that most HVAC parts cannot be rebuilt. Parts that can be rebuilt include the compressor, radiator, and water pump.

In many cases, the customer will come out ahead with a new or remanufactured assembly instead of paying to rebuild an old part. The price of the new or remanufactured part is often less than the charge to rebuild the old part. These parts often come with a limited warranty from the remanufacturer and the assurance the part was assembled in a clean, controlled environment. The technician will often come out ahead, since the labor time saved rebuilding the old part can be devoted to other work.

Along with the cost of repairs, another factor that must be considered is the necessity to retrofit the vehicle to use another refrigerant. The cost of retrofitting will have to be included, especially if the parts being replaced are not normally serviced during a retrofit.

Therefore, when deciding what to do to correct a problem, make sure that all parts that could contribute to the problem have been tested. In one form or another, every possible component and system should be tested. Then you can decide with assurance what components are defective.

Special Tools

Special tools are often needed to adjust or disassemble a complex assembly, such as a compressor. Often, the cost of the tool may exceed the price of a complete replacement assembly. However, special tools can be used again for the same type of repairs in the future, and may be a good investment. You should also figure in the initial cost of the tool versus the number of jobs that will be possible using that tool. If you expect to do a lot of the same type of repairs in the future, and the special tools are reasonably priced, they should be purchased.

Contacting the Owner about Needed Work

After determining the parts and labor necessary to correct the problem and before proceeding to actually make repairs, contact the vehicle owner and get authorization to perform the repairs. The best way is to show the owner the completed inspection form. *Never* assume the owner will want the work done. The owner may not have sufficient money for the repairs, may prefer to invest the money in another vehicle, or prefer to have someone else perform the repair work. The defective part or problem may be covered by the vehicle manufacturer's warranty or a guarantee given by another repair shop or chain of service centers. In these cases, the vehicle must be returned to an approved service facility for repairs. If your shop is not one of these approved facilities, you cannot expect to be reimbursed for any more than diagnosing the problem.

If the vehicle is leased, the leaseholder is the actual owner. Depending on the terms of the lease, the leaseholder may be the only one who can approve any expenses in connection with the vehicle. Be especially careful if the vehicle is covered by an extended warranty or service contract. Extended warranties and service contracts are a form of insurance, and like all types of insurance, it is necessary to file a claim for any expenses. In some cases, the owner can file a claim after repairs are completed, while in other cases, approval must be granted from the insurer before the repair work can begin. Sometimes, the insurer will send an adjuster to inspect the vehicle before approval is granted.

Before talking to the vehicle owner, leaseholder, or extended warranty company concerning authorization to perform needed repairs, you should make sure you can answer three questions that will be asked. First, be prepared to tell exactly what work needs to be done, and why. Next, have available a careful breakdown of both part and labor costs. Third, be ready to give an approximate time when the vehicle will be ready. If you suspect a problem that requires further disassembly, be sure the customer understands that further diagnosis (and costs) may be needed before an exact price is reached.

Customer Name:	Sylvia Smith		License No.:	TKD 330
Automobile Year/Make:	97 Dodge	Model: Neon	Engine Size:	2.0 L
Inspection Performed By:	J. Seeker		Date:	

Procedure	Recommendations	Estimated Cost of Repairs	
		Parts	Labor
VISUAL INSPECTION - Engine Compartment			
1) Hoses, tubing and connections (Suction, Discharge & Liquid Lines)	Leak at liquid line, near condenser	48.00	55.00
2) Compressor	OK		
3) Compressor Clutch	OK		
4) Service Ports	OK		
5) Condenser	OK		
6) Expansion Valve/Orifice Tube	OK		
7) Evaporator Pressure Regulator (POA, STV or VIR)	N/A		
8) Cabin Air (Evaporator) Filter (if equipped)	N/A		
9) Accumulator/Drier	OK		
10) Drive Belts, Pulleys and Tensioners	OK		
11) O-rings, Gaskets, Seals and Spring Locks			
12) Inline Filter	N/A		
13) Electric Fan, Fan Clutch & Fan Blade	OK		
14) Electrical Components	OK		
VISUAL INSPECTION - Passenger Compartment			
1) Air ducts, louvers, sensors, control knobs and cables	OK		
2) Control Head	OK		
3) Interior Condition	OK		
LEAK CHECK - Engine Compartment (NOTE: Engine must be off during this procedure)			
1) Refrigerant Check	40% charge	55.00	55.00
2) Results of Leak Check	Located leak in line near condenser		
Subtotal Of Estimated Repair Costs		103.00	110.00

Total Estimated Cost of Repairs Based On This Inspection:	213.00

The above inspection was done in accordance with the IMACA Code Of Professional Practice procedures manual. If repairs are recommended you will be provided an estimate and only the repairs authorized by you (the customer) will be made. If further repairs are necessary you will be informed of and approve the additional parts and labor costs before the repairs are performed.

Thank You for your business!

Manager: _____
Date: _____

INITIAL PERFORMANCE EVALUATION

Type of Refrigerant: R-134a Purity ☒Yes __100__ % ☐No High-Side Press.: 120 Low-Side Press.: 12

Louver Temperature: 66   Interior Temperature: 85   Ambient Temperature: 92

Amount of Refrigerant Added to system: 1.14 lbs   Amount of Refrigerant Recovered from system: – 67 lbs

FINAL PERFORMANCE EVALUATION

Type of Refrigerant: R-134a Purity ☒Yes __100__ % ☐No High-Side Press.: 190 Low-Side Press.: 36

Louver Temperature: 46   Interior Temperature: 68   Ambient Temperature: 92

Amount of Refrigerant Added to system: 1.89 lbs

Figure 15-26. *A typical filled out inspection form. Note how following the checking procedures will uncover all problems. This leads to the correct diagnosis. Parts and labor costs can then be figured to give the owner a clear idea of what is wrong and what will take to correct the problem. (IMACA)*

Step 6—Correct the Defect

In Step 6, you correct the defect by making system repairs as needed. This repair can be as simple as tightening a loose fitting or may require replacement of almost every part in the refrigeration system. For repairs, refer to the procedures in the following chapters.

Be sure to completely fix the problem. Do not, for instance, correct leaks and let the vehicle go with a worn compressor. Keep in mind that disassembling the HVAC and refrigeration systems often uncovers other problems. Be sure to inform the owner about additional charges and get an ok before starting repairs.

Step 7—Recheck System Operation

Recheck system operation by conducting another performance test, checking refrigeration pressures and output temperatures. Do not skip this step, since it allows you to determine whether the previous steps corrected the problem. If necessary, repeat Steps 1 through 6 until Step 7 indicates the problem has been fixed. If you are satisfied the problem has been corrected, road test the vehicle to ensure there are no other problems and the repair you made actually corrected the customer's problem.

Follow-up

Once the seven-step checking process has isolated and cured the immediate problem, your first impulse is park the vehicle and get on to the next job. However, it is worth your time to think for a minute and decide whether the defect you found is really the ultimate cause of the problem. This process is known as *follow-up.*

For example a customer brings a vehicle into the shop complaining of poor cooling and rapid compressor clutch cycling. The refrigerant is a little low, so you add about 1/2 pound and the clutch cycling returns to normal. Do not assume the vehicle is fixed until you ask yourself where that 1/2 pound of refrigerant went. In this case, there is most likely an undiscovered leak that may soon empty the refrigeration system. If you do not locate the real problem, the vehicle will be back soon, along with a dissatisfied customer.

Hidden defects are common, and may cause a vehicle to return again and again with the same defective part. Do not let the vehicle leave until you are reasonably sure the observed defect is the real source of the problem. Some hidden problems can be tricky, such as a high side seal that checks out ok with the engine off but leaks at high pressures, or a bad relay that ruins a series of HVAC control modules. This is where good diagnostic skills and customer feedback can be helpful.

Whenever you work on a refrigeration system, always try to determine the real cause of a failure, even when the problem appears to be simple.

Documentation of Repairs

Part of the follow-up process includes writing on the repair order what the problem was, and what was done to correct the problem. This is called *documentation* and it is a vital part of the diagnostic process. Every repair order line should have three things. These three things are:
- ❑ What the driver's complaint was.
- ❑ The cause of the complaint.
- ❑ What was done to correct the complaint.

This type of documentation not only allows the driver to clearly see what was done to correct the vehicle's problem, it also supplies a good history of what has been done, **Figure 15-26.** If the vehicle should come back with a similar problem, it gives you or the technician working on the vehicle a place to start looking, without having to repeat some of the steps you took to find the problem.

Remaining Calm

One of the hardest principles of diagnosis is to remain calm. Mastering your own emotions is often the hardest thing to do, especially if you meet with a series of dead ends while looking for a problem or are having to deal with an angry customer, but it is necessary. Nothing will be accomplished by losing your composure. If you lose your composure, you will waste valuable time and possibly upset the customer. If you have picked up a tendency to overreact to situations, you must unlearn this behavior and teach yourself to remain calm. Only a calm person can think logically.

Summary

When troubleshooting any refrigeration system or other HVAC system problem, always proceed logically. The seven step troubleshooting process enables the technician to quickly locate and correct refrigeration system problems.

The first step is to determine the exact problem. This usually involves questioning the driver. A series of questions is the best way to determine the exact problem. In Step 2 check for obvious problems, or problems that can be easily tested. At this time attach gauges or a charging station and check static pressure. Excessively high or low pressures in the system indicate a problem. The third step is to determine which refrigeration system components or systems could cause the problem. Do this by conducting a system performance test. In Step 4 put together the information you gathered in the first three steps. Begin by checking the components or systems that are the most likely sources of the problem. In step 5 double check the cause of the problem determined in Step 4. In Step 6 correct the defect and in Step 7 recheck system operation. Do this by conducting another performance test. If necessary, repeat Steps 1 through 6 until Step 7 indicates the problem has been fixed.

Locating leaks is one of the most common refrigeration system diagnosis jobs. Oil on the refrigeration system fittings, compressor shaft, or evaporator drain hole indicates a leak. If an obvious leak cannot be found, use one of the following leak testing methods. Before leak checking, always make sure there is some pressure in the system. Some technicians pressurize completely empty systems with nitrogen. At one time the flame type leak detector was widely used. Today, however, electronic leak detectors are more common. They are more refrigerant sensitive than flame leak detectors.

Another common leak detection method involves injecting a dye into the refrigeration system and allowing it to circulate. The dye will leak out with the refrigerant and stain the components at the leak. A soap solution test can be used to locate large leaks.

Review Questions—Chapter 15

Please do not write in this text. Write your answers on a separate sheet of paper.

1. Strategy-based diagnostics involves reasoning through a problem in a _____ of steps.

2. List *in order* the seven step troubleshooting process.

3. Try to talk to the vehicle driver to find out the _____ complaint.

4. If the brakes are almost completely inoperable, should you go on a road test?

5. If you do not locate the real HVAC system problem, what will happen?

6. The refrigerant system must be _____ for any leak testing procedure to work.

7. The _____ on some vehicles are disabled by the engine control computer if the refrigeration system loses its charge.

8. The most effective leak detector is the _____ type.

9. A troubleshooting _____ will often simplify diagnosis.

10. _____ is very important in the troubleshooting process.

ASE Certification-Type Questions

1. All of the following are good questions to ask the driver to help with diagnosis, *except:*
   (A) does it happen whenever the engine is running?
   (B) do you hear any unusual noises?
   (C) how do you plan to pay for this?
   (D) does the air come out of the vents or somewhere else?

2. Technician A says if the vehicle has been retrofitted, the service fittings should be different from the original fittings. Technician B says if the vehicle has been retrofitted, a retrofit label should be installed under the hood. Who is right?
   (A) A only.
   (B) B only.
   (C) Both A and B.
   (D) Neither A nor B.

3. A good refrigerant identifier will be able to check for all of the following, *except:*
   (A) unknown refrigerants.
   (B) unknown refrigerant oils.
   (C) R-22 blends.
   (D) contamination by unknown gases.

4. Technician A says high refrigeration system static pressure indicates the vehicle has been retrofitted. Technician B says low refrigeration system static pressure indicates a refrigeration system leak. Who is right?
   (A) A only.
   (B) B only.
   (C) Both A and B.
   (D) Neither A nor B.

5. All of the following are preliminary steps for the refrigeration system performance check, *except:*
   (A) set the parking brake.
   (B) install a temperature gauge in the vent nearest the evaporator.
   (C) turn the HVAC control panel settings to the *off* position.
   (D) open the front windows.

6. If during the refrigeration system performance test you notice the cooling fan(s) are not operating, which of the following should you do *next?*
   (A) Stop the performance test.
   (B) Lower the engine speed to idle.
   (C) Add coolant to the engine radiator.
   (D) Turn the HVAC switch to vent.

7. Technician A says disconnecting the blower on a system with an evaporator pressure control device should cause the evaporator pressure to drop to about 28-30 psi (193-207 kPa). Technician B says disconnecting the blower on a system with an evaporator pressure control device should make the clutch cycle off in about 30 seconds. Who is right?

   (A) A only.
   (B) B only.
   (C) Both A and B.
   (D) Neither A nor B.

8. Refrigeration system pressures are correct on a cycling clutch type refrigeration system. No cool air is coming from the vents. Which of the following is the *least likely* cause?

   (A) Blend door cable broken.
   (B) Dirt on the evaporator.
   (C) Disconnected vacuum line at the heater/air conditioner door.
   (D) Plugged orifice tube.

9. Halide torch refrigerant leak checkers are being discussed. Technician A says a disadvantage of these checkers is they only work on R-12 or R-22. Technician B says a disadvantage of these checkers is they produce a poisonous gas. Who is right?

   (A) A only.
   (B) B only.
   (C) Both A and B.
   (D) Neither A nor B.

10. The dye type detector uses a fluorescent dye. This dye will glow when exposed to:

   (A) black light.
   (B) orange dye.
   (C) soap solutions.
   (D) open flame.

Chapter 16

Refrigerant Recovery, Recycling, and Handling

After studying this chapter, you will be able to:

- ❑ Add refrigerant to an operating refrigeration system.
- ❑ Discharge a refrigeration system and recover refrigerant.
- ❑ Evacuate a refrigeration system.
- ❑ Flush a refrigeration system.
- ❑ Check oil level and add oil to a refrigeration system.
- ❑ Vacuum leak test a refrigeration system.
- ❑ Pressure leak test a refrigeration system.
- ❑ Recharge a refrigeration system with new or recycled refrigerant.
- ❑ Purge a refrigeration system.
- ❑ Install an inline system filter.

Technical Terms

Refrigerant identifiers	Evacuating	Vacuum checking
Refrigerant scale	Flushing	Nitrogen pressurizing
Refrigerant recovery	Open loop flushing	Noncondensible gases
De minimis	Reverse flushing	Purging
Refrigerant recycling	Closed loop flushing	Inline filter

This chapter covers general refrigerant service. In this chapter, you will learn how to recover, recycle, and recharge the refrigeration system. The chapter contains procedures for flushing contaminated systems, and adding oil when necessary. It also explains how to add refrigerant to an operating system. You will also learn how to maintain a refrigeration service station. This chapter contains some of the basic steps that are part of all the refrigeration system service covered in later chapters. Everything you learn in this chapter will be used in the following chapters.

Refrigerant Service

Even in the tightest sealed refrigeration system, a slight amount of refrigerant will seep from the hoses and seals over time. If a refrigeration system has not been serviced for several years, the refrigerant level may be low, even though there are no detectable leaks. This loss should not amount to more than 1 pound (2.2 kg) over 3 or 4 years. In some cases, a previous technician may not have fully recharged the system and it may be necessary to add more refrigerant.

Caution: Before adding refrigerant to an operating refrigeration system, make sure the system is not leaking. Adding refrigerant to a leaking system is _not a repair._ Be sure to add the proper kind of refrigerant to the system. Use a refrigerant identifier if there is any doubt as to the type of refrigerant in a system.

Rules for Handling Refrigerants

The following is a review of refrigerant handling procedures, which were outlined in Chapter 6. They apply to all mobile vehicle refrigeration systems.

❑ All refrigerants must be recovered. Venting refrigerants to the atmosphere is illegal.

❑ Technicians who service air conditioning systems must be certified in refrigerant recovery and recycling.

❑ If your shop is not equipped to handle a particular type of refrigerant, you must treat it as contaminated refrigerant or decline the job.

Figure 16-1. _Use a refrigerant identifier to determine the content of the air conditioning system. This is the best way to ensure no cross-contamination takes place._

Refrigerant	High-Side Service Port		Low-Side Service Port		Label
	Diameter (inches)	Thread direction	Diameter	Thread direction	
HFC-134a	Quick-connect				Sky Blue
CFC-12	6/16	Right	7/16	Right	White
CFC-12 (Pre 1987)	7/16	Right	7/16	Right	White
Free Zone/ RB-276	1/2	Right	9/16	Right	Light Green
Hot Shot	10/16	Left	10/16	Right	Medium Blue
GHG-X4/ Autofrost	.305	Right	.368	Right	Red
GHG-X5	1/2	Left	9/16	Left	Orange
R-406A	.305	Left	.368	Left	Black
Freeze 12	7/16	Left	1/2	Right	Yellow
FRIGC FR-12	Quick-connect, different from HFC-134a				Grey

Figure 16-2. _Service port sizes and label colors for each refrigerant._

❏ Do not add refrigerant to a system with a detectable leak.
❏ Do not mix different refrigerants (for example R-134a and R-12).

Using a Refrigerant Identifier

Refrigerant identifiers have a probe which is able to contact the refrigerant at the port. Most probe connectors consist of a hose that is attached directly to the port. To use a refrigerant identifier, turn it on and allow it to warm up. Make sure the display panel is operating, and, if necessary, calibrate the identifier. Always follow manufacturer's instructions for calibration. Then attach the identifier to one of the service ports. Most refrigerant identifiers will then display the percentages of R-134a, R-12, and unknown refrigerants. **Figure 16-1** shows a typical refrigerant identifier being used.

Adding Refrigerant to a Partially Charged System

Refrigerant can be added using a gauge manifold and a separate container of refrigerant. Refrigerant can also be added by one of several types of refrigeration service machines, including charging stations, recovery and recycling machines, refrigerant management centers, or refrigerant service center. For convenience, in this section we will refer to all of these machines as refrigerant service centers.

> **Caution: Adding refrigerant to a partially charged system is referred to as topping-off. It is only permissible to top-off systems that do not have a detectable leak. Be sure to leak check any partially charged system.**

To add refrigerant to an operating system, begin by installing a temperature gauge in the vent nearest the evaporator. Next, remove the refrigeration system service fitting dust caps. Refrigerant fitting sizes and label colors are listed in **Figure 16-2.** Then prepare the gauge manifold or refrigerant service center as necessary.

Using Refrigerant Service Centers

Turn the master switch to the *on* position. Attach the hose assemblies to the refrigeration system. Be sure the shutoff valves are in the closed position. Once connections are made, open both shutoff valves. **Figure 16-3** shows a typical refrigerant service center attached to the refrigeration system.

> **Note: When using a refrigerant service center, it is usually easier to recover the existing charge, then recharge the system with the exact amount. This procedure is outlined later in this chapter.**

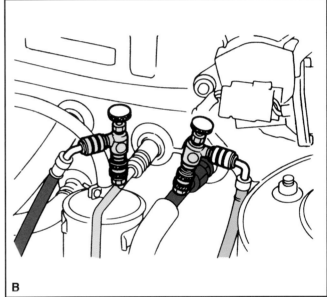

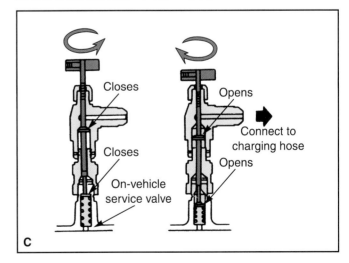

Figure 16-3. *A—R-134a refrigerant service center connected to a vehicle. B—These are refrigerant connectors for an R-134a system. C—The connector only depresses the service valve pin after the connection has been installed on the fitting. (Nissan)*

Using Gauge Manifolds

Make sure the hand valves are closed and attach the hoses to the service fittings. **See Figure 16-4.** Be sure to attach the high and low hose to the appropriate service fittings. Attach the center gauge manifold hose to the refrigerant container to be used. If a separate thirty pound cylinder is being used to add refrigerant, a *refrigerant scale* must be used. Begin by placing the cylinder on the scale.

Most refrigerant scales simply measure the total weight of the cylinder. To use these scales, you must subtract the amount to be added from the total weight of the cylinder. For instance, assume the scale shows a weight of 28 pounds (12.7 kg). You want to add 2.5 pounds (1.14 kg) of refrigerant to the system. Connect the cylinder through the gauge manifold and add refrigerant until the scale shows a weight of 25.5 pounds (11.6 kg) (28 - 2.5 = 25.5 or 12.7 kg - 1.14 kg = 11.6 kg).

Some refrigerant scales are designed so the technician can set the scale to the needed refrigerant weight and then add refrigerant to the system until the scale reading is zero. Before adding refrigerant, purge air from the hoses if necessary. Purging should not be necessary if the hose assemblies are equipped with shutoff valves.

Note: The following conditions apply on reasonably warm day, 75°F (24°C) degrees or above. At cooler temperatures, both of the conditions below may occur even if the system is fully charged.

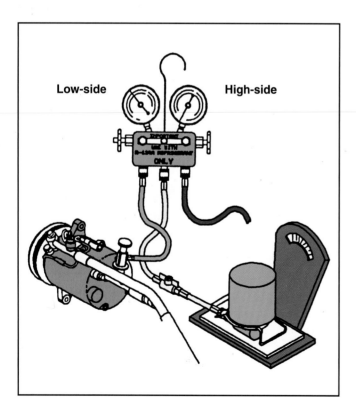

Figure 16-4. *Setup for adding refrigerant using manifold gauges.* (Saturn)

On an older refrigeration system with hand service valves, crack open the valves to connect them to the hoses. Once all hoses are connected, start the engine and turn on the HVAC system to maximum cooling. Set the fan to its lowest setting and allow the low side pressure to come down as much as possible. This is usually around 20-30 psi (135-200 kPa). Some refrigerant service centers can add refrigerant without the need to start the engine.

Once the low side pressure is at its minimum, open the gauge manifold low side valve or push the refrigerant service center charge button and open the low side hose valve. Allow about .5 pound (1.1 kg) of refrigerant to enter the system. Then check the refrigerant charge by monitoring the gauges and by observing the following conditions, as applicable:

❑ The temperature gauge installed in the dashboard vent should be about 30°F (17°C) less than outside air temperature. Check accumulator inlet and outlet temperature. If the inlet and outlet temperatures are not the same, the system charge is still low.

❑ Clutch cycling. If the clutch cycles excessively, the refrigeration level is still low.

❑ Sight glass. If the system has a sight glass, foaming indicates a low charge. Add refrigerant as necessary to obtain the right pressures and other indications of a full charge.

Warning: Closely watch high side pressures. Do not allow high side pressures to go over 350 psi (2412 kPa). If pressures rise over this value, stop adding refrigerant immediately and allow pressures to stabilize. If necessary, spray water on the condenser to lower pressure. If pressure does not stabilize at less than 350 psi, some refrigerant must be removed.

Once enough refrigerant has been added, shut off the air conditioner and engine. Remove the hoses from the refrigeration system. Reinstall the caps on the service fittings and return all equipment to storage.

Recovering Refrigerant

To perform any service on the refrigeration system, the refrigerant must be removed. It is wasteful and illegal to discharge refrigerant into the atmosphere; instead the refrigerant must be recovered. *Refrigerant recovery* consists of transferring the refrigerant from one container (the refrigerant system) to another (a temporary storage tank). The refrigerant can then be reused instead of being released to harm the environment.

To recover refrigerant, the shop must have a refrigerant recovery and recycling machine. Since different types of refrigerants cannot be mixed, the shop must have a separate dedicated machine for each type of refrigerant. Most shops have recycling and recovery machines for R-134a and R-12. Some newer units are combination types, with separate refrigerant recycling and recovery machines in a single cabinet. If the shop uses any kind of refrigerant blend, it must have its own recycling and recovery machine for each blend type. See **Figure 16-5.**

To use a recovery and recycling machine, the technician must be familiar with how it works. Note the relative complexity of the control panel shown in **Figure 16-6.**

Figure 16-5. *Every shop must have at least two refrigerant service centers, or one that can handle both R-134a and R-12. A third service center can be used to recover contaminated or blend refrigerants.*

Always study the manufacturer's instructions before trying to use a recovery and recycling machine.

A general procedure for using recovery and recycling machines is given in the following paragraphs. This procedure is applicable to most makes. Always consult the equipment manufacturer's service manual before starting the recovery and recycling procedure. The HVAC system and engine should be off during the recovery procedure.

Recovering R-134a or R-12

To recover R-134a, R-12, or other refrigerant, start by checking the refrigeration system with an identifier. Always use a refrigerant identifier before performing any recovery operations. Select the type of refrigerant recovery and recycling machine and hoses as indicated by the identifier, then turn the master switch to the *on* position. Attach the hose assemblies to the refrigeration system. Be sure all shutoff valves are in a closed position. Once the connections are made, double-check them, then open both shutoff valves.

Set the machine to recover the refrigerant charge. Most machines have pushbutton controls, **Figure 16-7.** A light will indicate the machine is drawing refrigerant from the system. Wait while the machine completely draws the refrigerant from the system. As the machine draws in the refrigerant, it will dry and filter it. This prepares the refrigerant for reuse. When the gauges read zero pressure or a vacuum, the system is empty. Wait about 10 minutes to be sure all refrigerant has been removed. When you are sure all refrigerant is removed, shut off the machine and proceed to perform other refrigeration system repair procedures.

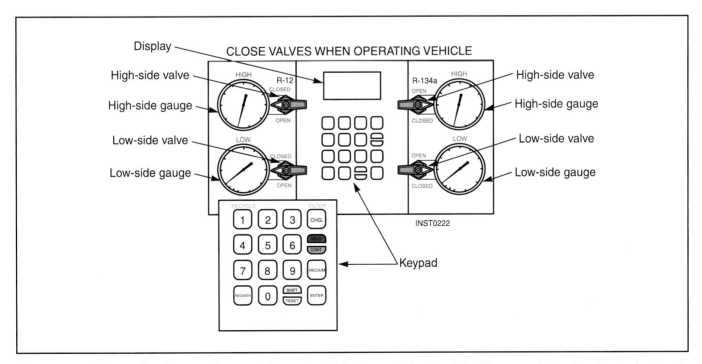

Figure 16-6. *Most recharging machine control panels will look something like this. These controls allow the technician to perform all refrigeration system service operations. (Robinair)*

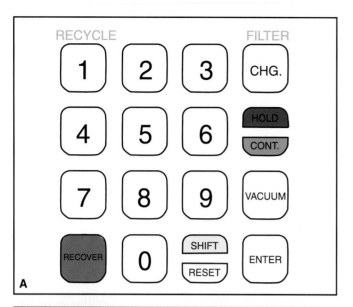

Figure 16-7. *A—Pushing the recover button causes the machine to draw the refrigerant out of the refrigeration system. B—The control panel will indicate the operational mode in use. Most machines have a light that signals when the recovery process is complete. (General Motors)*

Accidental Releases of Refrigerant

It is illegal to intentionally release refrigerants into the atmosphere. However, it is impossible to totally recover every molecule of refrigerant from a system. A small amount of refrigerant remains in all systems, even after a lengthy recovery. Any release of a small amount of refrigerant is referred to as a **De minimis** release. *De minimis* (minor) quantities of refrigerant are permitted to be released, so long as it is in the course of attempting to recapture and recycle or safely dispose of refrigerant. Other types of refrigerant releases permitted include:

❑ Refrigerants emitted from air conditioning and refrigeration equipment in the course of normal operation or as a result of mechanical purging or leaks.

❑ Releases of HFCs and CFCs not used as refrigerants.

❑ Small releases of refrigerant from purging hoses or from connecting or disconnecting hoses to charge or service appliances.

Recovering Blends and Contaminated Refrigerant

It is very important you do not mix refrigerant blends or contaminated refrigerant with pure R-134a or R-12. Refrigerants other than R-134a and R-12 must be recovered with a dedicated recovery and recycling machine or into a special waste cylinder.

Recovering Blended and Contaminated Refrigerants

To recover a refrigerant blend, the shop must have a refrigerant recovery and recycling machine dedicated to that blend. If the shop does not have a machine dedicated to the blend, or the blend is not recognizable, it must be treated as contaminated refrigerant.

Blended refrigerants containing flammable gases such as propane or butane can create an explosion or fire hazard. These refrigerants must be handled carefully. Many recovery and recycling machines contain internal arcing and sparking parts and should not be used to handle possibly flammable refrigerant blends. Some manufacturers now offer recovery stations designed to handle such refrigerants. To minimize the chance of cross-contamination, use a manifold gauge set to remove any unknown or contaminated refrigerants.

The shop must have a special container for contaminated or unknown refrigerants. Contaminated refrigerant containers are painted gray with a yellow top, **Figure 16-8.** The Department of Transportation (DOT) certifies this color for interstate shipment of waste. If your refrigerant identifier indicates the presence of contaminated refrigerant, or it cannot recognize the refrigerant, the refrigerant should be discharged into the shops' contaminated refrigerant container. Contaminated refrigerant should be shipped to a reclaiming facility for recycling or disposal.

 Caution: If you have no way to handle blended or contaminated refrigerants, turn down the job.

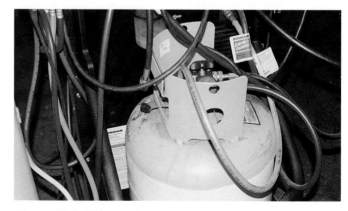

Figure 16-8. *This service center is set up to recover contaminated refrigerants. The gray tank with yellow top is used to store the refrigerant.*

Recycling Refrigerant

Refrigerant recycling means to reuse a portion or all the refrigerant taken from a refrigeration system. The refrigerant can be used to recharge the refrigeration system it came from, or in another system. Most recovery and recycling machines clean and dehumidify refrigerant for immediate reuse. The refrigerant may be stored in a standard 30 pound cylinder or in a separate charging tank. Only pure R-134a and R-12 should be recycled. Never attempt to recycle contaminated refrigerant. Current EPA regulations do not allow blended refrigerants to be recycled.

Evacuating Refrigeration Systems

The major reason for **evacuating**, or *pumping down* a refrigeration system, is to remove water. Water easily enters the refrigeration system in the form of vapor. It is attracted to the dry conditions in the system, just as it is attracted to a dry paper towel. Once in the system, water is hard to remove. Water mixes with the oil and refrigerant in the system, causing corrosion, expansion valve freezing and oil sludging.

To remove water, it must be turned back into a vapor and vacuumed out of the system. To turn water into vapor, the entire refrigeration system would have to be heated to above 212°F (100°C). However, this is impractical, so another method is used. If the atmospheric pressure can be lowered with a vacuum pump, the boiling point of water will drop to normal atmospheric temperatures. Note that vacuum can be measured in inches of mercury or microns. As the water changes into vapor, it expands and is drawn out of the system by the action of the vacuum pump. This is the process of evacuation. Evacuation also removes air (non-condensable gas) from the refrigeration system.

The chart in **Figure 16-9** gives the boiling point of water in relation to vacuum. Note that as vacuum increases, the boiling point temperature decreases. Lowering the boiling point ensures that any water in the system will turn to vapor and be exhausted through the vacuum pump.

 Warning: Never add alcohol or any alcohol based solutions to a refrigeration system to combat water. While adding alcohol may prevent expansion valve freezing, it will not prevent corrosion or oil sludging.

Evacuation Procedure

After refrigeration system repairs are complete, attach the gauge manifold or refrigerant service center hoses to the refrigeration system. If you are using a manifold gauge set, attach the center hose of the manifold to the vacuum pump. If using a refrigerant service center, push the appropriate buttons to connect the pump to the refrigeration system. Then start the vacuum pump and open the hose valves to the refrigeration system.

Observe the vacuum displayed on the low side gauge. The actual performance of various vacuum pumps varies widely. Some pumps will pull a vacuum of as much as 29.9 inches (250 microns). Other widely used vacuum pumps will produce a vacuum of only 29.6 inches (7500 microns). When the temperature is over 45°F (7°C), 29.6 inches of vacuum will remove most moisture from a refrigeration system if the pump is attached for a long period. At lower temperatures, this type of pump may not remove enough water.

The evacuation process does not occur instantly. As the water changes from a liquid to a vapor, its volume increases. This additional volume causes the vacuum to decrease. The pump must continue to operate to remove the vaporized water and again lower the vacuum. It will usually take at least five minutes for full vacuum to be obtained.

Boiling Point – Vacuum Chart		
Boiling Point of Water °F (°C)	**Inches of Mercury**	**Microns**
212 (100)	0.00	759,968
205 (96)	5.00	535,000
194 (90)	9.81	525,526
176 (80)	16.02	355,092
158 (70)	20.80	233,680
140 (60)	24.12	149,352
122 (50)	26.36	92,456
104 (40)	27.83	55,118
86 (30)	28.75	31,750
80 (26.6)	29.00	25,400
76 (24.5)	29.10	22,860
72 (22)	29.20	20,320
69 (20.5)	29.30	17,780
64 (18)	29.40	15,240
59 (15)	29.50	12,700
53 (11.6)	29.60	10,160
45 (7)	29.70	7,620
32 (0)	29.82	4,572
21 (-6)	29.90	2,540
6 (-14.5)	29.95	1,270
-24 (-31)	29.99	254
-35 (-37)	29.995	127

Figure 16-9. *Study this chart. It shows that lowering pressure, or creating a vacuum, causes the temperature at which water boils to decrease. Lowering the boiling point of water is much easier than raising the temperature of the refrigeration system above 212° F (100° C). (Robinair)*

Note: The 29.9 inches of vacuum is for a vehicle at sea level. Subtract 1 inch of vacuum for every 1000 feet of elevation above sea level.

The pump must be attached to the system for some time before all water is removed. Do not turn off the pump as soon as the vacuum reaches its maximum. Allow the pump to evacuate the system for at least fifteen minutes after maximum vacuum is obtained. While there is no maximum time limit on evacuation, longer is better, especially when the pump cannot produce a vacuum of 29.9 inches. Most shops evacuate the system for a minimum of 30 to 45 minutes.

 Note: If the vacuum does not approach 29 inches of vacuum after the pump has been operating for a few minutes, the system has a leak that is allowing air to enter. Shut off the pump and locate the source of the leak.

If the outside air temperature is relatively low, apply a slight amount of heat to the desiccant in the accumulator or receiver-drier. This will help to boil the water out of the desiccant as the vacuum pump operates. Heat can be applied to the outside of the accumulator or receiver-drier with a heat lamp. Do not heat the accumulator or receiver-drier over about 120°F (49°C), and do not use an open flame to apply heat.

Once the system has been evacuated for the proper amount of time, close the valves to the vacuum pump (if necessary). Then turn off the pump and vacuum leak test the refrigeration system.

Flushing Refrigeration Systems

Flushing can be used to save a condenser, evaporator, or hose assembly that is clogged with debris. This debris is usually caused by metal particles from a failed compressor mixing with system oil and pieces of rubber flaked off the inside of hoses. In some cases, the buildup can become large enough to plug the tubes of the evaporator or condenser.

Do not flush compressors, accumulators, receiver-driers, or hoses with built-in components. Expansion valves and orifice tubes should be removed from the system for cleaning or replacement. Do not flush rubber hoses being retrofitted to R-134a, as this will remove the oil that helps to seal in the refrigerant. Some systems are almost impossible to completely clean, and it may be necessary to install an inline filter in the system. Filter installation is covered in Chapter 19.

There are two kinds of flushing:

❏ *Open loop.* The flushing agent is blown through the individual components.

❏ *Closed loop.* The entire system is flushed with a solvent that is recycled.

These flushing procedures are discussed in the following paragraphs.

Open Loop Flushing

If the system is severely contaminated, the individual components should be open loop flushed. **Open loop flushing** is usually done with liquid solvents. In the past, technicians used mineral spirits. However, mineral spirits are flammable and slowly evaporate from the system. Mineral spirits left in the system will dilute the oil and possibly react with the refrigerant. For these reasons, vehicle manufacturers do not recommend flushing the refrigeration system with mineral spirits. Most shops now use commercial refrigeration system cleaning solvents such as R-141b or ester oil based solvents. These solvents are nonflammable, clean more thoroughly, are compatible with refrigerant oils, and evaporate quickly.

 Caution: Do not open flush accumulators and receiver-driers. Replace these components if the system suffered contamination.

To open loop flush, recover the refrigerant and open the inlet and outlet fittings of the component to be flushed. Sometimes the flushing is more effective if the component is removed from the vehicle. Attach a drain hose to the inlet opening of the component and place the other end in a container to trap debris. Then pour about one pint of liquid solvent into the outlet opening of the component. Use a rubber tip blowgun to blow into the outlet end of the component. Some technicians use a tank and hose design that injects the solvent into the system using air pressure,

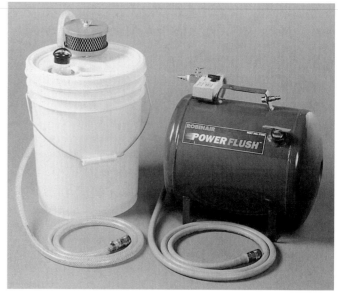

Figure 16-10. *A high pressure hose and tank can often be used to flush refrigeration system components. (Robinair)*

Figure 16-10. Make sure the compressed air source is dry, or use dry nitrogen. Air pressure should not exceed 100 psi (689 kPa). Always blow in the reverse direction to refrigerant flow to back (reverse) flush as much debris as possible. Add more solvent and repeat the process until only clean solvent comes out.

Switch the drain hose to the outlet opening and make a final pass in the forward direction using a small amount of solvent. Continue blowing air through the component until no more solvent comes out. Allow the component to dry thoroughly before reattaching the inlet and outlet fittings. Whenever a component has been flushed, evacuate the system for at least one hour to completely remove any solvent and air.

With either open or closed loop flushing, all traces of the solvent must be removed from the system before it is evacuated and recharged. .For this reason, you should never try to open flush receiver-driers, accumulators, filters, expansion valves, STV or POA valves, or any other part which could trap solvent. Compressors should never be flushed because refrigerant oil will be washed out of the compressor crankcase.

The most commonly flushed parts are evaporators and condensers. Hoses can also be flushed. It is recommended that non-barrier hoses not be flushed if the system is being retrofitted to R-134a. The oil on the hose walls helps to seal in the smaller R-134a molecules. Removing the oil by flushing may create leaks.

To begin flushing the components of a refrigeration system, recover the refrigerant and disconnect the refrigerant lines from the component to be flushed. Then decide in which direction to flush. Some technicians recommend beginning the flushing operation by reverse flushing. *Reverse flushing* is flushing in the opposite direction of normal refrigerant flow. Reverse flushing will dislodge any compacted debris. After reverse flushing, the connections are reversed and the flushing process is repeated in the direction of refrigerant flow to remove any remaining debris.

Some technicians use mineral spirits, but this is not recommended because of residual fluid left in the system, and because of fire danger or asphyxiation from the mineral spirits vapor. Many specific flushing solvents, such as R-141b are available. These are specifically designed for refrigeration system flushing and are non-hazardous.

If the flushing equipment does not use a flushing cylinder, pour the solvent into the component to be flushed. If a flushing cylinder is used, fill it with solvent. Run a hose from opening of the component to a recovery container. Then using no more than 60-100 psi (414-690 kPa) of air pressure, blow through the component.

Note: Do not exceed 125 psi (862 kPa) air pressure. This could damage refrigeration system components and may not allow the solvent to remain liquid for enough time to dissolve deposits.

Observe the condition of the solvent as it leaves the component. When the solvent no longer contains any oil or debris, the component is sufficiently cleaned. If it is necessary to clean an expansion valve or pressure control valve (POA, STV) remove it from the vehicle and clean it by immersing it in solvent. Then thoroughly dry the component with compressed air before reinstalling it. If possible, allow the part to air dry overnight before reinstallation.

Closed Loop Flushing

Closed loop flushing is usually done with refrigerants, or by dedicated liquid flushing machines. In the past, refrigerants such as R-11 and R-113 were used to open loop flush systems. Since both R-11 and R-113 are CFCs, these refrigerants can no longer be used. The refrigerant commonly used for flushing is R-134a.

This procedure uses either a special refrigerant cycling machine or a recovery and recycling machine to direct the solvent through the system. No refrigeration system fittings need to be disconnected to perform closed loop flushing. Closed loop flushing will not open completely blocked passages. It will remove oil and some contamination. Some vehicle manufacturers do not allow closed loop flushing.

To perform closed loop flushing, connect the machine to the system and follow instructions. Most closed loop machines will automatically perform the flushing operation. Refrigeration system flushing procedures are similar to flushing procedures for cooling system parts, or flushing any type of closed container.

Checking and Adding Oil

Refrigeration systems do not require periodic oil changes. Since the oil circulates with the refrigerant, however, it can leak out of the system with refrigerant. If the system is recharged without adding oil to make up for what has been lost, the compressor will be damaged from lack of lubrication.

If too much oil is added, damage may also occur. Too much oil causes compressor knocking and possible damage. The oil also coats the surfaces of the evaporator and condenser, leading to poor heat transfer. A symptom of this is excessive high and low side pressures. Sometimes excessive oil in the system causes the compressor oil level to be lower than it would be with the normal amount of oil. This happens when the mixture of cold refrigerant and oil from the low side enters the warmer compressor crankcase. The refrigerant boils and causes the oil to foam. The oil then circulates through the system with the refrigerant instead of staying in the compressor and providing lubrication for moving parts. Also, excess oil will not condense.

It is sometimes difficult to determine whether a system has too much oil. In many situations, the system must be discharged and the oil drained out and measured. Various oil checking procedures are explained in the following paragraphs.

Checking Compressor Oil Level

It is sometimes necessary to check the oil in the compressor crankcase. On most modern refrigeration systems, the only way to do this is to discharge the system, remove the compressor, and drain the oil into a measuring cup. Oil will tend to stay in the other system components, so if the system is low on oil, the amount of oil in the compressor will be less than specified.

To check compressor oil level using this method, first operate the refrigeration system at maximum capacity for about 15 minutes. This will cause any oil that has settled in other components to return to the compressor. Do not operate the system for a longer period, as the oil may begin to foam and leave the compressor. Shut off the engine, wait a few minutes for the oil to settle, then recover the system refrigerant and remove the compressor.

Open the crankcase oil fill port, or the suction port if no oil fill port is used. Tilt the compressor and pour the oil out. If you need to know how much oil the compressor contained at the time of removal, pour the oil into a measuring cup or graduated cylinder, **Figure 16-11.** Compare the amount of oil with the manufacturer's specifications to determine whether the system was low on oil. Add the correct amount of new oil as explained later in this chapter. Never reuse refrigerant oil.

A version of this method uses a dipstick to check the oil level, **Figure 16-12.** After operating the system for 15 minutes, then recovering the refrigerant, locate the oil checking plug on the upper part of the compressor crankcase. Remove the plug and insert a rod into the crankcase. Make sure the rod goes completely to the bottom of the crankcase. Remove the rod and measure the length from the bottom of the rod to the oil line. Compare this length with the manufacturer's specifications and add or remove oil as necessary.

Another method of checking does not require refrigerant removal. To use this method, the technician must first determine whether there is an oil inspection fitting on the

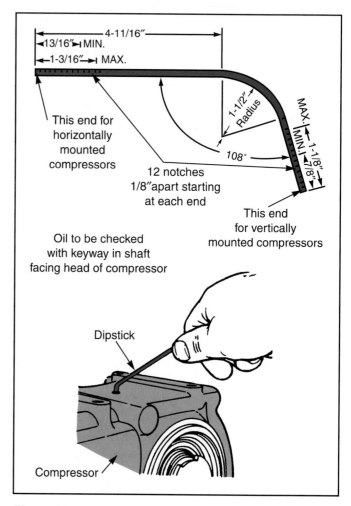

Figure 16-12. *The oil level of some compressors is checked using a dipstick. The dipstick can usually be made from a length of wire. Follow manufacturer's directions closely to avoid getting a false reading. (Sun)*

lower part of the compressor crankcase. Not all compressors have such a fitting. Operate the system for 15 minutes then allow it to sit for about 10 minutes. Then crack (slightly open) the fitting and observe whether oil or refrigerant comes out. If oil drips out, the oil level is sufficient. If only refrigerant comes out, the oil level is too low.

Adding Oil to a Charged System

To add oil to a charged refrigeration system, a special oil injector is needed. A common pump oil injector is shown in **Figure 16-13.** To add oil by this method, attach the pump to the high side service port of the system. The engine must be off when adding oil. Use the high side fitting to reduce the possibility of liquid oil entering the compressor. Pump in the needed amount of oil, remove the pump, and reinstall the service port cap. Then start the engine and turn the HVAC controls to maximum air conditioning to circulate the oil.

Other injectors are used with a gauge manifold, **Figure 16-14.** To use this type of injector, fill it with the proper oil and connect it to the gauge manifold as shown.

Figure 16-11. *Pour the compressor oil into a graduated cylinder to measure it. Also visually check the condition of the oil. Do not reuse this oil. (Subaru)*

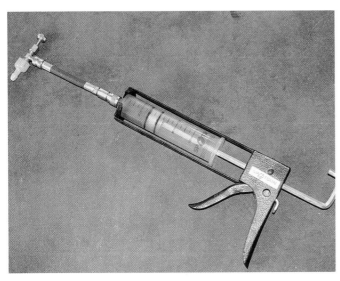

Figure 16-13. *Oil may be injected into the refrigeration system using a special screw type unit similar to those used to inject dye. Squeezing the handle forces oil through the fitting and into the system.*

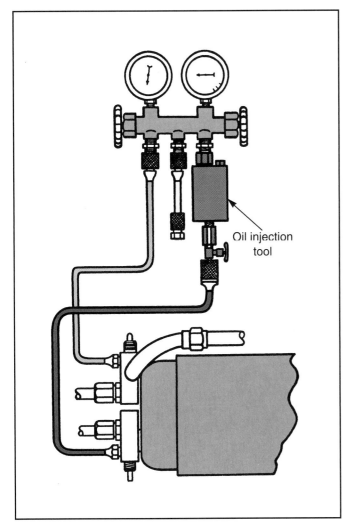

Oil injection tool

Figure 16-14. *This type of oil injector uses the pressure difference between the low and high sides of the system to cause oil to flow into the system. (General Motors)*

Then connect the gauge manifold to the system, start the engine and turn the HVAC controls to maximum air conditioning. Crack open the low and high side hand valves. This allows the high pressure to push the oil into the low side. Do not open the valves more than a small amount. Allow the oil to enter the system, then shut both valves. Allow the refrigeration system to operate for a few more minutes to distribute the oil. Then stop the engine, remove the hoses and replace the fitting caps.

Adding Oil to Replacement Components

Some of the circulating oil settles in major components during system operation. If the component is replaced, some of the remaining system oil will be trapped in the new component. This will reduce the amount of oil returning to the compressor and it will be starved for oil. Therefore, when any major component such as the condenser, evaporator, accumulator, or receiver-drier is replaced, oil must be added. Manufacturers publish specifications as to how much oil should be added to a replacement component. **Figure 16-15** is a chart showing typical amounts of oil to be added to various refrigeration system components. To add oil, simply pour the oil into the component before installing it.

 Caution: Do not add oil by immersing the refrigerant fill hose into a bottle of clean oil as the system is operating. This can lead to overcharging with oil as well as contamination of the oil remaining in the bottle.

Adding Oil to the Compressor

When the compressor is replaced, or the old compressor reinstalled, it must be filled with oil. Always check the oil level before installation. Then add the correct type and amount of new oil. See **Figure 16-16.** Do not reuse the original oil.

 Note: Remember to tightly close all refrigerant oil containers after use to prevent water entry.

Vacuum and Pressure Testing Refrigeration Systems

Before recharging the system, be sure that it does not leak. There are two ways to check for leaks, using **vacuum checking** or **nitrogen pressurizing.** These methods are explained in the following paragraph.

COMPONENT REPLACEMENT OIL CAPACITY

Accumulators	All	1 oz.		Evaporator Cores	Chrysler	2 oz.
Condensers	All Types	1 oz.			Ford - GM	3 oz.
Evaporator Cores	AMC	1 oz.			Imports	1 oz.
				Receiver–Driers	All Types	1 oz.

COMPRESSOR OIL CAPACITIES

UNIT MODEL:	REMARKS:		CAPACITIES*:
Chrysler RV2	Oil level indicator		1 5/8" - 2 3/8" - 8 oz.
Frigidaire/Harrison			
A6	Drain to check & flush system		11oz.
R4	Drain to check & flush system		6 oz.
DA. HR6, HR6HE, V5	Drain to check & flush system		8 oz.
Nippondenso			
Chrysler C171, A590, 6C17	Drain to check & flush system		9 oz.
6E171	Drain to check & flush system		13 oz.
10 PA17	Drain to check & flush system		8 oz.
Ford FS6, 6P148	Drain to check & flush system		10 oz.
FX15	Drain to check & flush system		7 oz.
10P15C, 10P15F, 10PA17C	Drain to check & flush system		8 oz.
GM10PA20, 10P15			8 oz.
Sankyo/Sanden			
SD508	Oil level indicator		6 oz.
SD510, 709	Oil level indicator		5 oz.
SD519HD, 10P15	Oil level indicator		7 oz.
Tecumseh			
HR980	Drain to check & flush system		
HG1000	Oil level indicator	-Vertical Mount	7/8" - 1 3/8" - 11 oz.
		- Horizontal Mount	7/8" - 1 5/8" - 11 oz.
York 206, 209, 210	Oil level indicator	-Vertical Mount	7/8" - 1 1/8" - 10 oz.
		- Horizontal Mount	15/16" - 1 3/16" - 10 oz.
Zexel/Diesel KIKI	Drain to check & flush system		5 oz.

*This measurement represents a system that has been completely discharged and thoroughly cleaned.

Figure 16-15. *Before installation, refill each component being replaced with the correct amount and type of oil. (Four Seasons)*

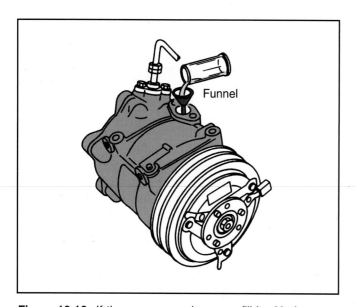

Figure 16-16. *If the compressor is new, refill it with the same amount of oil that came out. (Subaru)*

Vacuum Testing

This test should be performed as soon as the system has been evacuated. Begin by closing both high and low side valves then turning off the vacuum pump. Wait at least 20 minutes then observe the low side gauge. The vacuum should rise no more than .01 inches of mercury (700 microns). If the vacuum rises more than this, the system is leaking and the source of the leak must be located and corrected.

Nitrogen Pressure Testing

To perform nitrogen pressure testing, use dry nitrogen to pressurize the refrigeration system to 100 psi (689 kPa). Some service literature recommends using nitrogen at 150 psi (1034 kPa), but this should be done carefully since excessive pressures could rupture low side components such as the evaporator. After pressurizing the system, observe it for 20 minutes. The pressure may drop a few PSI due to temperature changes. More than a 5 psi (34 kPa) loss, however, indicates a leak.

Recharging Refrigeration Systems

After system repairs are made, the system must be refilled with the proper amount of refrigerant. Before recharging, make sure the system has the proper amount of oil, has been evacuated and thoroughly checked for leaks. Then determine the proper amount of refrigerant before starting the recharging procedure. Amounts vary from 1.5 pounds (3.3 kg) on a small car to as much as 8 pounds (17.6 kg) on a large vehicle with rear air conditioning. Also make sure you know for certain what type of refrigerant should be used.

 Note: Once the system has been evacuated, do not release the vacuum before charging. Allow the vacuum to draw in the refrigerant.

Recharging with Refrigerant Service Center

Turn the master switch to the *On* position. Attach the hose assemblies to the refrigeration system. Be sure the shutoff valves are in the closed position. Once connections are made, open both shutoff valves. Set the refrigerant service center controls to the proper amount of refrigerant. Consult the service material as needed for the correct amount of refrigerant. Without starting the engine, push the refrigerant service center charge button, **Figure 16-17.** Allow the service center to charge the system with the proper amount of refrigerant.

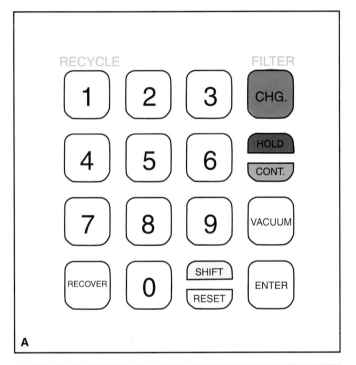

Figure 16-17. *Most manufacturers recommend recharging the refrigeration system by weight. A—If a charging machine is being used, set the correct amount of refrigerant and push the charge button. B—The machine will automatically add the proper amount of refrigerant. (General Motors)*

Then start the engine and HVAC system and set the blower speed to high. Wait about five minutes for pressures and temperatures to stabilize. If desired, place a temperature gauge in the vent closest to the evaporator to check outlet temperature. Observe the gauges to ensure the high and low side pressures are correct. Monitor outlet temperature and other operating conditions to ensure the refrigeration system is operating properly.

Charging with Gauge Manifold

Make sure the manifold hand valves are closed and attach the hoses to the appropriate service fittings. Then attach the center gauge manifold hose to the refrigerant container to be used. Purge air from the hoses if necessary. Once all hoses are connected and purged, open both gauge manifold valves. Allow refrigerant to enter both sides of the system with the engine off.

Once pressures have stabilized, close both the high and low side valves and start the engine. Turn the HVAC system to maximum cooling and set the blower to its lowest position. Slowly open the low side valve and add refrigerant to the system. Do not open the high side valve. Closely watch high side pressures as you add refrigerant and do not exceed the specified maximum amount. If pressures rise over 350 psi (2412 kPa), stop adding refrigerant and allow pressures to stabilize. If necessary, spray water on the condenser to lower pressure. If high side pressure does not stabilize at less than 350 psi (2412 kPa), some refrigerant must be removed.

Once enough refrigerant has been added, set the blower speed to high. Wait about five minutes for pressures and temperatures to stabilize. Then observe the gauges to ensure the system high and low side pressures are correct. Ensure the refrigeration system is operating properly, then remove the hoses from the refrigeration system and reinstall the caps on the service ports.

> **Caution:** If the system is being recharged with a blended refrigerant, the refrigerant must be charged into the system as a liquid. If a blended refrigerant is charged as a vapor, composition change may occur, resulting in system damage.

Refrigeration System Purging

Excess air (sometimes called **noncondensible gases**) raises pressures since it cannot condense at normal system pressures. Air interferes with normal operation of the refrigeration system since it cannot change state to transfer heat. The air also contains water vapor. Usually only a small amount of air gets into the refrigeration system. Occasionally improper charging procedures will allow large amounts of air into a system. Air enters refrigeration systems often enough that you should know how to remove it. Air removal is called **purging.**

The best indication of excess air in a system is high static (system not operating) pressures in a system that has not been overcharged. The simplest way to remove air is by slightly opening (cracking) a valve or hose fitting at the highest point in the system. Since air is lighter than the refrigerant, it will rise to this point. This procedure should be done after the system has been sitting overnight to allow the air to separate from the refrigerant. Watch the connection and close the valve or fitting as soon as refrigerant begins to come out of the connection. The disadvantage of this method is it releases some refrigerant to the atmosphere. Another problem is most refrigeration systems do not have a convenient fitting at the highest point in the system, and that air pockets may develop at two or more locations.

Another method uses a refrigerant recovery and recycling machine with an automatic air purging device. Use the machine to evacuate the system and allow the automatic purging device to remove air. The purged refrigerant can then be reinstalled in the system.

Two dedicated devices are used to purge air from the refrigeration system. One uses an oxygen sensor, while the other uses an infrared sensor. To use them, attach them to the system and follow instructions.

Installing an Inline Filter

Sometimes a refrigeration system becomes so contaminated it cannot be completely cleaned. In these cases, an *inline filter,* **Figure 16-18,** should be installed. Some manufacturers recommended installing a filter instead of attempting to flush the system. Inline filters are usually installed in the high pressure line just ahead of the orifice tube. Filters are usually not used on systems with an expansion valve, since the receiver drier already contains a filter.

Figure 16-18. *Use extreme caution when cutting the lines and attaching the fittings to install an inline filter. (General Motors)*

Various kinds of filters are available. Some filters contain a filter material only, while others have filters and desiccant to remove moisture. A few filters are designed with a built-in orifice tube to replace the existing tube.

Some filters are equipped with fittings that allow them to be installed at the high pressure line where it connects to the orifice tube. Other filters can only be installed by cutting the line and splicing in the filter. These procedures are covered in the next chapter. If the filter contains an orifice tube, the original tube must be removed.

Maintaining Refrigerant Service Centers

In addition to knowing how to use a refrigerant center to recover, recycle, and recharge refrigerant, you should know how to perform routine maintenance to keep the service center operating properly. The following section discusses various service procedures needed to keep a refrigerant service center operating. In most cases, the service center will give a code or other notice that maintenance is needed.

 Note: The procedures here are general and should not be used as a substitute for the refrigerant service center's manual.

Replacing the Filter-Drier

On most service centers, the filter-drier is used to trap dirt and particulates. It also removes any moisture present in refrigerant. The filter-drier has a service life of approximately 200-500 hours, depending on the manufacturer. Usually, the service center's display will tell you when the filter-drier needs to be replaced.

To replace the filter-drier, make sure the service center is not connected to a vehicle. Then follow the service center's manual to recover any refrigerant from the low side of the unit, Once all the refrigerant has been removed, the display on the center will notify you when the filter can be replaced. If the filter assembly uses gaskets, replace them at this time. Hand tighten any fasteners holding the filter-drier, **Figure 16-19.** Evacuate and refill the filter-drier using the procedure outlined in the manufacturer's literature.

Calibrating the Service Center's Scale

After changing the tank or other service, the refrigeration unit should be calibrated to ensure proper operation. The most important is the scale calibration. This ensures the proper amount of refrigerant is added to each vehicle. To calibrate the scale, follow the directions in the service center's manual.

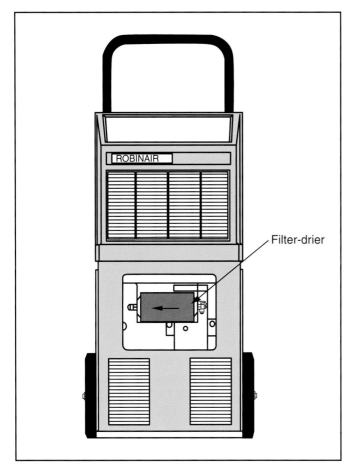

Figure 16-19. *Replace the filter-drier in a refrigerant service center every 200-500 hours of use. (Robinair)*

 Note: An exact weight is usually needed to properly calibrate the scale. Consult the manual for the required weight.

When the scale is properly calibrated, the reading of the exact weight should be within .04 lbs (.02 kg). If the scale does not show the proper weight, recalibrate the scale.

Changing the Vacuum Pump Oil

If the service center has a built-in vacuum pump, the oil in the pump must be changed to maintain proper operation. The oil is stored in a reservoir mounted on the pump itself. Most vacuum pumps will need approximately 6 ounces of oil. Follow the instructions in the service center's manual when changing the pump oil.

Checking for Leaks

You should check the refrigerant center every three months for leaks. Begin by turning off and unplugging the unit from the wall outlet. Remove any shrouds or covers from the unit. Search for leaks by using a leak detector. Tighten any fittings found to be leaking. Reassemble the unit once you are sure there are no leaks.

 Note: Some local and state laws specify a mandatory period that refrigerant service equipment must be leak tested.

Installing a New Tank

If the refrigerant storage tank is leaking or has been filled with contaminated refrigerant, it may be necessary to install a new tank. All new tanks come with a dry nitrogen charge.

Begin installation by venting the tank's dry nitrogen charge to the atmosphere. When replacing the tank, you may want to calibrate the scale. Place the new tank on the scale and secure it to the station. Attach the temperature probe, hoses, and complete installation following the directions in the service center manual.

Adding Refrigerant from a Disposable Tank

Service center tanks have two shut off valves versus one for disposable refrigerant tanks. After replacing a tank or when the center's tank is empty, it is necessary to refill the tank using a disposable cylinder.

Close the liquid valve on the unit tank. Connect one end of the blue tank hose to the disposable cylinder, **Figure 16-20.** Make sure the disposable cylinder is upside down to transfer liquid. Open the valve on the disposable

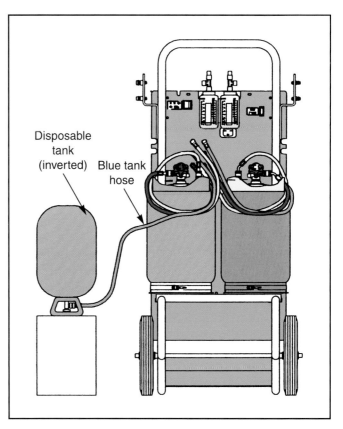

Figure 16-20. *Recharging service center tank using a disposable refrigerant cylinder. (Robinair)*

tank then open the red gas valve on the service center tank. Begin the transfer by pressing the appropriate keys on the service center. The service center should notify you when transfer is complete, which should take from 45 minutes to two hours, depending on atmospheric conditions.

 Note: Space is left in the unit tank after transfer (approximately 10 pounds). This allows space for refrigerant recovery.

Removing Air from Charging Cylinders

It is common to find air in charging cylinders. As with refrigeration systems, static pressures in a charging cylinder indicate the presence of air. Since air rises to the top of the cylinder, it can be removed by placing the cylinder upright and cracking the valve to allow the air to escape. Every refrigerant has a specified pressure at a given temperature. Removing air causes the refrigerant to boil and absorb heat. Therefore, to get an accurate reading after removing some of the air, the cylinder must be left overnight before removing more air.

The cylinder can also be attached to a refrigerant recovery and recycling machine with an automatic air purging device. Attach the cylinder to the machine and slowly crack open the fill valve to evacuate the system and allow the automatic purging device to remove air. Any refrigerant removed can then be returned to the cylinder. The dedicated purging devices described above can be used to purge air from the cylinder. Attach the hoses to the cylinder and follow instructions.

Summary

If a refrigeration system has not been serviced for several years, the refrigerant level may be low due to slight seepage. In this case, refrigerant may need to be added with no other service. Before adding refrigerant, make sure the system is not leaking and use a refrigerant identifier if there is any doubt about the type of refrigerant.

Refrigerant can be added with a gauge manifold and a separate container of refrigerant, or a refrigerant service center. With either method, install a temperature gauge in the vent nearest the evaporator. Next, attach the hoses to the refrigeration system. Open the gauge manifold low side valve or push the refrigerant service center charge button and allow refrigerant to enter the system. When enough refrigerant has been added, shut off the air conditioner and engine and remove the hoses from the refrigeration system.

To recover refrigerant, a refrigerant recovery and recycling machine is needed. The shop must have a refrigerant recovery and recycling machine for every kind of refrigerant, including blends. Contaminated or unfamiliar refrigerant must be placed in a special container and sent to a reclaiming facility. Most recycling machines also clean and dehumidify refrigerant for immediate reuse. Never attempt to recycle contaminated refrigerant.

Water enters the refrigeration system and mixes with the oil and refrigerant in the system. If not removed, it will cause corrosion, expansion valve freezing and oil sludging. Water is removed by evacuating the system with a vacuum pump. Evacuation also removes air from the refrigeration system. Refrigeration systems should be evacuated for at least 15 minutes. After evacuating the system, vacuum leak test the refrigeration system.

Sometimes refrigeration systems must be flushed. The two types of flushing are open loop and closed loop. Open loop flushing is usually done with liquids and closed loop flushing is usually done with refrigerants.

Refrigerant oil can leak out of the system along with refrigerant. Refrigerant collects in other system components, causing the compressor oil level to be low. Compressor oil level can be checked by pouring the oil into a measuring cup, using a dipstick, or by cracking a fitting at the underside of the compressor. Before checking the oil, operate the refrigeration system for about 15 minutes and wait a few minutes for the oil to settle. To add oil to an operating system, a special oil injector is needed. Oil should always be added to any replaced major components before installation. Refrigeration systems can be vacuum or pressure tested. Vacuum testing is done after the system is evacuated. Pressure testing is done with dry nitrogen.

After system repairs are made, the system must be recharged with the proper refrigerant. Recharging can be done with a gauge manifold or a refrigerant service center. Always connect the hoses to the proper high and low side connections and purge air if necessary. Be sure the system is recharged with the correct refrigerant. Charge through the low side only if the system is operating. Stop adding refrigerant if pressures rise over 350 psi (2412 kPa). Once charging is complete, make sure high and low side pressures are correct and ensure the refrigeration system is operating properly.

Sometimes a system must be purged of air. Recovery and recycling machines will evacuate the system and automatically purge the air. Two dedicated devices are available to purge air from the refrigeration system. Air must occasionally be removed from some charging cylinders.

Review Questions—Chapter 16

Please do not write in this text. Write your answers on a separate sheet of paper.

1. *True or False?* Some refrigerant will seep from hoses and seals.

2. The HVAC system should be off during _____.

3. *True or False?* Never recover two kinds of refrigerant using the same machine.

4. _____ should be sent to a reclaiming facility.

5. To remove water, it must be turned into a _____.

6. *True or False?* Refrigeration systems do not require periodic oil changes.

7. A _____ can sometimes be used to check compressor oil level.

8. The vacuum test should be performed as soon as the system has been _____.

9. Do not release the vacuum before _____.

10. Air in the refrigeration system is sometimes called _____ gas.

ASE Certification-Type Questions

1. When preparing to add refrigerant to an operating system, install a temperature gauge in the dashboard vent nearest the:
   (A) steering wheel.
   (B) evaporator.
   (C) right window.
   (D) compressor.

2. All of the following operating conditions are signs of a fully charged refrigeration system, *except:*
   (A) high and low side pressures about 350 psi (2412 kPa).
   (B) vent air temperature about 30°F (17°C) below outside air temperature.
   (C) accumulator inlet and outlet temperatures about the same.
   (D) compressor clutch cycling normal.

3. Refrigerant recovery is being discussed. Technician A says you should always start by checking the refrigerant with an identifier. Technician B says you should always recover blended refrigerants into an R-12 container. Who is right?
   (A) A only.
   (B) B only.
   (C) Both A and B.
   (D) Neither A nor B.

4. Contaminated refrigerant containers are painted:
   (A) white.
   (B) blue.
   (C) gray with a yellow top.
   (D) pink.

5. Water in the refrigerating system can cause all of the following conditions, *except:*
   (A) corrosion.
   (B) alcohol formation.
   (C) expansion valve freezing.
   (D) oil sludging.

6. Technician A says creating a vacuum in the refrigeration system causes the water in the system to boil. Technician B says creating a vacuum in the refrigeration system removes air as well as water. Who is right?
   (A) A only.
   (B) B only.
   (C) Both A and B.
   (D) Neither A nor B.

7. Which of the following is no longer used as a flushing agent?
   (A) Mineral spirits.
   (B) R-141b.
   (C) Ester oil solvents.
   (D) R-134a.

8. Technician A says used refrigerant oil can be reused if it is clean. Technician B says that, before checking the oil, the refrigeration system should be operated at maximum for about 15 minutes and allowed to sit for a few minutes. Who is right?
   (A) A only.
   (B) B only.
   (C) Both A and B.
   (D) Neither A nor B.

9. Excessive pressures developed during nitrogen pressure testing could rupture the:
   (A) condenser.
   (B) compressor.
   (C) accumulator.
   (D) evaporator.

10. The simplest way to remove air from a refrigeration system by cracking a valve or hose fitting at the:
   (A) lowest point in the system.
   (B) highest point in the system.
   (C) recovery cylinder.
   (D) low side hose.

The green on this vacuum pump is not moss, its leak trace dye. If dye is used to find a leak, some may be removed from the air conditioning system during evacuation.

Chapter 17

Hose, Line, Fitting, and O-ring Service

After studying this chapter, you will be able to:
- ❏ Remove and replace a compression fitting.
- ❏ Remove and replace a spring lock coupling.
- ❏ Remove and replace refrigeration system O-rings and gaskets.
- ❏ Remove and replace a refrigeration system hose.
- ❏ Make a new hose from stock hose lengths.
- ❏ Install a crimp fitting on a hose.
- ❏ Remove and replace a refrigeration system line.

Technical Terms

Spring lock coupling

Compression fittings

Crimped fittings

Hose clamps

Beadlock fitting

Refrigeration system hoses, lines, fittings, O-rings, and gaskets seldom cause refrigeration system performance problems. However, they are common causes of leaks. O-rings and gaskets must often be replaced as part of other refrigeration system service. It is easy to install the wrong kind of gasket or O-ring or damage hoses or lines during service operations. This chapter covers replacement of refrigeration system hoses, lines, and fittings. It also addresses the replacement of O-rings and gaskets. In this chapter, you will learn how to remove these components and replace them with new ones.

> **Warning: Before loosening any refrigeration system fitting, make sure the system is completely discharged. If necessary, place a vacuum on the system to ensure all pressure has been removed. If any refrigeration system component will be removed for more than 10 minutes, cap the refrigeration system openings to minimize water vapor collecting in the system.**

Servicing Fittings and O-rings

The refrigeration system fittings must be disassembled to replace leaking O-rings, to flush system components, or to change components. Seals and O-rings compress and harden over time and will not seal properly if they are reused. Lip seals and O-rings should always be replaced when the related fitting is removed. Always replace O-rings with the exact type and size. Some O-rings are replaced with updated seals that are a different color. If a different color O-ring is supplied, make sure it is the intended replacement. The following sections give general procedures for fitting and O-ring service.

> **Note: Compressor shaft seal replacement will be covered in Chapter 18.**

Begin service by identifying and recovering the system's refrigerant charge. This was covered in Chapter 16. Remember that even after the recovery process is complete, there is still a residual amount of refrigerant left.

Compression Fittings

Compression fittings consist of two lines threaded together with an O-ring between the two fittings. These fittings and their related O-rings can be replaced using hand tools. Use the proper size wrenches to avoid damaging the fittings. Whenever possible, use line wrenches to loosen the fittings.

Fitting Disassembly

To disassemble compression fittings, use two wrenches to loosen the fitting, one on each side of the fitting. **Figure 17-1** shows how the wrenches should be placed to prevent damage to the fitting. Do not try to loosen the fitting using only one wrench, as this will twist the tubing. After loosening and removing the nut, pull the fitting apart to expose the O-ring(s).

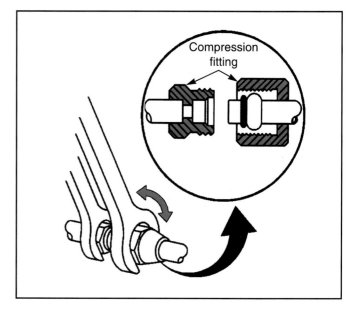

Figure 17-1. *Always use two wrenches to remove refrigeration line fittings. This will prevent damage to the lines. (General Motors)*

Fitting Reassembly

Before reassembling the fitting, lightly coat the O-ring with clean refrigerant oil, if it was not done during installation. Reinstall the fitting and tighten it until resistance is felt, indicating the O-ring is being compressed. Then turn the fitting another 1/4 turn. Do not overtighten the fitting. Overtightening will crush the O-ring or strip the fitting.

O-ring Replacement

O-rings can be replaced once the fitting is disassembled. Pull the old O-ring from the fitting and thoroughly clean the area with a rag and solvent. Do not use a wire brush or any cleaner that could damage the O-ring grooves. If the O-ring grooves show any sign of damage, corrosion, or deforming, the fitting should be replaced. Do not forget to check the inside sealing surface of the nut or line.

> **Caution: Do not use black nitrile O-rings on R-134a systems. They will allow refrigerant to leak.**

Before installing the new O-ring, make sure it is the right size for the O-ring groove and it is the right material for the system. There are two types of O-ring grooves, non-captured and captured, as shown in **Figure 17-2.** Coat each O-ring with clean refrigerant oil and seat it flat against the shoulder(s) of the fitting or groove, as shown in **Figure 17-3.** Improper installation will cause the O-ring to leak.

Spring Lock Couplings

Spring lock couplings are held together by a garter spring. A special tool, **Figure 17-4A,** must be used to disassemble spring lock couplings. This tool is used to retract the spring away from the inner tube, permitting the coupling to be pulled apart.

Coupling Disassembly

To use the tool, assemble it over the coupling, **Figure 17-4B.** Then push the tool into the coupling to expand the spring, **Figure 17-4C.** Once the tool is holding

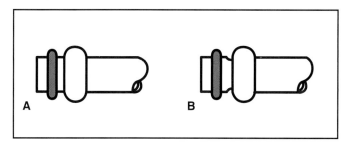

Figure 17-2. *A—Non-captured O-ring B—Captured O-ring. Dimensions for each type of O-ring vary, even though the line may be the same size. (General Motors)*

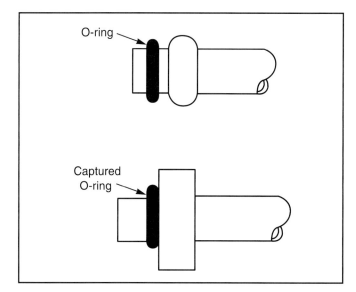

Figure 17-3. *Be sure O-rings are properly placed before reassembling the fittings. Improperly installed O-rings will be damaged when the fitting is installed and will leak. (General Motors)*

the spring in the released position, pull the coupling apart. Then open the tool and finish disassembling the coupling, **Figure 17-4D.**

O-Ring Replacement

Once the coupling is apart, the two O-rings can be changed. Carefully remove the O-rings using a wire or pick. Do not damage the O-ring grooves with the

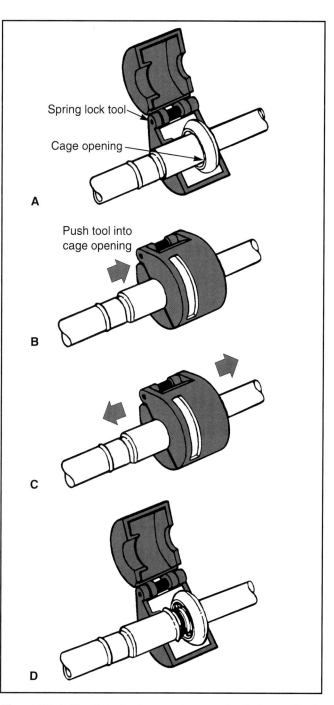

Figure 17-4. *The first step in removing a spring lock coupling is to make sure all refrigerant has been recovered from the system and the system has no pressure. A—Assemble the special tool over the coupling. B—Push the tool into the cage opening to release the female fitting. C—Pull the tool and the fittings apart. D—Remove the tool from the fitting and line. (Ford)*

removal tool. Clean the inside and outside sealing areas thoroughly. Run a piece of string through the O-ring groove to remove any old O-ring material. Then blow the fitting clean with compressed air. A damaged garter spring can be replaced, **Figure 17-5A.** If the grooves show any sign of damage, the fitting should be replaced.

Coupling Reassembly

To reassemble the coupling, lightly lubricate the O-rings with clean refrigerant oil, install the O-rings, and press the two halves of the coupling together using a light twisting motion. See **Figure 17-5B.** You will hear a click when the two halves are fully seated. Pull on the coupling to ensure it is locked in position.

Spring lock couplings tend to leak. To ensure leakage is kept to a minimum, reinstall any clips that were placed over the coupling. Special aftermarket clamps are available to tightly hold the spring lock coupling in place. **Figure 17-6** illustrates this type of clamp.

Service Fittings

While service fittings are rarely the cause of refrigerant leaks, they sometimes do require service. Usually, service is required due to damage to the fitting itself or the valve inside the service fitting.

Quick Disconnect Service Fittings

To remove a valve from a push-on service fitting, a special octagonal socket is usually needed. Recover any refrigerant charge still present, then use the socket to remove the valve, **Figure 17-7.** Installation is in the reverse order.

Valve Core Replacement

To replace a valve core, use a special tool to unscrew the core from the fitting, **Figure 17-8.** This tool resembles a tire valve core removal tool. To replace the core, first lubricate the core O-ring with the right type of lubricant. Most manufacturers recommend the same type of refrigerant oil that is used in the refrigeration system. Next, use the core removal tool to install and tighten the new core.

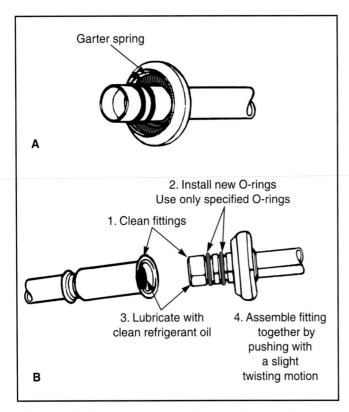

Figure 17-5. A—*Before reassembling make sure the spring is in place and n good condition. If the spring is bent or expanded, the coupling should be replaced. B—Replace the O-rings, and lubricate them thoroughly with the proper type of refrigerant oil. Then align the halves and push them together to reassemble the coupling. (Ford)*

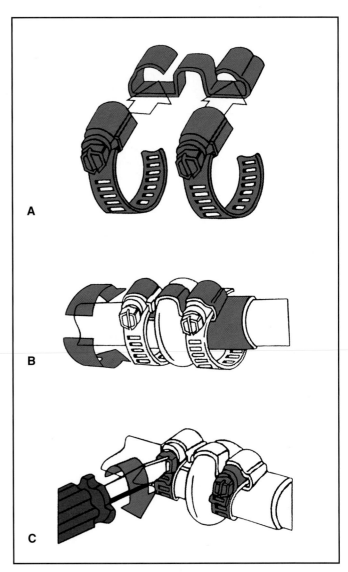

Figure 17-6. *Spring lock connections have a tendency to leak. Some aftermarket companies make clamps to help secure spring lock connectors. A—Assemble the clamps and bracket. B—Install the bracket and clamps on the refrigerant line. C—Tighten the clamps and bracket. These clamps hold the fitting halves tightly and place additional pressure on the O-ring seal area. (TDR)*

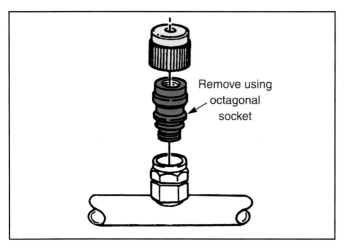

Figure 17-7. *Remove some R-134a fittings by using a socket to loosen and remove the fitting. (General Motors)*

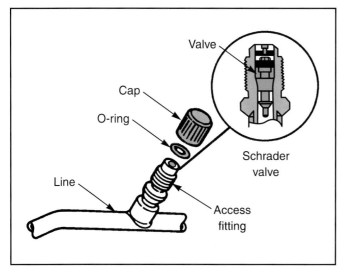

Figure 17-8. *Schrader valve assemblies can be removed like a bicycle tire valve. (General Motors)*

Retrofit fittings

When retrofitting an R-12 refrigeration system to R-134a, the original R-12 fittings must be replaced with R-134a fittings. To install this type of fitting, remove the valve core from the existing R-12 fitting. Then coat the threads of the male fitting with Locktite or other thread locking compound. This will prevent removal. Then thread the R-134a fitting onto the R-12 fitting, **Figure 17-9.** Other retrofitting procedures are covered in Chapter 23.

Servicing Hose Fittings

Hose fittings can be divided into hose clamps and crimp fittings. Hose clamp fittings are relatively easy to remove and install using a screwdriver. Crimp fittings cannot be removed without being destroyed, and new crimp fittings must be made with special tooling. Procedures for servicing these fittings are given in the following paragraphs.

Figure 17-9. *R-134 retrofit fittings screw over the R-12 service fittings already on the vehicle.*

Crimped Fittings

Crimped fittings are bands or rings formed (crimped) over the hose at the time of manufacture. They should only be removed if a new hose is being made up. A damaged crimp fitting cannot be replaced unless the hose is also replaced. See the section on making hose in this chapter for information on replacing crimped fittings.

Hose Clamps

Hose clamps, **Figure 17-10,** were often used to clamp a rubber hose to a metal line on older cars. These clamps are sometimes called *worm gear clamps.* They are used on some aftermarket air conditioners and sometimes to make

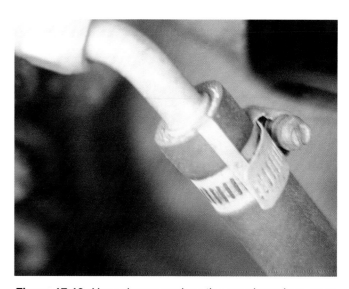

Figure 17-10. *Hose clamps, such as the one shown here, were once widely used in refrigeration systems. Other types of fittings have since replaced hose clamps. They should not be used on modern R-134a systems, or in combination with bead-lock fittings.*

hose repairs. Sometimes hose clamps do not seal adequately, especially when a clamp is reused. Whenever possible, hose clamps should be replaced with crimped or other types of fittings.

 Caution: Hose clamps should never be reused when an R-12 system is retrofitted to R-134a. Never use hose clamps with barrier hoses.

If a hose clamp is being used, ensure the fitting to be inserted into the hose is a barb fitting. Then insert the fitting fully into the hose, making sure the clamp will be over all the barbs. Tighten the clamp until the rubber just begins to show through the holes in the clamp. After tightening, try to twist the hose to make sure it cannot move in relation to the metal line. If the hose can be turned, it is not tight enough. If the hose cannot be tightened sufficiently, replace the clamp or use a crimp fitting.

Gasket Replacement

To replace a gasket, disassemble the mating parts by removing the fasteners. Then remove the gasket, carefully noting its position. **Figure 17-11** shows the placement of a gasket between an expansion valve and tubing flange.

Carefully scrape the mating surfaces to remove any remaining gasket material. Carefully check the condition of the mating surfaces. Compare the old and new gaskets to ensure the replacement gasket is correct. Then install the new gasket in the same position as the old one. Most refrigeration system gaskets do not require gasket sealer. Tighten the fasteners to the proper torque.

Hose Service

A slight amount of refrigerant will seep from the hoses and seals over time. This is not cause for hose replacement. If a hose is leaking severely or shows signs of damage, it should be replaced.

Hose Removal and Replacement

Begin hose replacement by removing any clamps that hold the hose in position. Then disassemble the fittings and remove the hose. Clean any fitting mating surfaces and make sure all related O-rings and gaskets are replaced. If a hose is being replaced with an exact replacement, compare the old and new hoses to make sure the new hose is correct. Make sure any plastic caps or plugs have been removed from the new hose before installing it.

Place the new hose in position and lightly install the fittings using new O-rings. Before tightening the fittings,

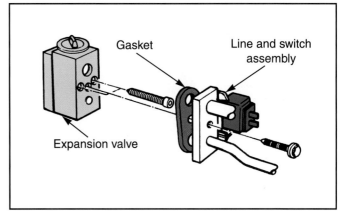

Figure 17-11. *Gaskets are sometimes used in refrigeration systems. This gasket is used between the expansion valve and the refrigeration lines. Refrigeration gaskets are made of special refrigerant sealing material and should not be replaced with standard gasket materials. (DaimlerChrysler)*

make sure the hose has no extreme bends or twists, and does not pass closer than 2 1/2 inches (6.35 cm) to any exhaust system part. Hoses should also clear the drive belts, cables, and other parts.

Once the hose is in place, reinstall any clamps. Tighten the hose fittings and leak test the system as necessary. Then evacuate and recharge the system as explained in Chapter 16.

Making Hoses

Many shops are equipped to make replacement hoses, or may have the hose made by a local company that specializes in hose fabrication. Making replacement hoses can save time and is usually cheaper than buying a new hose from the manufacturer. Making hoses is sometimes necessary when a retrofit is performed and the old hoses must be replaced with barrier hoses. Sometimes the exact replacement hose is no longer available from the manufacturer and a new hose must be made. Sometimes the fittings and header of a complex hose assembly are reused with new hoses.

Selecting Proper Hoses

It is important to use the proper kind of refrigeration hose and fittings. The hoses must be designed for refrigeration system service. Barrier hoses must be used when the system uses R-134a and R-22 blends. The hose must be the proper size to allow refrigerant to pass through the system with minimum restriction.

Shops making their own hoses will keep a stock of the most needed hose and fitting sizes. **Figure 17-12** shows the most common crimped hose fitting, the **beadlock fitting**. Barb fittings can also be used with a crimped fitting. However, crimping machines and their fittings are dedicated to each type and cannot be interchanged.

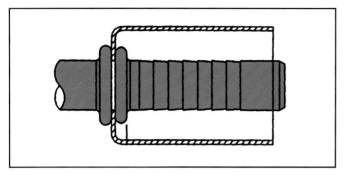

Figure 17-12. *Most modern refrigeration systems have bead-lock fittings. Some beadlock fittings do not have a self-contained outer covering. (Four Seasons)*

Remember that most manufacturers recommend using beadlock fittings on R-134a systems. Barb fittings can be used with older R-12 systems. Barb fittings should never be used to make ends on barrier hoses. Sometimes the original fittings or headers are used with new hoses.

To reuse an existing header or fitting, cut the tubing end with a tubing cutter. The cut must be square and as close to the original hose fitting as possible. Before cutting, be sure there is a sufficient straight run of tubing to install the fitting. Next, install the hose fitting over the end of the tubing and tighten to fit. Once all repairs to the headers and fittings have been made, attach the hoses.

Cutting Hose

When a hose is needed, the technician can cut the right length of proper size hose and attach the needed fittings. Hose length is important. If the hose is too short, it will be strained and may pull apart if there is too much movement between parts. If the hose is too long, it may come into contact with moving parts or exhaust components.

Once the correct hose has been obtained, cut it to the proper length. Always use a cutter that will make a clean square cut, **Figure 17-13.** Some hose makers recommend wrapping the hose with tape before cutting. This prevents damage to the hose surface.

 Note: Never use an abrasive wheel to cut hose. An abrasive wheel can damage the internal barrier hose and may leave debris inside of the hose.

Installing and Crimping Hose Fittings

After cutting, lightly lubricate the inside of the hose. Be sure to use the type of refrigerant oil that will be used in the system. Then push the fitting assemblies onto the hose ends. The hose may have to be installed a specific distance into the fitting, depending on the manufacturer's instructions. Next crimp the hose to the fitting using a crimping machine. **Figure 17-14** shows a hose being

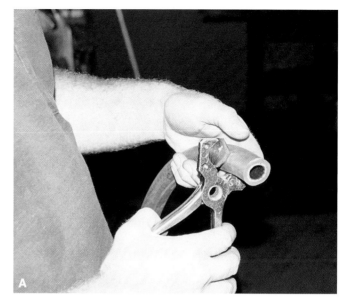

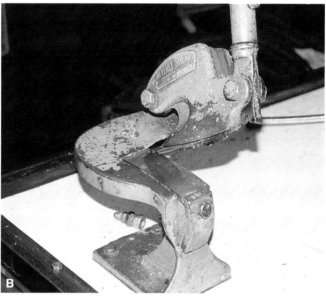

Figure 17-13. *Before a fitting can be installed, the hose must be cut cleanly and straight. A—A hand cutter can be used if it is properly positioned. B—A bench-mounted hose cutter is more effective in making straighter cuts. Some technicians prefer to use a rotary cutting blade.*

placed in a hydraulically operated crimping machine. In **Figure 17-15,** the technician has operated the hydraulic ram to crimp the fitting to the hose. **Figure 17-16** shows the crimped hose assembly. Types of hose crimping machines were discussed in Chapter 3.

 Note: Do not use hose clamps unless there is no other way to connect the hose and fitting. Never use a hose clamp with a barrier hose.

Once the hose has been assembled, if possible, test the hose for leaks, **Figure 17-17.** Install it as explained earlier in this chapter.

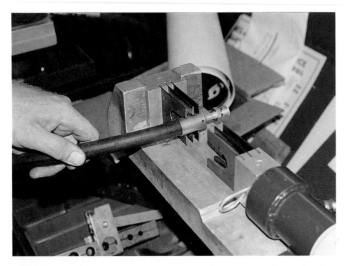

Figure 17-14. *To crimp a hose, place it in the crimping machine. Make sure the hose is fully inserted in the fitting, and the fitting is properly placed in the crimping machine.*

Figure 17-16. *The finished crimp. Note how it consists of three different crimped areas that extend completely around the fitting. Each crimp provides additional sealing.*

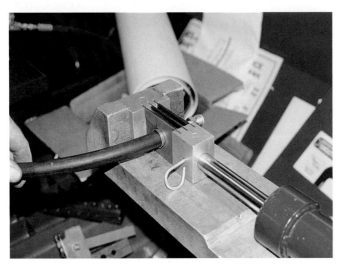

Figure 17-15. *This hose crimping machine uses hydraulic pressure to make the crimp. The crimp must extend around the fitting.*

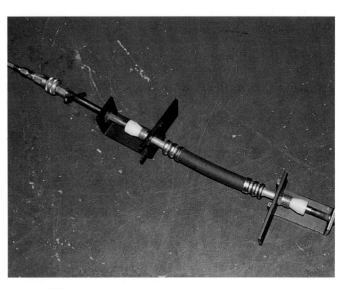

Figure 17-17. *After assembly, the new hose should be leak tested to ensure it will not leak.*

Line Service

The metal lines of the refrigeration system usually last the life of the vehicle. Unlike rubber hoses, they usually do not leak. If a line is damaged, it can be serviced in the same way as a hose.

Line Removal and Replacement

To remove a metal refrigeration line, remove any clamps holding the line to the vehicle, **Figure 17-18.** Loosen the fittings and allow any residual pressure to escape. Then unscrew the fittings and remove the line from the vehicle. Clean the fitting mating surfaces as necessary. Check that all plastic caps have been removed from the new line and there are no other obstructions in the line.

Then install the new line with new O-rings and any other seals and gaskets. Do not bend or kink the line during installation.

Making Lines

Metal lines can be simply made. However, it is important the right kind of material be used for lines. If you do not have the proper kind of tubing, purchase an exact replacement line. Sometimes lines can be replaced with rubber hoses, however, this is not recommended.

The line should be made of the same material as the original line. This is usually aluminum or steel. Stainless steel is often used to make replacement lines; especially those used on the high side of the refrigeration system. Any line used must be strong enough to withstand the pressures of the refrigeration system. Do not use tubing intended for

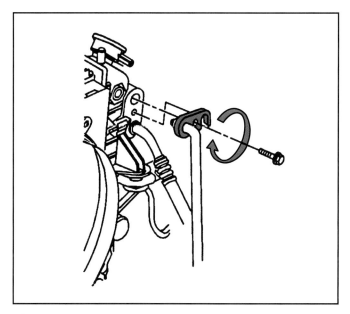

Figure 17-18. *Many line fittings are tightly clamped to the related refrigeration component by a bolt. Sometimes a threaded stud and nut are used. Always remove all refrigerant pressure from the system before removing the bolt or nut.*
(General Motors)

fuel systems or transmission coolers. This tubing is intended for use under relatively low pressures, and will burst under refrigeration system pressures. Never use low pressure or unknown tubing just because it is handy.

In the past, copper tubing was sometimes used in place of the original aluminum or steel lines. However, copper tubing cannot withstand high refrigerant pressures and is not recommended. If copper tubing is used anywhere in the refrigeration system, it is vital the system be free of water. Make sure the refrigeration system has been thoroughly evacuated if copper tubing is being used. Water will react with the refrigerant to form an acid that will dissolve copper from the tubing interior. Copper plating (transfer of copper to other metal surfaces) will occur throughout the system.

Once the proper line has been selected, uncoil it from the roll and use a tubing bender to form it into the shape of the line to be replaced. If the line is kinked during the bending procedure, discard it and start over. When the line has been formed, cut the line end square with a tubing cutter and install the line fittings. The line is now ready to install.

Summary

Hoses, lines, fittings, O-rings, and gaskets seldom cause refrigeration system performance problems. However, they often leak. The refrigeration system fittings must be disassembled to replace leaking O-rings, flush the system components, or change components. Always make sure all pressure has been removed from the system before loosening any fitting.

Lip seals and O-rings should always be replaced when the fitting is removed. Compression fittings can be removed using hand tools. Line wrenches should be used when available. O-rings can be replaced once the fitting is disassembled. Coat the O-ring with the proper type of refrigerant oil before reassembly.

A special tool is needed to disassemble spring couplings. After the coupling is disassembled check the seal grooves for damage. Replace the coupling if any damage is found. After installing new O-rings press the coupling together. Special locks are available to reduce leaks.

Screw type clamps were often used to clamp a rubber hose to a metal line on older cars. They should only be used when no alternative connecting method is possible. Crimped fittings are bands or rings that are formed over the hose at the time of manufacture. A damaged crimp fitting cannot be replaced unless the hose is also replaced.

Most refrigeration system gaskets can be easily replaced. Disassemble the mating parts and scrape the surfaces clean. Inspect for damage and then replace the gasket.

Replace a hose by removing any positioning clamps, then disassemble the fittings and remove the hose. Compare the old and new hoses then place the new hose in position. Loosely install the fittings using new O-rings. Once the hose is in place, reinstall any clamps.

Hose fabrication is relatively simple. However, the proper kind of hose must be used. The hoses must be designed for refrigeration systems. Shops making hoses keep a stock of the most needed hose and fitting sizes. Sometimes the fittings and header are reused with new hoses.

Lightly lubricate the inside of the hose. Then push the fitting assemblies onto the hose ends and crimp the hose to the fitting using a hose crimping machine. Do not use screw type hose clamps unless no other method is possible.

Metal usually lines do not leak. If a metal line is damaged, it can be easily replaced. Sometimes replacement lines can be made. If the proper tubing is not available, use an exact replacement line. Some lines can be replaced with rubber hoses.

Review Questions—Chapter 17

Please do not write in this text. Write your answers on a separate sheet of paper.

1. To disassemble a compression fitting, use two _____.

2. An updated O-ring may be a different _____ than the original.

3. What should be used to lubricate the O-ring(s)?

4. A spring lock coupling has _____ O-ring(s).

5. A hose clamp should only be used with a _____ type fitting.

6. Why should the technician compare the old and new gaskets?

7. A hose should not pass closer than 2 1/2 inches (63 mm) to any _____ part.

8. The two common types of hose fittings are the _____ and the _____ fitting.

9. Some hose makers recommend wrapping _____ around the hose before cutting.

10. *True or false?* One crimping machine can be used to crimp either type of the above fittings.

ASE Certification-Type Questions

1. All of the following procedures should be performed during an O-ring replacement procedure, *except:*
   (A) make sure all pressure has been removed from the system.
   (B) cap open system lines.
   (C) clean the fitting surfaces carefully.
   (D) replace the O-ring with one of the same color.

2. Which of the following service procedures can be performed with hand tools?
   (A) Replacing a compression fitting O-ring.
   (B) Replacing a crimped fitting.
   (C) Replacing a spring lock coupling O-ring.
   (D) Metal line fabrication.

3. A vehicle has a leak at a compression fitting. Technician A says you should use line wrenches to loosen the fitting. Technician B says the fitting must be replaced as it is damaged once loosened. Who is right?
   (A) A only.
   (B) B only.
   (C) Both A and B.
   (D) Neither A nor B.

4. Technician A says captured O-rings cannot be replaced. Technician B says non-captured O-rings cannot be replaced. Who is right?
   (A) A only.
   (B) B only.
   (C) Both A and B.
   (D) Neither A nor B.

5. All of the following should be done when installing fittings during a retrofit, *except:*
   (A) remove the existing R-12 valve core.
   (B) coat the threads of the R-12 fitting with Loctite.
   (C) coat the inside threads of the R-134a fitting with Teflon sealer.
   (D) thread an R-134a fitting over the R-12 fitting.

6. Technician A says when replacing a gasket, you should carefully scrape the old gasket from the part's mating surface. Technician B says you should use sealer on all refrigeration system gaskets. Who is right?
   (A) A only.
   (B) B only.
   (C) Both A and B.
   (D) Neither A nor B.

7. Which of the following should *never* be used with barrier hoses?
   (A) Compression fittings.
   (B) Hose clamps.
   (C) Crimp fittings.
   (D) Beadlock fittings.

8. Which of the following materials would make a good replacement metal refrigeration line?
   (A) Stainless steel.
   (B) Aluminum.
   (C) Copper.
   (D) Both A and B.

9. Technician A says when replacing metal lines, you should remove all clamps holding the line to the vehicle. Technician B says you should clean the mating surfaces before installing the new line. Who is right?
   (A) A only.
   (B) B only.
   (C) Both A and B.
   (D) Neither A nor B.

10. If a metal line is kinked, it should be:
    (A) straightened.
    (B) heated and straightened.
    (C) use with the kinked portion intact.
    (D) discarded.

Chapter 18

Compressor and Clutch Service

After studying this chapter, you will be able to:

❑ Diagnose compressor and clutch problems.
❑ Remove and replace a compressor clutch.
❑ Remove and replace a compressor clutch electromagnet.
❑ Remove and replace a compressor shaft seal.
❑ Remove and replace compressor gaskets and O-rings.
❑ Remove and replace a compressor valve plate and valve assembly.
❑ Remove and replace a compressor capacity control valve.
❑ Explain how to overhaul an older piston compressor.

Technical Terms

Chatter

Locks up

Compressor retrofitting

Air gap

Oil wick

Shell

Valve plates

Cylinder head

Split crankcase

Capacity control valves

The compressor is the pump for the refrigeration system. Any problems in the compressor will cause the system to malfunction. In most cases, technicians simply choose to replace the compressor. However, many compressor problems can often be corrected with simple part replacement. New and rebuilt compressors are available for all vehicles made within the last 20 years. Studying this chapter will enable you to determine compressor problems and make repairs. The information in this chapter will also enable you to tell when the problem is not in the compressor.

⚠ **Warning: Before loosening any refrigeration system fitting, make sure the system is completely discharged. If necessary, place a vacuum on the system to ensure all pressure has been removed. If any refrigeration system component will be removed for more than 10 minutes, cap the refrigeration system openings to minimize water vapor collecting in the system.**

Common Compressor and Clutch Problems

Compressor and clutch problems consist of noises and poor cooling. The following sections cover the causes of common compressor problems. **Figure 18-1** lists some causes of compressor noises. Keep in mind the compressor is not the only cause of poor cooling.

 Note: Check all other sources before condemning the compressor.

Common Causes of Compressor Noises	
High pitch squeal at all times	Belt slipping Defective belt tensioner
High pitched squeal only when compressor engaged	Compressor seized Clutch seized
Compressor makes rattling or knocking noise	Compressor damaged internally
Compressor clutch not engaging	Air gap too large Electromagnet weak or inoperative
Compressor clutch slipping	Loose belt Weak electromagnet
Compressor makes chattering noise	Clutch plate damaged or loose

Figure 18-1. *List of common causes of compressor noise.*

Noises

Before deciding the compressor is the source of a noise, check the engine condition. Loose engine or exhaust system parts, loose belts, and even vacuum leaks can sound like a noisy compressor. Check all belt driven accessories, especially the water pump and air pump, if used. In addition to checking the accessories themselves, check for loose mounting bolts.

Belt and Clutch Noises

The most common source of HVAC system noise is a slipping belt. The belt will make a loud squealing noise as it slips. If the slippage goes on for enough time, the belt will be ruined. A common cause of belt slippage on vehicles with serpentine belts is a defective belt tensioner.

 Note: Serpentine belts will make a squeaking or chirping noise if the pulleys are misaligned, contaminated with paint, belt dressing, oil, grease, or debris from the belt. Unless the pulleys are misaligned, the only way to correct this condition is to replace the belt and clean the pulleys.

If the clutch slips it may make noise, but in most cases slipping clutches are silent. Clutch slippage is often caused by a weak electromagnet. If the clutch is loose or damaged, it may chatter. *Chatter* is a rapid clicking or knocking noise. A chattering clutch can usually be seen vibrating as the system operates.

If the compressor *locks up* (stops turning), the belt will squeal excessively. This is hard to mistake for anything else. If the compressor remains locked up for more than a few seconds, the belt will begin to burn and smoke. Causes of intermittent compressor lockup are excessive high-side pressure caused by refrigerant overcharge or excessive condenser heat load. Sometimes the compressor is starved for oil or is binding internally.

Internal Compressor Noises

Rattles and knocking noises can be caused by internal problems in the compressor. A noisy compressor is usually about to fail and may already be unable to develop proper pressures. Adding about 2 oz. (60 ml) of refrigerant oil to the refrigeration system can sometimes quiet older compressors, especially axial piston types. This is usually done as a diagnostic procedure only, and is not considered a fix. The compressor is worn and will eventually need replacement.

Compressor Clutch Not Engaging

If the compressor clutch does not engage, the air gap may be too large, the electromagnet winding may be open or shorted, or there may be a control system problem such as a blown fuse, stuck high pressure switch, or clutch cycling switch. To check the electromagnet, bypass the control system by unplugging the clutch connector, **Figure 18-2.** Then directly energize the electromagnet using one jumper wire to ground and another to battery positive. Polarity is not important as long as the electromagnet is disconnected from the rest of the HVAC electrical system.

If the clutch engages, the electromagnet is good and the problem is in the control system. If the clutch does not engage, the electromagnet or clutch assembly is defective. The electromagnet may be tested with an ohmmeter without removing it from the vehicle. Turn the ohmmeter on and touch one probe to each electromagnet lead. If the electromagnet winding has zero or infinite resistance, it is defective.

Excessive Pressure

A compressor in good condition can produce very high pressures. The pressure switches and other components of the refrigeration system work to keep pressures within the proper limits. Overcharge or excessive temperatures (high heat load) can cause excessively high pressure. While it is often mistaken as such, this is not a compressor problem. To correct high pressure problems, remove refrigerant as necessary or correct the high temperature problem. Typical causes of high temperature are an inoperative fan, debris on the condenser, or an engine cooling system defect. Once the temperature problem has been corrected, recheck system operation.

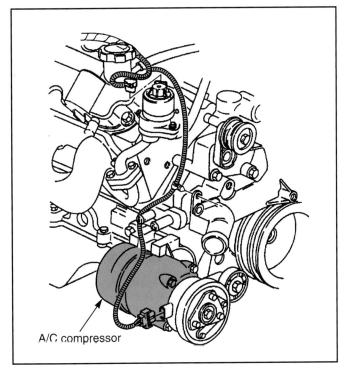

A/C compressor

Figure 18-2. *Many modern vehicles locate the compressor low on the engine because of hood clearance concerns. This makes the compressor somewhat harder to test and service. (General Motors)*

 Note: Extreme ambient temperatures can cause reduced refrigeration system performance. This should not be mistaken for a problem which causes extreme high pressure.

Constant Low Pressure

If the pressure is low under all operating conditions, the refrigerant level is low or the compressor has internal damage. If the refrigerant level is low, check for leaks, correct, and retest the system. If the system is properly charged, and both high and low sides are lower than normal, the compressor may be defective. One sign the compressor may be defective is if the gauge readings quickly equalize after the engine is shut off, **Figure 18-3.** Racing the engine with the HVAC system operating sometimes isolates a defective compressor. If the pressures do not approach normal readings when the engine is raced, the compressor reed valves are probably not sealing.

 Note: It is normal for system pressures to be lower when the ambient temperature is below 60°F (16°C).

A quick check of the compressor internal parts involves turning the compressor by hand. Stop the engine, remove the ignition key, and attempt to turn the clutch driven plate by hand, **Figure 18-4.** If the driven plate turns without binding, the compressor crankshaft and the parts attached to it (cylinders, vanes, or scrolls) are not damaged. If the driven plate is hard to turn or feels rough when it is turned, the compressor is probably defective. If the compressor turns, it may still have defective valves or another problem.

Occasional Low Pressure

Common causes of occasional low pressure are clutch or belt slipping. Slippage is usually accompanied by noise. Low refrigerant charge can cause excessive clutch cycling. Another cause of low pressures is the clutch cycling too often, even with a full refrigerant charge. Also check the HVAC control system for proper operation.

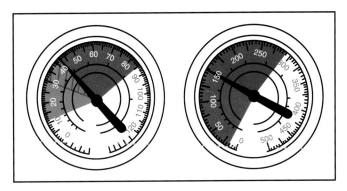

Figure 18-3. *Pressure gauge readings indicating a possible defective compressor.*

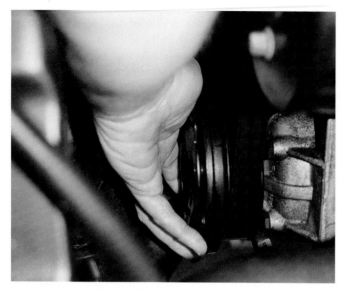

Figure 18-4. *After allowing the engine to cool off, try turning the driven clutch plate by hand. If the clutch cannot be turned, the compressor is seized. A common symptom of this is squealing when the air conditioner is turned on, along with belt glazing.*

Compressor Leaks

The compressor is sometimes a source of leaks. Body gaskets and O-rings can leak and may seep slightly. Slight oil seepage on the compressor body is not a sign of leakage. More serious leaks will cause oil to run down the compressor housing from the source of the leak.

Leaks from the compressor are usually around the shaft seal. A sign of shaft leakage is oil on the clutch assembly. Severe leaks will cause oil to be thrown on the underside of the hood. As with other seals, slight oil seepage is normal.

Leak Testing Compressors

It is possible to leak test some compressors off the vehicle. This is done through the use of a special adapter plate, which has two refrigerant fittings. Off-engine leak tests are performed on compressors that are difficult to reach where they are located on the engine or after a compressor has been rebuilt.

Begin the test by connecting the adapter plate to the compressor, **Figure 18-5.** Connect a manifold gauge or service center to the adapter plate. Evacuate and add a little refrigerant to the compressor. Using a leak detector, probe around the compressor. There should be no leaks at any point.

Compressor Removal and Replacement

Compressor removal and replacement is relatively easy, but the system's refrigerant charge must be recovered first. A few older vehicles will have hand valve service fittings. These fittings can be used to isolate the compressor

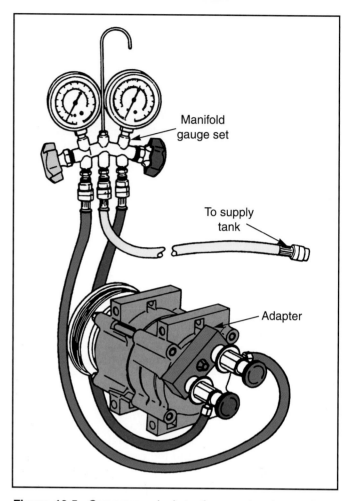

Figure 18-5. *Compressor leak testing can be done using a special adapter plate. (Ford)*

from the rest of the system. The refrigerant in the compressor should be recovered before the connections are removed. On some vehicles, the compressor is located at the bottom of the engine and the vehicle must be raised to gain access. This procedure assumes the clutch is preinstalled on the new compressor.

 Note: It may be necessary on some front-wheel drive vehicles to remove splash panels, one of the front wheels, or loosen the engine mounts and lift the engine slightly to remove the compressor.

Compressor Removal

Once the refrigerant has been recovered, remove and cap the compressor inlet and outlet lines. Loosen the belt adjuster and remove the belt. Remove or relocate any parts as needed to access the compressor. If the compressor is mounted low, you may need to raise the vehicle. Remove the clutch electrical connector, and connectors to any switches on the compressor body. Remove the bolts holding the compressor to the engine and remove the compressor from the engine compartment, **Figure 18-6.**

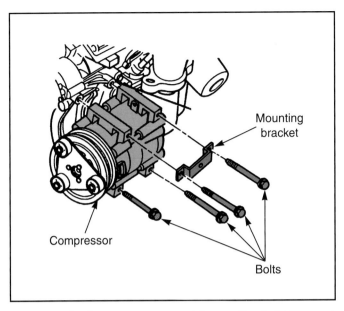

Figure 18-6. *Compressor removal is usually straightforward. On newer cars, you will probably have to raise the vehicle and remove other components to access the compressor. (Ford)*

Adding Oil to Compressors

When replacing the compressor, you must drain the oil out of the old compressor before installing the new unit. Remember that compressor repairs are not complete until the correct amount of oil has been added to the system. In some cases, checking the oil can be done with a dipstick, which was outlined in Chapter 16.

Once the compressor is removed, allow the oil to drain into a measuring cup or graduated cylinder, **Figure 18-7.** Measure the amount of oil recovered from the compressor and compare to specifications. This measurement is important if the compressor is being rebuilt. In some cases, the new compressor may come prefilled with refrigerant oil.

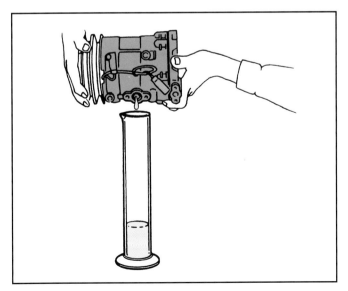

Figure 18-7. *Drain the oil from the old compressor. Be sure to add the correct amount of oil to the new compressor before installation. (Subaru)*

Be sure to check and add oil according to the instructions given in the service manual. Also be sure to use the proper kind of oil. Do not add oil to the compressor until you are ready to reinstall the unit.

 Note: You should replace the accumulator or receiver-drier anytime you replace or rebuild the compressor. On CCOT systems, you should also replace the orifice tube.

Compressor Installation

Place the compressor in position and loosely install the mounting bolts. Install all the bolts before tightening any of them. Reconnect the clutch and any pressure switch electrical connectors. Install the compressor refrigerant lines using new gaskets or O-rings. If needed, reinstall any other parts that were removed and lower the vehicle. Install the belt over the pulleys. Adjust the belt if needed, and retighten all compressor bolts. Evacuate and recharge the refrigeration system and check compressor operation.

 Caution: Make sure all bolts are installed properly and tightened to the correct torque. The compressor can become a source of vibration if not installed properly.

Compressor Retrofitting

Sometimes a newer compressor must be installed on an older refrigeration system. This is known as *compressor retrofitting.* Retrofitting is done when the original compressor model is no longer available. Compressor retrofitting is also done when some systems are being converted to use R-134a as some R-12 compressors are not compatible.

 Note: Compressor retrofitting generally means replacing the compressor on an older vehicle with one that was not original equipment. It should not be mistaken for modifying or resealing an older compressor so it can handle R-134a. Most newer compressors can handle either R-12 or R-134a without modification or resealing.

When retrofitting a compressor, additional parts, such as brackets, hoses, wiring harnesses, and relays may need to be replaced in order to use the new compressor. Special retrofitting kits are often used to install a retrofit compressor.

Compressor Clutch and Electromagnet Service

Sometimes the clutch can be replaced without removing the compressor from the vehicle. On most vehicles, engine compartment clearances are so close, the clutch cannot be serviced without removing the compressor. Always consult the service manual for exact clutch service procedures.

⚠️ **Warning: To prevent damage or injury, ensure the ignition switch is in the off position before beginning repairs. For added protection, the negative cable should be removed from the battery.**

Clutch Removal

Special tools may be needed to remove the compressor clutch. Begin disassembly by removing the belt and compressor (if needed) from the vehicle. Then remove the nut holding the driven plate (sometimes called the hub) to the compressor crankshaft, After removing the nut, remove the clutch hub with a puller designed for the job, **Figure 18-8.**

Once the driven plate has been removed, the drive plate and pulley assembly can be removed. Most pulley assemblies are held to the compressor nose by a large snap ring. Once the snap ring has been removed, the pulley and bearing assembly can be removed from the compressor. Some pulleys are pressed onto the compressor and must be removed with a puller. The clutch bearing is usually attached to the pulley and will be removed with the pulley.

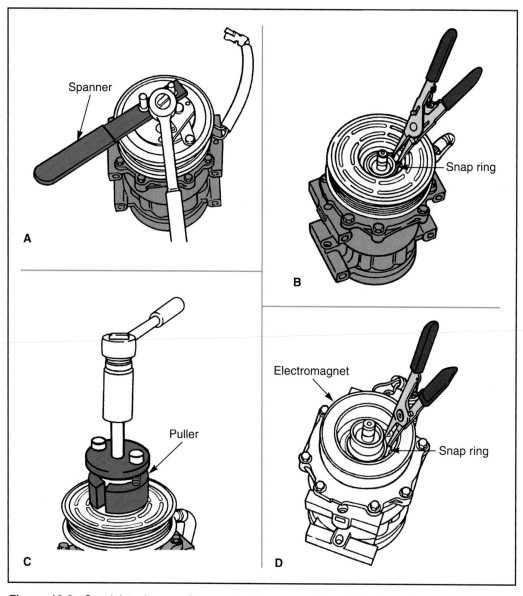

Figure 18-8. *Special tools are often needed to remove the clutch from the compressor. A—Removing the nut. B—A snap ring usually holds the pulley to the electromagnet. C—Pulley removal; this usually requires a special puller. D—Electromagnet removal. Do not try to pry the clutch from the compressor. (DaimlerChrysler)*

Electromagnet Removal

If the electromagnet failed any tests, it can be replaced once the clutch pulley has been removed. The electromagnet may be bolted to the front of the compressor, or may be held to the compressor nose by a snap ring. Remove the fasteners as necessary and remove the electromagnet.

Inspect the electromagnet for overheated or shorted windings. Also check for contact with the electromagnet and the pulley. If the electromagnet shows evidence of damage, replace it. Some older systems deliver current to the electromagnet through brushes. Inspect the brushes for wear and replace if necessary.

 Note: Some brushes can be removed without removing the clutch pulley assembly.

Clutch Inspection

The mating faces on the clutch should be checked for wear and scoring. Light wear and scoring are normal and the clutch does not need replacement. If the faces are heavily scored, worn, or damaged, they should be replaced, **Figure 18-9.**

Figure 18-9. A— A misaligned clutch plate indicates that the clutch assembly is badly damaged. B—This clutch is destroyed and must be replaced.

Also check the pulley grooves for wear and damage. If the driven plate uses vulcanized rubber to connect the inner and outer portions, check the rubber for signs of overheating. If the rubber section is melted or charred, the plate should be replaced. Check metal connections for breakage or bending. If one plate is defective, both clutch plates should be replaced.

Clutch Bearing Inspection

The clutch bearing can be inspected to ensure it is not damaged. Turn the bearing's inner race and note any roughness or binding. If the bearing does not turn smoothly, it should be replaced. Most bearings are held to the drive plate and pulley assembly with a snap ring. Remove the snap ring and pull the old bearing from the pulley assembly. Check the snap ring to ensure it is not bent or damaged. Slide the new bearing into position and reinstall the snap ring. Some clutch bearings must be installed using a special tool, **Figure 18-10.**

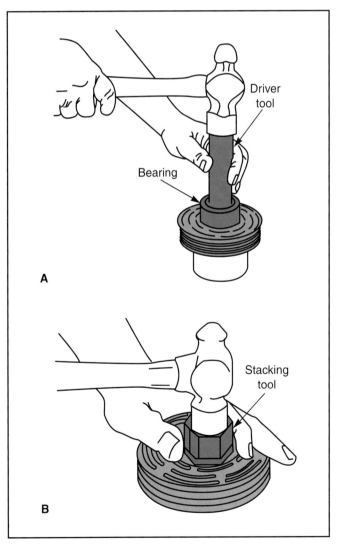

Figure 18-10. Bearings should be installed with the proper size bearing installer. Some bearings can be hammered into place, while others can only be installed using a press. (General Motors)

Electromagnet and Clutch Replacement

Before replacing the electromagnet and clutch assembly, make sure the mounting faces of the compressor are not damaged. Then follow the installation steps in the following paragraphs.

Electromagnet Replacement

To install the electromagnet, place it in position and install the fasteners. Be sure the electrical connectors are in the same position as originally. If the electromagnet mounting uses alignment pins, be sure to install the mounting flange over the pins before tightening any fasteners. Proper alignment is important to prevent contact between the electromagnet and pulley assembly. Once the electromagnet is installed, install the electrical connector.

Plate and Pulley Replacement

Some drive and driven plates and pulley assemblies must be installed using special tools. See **Figures 18-11** and **18-12**. Other assemblies can be slid into position or

Figure 18-11. *A—Make sure the key is in place on the compressor crankshaft, then align the key and keyway before reinstalling the clutch plate. B—Placing the clutch plate in place.*

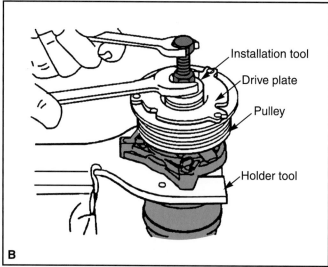

Figure 18-12. *A—The clutch plate is usually installed with a special tool similar to the one shown here. B—Use wrenches to turn the tool, which installs the clutch plate. Attempting to hammer the clutch plate into position will damage the compressor. (General Motors)*

lightly hammered into place. Once the pulley assembly is installed, make sure it turns freely without contacting the electromagnet. If there is any contact between parts, remove the pulley and correct the problem.

Most compressor shafts have a key and keyway. The keyway slot in the driven plate must be aligned with the key before the plate is installed. Many driven plates can be installed by tightening the retaining nut into place. Others require special installation tools.

Setting Air Gap

The *air gap* between the drive and driven plates is critical to clutch operation. A small air gap may cause the clutch to drag. Too large an air gap will cause the clutch to slip. Most air gap checking is done with a feeler gauge. Insert the proper thickness feeler gauge in the space between the plates, **Figure 18-13.** Then tighten the driven pulley until the feeler gauge lightly drags between the two pulleys. Check the clearance at several places around the pulley.

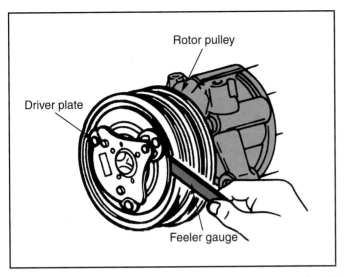

Figure 18-13. *Check clutch plate clearance as you install the plate. Most manufacturers suggest inserting a proper size feeler gauge between the clutch faces. When the gauge slides between the two faces with a slight drag, the clutch clearance is correct. (Nissan)*

Once the pulley has been installed and the air gap adjusted, reinstall the drive belt and tighten it to specifications. Belt service is covered in more detail in Chapter 20.

Compressor Overhaul

Most shops prefer to replace the entire compressor rather than perform any compressor repairs. Some compressors can be returned to service by performing the repairs in the following paragraphs. This is sometimes done when the replacement compressor is expensive, or when the compressor is otherwise known to be good.

The only time a compressor should not be rebuilt is if there is evidence of internal damage. If the compressor shaft cannot be turned by hand or with a wrench or socket, the compressor should be replaced. A compressor with internal damage almost always will cost more to rebuild than a replacement or retrofit compressor.

Shaft Seal Replacement

To replace the shaft seal, the clutch driven hub must be removed. Most shaft seals are serviced with the compressor assembled. On a few compressors, the compressor must be disassembled to change the seal. Replacement procedures are given in the following paragraphs.

 Note: In most cases, special tools are needed to replace the compressor shaft seal. Consult the service manual.

Shaft Seal Replacement with the Compressor on the Vehicle

Many shaft seals can be changed without removing the compressor from the vehicle. Once the driven hub is removed, the seal can be removed using special tools. A typical compressor shaft seal removal and replacement procedure is shown in **Figure 18-14.** Begin by removing the *oil wick.* The oil wick is a felt roll which absorbs slight amounts of oil that seep from the shaft seal. The wick keeps oil from reaching the pulley and belt.

 Note: The oil wick may be soaked with refrigerant oil. This is not necessarily a sign the seal is leaking excessively, as some oil seepage occurs even with a good seal.

Next, remove the shaft key. Then use the proper pliers to remove the internal snap ring holding the seal in place. Remove the seal assembly from the seal cavity, then remove the internal rubber O-ring.

Inspect all reusable parts for wear, then begin the process of installing the new seal. Lubricate all new parts with refrigerant oil. Do not touch any of the new seal parts with your hands. It is usually possible to handle the parts without touching them by only partially removing the wrappings. Once the sealing part has been placed on the special installation tool, completely remove the wrapping.

Install the O-ring using the special installation tool. Then install the seal. This also requires a special tool. Many manufacturers recommend placing a seal protector over the nose of the compressor crankshaft before installing the seal. Then install the snap ring, making sure it is fully seated. If the snap ring is not fully seated, refrigerant pressure will push the seal out of place. Some manufacturers recommend a special tool to fully install the snap ring.

After the snap ring is in place, place the oil wick in the seal cavity and install the clutch parts that were removed. Check the compressor oil level as explained in Chapter 16. Then evacuate and recharge the refrigeration system and check the new compressor seal for leaks.

Shaft Seal Replacement with the Compressor Removed

Some shaft seals can only be replaced after the compressor is removed from the vehicle and the end housing is removed. If possible, place the compressor in a holding fixture. If no holding fixture is available, place the compressor on a clean workbench. Drain the oil as explained in Chapter 16.

A typical compressor shaft seal removal and replacement procedure is shown in **Figure 18-15.** Once the clutch assembly has been removed, remove the key from the compressor shaft keyway. Then remove the bolts holding

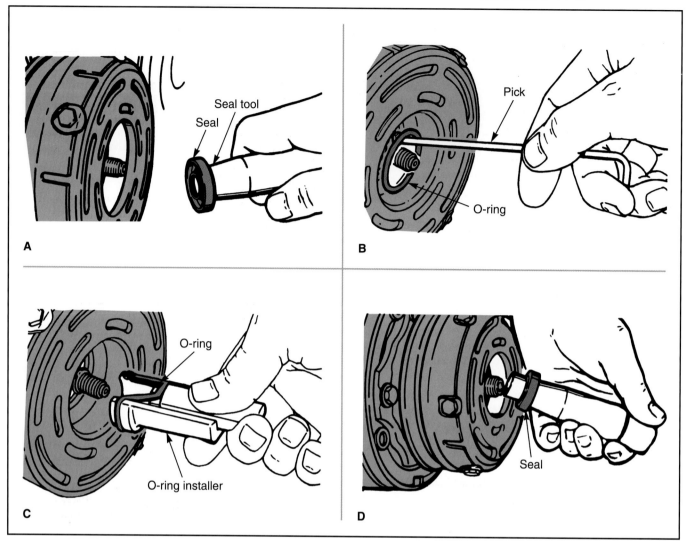

Figure 18-14. *Some compressors must be separated to gain access to the center O-ring. A—After removing the attaching bolts, lightly tap on one side of the housing with a plastic hammer to loosen the connection. B—Pull the two halves no more than 1 inch (2.54 cm) apart to reduce the chance of internal compressor damage. C—The O-ring can then be removed and replaced. D—Make sure you install the new seal completely to prevent leaks and damage. (DaimlerChrysler)*

the end housing to the compressor body. Slide the end housing from the compressor shaft and remove the O-ring from the front cover.

Remove the snap ring holding the seal in place and remove the seal from the cavity. Then remove the internal rubber O-ring from the seal cavity. Clean the seal bore and check for damage. Lightly lubricate the seal and O-ring with the proper refrigerant oil. Install a new O-ring, then install the seal. Use a socket to lightly tap the seal into position. Then install the snap ring, making sure it is fully seated in its groove.

After the snap ring is in place, install a new gasket or O-ring in the end housing. Place a seal protector over the nose of the compressor crankshaft and slide the end housing over the compressor shaft. Install and tighten the end housing bolts. Install the key in the keyway, install the oil wick in the seal cavity (if used), and install the clutch parts that were removed. Then reinstall the compressor, evacuate and recharge the refrigeration system, and check the front of the compressor for leaks.

Rebuilding Compressors

There are several rules which apply to the reconditioning and service of all compressors. Failure to follow these rules may result in poor compressor or refrigeration system operation. These rules apply to all compressor types:

❏ Replace the accumulator, receiver-drier, and orifice tube (if equipped) when performing a rebuild.

❏ If the compressor is from an R-134a system, evacuate the refrigeration system for a minimum of 45 minutes before recharging. PAG oil is hygroscopic and will absorb moisture very quickly.

❏ All parts, especially O-rings, should be lubricated with refrigerant oil before installation.

❏ Keep dirt and foreign material from getting into or on the compressor parts.

❏ Clean all parts with a nonpetroleum based solvent.

❏ Do not use compressed air to dry parts. Dry all parts using a clean, lint-free cloth.

Begin the overhaul procedure by recovering the refrigerant charge and removing the compressor from the vehicle. Once the compressor is off, proceed to drain the oil into a graduated cylinder or measuring cup. Begin draining by allowing as much oil as possible to drain from the suction and discharge ports. Remove the oil plug (if equipped) and drain the oil from the compressor housing.

Overhauling Radial Piston Compressors

A radial compressor can be removed and disassembled to replace most compressor body and housing gaskets and O-rings. However, most technicians prefer to simply replace a defective compressor. Gaskets and O-rings are often used to seal the compressor body and end housings. They may also be used to seal the compressor inlet and outlet fittings. Seals can be replaced with the compressor installed on the vehicle. Always place the compressor in a holding fixture or use a clean workbench to perform the following service operations.

Replacing Compressor Shell and O-rings

Some compressors have bodies that are covered by a sheet metal *shell.* O-rings seal the shell to the body. To service these valves, remove the shell using a special tool, **Figure 18-16.** Once the shell has been removed, the O-rings can be replaced and the shell reinstalled.

Front Head and O-ring

Begin by removing the clutch and electromagnet from the compressor. Remove and discard the shaft seal.

 Note: Mark the position of the head before removal.

Remove the front head mounting screws and remove the front head. You may need someone to help you with portions of this procedure. Discard all the old O-rings as

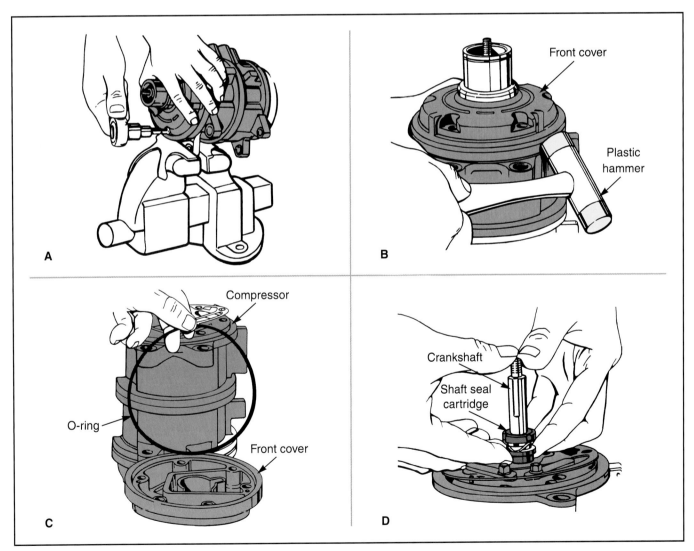

Figure 18-15. *This series of illustrations show the compressor shaft seal removal process when the compressor must be partially disassembled for replacement. A—Remove the compressor through bolts. B—Lightly tap on the front cover to loosen it. C—Remove and replace the O-ring sealing the front cover. D—Replace the shaft seal. (DaimlerChrysler)*

they will be replaced with new ones. When installing the resealed front head, check the gap between the rotor and new O-ring.

Replacing Compressor Main Bearing

Begin by removing the clutch and electromagnet, then, remove the front head assembly. Place the front head on blocks and use a special bearing removal tool to drive the main bearing out. To install the new bearing, place the front head with the neck end down on a solid flat surface. Place the new bearing and the installation tool on the front head. Carefully drive the new bearing on the front head. Reinstall the front head, electromagnet, and clutch on the compressor.

 Caution: Make sure you install the bearing in the correct direction. On some compressors, the bearing has a distinctive mark, telling you which side it should face.

Replacing Compressor Discharge Valves

Some radial compressors have individual valves located in the compressor body. To service these valves, remove the shell that covers the compressor body. Once the shell has been removed, determine which valve needs replacement. You may need to remove and examine each valve to locate the defective one(s).

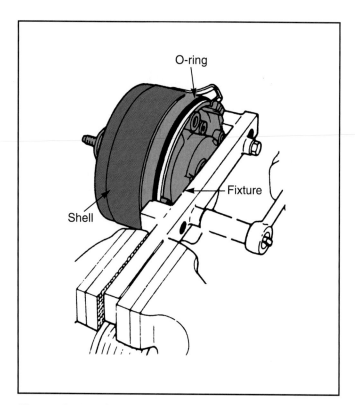

Figure 18-16. *On radial compressors, the cover must be pressed off. Clips retain the cover to the compressor. (General Motors)*

To remove the valve, use the proper size pliers to remove the snap ring holding the valve assembly in place, **Figure 18-17**. Remove the valve assembly from the compressor body. Check the valve assembly for bent reed valves and worn seats. If the valve shows signs of damage, it should be replaced. Compare the old and new valves then install the valve in the compressor body, **Figure 18-18**. Replace the snap ring. Reinstall the compressor body shell using new O-rings. Install the compressor on the engine and evacuate and recharge the refrigeration system. Start the engine and HVAC system and check compressor operation.

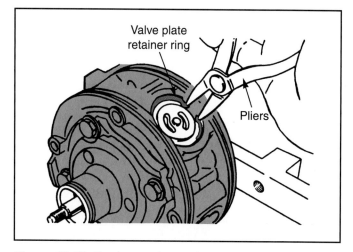

Figure 18-17. *To change the reed valve assembly on a four-cylinder radial compressor, first remove the shell then use snap ring pliers to remove the snap ring holding the valve assembly in place. (General Motors)*

Compressor Cylinder and Shaft Assembly

The compressor drive shaft and cylinders are replaced as a unit on most radial compressors. Start by removing the compressor from the vehicle and mounting it on a vise or other secure location. Remove the clutch, electromagnet, and shaft seal. Remove the front head, thrust washers, and shell next, followed by the discharge valve plates. Installation is in the reverse order of disassembly.

Overhauling Axial Piston Compressors

Rebuilding axial compressors is somewhat easier than radial compressors. However, procedures can vary greatly, due to the many different compressor designs. Be sure to follow manual procedures, as the rebuild in this text is generic, and does not apply to any one compressor.

Replacing Valve Plate and Cylinder Head

Valve plates are a flat assembly installed between the body and housing of the compressor. The valve plates are located on a single plate between the compressor body and the end housing, sometimes called the *cylinder head,* on some compressors.

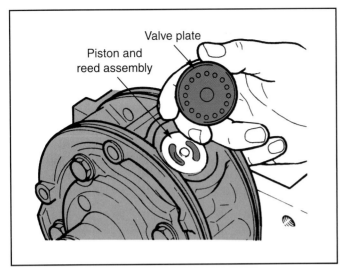

Figure 18-18. *Four-cylinder radial compressor reed valve assembly. As with other compressors, there are separate inlet and discharge valve plates. (General Motors)*

Some compressors have valve plates at both ends of the compressors. Others have only one plate, located at the end opposite the clutch. Most valve plates are replaced as an assembly. Other compressors have individual valves located in the compressor body.

Begin by removing the bolts holding the cylinder head to the compressor. After the bolts are removed, gently pry on the housing to split it from the body. Gently remove the housing and valve plate from the body. See **Figure 18-19.** Remove the gasket or O-ring between the cylinder head and valve plate. Using a small hammer and gasket scraper, separate the valve plate from the compressor. If a gasket was used, remove any material left on the compressor assembly.

 Caution: Do not damage the surface of th compressor components. Leaks could result if the surface is scratched.

Check the valve plate for damage. Look for bent reed valves and worn seats on the plate. If the plate shows signs of damage, it should be replaced. Compare the old and new valve plates to ensure the replacement plate is correct. See **Figure 18-20.** Install a new gasket on the compressor body, then install the plate. Make sure the valve plate is properly positioned and placed over any alignment dowels. Position and install a new gasket or O-ring, then install the cylinder head and tighten the attaching bolts. Install the compressor, evacuate and recharge the refrigeration system, and check the compressor for leaks.

Replacing Compressor Front Head, Valve Plate, and Seal

To change the front head, valve plate, gasket, or O-ring on an end housing, begin by removing the compressor clutch and coil. Then remove the bolts holding the

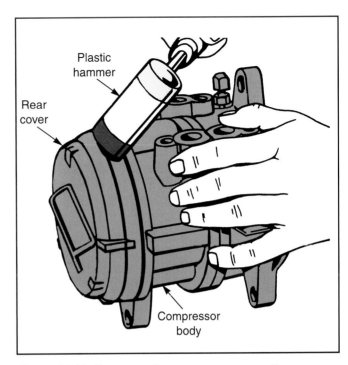

Figure 18-19. *To remove the rear cover, remove the compressor through bolts and lightly tap on the end cover while holding the body. (DaimlerChrysler)*

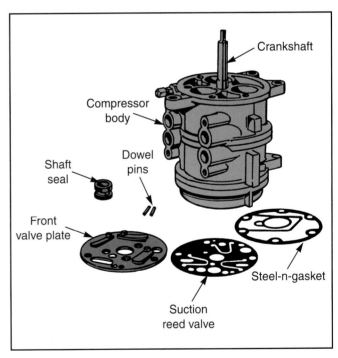

Figure 18-20. *Disassembled view of the front cover. When replacing the front seal on this type of compressor, change the O-rings and gaskets also. (DaimlerChrysler)*

end housing to the compressor. Gently pry on the front housing to split it from the body. Remove the housing from the body. On many compressors, the valve plate is installed between the end housing and compressor body. Be careful not to damage the valves. **Figure 18-21** illustrates a disassembled end housing.

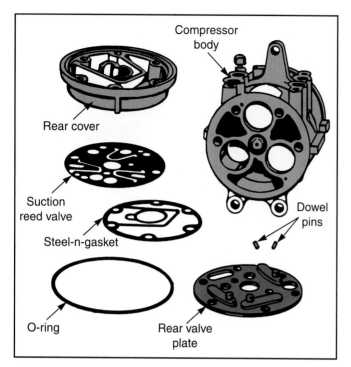

Figure 18-21. *Disassembled view of the rear cover. Replace all parts as needed. (DaimlerChrysler)*

Clean the gasket or O-ring mounting surfaces. Carefully check them for damage. If the mounting surfaces show any signs of nicks, deep scratches, or warping, the part must be replaced. Most refrigeration gaskets and O-rings do not require sealer.

Place the new gasket or O-ring in position. Most refrigeration gaskets and O-rings do not require sealer. Reassemble the compressor and torque the bolts to the proper specifications.

Replacing O-rings between Compressor Halves

A few compressors have a **split crankcase** with O-rings installed between the compressor halves. To service this type of O-ring, the compressor must be disassembled. Remove the front and rear compressor ends and valve plates. When servicing this type of compressor, do not separate the compressor halves any more than necessary to remove and replace the O-rings. **Figure 18-22** illustrates this procedure. Stretch the new O-rings as little as possible when installing them. Reassemble the compressor and leak test.

Overhauling Scroll Compressors

Because of the tight specifications of the scroll wheels, scroll compressors cannot be rebuilt in the field. The only service possible on scroll compressors is clutch replacement. Follow manufacturer recommendations for scroll compressor service.

Capacity Control Valve Removal and Replacement

Many **capacity control valves** are not serviced separately and a defective valve means the compressor assembly must be replaced. Some capacity control valves can be removed by the following procedure.

Recover the refrigeration system's charge. Next, locate the control valve snap ring at the rear of the compressor. Thoroughly clean the area around the snap ring. Then remove the snap ring and slide the valve from the compressor body. See **Figure 18-23.**

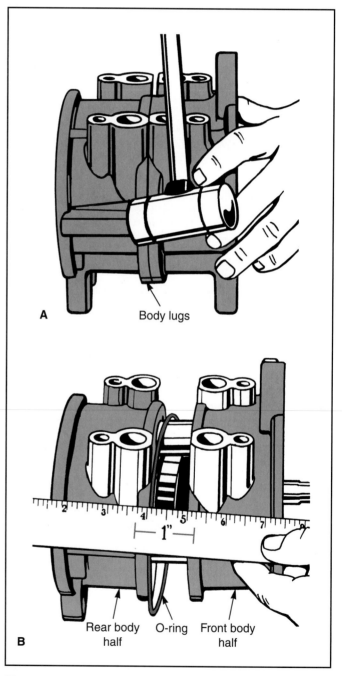

Figure 18-22. *A—Lightly tap on the compressor body to loosen it. B—Separate the body only far enough to replace the O-ring seal. (DaimlerChrysler)*

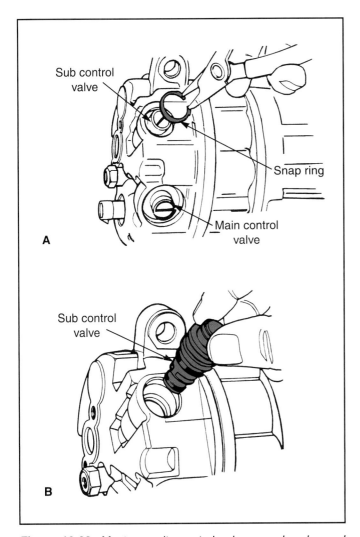

Figure 18-23. *Most capacity control valves can be changed without disassembling the compressor. On some vehicles, the compressor can be left installed. However, the refrigeration system must be discharged first. A—Remove the snap ring holding the valve in place. B—Slide the valve from the compressor housing. (DaimlerChrysler)*

To install the new valve, coat the valve O-rings with the proper refrigerant oil. Push the valve into the compressor body. The valve should slide in with hand pressure. If it does not slide in easily, do not force it. Remove the valve and check for nicks, incorrect valve, or incorrect O-rings.

Once the valve is installed, install the snap ring and make sure it is seated in its groove. Then evacuate and recharge the refrigeration system and check control valve and compressor operation.

Replacing Other Compressor Components

Some other components are installed in the compressor. Removal and replacement of these components is relatively easy. However, the refrigeration system must be discharged to perform them.

Suction Throttling Valves in Compressor

A few refrigeration systems locate the suction throttling valve in the compressor. The valve is located in the compressor inlet port. To remove the valve, recover the system charge and remove the inlet fitting from the compressor. The valve will be visible in the compressor inlet. A snap ring may be used to hold the valve in the compressor. Remove the ring and slide the valve out of the compressor inlet port.

Coat the O-rings of the new valve with the proper refrigerant oil. Push the valve into the compressor body. If the valve does not slide in with hand pressure, remove it and check for nicks or an incorrect part.

If a snap ring is used, install it and make sure it is seated in its groove. Then evacuate and recharge the refrigeration system and check system operation.

Compressor Switches

Several kinds of switches are installed on compressors. **Figure 18-24** shows two switches installed on the rear end housing of a common compressor. To change a switch, discharge the system and locate the switch on the compressor.

> **Caution: Be sure you know which switch should be changed. Most modern compressors are equipped with more than one switch.**

Unscrew the switch (a special socket may be necessary) and remove it from the compressor. Compare the old and new switches to ensure they match. Then install the new switch, evacuate and recharge the refrigeration system, and check operation.

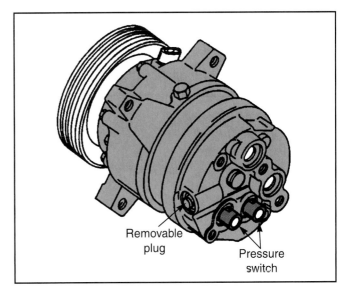

Figure 18-24. *Common locations of pressure relief and other control valves. Some are removed by unscrewing them from the compressor body, while others are held in place with snap rings. (General Motors)*

Many switches are held to the compressor by snap rings. To remove one of these switches, use a pair of snap ring pliers to remove the ring holding the switch. Carefully pull the old switch from the compressor.

 Note: It may be necessary in some cases to recover the refrigerant charge or remove the compressor from the vehicle when replacing one of these switches.

Install the new switch and a new snap ring on the compressor. If any refrigerant oil was lost, be sure to replace it when you recharge the system.

Pressure Relief Valves

Some older compressors use a pressure relief valve. To change a pressure relief valve, recover the system charge and unscrew the valve. Most pressure relief valves can be removed with standard wrenches or sockets. Compare the old and new valves, and then install the new valve. Evacuate and recharge the refrigeration system and check the valve to ensure it is not leaking.

Overhauling Older Compressors

The pistons and cylinders of all modern compressors are serviced as an assembly. Many manufacturers do not supply the internal parts for modern compressors. Therefore, many compressors cannot be rebuilt. The piston rings are often made of plastic molded on the piston during assembly. There is no way to duplicate the factory process for installing these piston rings.

Compressor remanufacturers rebuild compressors on an assembly line to increase quality and reduce costs. Buying a remanufactured compressor is usually cheaper and more reliable than investing parts and labor rebuilding a compressor in the shop. Most shops prefer to replace the entire compressor rather than perform any compressor repairs.

It may be necessary to rebuild the pistons and cylinders in an older compressor if the owner wants to keep the original equipment appearance or when no replacement is available. To rebuild a compressor, remove it from the vehicle and remove the end housings as discussed earlier in this chapter. If the compressor is an axial type, split the two halves of the compressor and pull them apart. The pistons and cylinders will be removed when the halves are split.

 Note: Replace all rubber and plastic seals in an older compressor. Many older compressors have seals made of Viton, which will swell if it contacts R-134a or its oil.

If the compressor is a radial type, unbolt the connecting rods from the crankcase and remove them from the cylinder bores. Then remove the bearing caps and remove the crankshaft from the compressor body.

Inspect all parts for damage and obtain new parts as necessary. Then lubricate the new piston rings with the proper kind of refrigerant oil. Install the new rings on the pistons. Lubricate the crankshaft and related parts. Reinstall the pistons in the cylinder bores. Install the pistons on the crankshaft and wobble plate (axial compressor only). Then reassemble the compressor cylinder assembly and install other parts as necessary.

Summary

The compressor moves refrigerant through the refrigeration system and provides pressure changes necessary to cause a change of state. Many compressor problems can be corrected with simple parts replacement. Most technicians and shops prefer to correct compressor problems by replacing the compressor assembly.

Compressor and clutch problems fall into two categories: poor cooling and noises. If the compressor clutch does not engage, check for current to the electromagnet. If no current is available to the electromagnet, check the fuses and control system. To check the compressor internals, turn the compressor by hand. If the compressor does not turn, it is defective. If compressor turns, it may have defective valves or another problem.

A loose belt or clutch will usually make a squealing noise. If the compressor locks up, the belt will squeal excessively. Internal compressor noises consist of rattles and knocking noises.

Special tools may be needed to remove the compressor clutch. Sometimes the compressor clutch can be removed without removing the compressor from the vehicle. The driven plate must be removed before the drive plate and pulley assembly can be removed. Most pulley assemblies are held to the compressor nose by a large snap ring. Once the snap ring has been removed, the pulley and bearing assembly can be removed from the compressor. A special puller may be needed.

Check the clutch faces for scoring and damage. If one plate is defective, both clutch plates should be replaced. Check the clutch bearing for roughness and binding.

To remove the electromagnet, remove the fasteners as necessary. Check the electromagnet for physical damage and incorrect resistance. Some electromagnets use brushes to deliver current.

To remove the compressor, remove the clutch electrical connector, and connectors to any switches on the compressor body. Remove the bolts holding the compressor to the engine and remove the compressor from the engine compartment. To install the compressor, place it in position and loosely install the compressor mounting bolts. Install the compressor refrigerant lines using new gaskets or O-rings. Adjust the belt and tighten all compressor bolts. Evacuate and recharge the refrigeration system and check compressor operation.

Sometimes a newer compressor must be installed on an older refrigeration system. This is called retrofitting. Most shops prefer to replace the entire compressor rather than perform any compressor repairs. However, some compressor service can be performed. Compressor gaskets and O-rings can be replaced on many compressors. Some gaskets and seals can be replaced with the compressor installed on the vehicle. The compressor must be removed and disassembled to replace most compressor body and housing gaskets and O-rings.

On a few compressors, the compressor must be disassembled to change the seal. Most shaft seals are serviced with the compressor assembled, and some shaft seals can be changed without removing the compressor from the vehicle. To replace the compressor shaft seal, remove the clutch driven hub and shaft key. Then remove the seal assembly from the seal cavity using a special tool. Remove the internal rubber O-ring. Install the new O-ring and then the shaft seal. Install all other parts, evacuate and recharge the refrigeration system, and check the new compressor seal for leaks.

To replace the seal on a compressor requiring disassembly, discharge the system and remove the compressor from the vehicle. Remove the clutch assembly and remove the key from the shaft keyway. Remove the bolts holding the end housing to the compressor body and slide the housing over the compressor shaft. Remove the snap ring holding the seal in place and remove the seal from the seal cavity. Install a new O-ring and seal. Then reinstall the compressor, evacuate and recharge the system, and leak check the compressor.

Most compressor valves are located in a flat assembly installed between the compressor body and housing. A few compressors have individual valves located in the compressor body. Valve replacement involves partial compressor disassembly. Some capacity control valves can be replaced. Other replaceable compressor components include suction throttling valves and several kinds of electrical switches.

Most compressors are replaced with new or rebuilt units. In rare instances, it may be necessary to overhaul an older compressor. After any compressor repairs, always check oil level and add oil as necessary.

Review Questions—Chapter 18

Please do not write in this text. Write your answers on a separate sheet of paper.

1. The two major classes of compressor and clutch problems are _____ and _____.

2. If the compressor clutch does not engage, the _____ may be too large.

3. An _____ can be tested without removing it from the vehicle.

4. The most common cause of squealing noises is a _____.

5. A loose or damaged clutch may _____.

6. Many clutch parts are held to the compressor by _____.

7. Most compressor _____ have a key and keyway.

8. Most _____ checking is done with a feeler gauge.

9. In what two situations is compressor retrofitting performed?

10. When a compressor develops a problem, what will most shops do?

11. If seal mounting surfaces have any nicks, deep scratches, or warping, the part must be _____.

12. What is an oil wick?

13. Which of the following has to be removed to remove a valve plate?
    (A) Shaft seal.
    (B) Compressor shell.
    (C) Compressor pistons.
    (D) End housing.

14. The refrigeration system must be discharged to remove which of the following?
    (A) Driven plate.
    (B) Pulley.
    (C) Clutch bearing.
    (D) Shaft seal.

15. In what circumstances would an older compressor be overhauled?

ASE Certification-Type Questions

1. To check an electromagnet winding, the technician can do all of the following, *except:*
   (A) energize the electromagnet using jumper wires.
   (B) check for a blown fuse.
   (C) use an ohmmeter to check for shorts.
   (D) use an ohmmeter to check for opens.

2. All of the following statements about excessive pressures are true, *except:*
   (A) excessive temperatures due to high heat load is a compressor defect.
   (B) other components of the refrigeration system keep pressures within limits.
   (C) a refrigerant overcharge can cause high pressures.
   (D) a compressor in good condition can produce very high pressures.

3. Compressor lockup can be caused by:
   (A) low refrigerant pressures.
   (B) low ambient air pressures.
   (C) insufficient refrigerant.
   (D) internal compressor damage.

4. Technician A says clutch driven plate removal often requires a special tool. Technician B says clutch pulley removal often requires a special tool. Who is right?
   (A) A only.
   (B) B only.
   (C) Both A and B.
   (D) Neither A nor B.

5. Where is the compressor clutch air gap located?
   (A) Between the drive and driven plates.
   (B) Between the clutch pulley and electromagnet.
   (C) Between the driven plate and the compressor shaft.
   (D) Between the clutch bearing and the compressor body.

6. Air gap is set with a(n):
   (A) ohmmeter.
   (B) special tool.
   (C) feeler gauge.
   (D) dial indicator.

7. Gaskets and O-rings are used to seal all of the following compressor parts, *except:*
   (A) body and end housings.
   (B) body and shell.
   (C) inlet and outlet fittings.
   (D) shaft.

8. Where are valve plates installed?
   (A) Between the body and end housing.
   (B) In the center of the compressor.
   (C) In the compressor body.
   (D) All of the above.

9. Technician A says the capacity control valve can be removed from the compressor without discharging the system. Technician B says the replacement capacity control valve should slide into place with hand pressure. Who is right?
   (A) A only.
   (B) B only.
   (C) Both A and B.
   (D) Neither A nor B.

10. Which valve is *never* installed in the compressor?
    (A) Pressure relief valve.
    (B) Expansion valve.
    (C) Capacity control valve.
    (D) Suction throttling valve.

Chapter 19

Valve, Evaporator, Condenser, and Related Parts Service

After studying this chapter, you will be able to:
- ❏ Diagnose expansion valve and orifice tube problems.
- ❏ Diagnose evaporator pressure control valve problems.
- ❏ Diagnose compressor cycling switch problems.
- ❏ Diagnose evaporator and condenser problems.
- ❏ Diagnose accumulator and receiver-drier problems.
- ❏ Remove and replace expansion valves and orifice tubes.
- ❏ Remove and replace evaporator pressure control valves.
- ❏ Remove and replace compressor cycling switches.
- ❏ Remove and replace evaporators and condensers.
- ❏ Remove and replace accumulators and receiver-driers.

Technical Terms

Inlet screen

Expansion valve freezing

Oil sludging

Variable orifice valve

Biocides

Energy module

This chapter covers the diagnosis and replacement of expansion valves, compressor cycling switches, orifice tubes, evaporator pressure control valves, evaporators, condensers, accumulators, and receiver-driers. Servicing of related screens, filters, and seals is also discussed. Studying this chapter will complete your knowledge of refrigeration system component service. This is the last chapter covering service on the refrigeration system. Air conditioning system installation and retrofitting will be discussed in Chapter 23.

Common Problems

Expansion valves, evaporator pressure control valves, and compressor cycling switches can fail in ways that cause reduced system performance, or a total lack of cooling. Evaporators, orifice tubes, and condensers have no moving parts to wear out. However, they can develop leaks, clogging, and other problems. These failures can cause excessive temperatures which can result in damage to other parts of the refrigeration system. Accumulators and receiver-driers also have no moving parts. Failure of the desiccant, however, can cause severe damage to the other parts of the refrigeration system.

The following sections cover the causes of common problems in expansion valves, evaporator pressure control valves, orifice tubes, compressor cycling switches, evaporators, condensers, accumulators, and receiver-driers. Problems are listed by component.

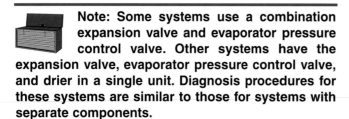 **Note: Some systems use a combination expansion valve and evaporator pressure control valve. Other systems have the expansion valve, evaporator pressure control valve, and drier in a single unit. Diagnosis procedures for these systems are similar to those for systems with separate components.**

Expansion Valve Failures

Expansion valve failures are always related to their open or closed states. If the valve sticks open, the evaporator will be flooded with refrigerant. This will cause evaporator sweating or frost formation, but with very little cool air output. A pressure gauge installed on the system will show excessive low side pressure. Loss of the refrigerant charge in the sensing bulb will cause the valve to stick open. If the expansion valve sticks closed, there will be no cooling, and pressure gauge readings will show a vacuum on the low side, **Figure 19-1.** Check expansion valve operation by heating and cooling the sensing bulb. One method is to remove the bulb from the evaporator and squeeze it in your hand. If heating or cooling the bulb has no effect on low side pressure, you can assume the valve is stuck or the bulb has lost its charge. In either case, the

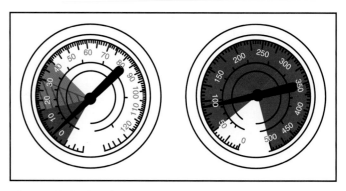

Figure 19-1. *A defective or restricted expansion valve or orifice tube will be indicated by the manifold gauge readings shown here. The red indicates the areas where the gauge needles will fall.*

valve must be replaced. A clogged inlet screen, discussed in the next paragraph, can also cause these symptoms.

Almost all expansion valves have an **inlet screen** built into the valve assembly. **Figure 19-2** shows a cross-section of the inlet screen. If the screen becomes clogged, it will reduce flow through the expansion valve and evaporator. A partially clogged screen will cause reduced flow leading to low pressure and icing. Sometimes the expansion valve assembly will become coated with ice. A completely clogged screen can be identified by total loss of cooling and the presence of a vacuum on the low side of the system.

Water in the refrigeration system can often cause **expansion valve freezing.** The water turns to ice as it passes through the expansion valve. This causes a no cooling condition, usually accompanied by the presence of vacuum on the low side. The lack of flow causes the expansion valve to warm up and the ice melts. The system operates normally until ice forms again at the expansion valve. The symptom of this is intermittent loss of cooling. The system may operate normally for several minutes between episodes of cooling loss. This problem can be corrected by evacuating the system to remove water, and changing the accumulator or receiver-drier. The excess water, however, may have caused other damage to the system, especially if it has been in the system for a long period of time.

Orifice Tubes

Because they have no moving parts, orifice tubes are usually trouble free. If other problems occur in the system, the screen will likely become plugged with debris, **Figure 19-3.** Low pressure or a vacuum on the low side of the system is a symptom of screen plugging. Once the cause has been identified, the orifice tube and screen should be replaced.

Evaporator Pressure Control Valves

Most evaporator pressure control valve problems are caused by the failure of the valve to prevent evaporator icing at higher vehicle speeds. Usually, the valve sticks open, or its set point is too low. Attaching a pressure gauge to the system should reveal pressures above the valve set point at idle, since the compressor pumping capacity is limited at idle. At high speeds, the compressor draws more refrigerant from the low side and evaporator pressure drops. If the valve is defective, pressures will fall below the normal valve set point, and the evaporator will begin icing.

Sometimes the evaporator pressure control valve setting is only slightly lower than normal. In this case, the evaporator will freeze up when the vehicle is driven on cool days with the blower speed set on low. At higher blower speeds,

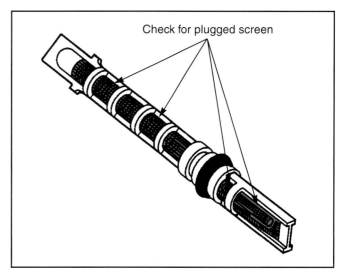

Figure 19-3. *Orifice tube screens also become plugged. An orifice tube with a plugged screen is usually replaced. (General Motors)*

or at engine idle, there is enough heat load to prevent freeze up, even if the control valve setting is slightly off.

In rare instances, the evaporator pressure control valve may stick closed or become clogged, causing no flow through the evaporator. If an oil return line is used with the pressure control valve, it may become frosty. Usually when there is enough debris in the system to make the pressure control valve stick, the expansion valve also becomes clogged.

Compressor Cycling Switches

Compressor cycling switches are usually installed on the accumulator. If the switch fails or sticks in the closed position, the compressor clutch will be engaged at all times. Since compressor clutch disengagement is what controls evaporator pressure, the evaporator will freeze up. If the switch sticks open, the compressor will not engage and the refrigeration system will be inoperative. Unplugging the switch with the HVAC system operating, tests for a switch stuck closed. If the clutch disengages, the switch is defective. If the clutch does not disengage, the problem is elsewhere in the system. Check for shorted wires or a control system problem.

If the clutch will not engage, first check for a blown fuse or a control system problem. Use jumper wires to bypass the switch. Make sure you are testing the proper switch, since switches for different purposes are sometimes installed side by side, **Figure 19-4.** The HVAC system should be on. If the clutch engages, the switch is defective.

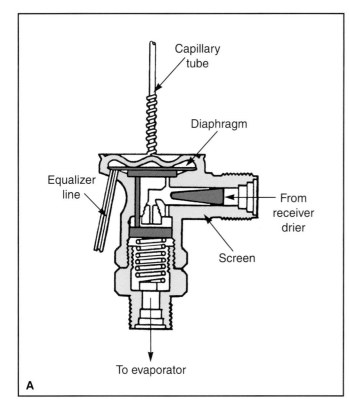

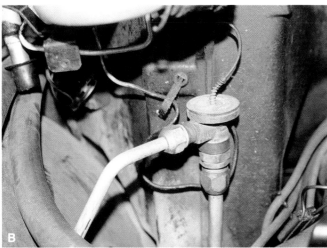

Figure 19-2. *A—This cutaway of an expansion valve shows the screen installed in the inlet. The screen can become plugged and keep the expansion valve from operating properly. The screen can be cleaned, although many technicians prefer to replace the entire valve. B—Relative location of the valve on many systems. (General Motors)*

 Caution: Some electronically controlled HVAC systems cannot be tested this way without damaging the control module. Always check the service manual before bypassing any HVAC switch.

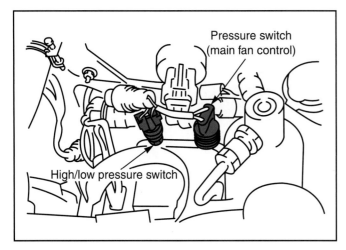

Figure 19-4. *It is important to identify refrigeration switches. Many switches are placed close together on refrigeration lines or components. Testing the wrong switch will lead to wasted effort and an improper diagnosis. (Subaru)*

Evaporator

Other than leaks, evaporator problems usually consist of poor heat transfer or odors. Blocked internal passages or excessive oil can cause poor flow through the evaporator. Oil checking, removal and flushing procedures were discussed in Chapter 16. Blocked external fins are usually caused by a buildup of dirt, leaves, or other debris in the air intake. See **Figure 19-5.** The system pressures will be ok, but the outgoing air will be warm. In cases of severe blockage, airflow will be sluggish. The buildup of dirt and water on the evaporator can sometimes cause microorganisms to grow on the surface of the evaporator. These growths can cause a musty odor to be present in the passenger compartment, especially when the HVAC system is first turned on.

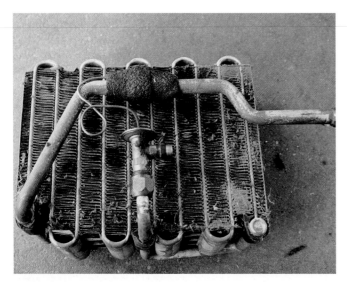

Figure 19-5. *The dirt and leaf particles on this evaporator can restrict airflow, leading to decreased cooling.*

Sometimes debris on the fins can be observed by removing the resistor assembly or blower motor. This debris can be removed by careful cleaning. If the vehicle has a cabin filter, check it for clogging. Refer to the vehicle owner's manual for the location of this filter.

Condenser

The condenser can contribute to poor cooling by failing to transfer heat from the refrigerant to the outside air. One possible cause is blocked internal passages by debris from compressor failure or excessive oil. A restricted condenser will cause excessive low and high side pressures, **Figure 19-6.** The external fins can become blocked with dirt and road debris. The condenser can also leak, although this is rare when the condenser has not been damaged in a collision.

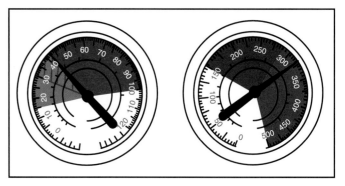

Figure 19-6. *Condenser restrictions will be indicated by higher than normal pressures. Make sure the restriction is not caused by external debris, which can be cleaned before condemning the condenser.*

Accumulator

Problems with the accumulator include leaks, failure of the desiccant, and oil line clogging. If the desiccant becomes saturated with water, any extra water will circulate through the system with the refrigerant. This will lead to corrosion and *oil sludging.* If the desiccant bag breaks, the desiccant will be carried throughout the system, causing compressor damage and plugging various screens and openings in the system. Desiccant failure cannot be spotted until another malfunction has taken place. Therefore, the accumulator should be changed if there is any doubt about its condition.

If the oil return hole becomes plugged, oil will collect in the bottom of the accumulator instead of being returned to the compressor. The compressor may become starved for oil and wear out prematurely. If the accumulator has more than about 5 ounces (150 ml) of oil, the oil return hole is probably plugged and the accumulator should be replaced.

Receiver-Drier

Like the accumulator, the receiver-drier can develop a leak or the desiccant can fail. If the desiccant becomes saturated with water, unabsorbed water will circulate throughout the system, leading to oil sludging and corrosion. If the desiccant bag breaks, the desiccant will be carried to the expansion valve clogging it and causing poor system cooling. As with the accumulator, there is no way to tell when the desiccant has failed until it is too late. The receiver-drier should be changed whenever there is any doubt as to its condition. Anytime a compressor is replaced, the receiver-drier should also be replaced.

Receiver-driers also have an internal debris screen to catch any metal shavings, rubber particles, or other debris generated by refrigeration system operation. If the screen becomes plugged, refrigerant cannot flow through the system, and no cooling will occur. A partially clogged screen can be detected by observing the line exiting the receiver-drier. The line should be hot after the system has been operating for a few minutes. If it is cool, or if frost forms, the screen is restricted.

Component Service

The components serviced in this chapter are replaced rather than repaired. Some related service operations, such as flushing and removing excess oil, were covered in other chapters. The following service procedures are grouped by component.

Expansion Valves

To replace an expansion valve, first recover the refrigerant. Then remove the sensing bulb from its attachment point. Use two wrenches to loosen the expansion valve fittings at the high pressure line and evaporator. If the expansion valve uses an external equalizer line, remove it also. Then remove the expansion valve from the vehicle.

 Note: On some vehicles, it may be necessary to disassemble part of the blower case to access the expansion valve.

Compare the old and new valves, then place the valve in the system. Lightly install all fittings using new O-rings. Once all fittings have been started, tighten the fittings. Carefully install the sensor tube and bulb, being careful not to kink the tube. Sensor tube and bulb placement is critical. If the expansion valve bulb fits into a well in the evaporator outlet tube, insert the bulb completely in the tube. If the tube well contains debris or oil, clean it out with spray carburetor cleaner before inserting the bulb.

All sensor bulbs should be solidly attached to the evaporator outlet and thoroughly covered with insulating tape. **Figure 19-7** shows a typical expansion valve and

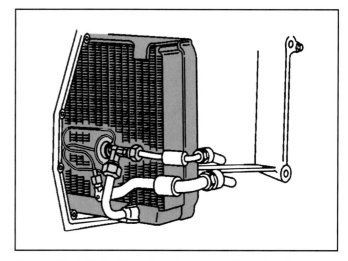

Figure 19-7. *Typical location of an expansion valve. Ducts and case components must be removed to gain access to this valve. (General Motors)*

sensing bulb installation. If the sensing bulb tube is much longer than the original, make sure the extra length of tubing does not contact the engine or exhaust system.

If the vehicle refrigeration system uses an H-block expansion valve, remove it by screws holding it to the evaporator and hose assembly. Most block type expansion valves are attached to the mating parts with screws that pass through the valve assembly and the flanges of the mating parts. See **Figure 19-8.** Always use new gaskets or O-rings as necessary.

Evaporator Pressure Control Valves

Begin by discharging the system and recovering the refrigerant. Then remove the pressure control valve. If the valve is installed in the evaporator outlet line, use two

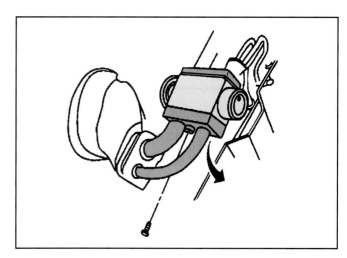

Figure 19-8. *H-valve or block expansion valves are usually attached with bolts or screws. The H-block expansion valve is often installed under the hood and is relatively easy to reach. Once the fasteners are removed, the valve can be removed. Always replace the O-rings or gaskets when changing this type of valve. (General Motors)*

wrenches to loosen each fitting at the evaporator and suction lines. If the system has external oil return and equalizer lines, remove them also. Then gently pull the fittings from the body of the valve and remove the valve from the vehicle. Make sure the new valve has the right O-rings. Lightly lubricate the O-rings with the correct refrigerant oil. Put the valve into position and loosely install all lines and fittings. Once all lines are installed, tighten the fittings to the proper torque. Evacuate and recharge the system as needed and recheck valve operation.

If the evaporator pressure control valve is installed in the compressor, remove the inlet fitting fasteners to expose the valve. Remove the snap ring if used and slide the valve out of the compressor inlet port. Push the new valve into the compressor body after coating the O-rings with refrigerant oil. Reinstall the snap ring if used. Install the fitting using a new gasket. Evacuate and recharge the system as needed and recheck valve operation.

Orifice Tubes

Orifice tubes are always replaced rather than being cleaned. The relative low cost of orifice tubes makes cleaning impractical. Also, cleaning can alter the size of the orifice opening and destroy evaporator flow calibration. Cleaning may also damage the screen.

To change the orifice tube, the system must be discharged. Once all pressure has been removed, disconnect the high pressure line fitting at the evaporator. Then remove the orifice tube using the correct tool, **Figure 19-9.** On some vehicles, a special tool must be used to remove the orifice tube. Lubricate the new orifice tube and slide it into position. Install the high pressure line fitting using a new O-ring. Then evacuate and recharge the system. Recheck system operation to ensure the new orifice tube is working properly.

Some aftermarket parts manufacturers offer a **variable orifice valve,** sometimes called a *smart orifice tube.* The variable orifice valve is installed in place of the original tube. It varies the orifice size depending on temperature and pressure changes. This increases system efficiency and lowers evaporator temperatures at low vehicle speeds. Variable orifice valves are sometimes used when the system is converted from R-12 to R-134a. Installation procedures are identical to those for a conventional orifice tube.

Compressor Cycling Switches

Compressor cycling switches are installed in the accumulator. Many compressor cycling switches are installed over a Schrader valve and can be changed without removing the refrigerant from the system, **Figure 19-10.**

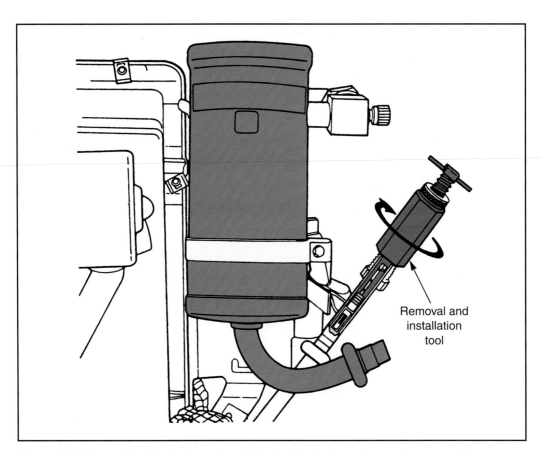

Removal and installation tool

Figure 19-9. *The orifice tube is always located in the line between the condenser and the evaporator. Removal is relatively easy, but a special tool is needed on some vehicles. (Ford)*

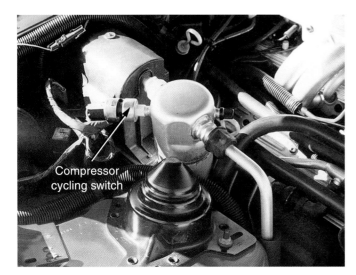

Figure 19-10. *The compressor cycling switch is usually located on or near the accumulator. Make sure the switch is attached to a Schrader valve. If it is not, you will have to recover the system charge before replacing the switch.*

 Warning: Find out whether the switch is connected to a Schrader valve before beginning removal. If the switch does not connect to a Schrader valve, the refrigerant charge will be lost when the switch is removed.

To remove the switch, remove the electrical connector and unscrew the switch from the accumulator. Compare the old and new switches to ensure they match. Install the new switch, evacuate and recharge the refrigeration system (if necessary), and check switch operation.

Evaporators

The evaporator has no moving parts, and there are no adjustments or overhaul procedures. Evaporator service is confined to cleaning and replacement. These procedures are discussed in the following paragraphs.

Evaporator Cleaning

Evaporators can become clogged internally or externally. Internal flushing and oil removal was discussed in Chapter 16. To clean the exterior of the evaporator, several commercial cleaners are available. On many vehicles, the evaporator fins can be reached by removing the blower resistor or control module assembly. Other evaporators can be reached by removing the blower motor assembly. On some vehicles, the only way to clean the evaporator is to remove it from the vehicle, or remove enough ductwork or other parts to expose the finned surface. One procedure calls for drilling a small hole in the case, and inserting the

nozzle of a can of cleaner in the hole. The cleaner will exit from the nozzle and dissolve deposits, which will drip out of the evaporator case drain. The hole is then plugged with a small cap.

To clean the evaporator, spray the detergent solution on the fins and allow it to work for 10 to 15 minutes. Then hose the fins with a strong stream of water. Make sure water flows out of the drain hole. Unplug the drain hole if necessary. Then reinstall any parts that were removed and check HVAC system operation.

Disinfecting Evaporator Core Using Biocide

If the evaporator is being cleaned of microorganisms, special **biocides** (microorganism killers) should be used. Some biocides come in two bottles that must be mixed together before use. Wear protective gloves and safety goggles while performing this procedure. Be sure to perform this procedure in a well-ventilated area.

 Warning: Most biocides can cause irritation of the eyes. If the disinfectant gets into your eyes, flush with water for 15 minutes and consult a physician.

Mix the biocide if needed. Remove the blower resistor or module to access the evaporator core; do not disconnect any wiring. Shine a light into the evaporator core and check for debris; remove as needed. If the debris is heavy or impacted on the evaporator surface, you will have to remove the core. If a large amount of debris is present, check the air inlet screen and make sure it is intact. Place a drain pan below the evaporator drain.

Caution: Do not allow the resistor pack or the module heat sink to touch ground. This can cause damage to the HVAC circuitry.

Open all vehicle windows and doors and turn the ignition key on, but do not start the engine. Turn the blower motor on to low and mode to vent. Using a cleaning spray gun, insert the siphon hose into the disinfectant and the nozzle into the evaporator case, **Figure 19-11.** Spray the disinfectant into the evaporator core, making sure the entire core is treated. During this process, you should use the entire bottle of disinfectant. You should also spray a little into the vents to kill any bacteria that may have migrated into the ductwork, **Figure 19-12.** Shut off the ignition and allow the disinfectant to stay on the core for at least five minutes. Then, turn the ignition key back to on. Using the spray gun with a quart container of water, thoroughly rinse the core of all disinfectant. Turn off the ignition and reinstall the resistor or power module.

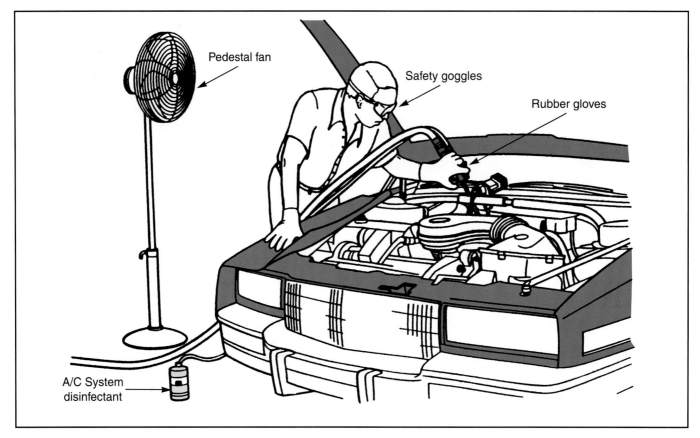

Figure 19-11. *Disinfect the evaporator core by spraying a biocide through the core using a pressure spray gun and disinfectant. (General Motors)*

Figure 19-12. *Spray bottles of biocide are available. When disinfecting an evaporator, spray a little down into the vents, as bacteria may have migrated into the ductwork. (Jack Klasey)*

Evaporator Replacement

If an evaporator leaks refrigerant, it should be replaced. Leak repairs to an evaporator are usually unsuccessful. To remove the evaporator, recover the system refrigerant and ensure no residual pressure is left in the system. Then remove the inlet and outlet fittings. Also remove the oil return line and temperature probe if used.

 Caution: Evaporator removal is often complex, requiring the removal of the dashboard, interior components, or underhood parts. Always refer to the appropriate service manual before beginning evaporator removal.

Disabling Air Bag System

If the vehicle is equipped with air bags, you will need to disarm the system when removing the evaporator core in most vehicles. Failure to do so could result in accidental deployment.

Warning: Most air bag systems have an *energy module* that stores power to deploy the air bags, even if the battery is disconnected. This power reserve can remain for up to 30 minutes after battery power is disconnected.

The first step in air bag removal is to turn the ignition switch to the locked position and remove the key. The key must be removed before starting any air bag disarming procedure. In some cases, you may need a scan tool to disarm the air bag system. After removing the key, disconnect the air bag system's source of power. This can be accomplished by removing the negative battery cable or pulling one or more fuses. Once power has been disconnected, remove the yellow connectors to the air bag module, unbolt the air bag, and place it in a clean area, with the inflator side facing up. Always follow the manufacturer's instructions for disarming the air bag system.

Evaporator Core Service—Blower Case in-Vehicle

Some evaporators are removed by removing one side of the evaporator case to expose the evaporator. Other evaporators can be removed by removing a cover in the engine compartment and sliding the evaporator from its case, **Figure 19-13.**

Begin by determining what covers and other vehicle components must be removed to gain access to the evaporator. Many older vehicles had a front or side cover that could be removed to expose the evaporator. Newer vehicles with a split case have a removable top cover over the evaporator. In some cases this cover must also be removed to gain access to the heater core. A few front-wheel drive models with a split case have a bottom mounted cover. To reach the bottom cover, the engine may have to be moved forward by loosening the engine mounts and prying on the engine block.

Once you have determined the evaporator core removal method, recover the refrigerant. On some vehicles with top mounted covers, it may be necessary to take off the intake grille at the windshield. Also remove the intake air filter if used.

Remove any electrical connectors attached to electrical devices on the evaporator cover and remove the evaporator tubing connections. On some vehicles, it may be necessary to remove the heater hoses. Then remove any ductwork as needed to gain access to the evaporator. Remove the fasteners holding the evaporator cover and remove the evaporator cover. This will expose the evaporator.

Remove the temperature probe if applicable then remove any evaporator brackets. Not all vehicles have brackets, relying on the case itself to hold the evaporator in position. Slightly twist the evaporator to break the gasket seals, then remove the evaporator from the case. The evaporator may be pulled up, sideway, or through the bottom of the case depending on the design.

After the evaporator has been removed, clean the inside of the case and check it for damage. Replace any internal seals if necessary. Check that the evaporator drain hole(s) are open.

Compare the old and new evaporators and place the new evaporator into position. Reinstall the brackets and temperature probe if applicable. Then replace the evaporator cover. Reinstall the evaporator connections using new O-rings and/or gaskets where needed. Reconnect all electrical wiring and install any other vehicle parts that were removed. Evacuate and recharge the refrigeration system. Start the HVAC system and check the evaporator and fittings for leaks.

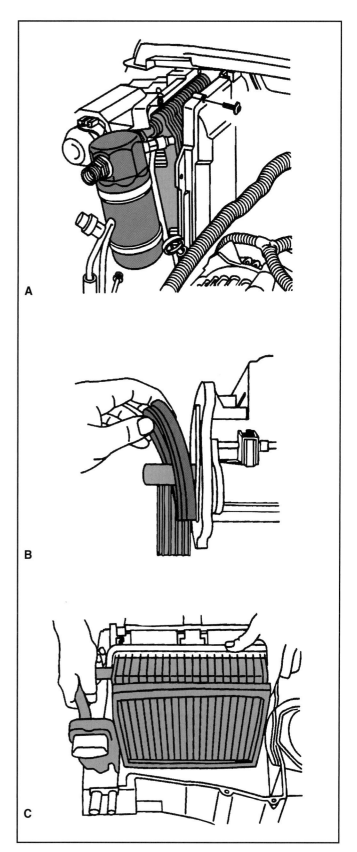

Figure 19-13. *A—On many vehicles the two halves of the evaporator case must be separated, or split, to remove the evaporator. Usually one side is loosened and pulled away to gain sufficient clearance for the evaporator to be pulled out. B—Be sure to reinstall all seals and grommets. C—Remove the old evaporator, clean out any debris before installing the new core. (General Motors)*

Evaporator Core Service—Blower Case Removal and Installation

In some cases, the only way to access the evaporator core is to remove the entire blower case from the vehicle. Depending on the vehicle, this can be a very complex and time-consuming task. Begin by disabling the vehicle's air bag system.

After disarming the air bag system, disconnect the negative battery terminal to prevent draining the battery. Check the service manual to see what components may need to be removed, relocated, or loosened to access the blower case, **Figure 19-14.** On most vehicles that require case removal for evaporator replacement, the following interior components and systems are often involved:

❑ Dashboard (upper and lower).
❑ HVAC ductwork.
❑ Front carpet.
❑ Center console (if equipped).
❑ Steering column.
❑ Air bag system.
❑ Front seat(s).
❑ Instrument cluster.
❑ Dashboard wiring harness.

Follow service manual directions for the removal of each interior component as needed. Once the blower case is accessible, remove all mounting bolts, **Figure 19-15.** Some blower case bolts may need to be accessed from the engine compartment. Carefully pull the case back and remove it from the vehicle. Place the case flat on the shop floor, with a fender cover to prevent damage.

In most cases, the case will have several bolts to remove in order to access the evaporator. However, a few evaporator cases require the technician to cut a hole in the case to access the core. Lift the evaporator from the duct-work and place it on a bench. Observe the evaporator for confirmation it is leaking, **Figure 19-16.**

Compare the old and new evaporators, then add the recommended amount of refrigerant oil to the new evaporator. Clean out any leaves and debris from the evaporator and heater areas. Install the new evaporator core and reseal the blower case.

 Note: If the case was cut open, you will need gasket cement, RTV, or weatherstrip sealer to reseal the case properly. Some require gaskets. Follow service manual instructions.

Afterblow Module Installation

Some vehicles are equipped with an afterblow module. The afterblow module operates the blower motor for a short time after the engine is shut off. This does not occur right away, but usually within an hour after the vehicle is shut off. This helps prevent the accumulation of mold on the evaporator core.

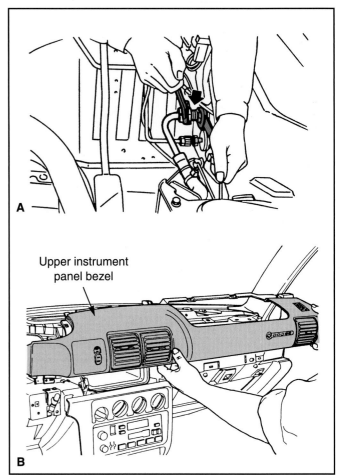

Upper instrument panel bezel

Figure 19-14. *A—Loosen the evaporator lines. B—To replace the evaporator on most vehicles, the entire blower case must be removed. Carefully disassemble the dash and other components as needed. C—A panel bezel with the dash removed. In some cases, you may have to remove such components to access the blower case. (Subaru, DaimlerChrysler, General Motors)*

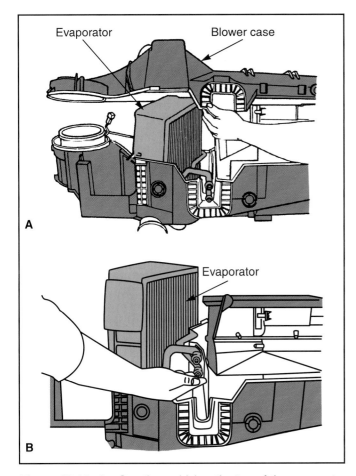

Figure 19-15. *A—On other vehicles, the top of the evaporator case is removed and the evaporator is lifted from the case. B—Lift the evaporator from the case, taking care not to displace any of the doors. Be sure to clean any debris before reassembling the case. (DaimlerChrysler)*

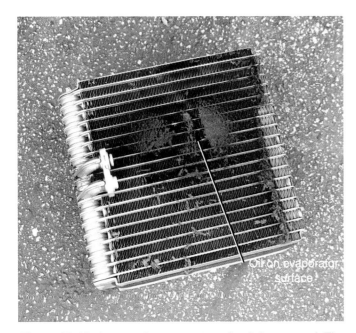

Figure 19-16. *Inspect the evaporator after it is removed. The presence of oil in the center of this evaporator confirms that it was leaking.*

Afterblow modules are sometimes installed on vehicles with persistent evaporator odor problems. Follow the manufacturer's recommendations for module installation, **Figure 19-17.** Be sure to advise the vehicle owner of the change in blower motor operation.

 Note: Afterblow modules can cause problems on vehicles with alarm systems that can sense battery voltage changes or current drains.

Condensers

Like evaporators, condensers have no moving parts. Condensers can be cleaned or replaced. Brazing occasionally can repair a leaking condenser. Internal flushing and oil removal was discussed in Chapter 16. External cleaning and replacement procedures are discussed in the following paragraphs.

Condenser Cleaning

Internal condenser flushing and oil removal was discussed in Chapter 16. To clean the exterior of the condenser, locate the condenser in front of the engine compartment. Cover any exposed vehicle paint or trim, then spray detergent solution on the fins. Allow the detergent to work for 10 to 15 minutes, then hose the fins with a strong stream of water.

Condenser Replacement

On some vehicles, the radiator may need to be relocated or removed to gain access to the condenser. To remove the condenser, recover the system refrigerant and ensure all residual pressure has been removed. Then remove the inlet and outlet fittings. Remove the bracket fasteners and remove the condenser. See **Figure 19-18.**

Add the recommended amount of refrigerant oil to the new condenser. Position the new condenser and reinstall the bracket fasteners. Connect the inlet and outlet lines using new O-rings. Reinstall the radiator if it was relocated or removed. Evacuate and recharge the refrigeration system. Start the HVAC system and check the condenser and fittings for leaks.

Accumulators

Defective accumulators are replaced. The parts in an accumulator cannot be reached without cutting into the accumulator. The clutch cycling switch can be removed from the accumulator.

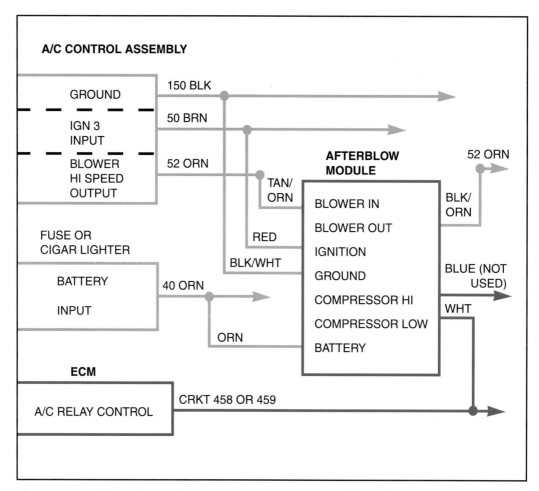

Figure 19-17. *Wiring diagram for an afterblow module. (General Motors)*

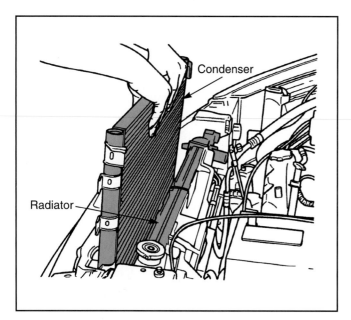

Figure 19-18. *Many condensers can be removed from the vehicle by pulling them out of the top. Others must be removed from the bottom. Usually air dams and other shielding must be removed from the front of the vehicle to get at the condenser fasteners and fittings. (DaimlerChrysler)*

Replacement Guidelines

Accumulators should be replaced when they are leaking, the desiccant becomes water saturated, or there is evidence the oil return hole is plugged. Often the accumulator is replaced as a preventive measure when the system is opened or other components are replaced. It is always safer to replace the accumulator when you have any doubts about its condition.

Accumulator Replacement

To remove the accumulator, first recover the system refrigerant and remove all residual pressure. Then use two wrenches to remove the inlet and outlet fittings. Remove the single screw holding the bracket around the accumulator and remove the accumulator. Transfer the clutch cycling switch and any other switch present from the old to the new accumulator as necessary. Add the recommended amount of refrigerant oil to the new accumulator, **Figure 19-19.** Then position the new accumulator on the bracket. Loosely install the inlet and outlet fittings using new O-rings. Reinstall and tighten the bracket fastener. Tighten the inlet and outlet lines, then evacuate and recharge the refrigeration system. Start the HVAC system and check the accumulator and fittings for leaks.

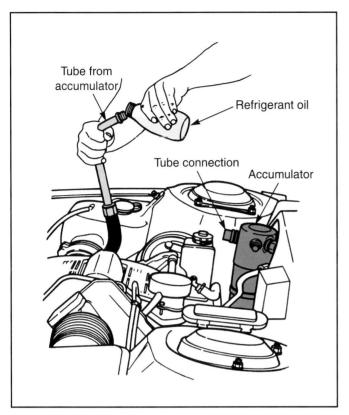

Figure 19-19. *Whenever a refrigeration system component is replaced, the proper amount of oil must be added. Usually the oil can be poured into one of the disconnected hoses. (General Motors)*

Receiver-Driers

Defective receiver-driers are usually replaced as a unit. The sight glass, pressure relief plug, and desiccant on some older receiver-driers can be replaced. In the past, some technicians would bake a used receiver-drier in an oven to remove water, but most modern shops prefer to replace the entire unit. Pressure switches, when used, can be changed without replacing the receiver-drier.

Replacement Guidelines

The receiver-drier is usually replaced as a preventive measure when the system is opened or other components are replaced. Receiver-driers are also replaced when they are leaking, the desiccant becomes water saturated, or there is evidence the internal screen is plugged. Another reason for replacing the receiver-drier is a clouded (dirty) sight glass. This is evidence of internal contamination.

Receiver-Drier Replacement

Receiver-drier removal usually involves removing the inlet and outlet lines and the attaching bracket, **Figure 19-20.** To remove the receiver-drier, first recover the system refrigerant and remove all residual pressure.

Use two wrenches to remove the inlet and outlet fittings. Remove the tensioning screws holding the brackets around the receiver-drier and remove the receiver-drier. Transfer any pressure switches to the replacement receiver-drier if necessary. Add the recommended amount of refrigerant oil to the new receiver-drier and position it in the brackets. Loosely install the inlet and outlet fittings using new O-rings. Reinstall and tighten the bracket tensioning screws. Tighten the inlet and outlet lines, then evacuate and recharge the refrigeration system. Start the HVAC system and check for leaks.

Systems with Multiple Component Units

On some refrigeration systems, two or more components are combined into a single unit. Service on these systems is somewhat different from service on systems using separate components for each function. Some systems have an H-valve, which is a combination expansion valve and evaporator pressure control valve. It can be removed from the system in the same manner as a block valve and the individual components serviced. However, most shops replace the entire assembly.

On some vehicles, the expansion valve, evaporator pressure control valve, and drier are combined into a single unit. This is a VIR or valves-in-receiver unit. This unit can also be removed from the system and the individual components serviced, **Figure 19-21.** However, most shops replace the entire assembly.

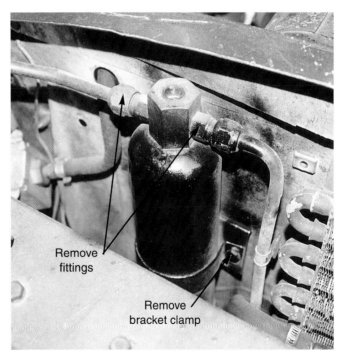

Remove fittings

Remove bracket clamp

Figure 19-20. *Receiver-driers are located at the front of the engine compartment, near the condenser. Always use two wrenches to loosen a fitting.*

Figure 19-21. *If desired, this older VIR unit can be disassembled to replace the internal components. Most technicians prefer to replace the entire assembly.*

Summary

Expansion valves, evaporator pressure control valves, and compressor cycling switches all have moving parts and can fail in ways that cause cooling problems. Evaporators, condensers, accumulators, and receiver-driers have no moving parts but can develop leaks or failures that affect the other parts of the refrigeration system. Some systems have combination units that contain two or more of the above components.

Expansion valves can stick open or closed. Water in the system can cause intermittent expansion valve sticking. Evaporator pressure control valve problems are usually caused by failure of the valve to prevent evaporator icing at higher vehicle speeds. Occasionally the valve will stick completely closed. Orifice tubes may clog up when excessive debris is present in the system. Compressor cycling switches can cause the compressor clutch to be either on or off at all times. Evaporators and condensers can develop leaks or become clogged with debris. Accumulators and receiver-driers are often replaced when the desiccant becomes saturated with water. If the desiccant bag breaks, the desiccant powder will be carried to system valves causing clogging and loss of system cooling.

To replace any of the parts covered in this chapter, first discharge the system and recover the refrigerant. Most parts can be removed and replaced with simple procedures. Removing the evaporator may require disassembly of the dashboard or other components. The external surfaces of evaporators and condensers can be cleaned with commonly available solvents. After changing parts, evacuate and recharge the system and check for leaks and overall system operation.

Some systems have multiple component units. Service procedures are similar to those for individual components.

Review Questions—Chapter 19

Please do not write in this text. Write your answers on a separate sheet of paper.

1. Match the symptom with the most likely problem.

Low airflow through the evaporator. _____	(A) Water in expansion valve.
Vacuum on low side. _____	(B) Restricted condenser fins.
Intermittent loss of cooling. _____	(C) Restricted receiver-drier.
Excessive high-side pressure. _____	(D) Restricted evaporator fins.
Evaporator icing when cruising. _____	(E) Expansion valve stuck open.
Icing on high pressure line. _____	(F) Expansion valve stuck closed.
Compressor clutch won't engage. _____	(G) Evaporator pressure control valve improperly set.
	(H) Cycling switch stuck open.
	(I) Cycling switch stuck closed.

2. Changing a cycling clutch control switch may not require removing the refrigerant from the system, since it may have a _____.

3. Which of the following refrigeration components is often changed as a safety precaution?
   (A) Evaporator.
   (B) Condenser.
   (C) Accumulator.
   (D) Expansion valve.

4. If the desiccant bag breaks, the desiccant will cause valve _____.

5. On some vehicles, it may be necessary to disassemble part of the _____ to access the expansion valve.

6. Why should fixed orifice tubes *not* be cleaned?

7. A _____ valve is sometimes used when an R-12 system is converted to R-134a.

8. Some technicians have baked used _____ in an oven to remove water.

9. Condensers can be cleaned with a strong stream of _____.

10. A good indication of an internally contaminated receiver-drier is a _____.

ASE Certification-Type Questions

1. If an expansion valve sticks open, the air conditioning system will have all the following symptoms, *except:*
   (A) sweating.
   (B) frost formation.
   (C) excessive high-side pressure.
   (D) very little cool air output.

2. Technician A says almost all expansion valves have an inlet screen. Technician B says almost all evaporator pressure control valves have an inlet screen. Who is right?
   (A) A only.
   (B) B only.
   (C) Both A and B.
   (D) Neither A nor B.

3. Heating an expansion valve sensing bulb causes low-side pressure to increase. Technician A says the sensing bulb has lost its charge. Technician B says cooling the sensing bulb should cause pressure to decrease. Who is right?
   (A) A only.
   (B) B only.
   (C) Both A and B.
   (D) Neither A nor B.

4. The compressor clutch engages only when the compressor cycling switch is bypassed. Technician A says the switch is defective. Technician B says the ECM is preventing clutch engagement. Who is right?
   (A) A only.
   (B) B only.
   (C) Both A and B.
   (D) Neither A nor B.

5. The expansion valve sensing bulb should be attached to the evaporator outlet and covered with _____.
   (A) dum dum
   (B) insulating tape
   (C) asbestos
   (D) plastic

6. If an evaporator is leaking, it should be:
   (A) cleaned.
   (B) soldered.
   (C) sealed with epoxy.
   (D) replaced.

7. All of the following interior components may need to be removed to remove a blower case, *except:*
   (A) dashboard.
   (B) engine.
   (C) center console.
   (D) front seats.

8. Technician A says brazing can sometimes repair a leaking condenser. Technician B says says a leaking condenser should be arc welded. Who is right?
   (A) A only.
   (B) B only.
   (C) Both A and B.
   (D) Neither A nor B.

9. An accumulator is removed from the system and is found to contain 7 ounces (210 ml) of oil. What is the *most likely* problem?
   (A) The desiccant bag is saturated.
   (B) The desiccant bag is broken.
   (C) The oil return hole is plugged.
   (D) The vapor return port is plugged.

10. All of the following are reasons to replace a receiver-drier, *except:*
   (A) evidence of excess water in the system.
   (B) clouded sight glass.
   (C) frost forms on the receiver-drier outlet.
   (D) pressure switch inoperative.

Replacement condensers have equal and in many cases, greater capacity than original equipment condensers. (Modine)

Chapter 20

Heater and Engine Cooling System Service

After studying this chapter, you will be able to:

- ❑ Diagnose heating system problems.
- ❑ Diagnose engine cooling system problems.
- ❑ Check coolant level and freezing point.
- ❑ Remove and replace heater cores and heater hoses.
- ❑ Remove and replace heater shutoff valves.
- ❑ Check and replace cooling system belts and hoses.
- ❑ Flush a cooling system and heater core.
- ❑ Remove and replace coolant pumps.
- ❑ Remove and replace cooling system thermostats.
- ❑ Remove and replace cooling fans, fan clutches, and fan motors.
- ❑ Remove and replace radiators.

Technical Terms

Depressurized

Petcock

Alkalinity

pH level

Electrochemical degradation (ECD)

Overheating

Coolant leaks

Overcooling

Reverse fill

Reverse flushing

The heater and its source of heat, the engine, are often overlooked. However, they are an important contributor to HVAC system operation. A malfunction in the cooling system or engine can cause air conditioning and heater system problems. Carelessness in servicing the heater components can cause loss of coolant and severe engine damage. This chapter covers diagnosis and service of the heater and engine cooling system. Since most vehicles use a liquid cooling system, emphasis will be placed on this system. Studying this chapter will add to your knowledge of HVAC operation.

Safety Considerations

Hot coolant can scald or blind. Before performing *any* service on the cooling system, make sure the cooling system has been **depressurized.** The cooling system is depressurized when all pressure has been removed from the system. A safer way to check for system pressure is to lightly squeeze the upper radiator hose, **Figure 20-1.** If the hose does not give easily under the pressure of your fingers, the system contains pressure. Do not keep your hand on the upper hose for an extended period of time.

 Note: Some upper radiator hoses contain reinforcing wire.

Do not remove the pressure cap or any fitting until you are sure no pressure remains in the system. The best way to depressurize the cooling system is to allow the engine to cool off completely. However, if this is not possible, the following procedure can be used:

- ❏ Allow the engine to cool off for at least 20 minutes. If necessary, open the hood and use a shop fan to blow cool air through the radiator and across the engine surface.
- ❏ Using several shop towels to protect your hands, crack the pressure cap loose (no more than 1/8 turn) and allow pressure to escape.
- ❏ Wait at least five minutes.
- ❏ Using the shop towels, turn the cap another 1/4 turn and make sure no pressure remains in the system.
- ❏ Remove the pressure cap.

Once the pressure cap has been removed, check the coolant and proceed with any needed service, keeping in mind the coolant is still very hot. Also keep in mind antifreeze is poisonous. If coolant must be removed, most radiators have a drain device, usually called a **petcock** at the bottom of the radiator, **Figure 20-2.** Do not allow antifreeze solutions to drain onto the ground. Make sure any containers of antifreeze are covered to keep animals from drinking it and being poisoned. Store used antifreeze in marked containers for recycling.

Figure 20-1. *Quickly grip and squeeze the upper radiator hose to see if the cooling system is under pressure. Only grasp the hose for a moment. (Jack Klasey)*

Figure 20-2. *This drain petcock is typical of those used on older vehicles.*

Cooling System Inspection

The cooling system is often overlooked until a problem occurs. The following sections briefly cover the inspection procedures that should be performed on a cooling system to keep it in good condition.

Cooling System Checks

Several checks should be made to every liquid cooling system. Make these checks whenever the engine oil is changed, or at least every two months. On an air-cooled engine, the only checks needed are to check the blower belt for tightness and condition and clean the cooling fins if necessary.

Coolant Level

⚠️ **Warning: Follow the safety precautions described earlier before removing the pressure cap. Newer vehicles use pressurized reservoirs like the one shown in Figure 20-3. It should be removed with the same caution as a standard pressure cap.**

For the most accurate level reading, check the coolant when the engine is cold. The coolant level should be at the top of the radiator filler neck if the vehicle has a coolant recovery system. The coolant level should be about 1-2″ (25-51 mm) below the neck if there is no recovery system. Also check the reservoir bottle for the proper level of coolant, **Figure 20-4.** If necessary, add coolant. Be sure to use the proper kind of antifreeze as many newer vehicles use long-life antifreeze. Adding conventional antifreeze to a system containing long-life antifreeze will shorten the service life of the coolant.

Figure 20-3. *Many vehicles have pressurized filler caps on the coolant reservoir instead of the radiator. Remove these caps as carefully as you would a radiator pressure cap.*

📦 **Note: Keep in mind the color of coolants (green, blue, red, yellow, orange, or pink) has no bearing on its composition or service life (either conventional or long-life). The various dyes are used to give antifreeze its color. Check the service manual for the proper type to use.**

Coolant Condition and Freezing Point

Coolant condition can be initially checked by visual observation. Remove the pressure cap after following safety precautions and observe the coolant in the system, **Figure 20-5.** If the coolant is clear to dark in color (dark

green or dark blue, for example), it is in good condition. If the coolant is rust colored, it must be changed and the system flushed. If oil is floating on the top of the coolant, engine oil or transmission fluid may be leaking into the cooling system. Check for milky oil in the engine and transmission.

🔧 **Caution: Coolant contaminated transmission oil will have the same color as a strawberry milkshake. If coolant has contaminated the transmission oil, repair or replacement of the radiator and a complete overhaul of the transmission/transaxle is required.**

Even if the coolant looks good, there may not be enough antifreeze present to keep the system from freezing in cold weather. The only sure way to check the freezing point of the coolant is to use a hydrometer or refractometer.

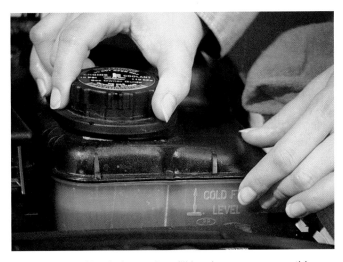

Figure 20-4. *Check the coolant fill level, on newer cars this can be done by checking the level of coolant in the resevoir.*

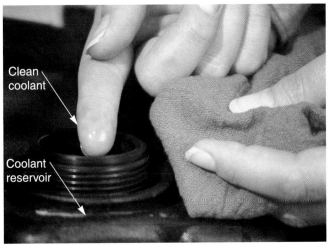

Figure 20-5. *Coolant should be clean with no corrosion or oil residue.*

Using a Hydrometer

To use a hydrometer, place the hose in the filler neck. Squeeze and release the bulb to draw in a sample of the coolant. Allow the pointer, floats, or balls in the hydrometer to stabilize, then read the temperature on the hydrometer, **Figure 20-6.** Some hydrometers have a temperature correcting feature and it may be necessary to add or subtract from the reading, depending on the temperature of the coolant.

Using a Refractometer

To use a refractometer, take a small sample of coolant from the radiator. Place a few drops on the refractometer lens. Look through the refractometer to get the freeze point reading.

Using Test Strips

Some shops now use test strips to quickly check engine coolant. Chemical test strips can check coolant for freeze protection, as well as its **alkalinity (acidity)** or **pH level.** However, most test strips can only be used on conventional coolants.

To use the test strips, remove one strip from the container. Be sure to close the container, as the color chart is often on the side. Dip the test strip in the engine coolant, making sure both test spots are immersed. Remove the test strip and wait approximately 30 seconds. Compare the two test sections to the color chart that comes with the strips, **Figure 20-7.**

Hose Condition

Observe the radiator and heater hose condition. Look for bulges and swelled spots, cuts and abrasions, cracks, and leaks. Allow the engine to cool off and observe the hoses. Any hose that collapses as the engine cools is soft and should be replaced. As a final check, release system pressure and squeeze the hoses near the clamps to check for soft spots. See **Figure 20-8.**

Some hose damage may not be evident to the naked eye. Hoses are weakened over time by **electrochemical degradation (ECD),** which is a reaction between the chemicals in the coolant and the metals in the engine and radiator, **Figure 20-9.** Often, this reaction can cause hose defects. Because of this, coolant hoses should be changed if they are more than three years old. ECD has the greatest effect on heater hoses, bypass hoses, and the upper radiator hose.

Checking for Cooling System Electrolysis

Improper grounding of electrical circuits can cause **electrolysis.** Electrolysis is the creation of an electric current in the cooling system. A multimeter can be used to check for electrolysis in the cooling system. Set the multimeter to the DC volts scale and connect the negative probe to the negative terminal of the vehicle's battery. Place the positive probe into the coolant at the radiator filler neck and note the reading. As a general rule, the voltage reading should be between 0.1 and 0.4 volts. Any reading over 0.8 volts indicates a problem that should be fixed immediately.

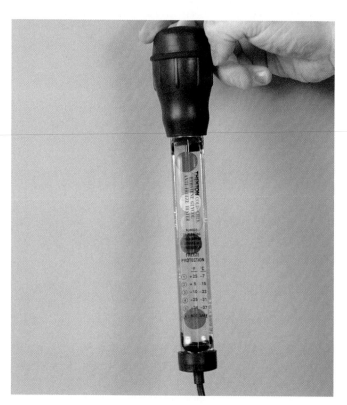

Figure 20-6. *A hydrometer can be used to check the coolant's ability to protect against freezing. Most cooling system hydrometers use floating balls or disc. (Jack Klasey)*

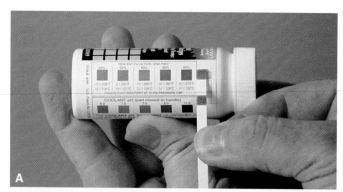

Figure 20-7. *Chemical test strips are now available and can quickly determine the cooling system's freeze protection capability, as well as its pH level. A—Test for freeze protection. B—Test for alkalinity or pH. (Jack Klasey)*

Figure 20-8. *Squeezing a hose near the clamps will determine whether the hose is soft or cracked. Allow the engine to cool off before making this test.*

Figure 20-9. *Hoses are subjected to electrochemical degradation (ECD), which can deteriorate hoses from the inside. (Gates)*

If the voltage reading is too high, remove one fuse at a time while monitoring the voltage reading. If voltage drops when a particular fuse is removed, the problem is in that circuit. Also check for missing ground wires. In some cases, it may be necessary to operate electrical components while watching the multimeter. If the voltage jumps when a component is turned on, the component is not properly grounded.

Belt Tension and Condition

Observe the belt condition and tension. Check the belt for cracks, splits, and frayed edges, **Figure 20-10.** If the vehicle uses a serpentine belt, check the pulleys for debris or foreign material. The belt should make no noise when the engine is operating.

Use a tension gauge to check belt tightness. If a tension gauge is not available, a general rule is the belt should not deflect more than 1/4″ (7 mm) under light thumb pressure. Vehicles with serpentine belts usually have self-tensioning devices. The tensioner should be checked to ensure it is supplying the correct tension.

Figure 20-10. *Inspect all drive belts for defects such as cracks, splits, frayed edges, and missing chunks. You can also perform a preliminary tension check during the inspection, before performing an actual test with a tension gauge. (Jack Klasey)*

Coolant Leaks

Make a visual check for coolant on the engine and radiator, or streaks that indicate coolant has been leaking. Lift the vehicle and check the freeze plugs for leaks and corrosion. **Figure 20-11** shows typical radiator failures.

Pressure Testing

If leaks are indicated or the coolant level is low, perform a pressure test. Leaks can usually be located quickly by pressure testing the system. Before attaching the tester, add coolant or water to bring the level to the top of the filler neck. Then use the pressure tester to pressurize the system, being careful not to exceed the rating of the pressure cap, **Figure 20-12.** Once the system is pressurized, check for a drop in pressure and obvious leaks in the cooling system. If the system holds pressure for 10 minutes, and no dripping coolant is found, it can be assumed the system is ok.

The pressure cap can also be tested with the pressure tester. Attach the cap to the tester using the proper adapters. Then pressurize the cap to the rating stamped on top of the cap. The cap should hold at its rated pressure for a minimum of two minutes.

Solder bloom—Solder corrosion caused by degradation of antifreeze rust inhibitors. Tube-to-header joints are weakened and corrosion can restrict coolant flow.

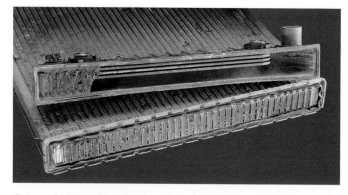

Internal deposits—Rust and leak inhibitors (Stop-leak) can form solids that collect in the cooling system and restrict flow.

Tube-to-header leaks—Solder joint failure resulting in coolant loss.

Leaky tank-to-header seam—Solder joint failure or a cracked header, usually the result of pressure-cycle fatigue.

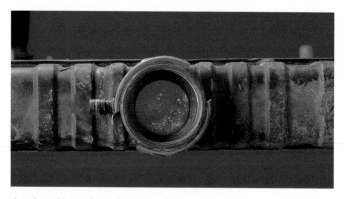

Leaky oil cooler—Coolant shows traces of oil. Transmission/ transaxle or engine damage can result.

Leaky inlet/outlet fitting—Leaks in this area can be caused by fatigue or solder joint corrosion.

Electrolysis—Electrical current created by the chemical reaction between coolant and two dissimilar metals. Causes corrosion of metal components.

Electrolysis—Electrical current created by the chemical reaction between coolant and two dissimilar metals. Will produce voids in tubes.

Figure 20-11. *Radiator failures are the very common, these are the ones you are most likely to see. (Modine)*

Fin deterioration—Chemical deterioration of the fins, caused by road salt or sea water.

Fin bond failure—A loss of solder bond between fins and tubes. Fins will be loose in the core.

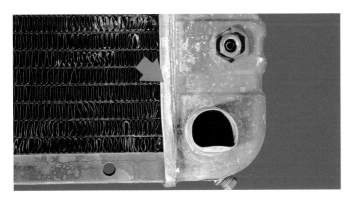

Blown tank-to-header seam—Caused by extreme cooling system pressure, usually as a result of exhaust leaking into the cooling system.

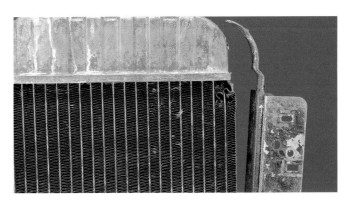

Loose side piece—Can lead to flexing of the core and radiator-tube failure.

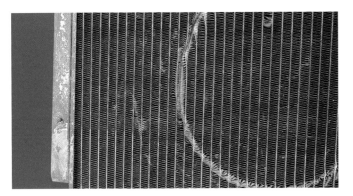

Fan damage—A minor collision, failed water pump, or loose fan support can result in radiator damage.

Overpressurization—Excessive pressure in the radiator caused by a defective pressure cap or engine exhaust leak.

Cracked plastic tanks—High stress in the radiator can cause premature plastic tank-failure.

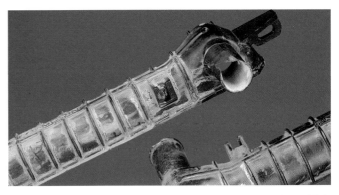

Steam erosion—Steam can break down plastic tanks, which will produce thinning and eventually, holes in the tanks. White deposits are often found.

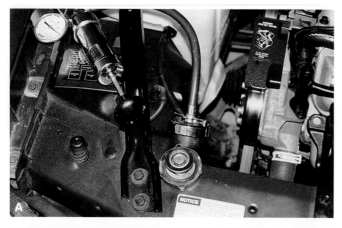

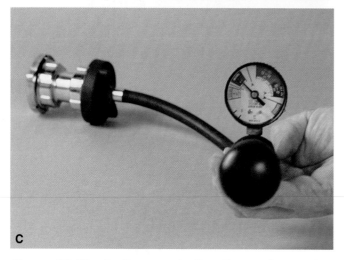

Figure 20-12. *A—Pressure testing the cooling system B—A special adapter is needed to pressure test a radiator cap. C—Pressure testing radiator cap. (Jack Klasey)*

Combustion Leak Testing

Most combustion leak detectors use a special solution that reacts with exhaust gases. To use the combustion leak tester, prepare the testing solution and pour it into the tester. Start the engine and allow it to operate for 10 to 15 minutes. Then install the tester on the radiator cap opening, **Figure 20-13.** Use the actuating bulb to draw gases from the cooling system into the tester. Closely monitor the testing solution. If the solution changes color, usually to a dirty yellow, exhaust gases are leaking into the cooling system.

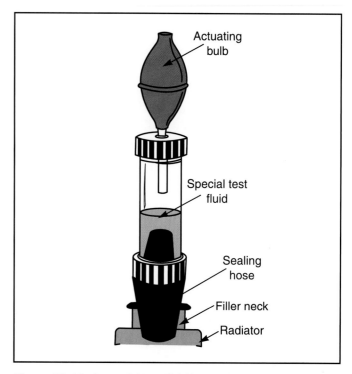

Figure 20-13. *A special test fluid is used to check for combustion leaks in the cooling system. If the fluid changes color, exhaust gases are present, and engine repair is needed. (P and G Manufacturing Co.)*

Another method of checking for combustion leaks is to use an exhaust gas analyzer. Depressurize the system and remove the radiator pressure cap. Start the exhaust gas analyzer, allow it to warm up, and calibrate it as necessary. Then place the probe of an exhaust gas analyzer in the radiator filler opening.

 Caution: Do not allow the probe to contact the coolant.

Start the engine and allow it to idle. Closely monitor the exhaust gas analyzer. If analyzer readings indicate that exhaust gases are present in the radiator, there is an exhaust leak in the cooling system.

Dye Testing Using Black Light

To locate hard to find leak sources, a dye can be added to the cooling system. After the dye has circulated through the cooling system, a black light is then used to locate the source of the leak. This is similar to refrigeration system leak testing using a dye, which was discussed earlier.

Heating System Problems

Unlike the refrigeration system, heater operation does not depend on change of state. In a properly operating liquid-cooled system, the coolant always remains a liquid.

Most heating problems result from failures involving the cooling system. Most problems on an air-cooled engine are caused by failure of the heated air to reach the passenger compartment. The following sections identify the various problems and their causes.

Poor or No Heating

Poor or no heat is the failure of heated coolant to transfer its heat to the passenger compartment. The most common causes are a clogged heater core; a coolant leak; a shutoff valve that does not open; collapsed or swollen heater hose; or an engine problem that does not allow the water pump to circulate coolant through the heater core. One indication of a clogged heater core is the presence of corrosion and deposits in the cooling system. Wait for the engine to cool, then remove the pressure cap and check for sludge and rust in the system. To check for a clogged heater core, disconnect the heater hoses and attempt to direct a low pressure stream of water through the heater core. Low flow indicates a clogged core. Also make sure the exterior of the core has not become plugged with debris.

 Note: Normal municipal water system pressure is about 35-40 psi (241-276 kPa). Do not allow the pressure in the heater core to reach this amount, or the core will be ruptured.

A stuck shutoff valve is a common cause of no heat. Place your hand on the linkage at the shutoff valve. Then have an assistant move the heater controls through various modes. If the shutoff valve linkage does not move as the modes and temperatures are changed, the shutoff valve is stuck. Some shutoff valves can be cleaned to restore their operation, but most technicians prefer to replace the unit. If a vacuum diaphragm operates the shutoff valve, check the diaphragm with a vacuum pump. If the diaphragm does not move the linkage, or will not hold a vacuum, it is defective.

A problem that prevents coolant flow through the heater core is sometimes hard to track down. An air bubble in the engine water jackets can sometimes cause flow problems. Air bubbles are common on many modern vehicles since the radiator filler neck is not at the highest point in the system. Other causes of poor heating are blown, out of position, or incorrect cylinder head gaskets, incorrect or damaged coolant pump impeller, or collapsed coolant hoses. Checking for these problems usually involves partial engine disassembly.

If the heater produces heat only after a long drive, the engine thermostat may be stuck open or may have been removed.

Too Much Heat

There are essentially two reasons for too much heat in the passenger compartment, an inoperative blend door or engine overheating. The problem is not in the heater core

itself. On some vehicles, the heater shutoff valve can cause overheating if it fails to close. Be sure to check shutoff valve operation.

Coolant Leaks

The most obvious heating system problem is coolant leaking into the passenger compartment. Leaks in the passenger compartment usually show themselves as a smell of antifreeze or the presence of steam, an oily or slimy film on the windshield, or coolant on the passenger compartment carpet. Usually, leaks will not cause a significant drop in coolant level before the other symptoms are noticed. Pressure testing may reveal the leak, but the surest way to tell is by the presence of coolant on the front right carpet or in the ducts. The heater core can be pressure tested on the vehicle, **Figure 20-14.** Once the heater core has been removed, the leak can be confirmed by bench pressure testing as shown in **Figure 20-15.**

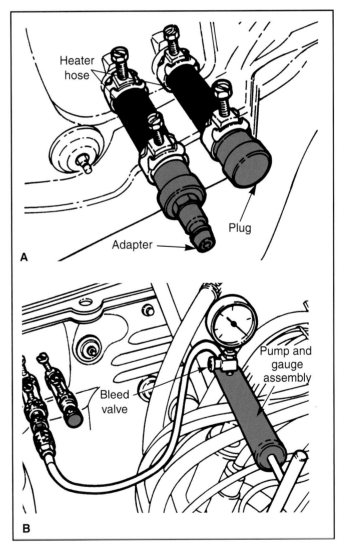

Figure 20-14. A—Pressure testing a heater core on the vehicle. This will reveal a heater core leak, but cannot be used to determine whether the core is plugged. B—Since a leak may not be visible with the core installed, carefully monitor the gauge for a drop in pressure. (Ford)

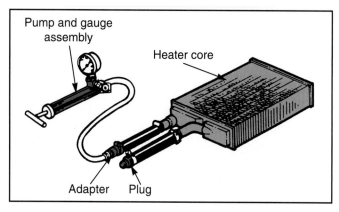

Figure 20-15. *Pressure testing a heater core that has been removed from the vehicle. Any leaks should be visible. (Ford)*

Heater Problems—Air-cooled Engine

Missing shrouds or ducts, closed heater doors, or failure of the thermostat when used is the cause of most heater system problems on vehicles with air-cooled engines. Visual inspection is the easiest way to locate a problem. If the blower drive belt is loose, air may not be forced into the system. This is usually accompanied by engine overheating.

Smells in the passenger compartment are caused by out of position shrouds, or by engine oil or exhaust leaks. Any odors in the passenger compartment should be investigated immediately to prevent carbon monoxide poisoning.

Engine Cooling System Problems

Since most modern vehicles use liquid cooling systems, the most common cooling system problem on most vehicles is overheating. Sometimes, the engine will be overcooled. Leaks commonly occur on liquid-cooled engines. Causes of various problems are discussed in the following paragraphs.

Overheating

Overheating occurs when the engine exceeds its maximum safe operating temperature. Overheating is usually indicated by an illuminated dashboard warning light or a gauge needle in the Hot position. On a liquid-cooled engine, steam may be coming from under the hood. An overheating engine may also knock excessively when accelerated.

> **Note:** If the warning light or gauge indicates overheating and no other symptoms are noted, check the light or gauge sender and wiring before proceeding.

An engine related problem might also cause overheating even when the cooling system is in good condition. Typical engine problems include retarded ignition timing, a lean mixture, or a restricted exhaust system. Even if you suspect one of these problems, check the cooling system first.

A good way to check the actual temperature of the cooling system is with an infrared temperature tester. Point the tester at the radiator outlet hose to determine the temperature of the coolant as it leaves the engine. If the temperature exceeds 250°F (121°C), the engine is overheating.

Causes of Liquid-cooled Engine Overheating

The most common causes of overheating on a liquid-cooled engine are a low coolant level, slipping fan belt, defective fan clutch or motor, defective pressure cap, clogged radiator, or stuck thermostat. Any of these defects cause the coolant to fail to pick up heat from the engine, or to fail to pass the heat into the atmosphere.

Coolant Leaks

Coolant leaks, or loss of coolant from the cooling system, can occur at several locations. Common leak locations are radiator heater hoses, the coolant pump, the heater core, and the radiator. Low coolant level can be checked at the radiator fill. Follow the precautions at the beginning of this chapter before removing the cap. If the coolant level is low, be sure to pressure test the cooling system.

Less common leaks are caused by cracks in the head or block or a blown head gasket. An indicator of an internal engine leak is the presence of coolant in the engine oil or excessive water or steam exiting from the exhaust pipe, coupled with loss of coolant from the cooling system. These defects can also cause exhaust gases to enter the cooling system, which can superheat the coolant very quickly. These problems are difficult to locate and require special testing procedures and in some cases, partial engine disassembly.

Slipping Belts

A slipping fan belt can usually be determined by listening to the engine as it is running or by visually inspecting the belt. Typical belt defects are shown in **Figure 20-16.** Loose or damaged belts often squeal when the engine is accelerated. To check a belt problem, squirt a small amount of water or penetrating oil on a noisy belt. If the noise disappears for a few seconds, then returns, the belt or pulleys are contaminated or misaligned. If the belt becomes noisier as soon as the water or oil is applied, the belt is slipping.

A belt that is loose but otherwise in good condition can be tightened. A glazed, cracked, or otherwise damaged belt should be replaced. If the vehicle uses a serpentine belt, the cause is either a defective tensioner or a stretched belt.

Figure 20-16. *Checking belts for wear A—This serpentine belt had chunks missing from the belt. B—This V-belt is cracked and frayed. Both belts should be replaced.*

Defective Fan Clutch or Motor

To check the fan clutch, allow the engine to reach operating temperature. The radiator should be hot. With the hood open, have an assistant stop the engine as you observe the fan. The fan should stop turning within one revolution. If the fan continues to spin with the engine off, the clutch is defective. Check the fan for broken or missing blades.

Checking for a defective fan motor or control system also requires the engine be warmed up. The fan should come on at a certain temperature. If possible, use a temperature probe or infrared tester to determine the exact temperature at which the fan turns on. If the fan motor does not start turning at the rated temperature, either the motor or a control unit is defective. See electrical checking in the next chapter.

Defective Thermostat

The best way to isolate a stuck thermostat is to allow the engine to cool off completely, then add coolant and start the engine. If the thermostat is stuck, the engine will begin overheating and steaming within five minutes.

 Caution: Do not try to cure an overheating problem by removing the thermostat. This will not correct the real problem, and may make the overheating condition worse.

Other Causes of Overheating

A clogged radiator is sometimes hard to spot. It is often necessary to remove the radiator and remove the tanks to check for clogged tubes. If the system is extremely dirty, it can be assumed some tubes are clogged.

Coolant sometimes gets into the transmission from a leak in the transmission oil cooler. Coolant in the transmission fluid will cause the fluid to look milky. In rare cases, failure of the coolant pump impeller or internal circulation baffles may cause overheating.

Causes of Air-cooled Engine Overheating

The usual causes of air-cooled engine overheating are a slipping blower belt or engine shrouds that are out of position. If the air-cooled engine uses a thermostat, it may have failed closed. All of these problems can be detected by visual inspection.

Overcooling

Overcooling is the failure of the engine to reach its normal operating temperature. Overcooling can reduce the amount of heat from the heater since the coolant never reaches its normal temperature. Overcooling will also damage the engine as water and unburned gasoline collect in the crankcase. The only cause of overcooling is a stuck open or missing thermostat. This applies to both liquid and air-cooled engines. A thermostat problem may not be noticed in warm weather. In cold weather, however, overcooling will usually be noticed quickly. In very cold climates, the engine may never reach normal operating temperatures. Remove the thermostat housing to check the thermostat.

Heater Service

The heater core, hoses, and their related parts require service, either as a part of HVAC maintenance or to correct a leak that affects the operation of the heating and cooling system. Basic procedures are covered in the following paragraphs.

Changing Heater Hoses and Shutoff Valves

Hoses and shutoff valves are usually located inside the engine compartment. They are generally easy to find and replace. To replace a defective hose or shutoff valve, depressurize and drain the cooling system below the level of the hose or valve.

 Note: When installing a new heater or radiator hose, spraying some aerosol silicone spray into the opening of the new hose will greatly ease installation.

Standard Hoses with Hose Clamps

Begin by loosening and removing the hose clamps, **Figure 20-17.** Then gently twist the hose in relation to the fitting to break it loose. If the hose will not come loose with gentle twisting, use a box cutter or other knife to slice the hose at the fitting. Then peel the hose away from the fitting. Cut a new length of the proper diameter hose and slip the clamps over each end. If desired, place a small amount of nonhardening sealer on the fitting nipple. Then place the hose ends over the fitting nipples. Tighten the hose clamps, add coolant, and check for leaks.

 Caution: Nonhardening sealers should never be placed inside a hose before installation. Doing this can result in clogging, especially in the case of heater hoses.

Quick Disconnect Hoses

Some heater hoses are one piece quick disconnect hoses. Quick disconnect hoses can be replaced with an identical replacement or repaired. Some quick disconnect fittings can be removed by unsnapping them from the

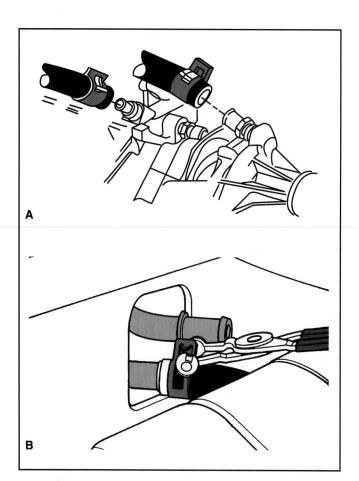

Figure 20-17. To remove heater hoses, depressurize the cooling system first. A—Remove the hoses from the engine side. B—Remove the hose clamps, then pull the hoses from the heater core. (General Motors, DaimlerChrysler)

attaching nipple. Others must be cut from the nipple, **Figure 20-18.** Then, the replacement hose is installed over the nipple and secured with a standard hose clamp. Nonhardening sealer can be used if desired.

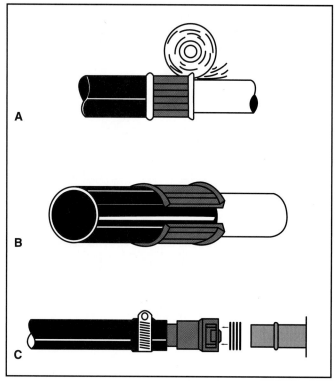

Figure 20-18. A—It is often necessary to cut the hose sleeve to make removal easier. B— Once a quick disconnect hose has been removed, cut off the sleeve around the hose. C— A quick disconnect hose can be replaced by clamping the replacement hose to the original coupling after the sleeve has been cut off. Another method is to clamp the replacement hose directly to the nipple of the hose fitting. (Gates)

Replacing Heater Shutoff Valves

There are many kinds of heater shutoff valves, with two, three, or four connections. The following procedure is a general guide only. Before replacing any shutoff valve, depressurize and drain the cooling system and loosen the hose clamps. Then remove the linkage or vacuum line and loosen the hose clamps. If the valve is attached to a bracket, remove the fasteners. Then gently twist the hoses and pull the valve nipples loose from the hoses, **Figure 20-19.** If the shutoff valve is threaded into the engine or heater core, remove all attaching parts and use the appropriate wrench to loosen and remove the valve.

Note: You may need to cut off the old hoses to remove the valve. If the shutoff valve is connected to the heater core by a short length of hose (less than 6 inches or 15 cm), replace the hose.

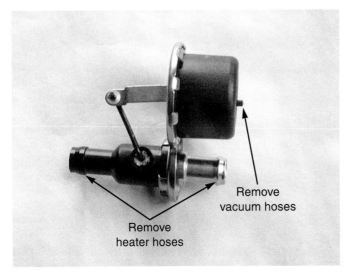

Figure 20-19. *Vacuum-operated heater valve removed from the cooling system.*

Blow through both the old and new valves to ensure they operate in the same manner. If you can blow through one valve and not the other, the replacement is incorrect. If the valve is linkage operated, be sure both levers are in the same position before making this test.

Place the new valve in position on the heater hoses. Use a small amount of nonhardening sealer on the hose nipples if necessary. Connect the linkage or vacuum line. Install the valve on the bracket if necessary. Tighten the hose clamps, add coolant, bleed the system, and check heater operation.

Changing Heater Cores

Heater core removal can be simple or extremely complex, depending on the vehicle. Most heater cores are installed in the blower case under the dashboard or against the firewall at the rear of the engine compartment. Before beginning heater core removal, depressurize and drain the cooling system below the level of the heater core.

 Note: Some technicians clamp off the hoses to the heater core before removal to minimize coolant loss and the entrance of air into the engine.

Remove the hoses from the heater core. Carefully blow air through the heater core to remove as much coolant as possible. This reduces spillage into the ductwork or passenger compartment when the heater core is removed. There are many variations of heater core placement and installation. Always consult the proper service manual before beginning removal of any unfamiliar heater core.

Disabling Air Bag System

You may need to disarm the air bag system when removing the heater core on some vehicles. Failure to do so could result in accidental deployment.

⚠ **Warning: Most air bag systems have an energy reserve module that stores power to deploy the air bags in the event battery power is lost in a collision. This power reserve can last for up to 30 minutes after battery power is disconnected.**

The first step in air bag removal is to turn the ignition switch to the locked position and remove the key. The key must be removed before starting most air bag disarming procedures. In some cases, you may need to leave the ignition on to use a scan tool to disarm the air bag system. After removing the key, disconnect the air bag system's source of power. This can be accomplished by removing the negative battery cable or pulling one or more fuses. If the system is equipped with an energy reserve module, wait 30 minutes before removing the air bag. Once power has been disconnected, remove the yellow connectors to the air bag module, unbolt the air bag, and place it in a clean area with the inflater side facing up. Always follow the manufacturer's instructions for disarming the air bag system.

Heater Core Removal from Engine Compartment

Heater cores mounted under the hood are usually easier to replace. On some vehicles with a split case surrounding the firewall, part of the inside section may need to be removed to gain access to the heater core. Disassemble any ductwork as necessary. Always refer to the appropriate service manual before beginning heater core removal.

Once the case or cover has been disassembled to expose the heater core, remove any brackets that hold the core to the case and remove the core. Compare the old and new core to ensure the replacement core is correct. Some heater cores are universal replacements, **Figure 20-20.** Clean any standing coolant, leaves, and debris from the case. Before installation, make sure the inlet tubes of the new core are almost exactly at the same position as the old core. Then install the core in the case. Install and tighten the brackets. Then reinstall the case components or cover on the vehicle. Reattach the heater hoses and refill the cooling system. Start the engine, bleed the system, and check heater operation.

Heater Core Removal from under Dash

If the heater is mounted under the dash, in some cases, it may be possible to remove a cover and lift the heater core out of its well, **Figure 20-21.** On other vehicles

Figure 20-20. *A—This heater core was leaking, note the coolant on the lower surface. B— This heater core is a universal type, made to fit many vehicles. The tubes can be moved to match the original heater core configuration. Other cores are designed for specific vehicles.*

with a one-piece case, the entire case must be removed and disassembled to get to the heater core, **Figure 20-22.** Removal procedures for heater cores mounted under the dash usually involve removing dashboard components. In some cases, the entire dashboard must be removed.

After disarming the air bag system, disconnect the negative battery terminal if needed to prevent draining the battery. Check the service manual to see what components may need to be removed, relocated, or loosened to access the case. On vehicles requiring case removal, the following interior components and systems are often involved:

❑ Dashboard (upper and lower).
❑ Refrigeration system charge recovery.
❑ HVAC duct work.
❑ Front carpet.
❑ Center console (if equipped).
❑ Steering column.
❑ Air bag system.
❑ Passenger front seat.
❑ Instrument cluster.
❑ Dashboard wiring harness.

Follow service manual directions for the removal of each interior component as needed. Once the case is accessible, remove all mounting bolts. Some blower case

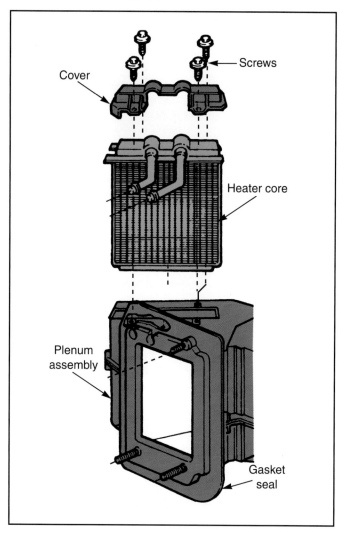

Figure 20-21. *Some heater cores can be removed by removing a cover under the engine compartment and pulling the core upward to clear the housing. (Ford)*

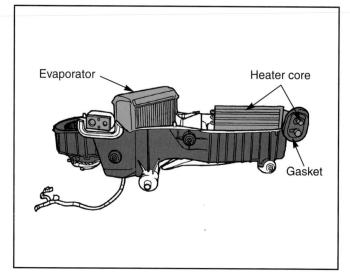

Figure 20-22. *This heater core can be removed after the case is split. Note the gasket that seals off the heater compartment from the outside. This gasket must be replaced to prevent air loss. (DaimlerChrysler)*

bolts may need to be accessed from the engine compartment. Carefully pull the case back and remove it from the vehicle. Place the case flat on the shop floor, with a fender cover or blanket to prevent damage.

In most cases, the case will have several bolts to remove in order to access the heater core. Lift the heater core from the ductwork and place it on a bench. Compare the old and new heater cores, then clean out any standing coolant, leaves, and debris from the evaporator and heater areas. Install the new heater core, reseal, and reinstall the blower case. Reattach the heater hoses and refill the cooling system. Start the engine, bleed the system, and check heater operation.

Liquid Cooling System Service

The liquid cooling system is used in over 97% of the vehicles on the road today. All liquid cooling systems have common failure points. The following general procedures are designed to efficiently service these points. If you are removing any cooling system part or fitting, drain the coolant to below the level of the part to be removed.

Changing Coolant and Flushing the Cooling System

Coolant should be changed according to manufacturers' instructions or at least every two years. It may also be necessary to flush the system if the coolant appears rusty or sludge is observed on the pressure cap or in the radiator.

Changing Coolant

A simple changing procedure is to operate the engine until the thermostat is open (the top radiator hose is hot). Place a drain pan under the petcock, then open the petcock. Allow the coolant to drain into the pan. When coolant stops flowing from the petcock, the system pressure has been removed. Remove the pressure cap and allow the coolant to drain completely. Then perform other service operations as necessary. Allow the old coolant to cool, then dispose of it properly.

Close the petcock, and fill the radiator with a 50-50 mixture of the proper antifreeze and water. Do not completely fill the radiator, as there must be room for the hot coolant to expand. Then bleed the cooling system as explained in the next section.

> **Caution: Do not install long-life coolant into a vehicle that uses conventional coolant after replacing a major cooling system part such as a heater core, water pump, or radiator. A possible loss in corrosion protection may result.**

Bleeding the Cooling System

Air pockets often form in the engine when the coolant is drained. These air pockets are hard to remove on many modern vehicles since the radiator filler neck is not the highest point in the system. Since air is lighter than coolant, it rises to the highest point in the engine. On older rear-wheel drive vehicles, the filler neck is above the engine block and air pockets usually remove themselves from the system as the coolant circulates. The next paragraphs explain bleeding procedures according to filler neck placement

System with Filler Neck above Engine Block

Start the engine and allow it to warm up. Turn the heater on full (mode and temperatures switches). Closely monitor the coolant level, and keep it to within about 3" (76 mm) of the filler neck. On most vehicles, the coolant level will slowly rise until the thermostat opens. The coolant may surge as the engine warms up. When the thermostat opens, the level in the radiator will drop suddenly. Add coolant to bring the system up to its normal full level. Monitor the level for a few more minutes to ensure all air has been removed. Then install the pressure cap.

System with Filler Neck below Engine Block

When the filler neck is below the engine block, the engine may be equipped with a bleeder valve. If the engine has a bleeder valve, locate the valve on the engine. It is usually at or near the thermostat housing, **Figure 20-23.** Open the bleeder valve and add coolant to the radiator until it begins to exit from the valve. Lightly close the bleeder valve. Start the engine, turn the heater on full, and allow the engine to warm up. Monitor the coolant level, and keep it to within about 3" (76 mm) of the filler neck. Open the bleeder valve occasionally to allow air to escape. When only coolant escapes from the bleeder valve, top off the radiator and install the cap. Allow the engine to cool, then recheck the level.

Figure 20-23. *Cooling system bleed valve location, usually they are close to the thermostat housing.*

Some older vehicles do not have a bleeder valve. To bleed these vehicles, begin by filling the cooling system to the top of the filler neck. Some technicians choose to *reverse fill* the engine with coolant to minimize the amount of trapped air in the water jackets. This is done by removing the upper radiator hose from the radiator and adding coolant to the engine through the upper hose.

Start the engine and turn the heater on full. As the engine warms up, monitor both the coolant level and engine temperature. Do not allow the engine to overheat. If the engine begins to overheat, shut it off and run a shop fan over the engine to cool it down. Once the air bubble comes out of the engine, the coolant level will drop significantly. When this occurs, fill the radiator to the top of the filler neck and run the engine to ensure the cooling system has been completely bled. It may be necessary in some cases to raise the front of the vehicle to remove all the air from the cooling system.

Flushing the Cooling System

There are several methods of flushing the cooling system. The easiest method is to drain the coolant, add a chemical cleaner, top off the system with water, and operate the engine for a specific period of time. After the chemical cleaner has dissolved as much of the deposits as possible, the system can be drained. Then the cooling system should be refilled with water and the engine operated for a few minutes. Repeat the draining and refilling until the water runs clear.

Reverse Flushing

If the system is very dirty, it may need to be cleaned by a process called *reverse flushing.* Reverse flushing is performed using water pressure to push water through the cooling system in the opposite direction to normal flow. Reverse flushing is often used as part of a chemical flushing procedure. Causing water to flow in the reverse direction allows it to get behind system deposits and loosen them. Reverse flushing is sometimes the only way to clean very dirty cooling systems. Simple reverse flushing can often be accomplished with a water hose. Combination water and air pressure flushing guns are also available.

To perform reverse flushing, depressurize the cooling system and remove the thermostat. Flushing with cool water will cause the thermostat to close. Then connect the water hose or flushing gun to the cooling system in a way that allows water to flow in the opposite direction of normal coolant flow. The most common way to make the connection is to remove the top (or inlet) radiator hose and direct water into the engine block. Adapters are available to allow for easy connections and to minimize pressure loss. On some vehicles one of the heater hoses can be used for flushing.

 Note: Always perform the flushing operation in an area that has adequate drainage. Some localities require the water flushed from the cooling system be recovered and reprocessed with other antifreeze solutions.

Once the connections are made, slowly open the valves and allow water (and air if used) to flow through the cooling system. Gradually increase flow as the water begins to exit through the radiator inlet hose and filler openings.

 Caution: Do not increase pressure in the cooling system beyond the cap rating. Excessive pressure may damage cooling system parts.

When only clear water flows from the radiator outlets, the system is flushed. Turn off the water and air, and reconnect the cooling system hose(s). If this method does not remove all sludge, the individual cooling system components must be removed for cleaning. In some cases, the radiator has become so plugged it must be removed for replacement or cleaning by a specialty radiator shop. Radiator removal is discussed later in this chapter.

Belt and Hose Replacement

Belt and hose replacements are among the most common cooling system jobs. Two kinds of belt designs, the V-belt and the serpentine belt, are used on modern vehicles. Follow the procedures for each kind of belt.

Serpentine Belts

To replace a serpentine belt, locate the automatic tensioning device. Before removing the belt, note the pattern of the belt. Usually, a sticker indicating the belt routing pattern is located somewhere in the engine compartment, **Figure 20-24.** Use a pull handle or pry bar to pull the tensioning device away from the belt. See **Figure 20-24.** In some cases, the pulley may have a fixed tensioner screw assembly. When the tension has been reduced, slide the belt over one of the pulleys and remove it from the engine. Install the new belt by placing it over all but one of the pulleys. Then move the tensioning device to allow the belt to be placed over the last pulley. Start the engine and check belt operation.

V-belts

To replace a V-belt, determine which driven accessories (air conditioning compressor, alternator, power steering pump, and air pump) need to be moved to remove the belt. In some cases, it may be necessary to remove more than one belt to access the compressor drive belt. Loosen the fasteners at each accessory and

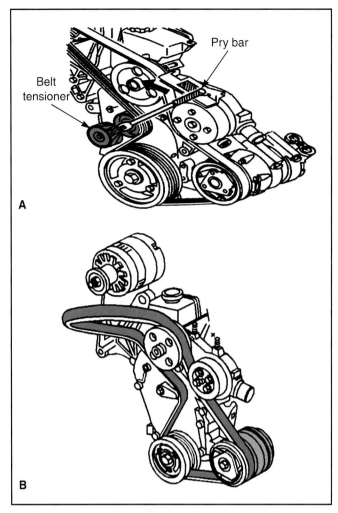

Figure 20-24. *A—Serpentine belts usually have an automatic tensioner. Use a wrench to push the tensioner away from the belt. B—The belt can then be slipped from around the pulleys. (General Motors)*

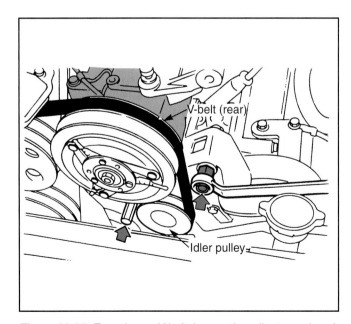

Figure 20-25. *To replace a V-belt, loosen the adjuster and push it toward the engine. The belt can then be removed. (Subaru)*

push it inward toward the engine. See **Figure 20-25.** Once enough slack is available, slide the belt over the pulley and remove it from the engine. To install the new belt, place it over the pulleys and pull the movable accessory outward to tension it. Check belt tightness with a tension gauge before fully tightening the bolts. Start the engine and check belt operation. If required by the belt manufacturer, readjust belt tension after about 10 minutes of operation.

Radiator Hose Replacement

Radiator hose replacement is similar to heater hose replacement. Begin by depressurizing and draining the cooling system below the hose fittings. Then loosen the hose clamps. Gently twist the hose in relation to the radiator or engine to break it loose. If the hose is stuck to the fitting, use a hose cutter or other knife to slice the hose at the fitting, **Figure 20-26.** Carefully peel the split hose away from the fitting.

Obtain the new hose and compare it with the original hose. Some hoses are made to fit several engines, and it may be necessary to cut a portion of the hose to allow it to fit without kinking. Before installing the new hose, slip the clamps over each end. It is usually advisable to use new clamps, especially if the original clamps were spring types, **Figure 20-26.** If desired, place a small amount of nonhardening sealer on the fitting (not inside the hose), then place the hose ends over the fitting nipples. Tighten the hose clamps and add coolant. Then start the engine and remove air from the system, and check for leaks.

Thermostat Replacement

To replace a thermostat, first locate the thermostat housing on the engine. Next, depressurize and drain the cooling system to a level below the thermostat housing. Remove the bolts holding the thermostat housing to the engine and remove the housing and thermostat.

Scrape the gasket material from the engine and thermostat housing. Compare the old and new thermostats to ensure the new unit is the correct replacement. Also make sure the new thermostat has the same temperature rating as the original. Using a thermostat that opens at a lower temperature will reduce heater output and may violate local emissions laws.

Place the new thermostat in position, making sure the temperature sensing element faces the engine. Some thermostats must be placed in a certain position to work properly, **Figure 20-27.** Install the thermostat housing using a new gasket. Install and tighten the attaching bolts. Add coolant and start the engine. Bleed the system, and ensure the thermostat opens as the engine warms up.

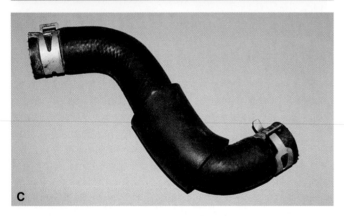

Figure 20-26. *A—Remove a radiator hose by first depressurizing the cooling system, then loosening and removing the clamps. B—Cut the hose from the engine and radiator, if needed. C—Make sure the clamps are placed on the new hose before installation.*

Testing Thermostats

To test a thermostat that has been removed from the engine, it must be suspended in a container of water as shown in **Figure 20-28.** To keep it from touching the sides of the container, the thermostat must be suspended from a wire. The container, must be placed over a burner or

other method of raising temperature. A temperature gauge capable of reading at least 250°F (121°C) should be placed in the water. Gradually raise the temperature of the water and observe the temperature at which the thermostat begins to open. Note whether this corresponds to the opening temperature stamped on the thermostat. After the thermostat begins to open, continue to observe it as the temperature continues to rise. The thermostat should be fully open at about 20-25°F (11-14°C) above the temperature at which it just starts to open. If the thermostat does not pass these tests, it should be replaced.

Cooling Fan Replacement

There are two kinds of cooling fans, the engine driven fan with a fan clutch and the electric motor driven fan. Replacement procedures for each are given in the following paragraphs.

Engine Driven Fan with Clutch

To replace an engine driven fan and fan clutch, disconnect the battery negative cable to prevent accidental engine starting. If needed, loosen and remove or relocate the fan shroud to allow clearance to remove the fan and clutch. Next, remove the bolts holding the fan and fan clutch to the drive pulley.

 Note: If possible, do not remove the belt(s). Leaving the belt(s) in place makes fan removal and installation easier.

Remove the fan and fan clutch. Remove the bolts holding the fan to the fan clutch and separate the two parts. See **Figure 20-29.** Reassemble with new parts as necessary, being sure the fan blades point in the proper direction. Then place the assembly on the pulley, and install the bolts. Start all bolts before tightening any bolt. Ensure the drive belts are tight, then reinstall the battery negative cable. Start the engine and check fan operation.

Electric Motor Driven Fan

To replace the cooling fan and motor, disconnect the battery negative cable. Then remove the electrical connector at the fan motor, **Figure 20-30A.** Remove the bolts holding the fan assembly to the radiator support, **Figure 20-30B.** Once the fan assembly is out of the vehicle, place the assembly on a bench and remove parts as necessary.

 Note: In some cases, you may need to replace both the fan and motor.

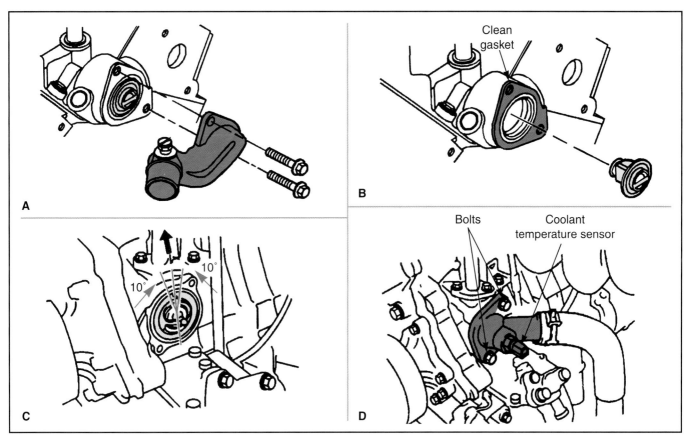

Figure 20-27. *A—To remove a thermostat, first locate the thermostat housing. Then remove the hose and the bolts holding the thermostat in position. B—Remove the old thermostat and clean the gasket surface. Some thermostats use O-rings. C—Some thermostats must be accurately positioned during installation. D—Install the housing, taking care not to damage any sensors or bleed valves close to the housing. (General Motors)*

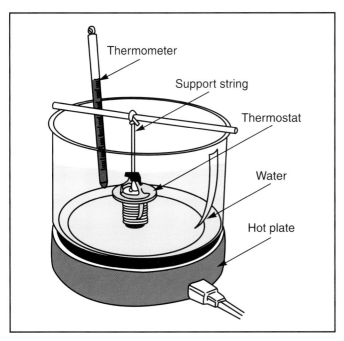

Figure 20-28. *Checking thermostat opening temperature. Keep the thermostat and thermometer from touching the sides of the container. (Honda)*

Install replacement parts and put the assembly into position on the radiator support. Install and tighten the bolts. Reattach the motor electrical connector and battery negative cable. Then start the engine and ensure the fan motor energizes at the proper temperature.

Coolant Pump Replacement

All coolant pumps operate in the same manner. The shape and place of coolant pumps, however, is different on every engine. The following general steps cover replacement of most coolant pumps.

Begin pump replacement by depressurizing and draining the cooling system. Then remove the fan and fan clutch if they are attached to the pump pulley. If necessary, remove the fan shroud to gain additional clearance. Then remove the drive belt(s) as necessary and remove the pump pulley. Remove any accessories and brackets as needed to access the pump.

 Note: On some engines where the pulley is bolted to the pump hub it is easier to loosen the pulley-to-hub bolts while the belt(s) still hold the pulley stationary.

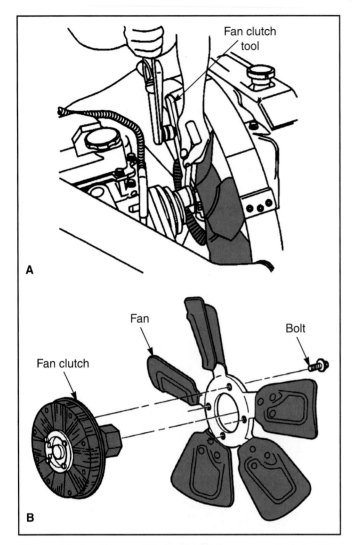

Figure 20-29. *Almost all belt driven fans are equipped with a fan clutch. A—Hold the fan as you remove the attaching bolts. B—This exploded view shows the relationship of the fan, fan clutch and drive pulley. Remove the fan and fan clutch from the pulley as an assembly, then separate the fan and fan clutch. (General Motors, DaimerChrysler)*

If any hoses are attached to the pump, remove them. If the pump uses bypass hoses, now is a good time to replace them. Remove the bolts and nuts holding the pump to the engine and remove the pump, **Figure 20-31.** Scrape all old gasket material from the pump mating surfaces on the engine. Inspect any baffles or channel plates in the pump cavity. Replace any plate that is bent or corroded. In some cases, you may need to transfer studs or fittings from the old pump to the new one.

Note: Compare the impellers on the old and new pumps. Both sets of impeller blades should be about the same size, have the same pitch direction, and number of blades. If the impellers do not match, do not use the pump.

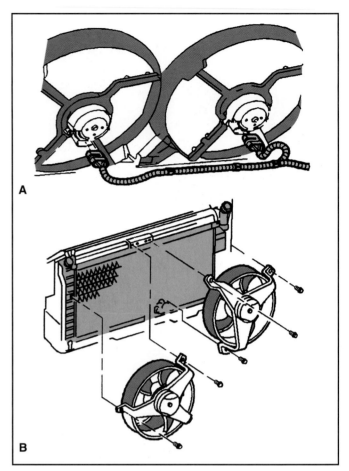

Figure 20-30. *A—Always remove the electrical connector before working on the electric fan. This prevents accidental fan operation. B—Most electric fans are installed in frames attached to the radiator housing or front radiator support. (General Motors)*

Place the gasket on the pump using a small amount of sealer or spray adhesive. Use only enough to keep the gasket in place. Then place the pump on the engine. Install and tighten the fasteners and install any hoses that were removed. Install the pulley and start the attaching bolts. Replace the belt(s) and install the fan assembly if necessary. Refill the cooling system and start the engine. Add more coolant, bleed the system, and check pump operation.

Radiator Replacement

Removing and replacing a radiator is usually simple, but must be done carefully to prevent damage to the radiator, condenser, and surrounding parts. Begin by depressurizing and draining the cooling system. Then remove the radiator hoses. Also remove the transmission and engine oil cooler lines when used. See **Figure 20-32.** Then remove the fan shroud or fan and motor assembly. Next, remove the bolts holding the radiator to the radiator support. On some vehicles, the radiator is held by rubber supports, and removing the upper radiator stiffener allows the radiator to be removed. Once all fasteners are removed, lift the radiator from the vehicle.

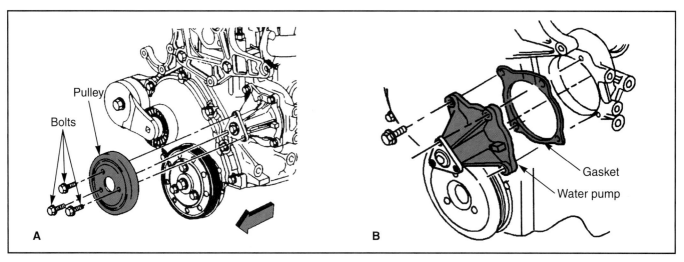

Figure 20-31. *A—Coolant pumps are commonly attached to the block or to an adapter on the front of the engine. Remove the bolts holding the pulley to the pump. B—Remove the bolts holding the pump to the engine and remove the pump. Scrape all of the gasket material from the sealing surface. (General Motors)*

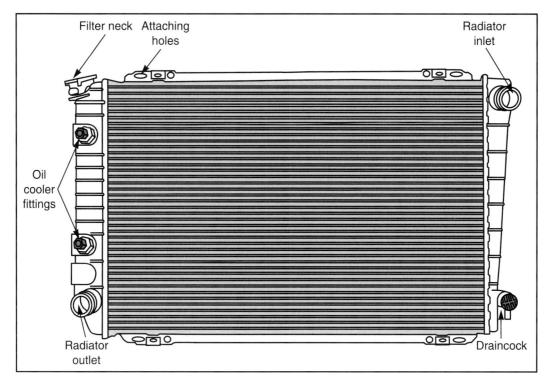

Figure 20-32. *This view of a radiator shows the connections that must be removed, radiator hoses oil cooler lines, and mounting bolts. Some radiators have electrical connectors to various radiator mounted sensors and switches. (Ford)*

 Caution: Proper radiator flow on newer cars is very critical. Make sure any replacement radiator comes from a reputable manufacturer.

Compare the old and new radiators, and transfer parts as necessary. **Figure 20-33** shows some typical electrical devices that may be attached to the radiator. Many replacement radiators use pipe plugs to seal the openings. These pipe plugs are left in any fitting not used so make sure they are tight before radiator installation. Then place the new radiator in position and install the fasteners. Install the fan shroud, hoses, and cooler lines. Then add coolant, start the engine and bleed the system. After all air has been removed, check the radiator and hose connections for leaks. Check and add transmission fluid, if necessary.

Air Cooling System Service

Although vehicles with air-cooled engines are rare, you may eventually encounter a vehicle so equipped. On the next page are repair operations that can be performed on an air-cooled engine.

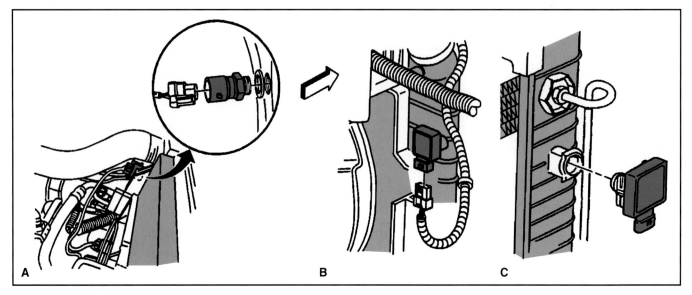

Figure 20-33. *Before replacing cooling system control switches always depressurize and drain the cooling system. A—Fan control switch. B—Disconnected low coolant level module. C—Removing low coolant module. (General Motors)*

Cleaning Fins On an Air-cooled Engine

The air-cooled engine uses fins to remove heat from the engine. These fins can become plugged with dirt, plant material, or other debris. To clean the fins, first allow the engine to cool completely. Then remove the engine shrouds as necessary and use compressed air to blow loose material away from the fins. If the fins are extremely dirty or oil soaked, spray a solution of detergent and water on the fins.

 Caution: The engine must be allowed to thoroughly cool before spraying any water or detergent on the fins.

Allow the detergent to sit on the fins for about 15 minutes, then spray the fins with a strong stream of water. Do not spray water on ignition or fuel system components. Once the fins are clean, reinstall the shrouds and start the engine to blow out any remaining water.

Replacing Blower Belt

To replace the blower belt, determine which of the belt driven accessories can be moved to remove the belt, **Figure 20-34.** Loosen the fasteners at that accessory and push it inward. Once the belt has enough slack, slide it over the pulley and remove it from the engine. To install the new belt, place it over the pulleys and pull the movable accessory outward to tension it. Check belt tightness with a tension gauge before fully tightening the bolts.

Replacing Blower and Bearing

To replace the blower and blower bearing, remove the belt as explained earlier. Then remove the portion of

Figure 20-34. *Location of the belt adjustment on an air-cooled engine.*

the shroud holding the bearing assembly. Remove the bearing and blower from the shroud. Some bearings must be pressed from the shroud assembly bracket. Install the new parts on the shroud and reinstall the shroud on the engine. Reinstall and tighten the belt and recheck blower operation.

Replacing Thermostat

To replace an air-cooled engine thermostat, locate the thermostat and remove the surrounding shrouds. Remove the link from the thermostat door and remove the fasteners holding it to the engine. Install the new thermostat and reattach it to the door. Reinstall shrouds as necessary. Start the engine and check the thermostat opens at the right temperature.

Summary

The heater and cooling system are interrelated and problems in one can cause problems in the other. Carelessness in servicing heater components can cause loss of coolant and engine damage. Before performing any cooling system or heater service, including removing the pressure cap, remove the system pressure. The best way to depressurize the cooling system is to allow the engine to cool off completely. Also remember coolant can be hot, and is poisonous. Before removing any cooling system part or fitting, drain the coolant to below the level of the part to be removed.

Coolant level should be checked at every oil change. If necessary, coolant should be added. Coolant condition and freezing point should also be checked. Remove the radiator and observe the coolant in the radiator. If the coolant is rusty, it should be changed and the system flushed. If oil is floating on the top of the coolant, engine oil or transmission fluid may be leaking into the cooling system. Also check the radiator and heater hose condition and drive belt tension and condition. To check for coolant leaks, start with a visual check for coolant on the engine or radiator, or streaks indicating leaking. Pressure test the system as necessary. Also use the pressure tester to check the pressure cap.

To change coolant, open the petcock and allow the system to drain into a pan. Do not remove the pressure cap until all pressure has been removed. After adding new coolant, bleed the cooling system depending on the location of the filler neck in relation to the engine. Once all air is removed from the cooling system, install the pressure cap.

The cooling system can be flushed with chemical cleaners or by reverse flushing. The fins on an air-cooled engine can be cleaned with common detergents. The engine should be allowed to cool completely before spraying the fins with water.

Heating system problems are usually caused by low coolant flow or reduced airflow. A clogged heater core, stuck shutoff valve, or cooling system problems are common causes of low coolant flow. The most common heating system problem is a leaking heater core. Missing shrouds or ducts, closed heater doors, or failure of the thermostat cause most heater problems on an air-cooled engine. If the blower drive belt is loose, air may not be forced into the system.

The most common causes of liquid-cooled engine overheating are a low coolant level, slipping fan belt, defective fan clutch or motor, defective pressure cap, clogged radiator, or stuck thermostat. Low coolant level can be checked at the radiator fill. The fan clutch should be checked when the engine is at operating temperature. The most common causes of air-cooled engine overheating are a slipping blower belt or engine shrouds out of position. The thermostat, if used, may have failed closed.

Coolant leaks can occur at several locations in the engine, including radiator heater hoses, the coolant pump, and the radiator. Internal leaks can occur due to defective head gaskets, or a cracked head or block. Pressure testing may be necessary to locate some leaks.

Heater service includes changing heater hoses and shutoff valves, and replacing the heater core. The cooling system should always be depressurized and drained before beginning any of these repairs. Removal procedures for heater cores involve removing ductwork and/or dashboard components. After removing the heater hoses, carefully blow air through the heater core to remove as much coolant as possible.

Typical liquid cooling systems service procedures include replacement of belts and hoses, radiator hoses, thermostats, cooling fans, coolant pumps, and radiators. Air cooling system service includes replacing blower belt or thermostat.

Review Questions—Chapter 20

Please do not write in this text. Write your answers on a separate sheet of paper.

1. The first step in cooling system service is to _____ the system.

2. A hydrometer is used to check the _____ of the coolant.

3. A pressure tester should be used to pressurize the cap to _____.
   (A) 5 psi (34 kPa)
   (B) 10 psi (70 kPa)
   (C) 15 psi (103 kPa)
   (D) Its rated pressure

4. If the coolant appears rusty, the system should be _____.

5. What is the purpose of the petcock?

6. If the heater cannot transfer its heat to the passenger compartment, which of the following is *not* a possible cause?
   (A) Clogged heater core.
   (B) Shutoff valve stuck closed.
   (C) Engine thermostat stuck closed.
   (D) Collapsed heater hose.

7. On some vehicles, the heater shutoff valve can cause overheating if it fails in the _____ position.

8. If a hose will not easily twist off of the fitting, the hose should be _____ and peeled away from the fitting.

9. Most _____ use a self-tensioning device.

10. Some replacement _____ must be cut to fit.

11. _____ should never be placed inside of the hose.

12. A low temperature thermostat may reduce _____.

13. The thermostat _____ should face the engine.

14. An engine-driven fan that turns several times after the engine is shut off has a defective _____.

15. Some air-cooled engines use a _____.

ASE Certification-Type Questions

1. Technician A says air-cooled engines are rare. Technician B says air-cooled engines need no maintenance. Who is right?
   (A) A only.
   (B) B only.
   (C) Both A and B.
   (D) Neither A nor B.

2. Radiator coolant levels are being discussed. Technician A says a radiator used with no coolant recovery system should be full to the bottom of the filler neck. Technician B says a radiator used with a coolant recovery system should be full to within 1-2 inches (30-60 mm) below the filler neck. Who is right?
   (A) A only.
   (B) B only.
   (C) Both A and B.
   (D) Neither A nor B.

3. All of the following coolant conditions require some service attention, *except:*
   (A) coolant rusty.
   (B) coolant green.
   (C) coolant contains oil.
   (D) coolant freezing point is 32°F (0°C).

4. A pressure tester can check all of the following, *except:*
   (A) radiator leak.
   (B) defective pressure cap.
   (C) water pump leak.
   (D) stuck shutoff valve.

5. To bleed a system with the filler neck below the engine, which of the following should be done *last?*
   (A) Add coolant as the level drops.
   (B) Open the bleed valve.
   (C) Close the bleed valve.
   (D) Install the pressure cap.

6. All of the following can be used to successfully clean the interior of a radiator, *except:*
   (A) cleaning with detergent.
   (B) cleaning with chemical cleaner.
   (C) reverse flushing.
   (D) removing the radiator for cleaning.

7. If you can blow through the replacement heater shut-off valve and not the original valve, the new valve is:
   (A) the correct replacement.
   (B) not the correct replacement.
   (C) an updated replacement.
   (D) cannot tell from this test.

8. All of the following statements about thermostat replacement are correct, *except:*
   (A) using a thermostat that opens at a lower temperature may violate local emissions laws.
   (B) the temperature sensing element should face the radiator.
   (C) some thermostats must be placed in a certain position to work properly.
   (D) add coolant after fully tightening the thermostat housing.

9. An electric fan and motor assembly is being replaced. Which of the following steps should be performed *last?*
   (A) Install battery negative cable.
   (B) Install motor electrical connector.
   (C) Install fan assembly on radiator.
   (D) Install fan on fan motor.

10. A coolant pump is being replaced. Which of the following steps should be performed *first?*
    (A) Remove any hoses attached to the pump.
    (B) Inspect any baffles in the pump cavity.
    (C) Remove the drive belts.
    (D) Remove the pump to engine bolts.

Chapter 21

Air Delivery and Manual HVAC Control Service

After studying this chapter, you will be able to:

❑ Diagnose blower problems.
❑ Diagnose diverter and blend door problems.
❑ Diagnose vacuum supply problems.
❑ Diagnose manual HVAC control panel problems.
❑ Remove and replace blower motors and related parts.
❑ Service air control doors and controls.
❑ Remove and replace control system components.

Technical Terms

Noise	Turnbuckle
Metal fatigue	Plastic connector
Adjusting slot	Clips

Problems with the air delivery and HVAC control system are not as common as problems in the refrigeration system. However, they occur often enough that you should know how to diagnose and repair them. Improper diagnosis can result in the needless replacement of expensive parts and, a dissatisfied customer. Carelessness when servicing these components can damage them. This chapter covers diagnosis and service of the air delivery and control system components. It will also prepare you for Chapter 22, which covers automatic temperature control systems.

Air Delivery and Manual Control System Problems

Most air delivery and temperature control problems are caused by very basic defects. Although the defects are usually simple, finding the exact defect is sometimes difficult, as many of the control system components are located under the dash. Always make sure the basic refrigeration and heating systems are working before deciding the problem is in the air delivery or control system. The following problems are listed by major system.

Blower Motors and Related Parts

Without the blower, cooled or heated air cannot be delivered to the passenger compartment. Common blower problems are shorted or open blower windings, a blown fuse, defective switch, noisy blower motor, or a defect in the blower resistor assembly or related relays.

No Blower at Any Speed

If the blower does not run at any speed setting, the blower motor may be defective, or a system fuse may be open. If the other HVAC system components are working, the fuses are probably good. After checking the fuses, use jumper wires to operate the blower directly. See **Figure 21-1.** Be sure to disconnect the blower lead to prevent damage to the control system.

 Caution: Operate the blower using jumper wires for short periods of time. Do not leave the jumper wires connected to the blower motor, as an electrical fire could result.

If the blower operates when the jumper wires are connected, the problem is in the control system or wiring. If the blower does not run, check for a proper ground. Since most cases are made of plastic, the blower motor has a separate ground. Many grounds are installed under one of the mounting screws. Check the blower motor ground for continuity with the battery negative cable. The resistance should be almost zero. If there is no continuity or the resistance is more than .25 ohms, repair the ground as necessary. If the ground is good and the blower still does not run, replace the blower.

 Caution: This test is for manual control systems only. Be sure to check the service manual before attempting any ohmmeter test on an automatic climate control system. More information on automatic control system diagnosis is in Chapter 22.

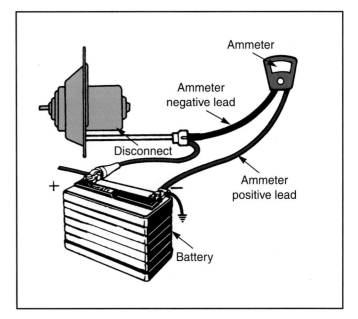

Figure 21-1. *A setup for bypassing the HVAC control system and operating the blower motor directly from the battery. This test can be performed with the blower installed in the vehicle. (Ford)*

Blower Inoperative at Some Speed Settings

On many HVAC systems, the highest blower speed is energized directly from the battery or alternator. If the blower operates on some speeds but not others, inspect the resistor assembly. If no obvious burned resistors are seen, check each resistor using an ohmmeter. Also check the thermal limiter (an internal resistor fuse), if used. **Figure 21-2** shows the location of the resistor assembly on a modern vehicle.

If the blower operates in high speed but no other, the problem is in the controls or resistor pack. If the blower runs in all speeds except high, the problem is in the high speed relay or the high speed fuse.

Most relays can be bypassed with a fused jumper wire. If bypassing the relay causes the blower to run at high speed, the relay is defective. Always consult the manufacturer's schematics before attempting to bypass the relay.

Blower Noises

Noise is the most common complaint regarding blower operation. A squealing noise from the blower is usually caused by dry motor bearings. Since the bearings cannot be serviced, the blower motor should be replaced. Rattling or beating noises from the blower may be caused by a loose blower wheel or debris caught in the blower wheel. To

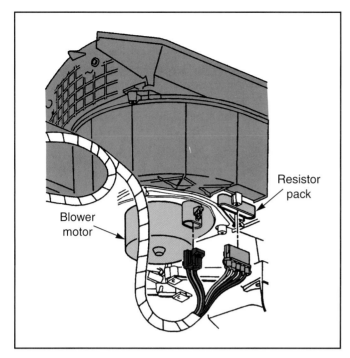

Figure 21-2. *Typical blower motor and resistor installation. On many older vehicles, the blower was mounted under the hood. However, tight engine compartment clearances on modern vehicles have forced manufacturers to mount the blower under the dashboard. (General Motors)*

check, remove the blower motor assembly as explained later in this chapter. Observe the wheel for tightness on the shaft and the presence of debris. Also check the inside of the blower cavity for leaves, dirt, or paper fragments. If the noise occurs after a blower is replaced, the shaft or wheel is incorrect or the wheel has been improperly installed.

Cabin Filter/Intake Air Filter

If the airflow is low at all blower speeds and the blower appears to be operating properly, the cabin filter may be restricted. Cabin filters should be changed every 12,000–15,000 miles (19,000–24,000 km), but they are often overlooked. Many technicians are not aware that some vehicles have cabin filters. Cabin filter replacement is discussed later in this chapter.

 Note: Low airflow accompanied by a musty smell in the passenger compartment is caused by mold growing on the evaporator fins. Removal of evaporator mold is covered in Chapter 19.

Ducts and Air Doors

Ductwork is seldom a cause of problems. Ducts can get out of position or be damaged by careless handling. The most common duct problem is a flexible duct that comes loose from the vent outlet. Other problems include misalignment of ducts, leading to air leaks, and plugging of drain holes, leading to water dripping onto the carpet. Occasionally, a previous technician may have worked on the system and carelessly mispositioned or forgot to reinstall a section of ductwork. You can usually spot problems by closely observing system operation during the functional and performance tests.

Uneven Airflow

Sometimes airflow from the vents is uneven. If airflow is normal at some times and too low at other times, the usual causes are a blower problem or evaporator icing. If airflow seems to vary in a random pattern, the blower motor may be wearing out, or a blower motor electrical connection may be bad. Sometimes a blower connection will become overheated or loose. This will increase the electrical resistance at the connection. A high or varying resistance connection will cause the blower speed to vary. Some overheated connections lose contact completely or cause the blower to stop when the vehicle hits a bump.

Another cause of erratic airflow is evaporator icing. If the airflow is ok when the air conditioner is first turned on, then gradually decreases as the system operates, the evaporator may be icing. To confirm icing, turn the system off and allow it to sit for about 10-15 minutes. If airflow is normal when the system is started up again, icing is probable.

Airflow often varies slightly between each outlet vent. This is due to different lengths, bends and sizes of the ducts, and is not a cause for concern. If the airflow from one or more vents is much lower than from the other vents, the vent may be disconnected or misaligned. Occasionally a vent will become plugged with debris. Carefully inspect under the dashboard for duct problems.

No, Slow, or Partial Change In Airflow

When this occurs, the air blows out of one set of outlets only, usually the heater ducts. This may be caused by a complete loss of vacuum or electric power, depending on the control system used. If the airflow changes slowly, low vacuum or a vacuum restriction is the usual cause. If the air control system works partially, an individual stepper motor may be defective or a vacuum diaphragm may have lost its vacuum source. On rare occasions, the air door linkage has broken or come loose from the diaphragm. Diagnosis procedures vary depending on the type of door control system.

 Note: Before deciding the vacuum system or stepper motors are defective, disconnect the door linkage and ensure the door itself is not stuck. If the vehicle has a cabin filter, make sure it is not plugged.

Vacuum Controlled System

If the system airflow does not change when different modes are selected, the problem may be a loose vacuum connection. Total lack of response indicates the system has lost vacuum at the source. Begin searching at the vacuum connection on the engine, **Figure 21-3.** An engine backfire or careless service procedures may have caused the hose to come loose from the intake manifold.

If the hose is attached to the manifold, trace it back toward the control panel. Look for loose connections and a collapsed or split hose. Closely examine the vacuum reservoir when used, and make sure the check valve is not clogged. Also check any vacuum restrictors. If you cannot blow through a vacuum restrictor, it is plugged and should be cleaned or replaced. If the hose problem cannot be found, start the engine and listen for a hissing noise as the engine idles. The hissing noise may be the source of the leak. Common leak points are cracked hoses and leaking diaphragms.

If vacuum does not seem to be reaching a component, use a vacuum gauge to confirm lack of vacuum. Trace the line back until vacuum is found. The clog or loose line should be located just past this point.

If the vacuum diaphragms are controlled by solenoids, their condition can be determined by checking the vacuum going in and out of the solenoids. A vacuum circuit schematic for each mode is sometimes provided in the service manual. If the solenoid does not transfer vacuum properly, and it is properly energized by the control system, it is defective.

Stepper Motor System

If the system uses stepper motors to operate the blend or diverter doors they can be checked with a voltmeter. If the voltages to the motor are correct, and the door is not jammed, the motor is defective.

Poor Heating or Cooling (Airflow Ok)

 Note: Check the refrigeration, heating, and engine cooling systems first.

Poor heating or cooling can be caused by a stuck blend door. The door itself may have jammed, although this problem is uncommon. The most likely cause is a defect in the actuating system. The following sections are grouped by actuating system: the manual cable and the electric motor. When used, vacuum diaphragm blend door actuators are part of an automatic temperature control system. Automatic temperature control systems are discussed in Chapter 22.

Cable Operated Blend Door

If a cable controls the blend door, heating problems often occur when the cable sticks, breaks, or goes out of adjustment. If the temperature lever moves very easily, the cable may be broken. If the cable lever does not move at all, the cable may be stuck. If the lever is hard to move in one spot only, the cable is kinked. Remove the dashboard components as necessary and observe cable operation. In the case of a stuck or hard lever, make sure the door is not stuck by disconnecting the cable and attempting to move the door lever.

Sometimes the cable has simply come loose from the door or lever. A slightly sticking cable can be lubricated. Most of the time, however, the cable will be damaged in some way. A damaged cable should be replaced.

If the temperature lever moves with a normal amount of resistance, check for a solid thumping noise when the lever is moved to the end of its travel in one direction, usually toward full heat. This noise indicates the door has fully closed. Some systems may make this noise in either direction. If moving the cable firmly to the end of its travel does not produce a solid thump, the cable may be out of adjustment. Refer to the adjustment procedures later in this chapter.

Electric Motor Operated Blend Door

A stepper motor operates many later model blend doors. To check the operation of a motor operated blend door, turn the ignition switch to the on position and turn off the radio. Place the HVAC system in Normal A/C mode and turn the blower switch to its lowest position. Then move the temperature lever and listen for motor operation. If the motor cannot be heard, the motor or the control panel switch is defective. Some stepper motors can be checked with a voltmeter.

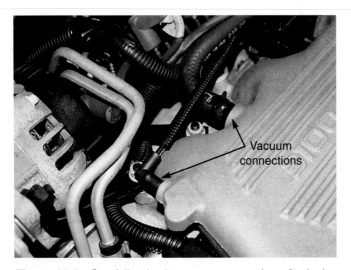

Figure 21-3. *Carefully check vacuum connections for leaks. Modern engines have many vacuum connections. In addition, intake plenum manifolds are often made in two or more pieces, increasing the chance of gasket leaks.*

No Defrost

Sometimes the system does not defrost the windshield. This complaint can take two forms: there is no airflow through the defroster ducts, or the air exiting the ducts does not clear the windshield. If air does not exit from the defroster ducts, check for disconnected ducts or an air door problem. If the airflow from the ducts is sufficient but does not remove moisture from the windshield, make sure the refrigeration system is operating and the blend door is not stuck. If the airflow is good and the windshield is not being defrosted, the customer will usually have other HVAC system complaints caused by refrigeration system or blend door troubles.

Odor from Air Ducts

The driver may complain of one of several kinds of odors from the HVAC system ducts. Coolant odors indicate the heater core is leaking and should be replaced. An engine or drivetrain oil leak or an exhaust gas leak is sometimes drawn into the air intake system. If the driver complains of this type of smell, check the hood-to-cowl seal and also make sure that there are no engine or drivetrain leaks. A refrigeration oil smell indicates a leaking evaporator, which must be replaced.

A musty odor is usually caused by the buildup of mold or other microorganisms on the evaporator. Evaporator cleaning was discussed in Chapter 19. Sometimes an afterblow module must be installed to keep the microorganisms from growing back. The afterblow module energizes the blower motor some time after the engine is turned off to dry out the evaporator.

Manual Control Panels

The control panel can be the source of several problems. Most common is breakage of levers or switches. Control panels use a variety of devices to operate the HVAC system, including vacuum switches, electrical switches, and cable levers. Many modern control panels are completely electrical or electronic, and can only be serviced by replacing the entire control unit. The panel may be lighted by bulbs or may contain light emitting diodes (LEDs). Before condemning the panel controls, make sure the controlled units are in working order. The following sections cover switch and cable problems and loss of illumination.

Vacuum Switches

A defective vacuum switch will fail to operate the vacuum devices, and may make a hissing noise. Older rotary and sliding type vacuum switches sometimes become stuck. Often they can be disassembled and cleaned. If cleaning does not restore normal operation, the switch should be replaced. Vacuum switches installed inside of another control assembly should be replaced when they develop problems. Other problems that should

be considered are a leaking vacuum reserve tank, inoperative vacuum assist pump, or low vacuum due to a problem with the engine itself.

Electrical Switches

Electrical switches usually work or do not work. Therefore, if operating the switch does not cause the controlled unit to operate, the switch is probably bad. Before condemning the switch, make sure the switch has power and the unit being operated is not defective. Bypass the switch using jumper wires to confirm it is defective. If the unit operates when the switch is bypassed, the switch is defective.

When only one function of the switch does not work, remember to check the operated unit, and in the case of blower speeds, the resistor assembly, before condemning the switch. If wiggling the switch makes the controlled unit operate, the switch contacts are worn or dirty. The switch can sometimes be cleaned, but may fail again. The best way to avoid the chance of a repeat failure is to install a new switch. Some modern control panels can only be replaced as a unit.

Cable Levers

The most common cable lever problem is **metal fatigue** of the panel lever, causing it to bend. The lever may have come loose at the pivot point. To check these problems, the panel must be removed from the dashboard, **Figure 21-4.** If a bent lever is found, make sure the cable is not sticking. Often a binding cable can cause the lever to bend.

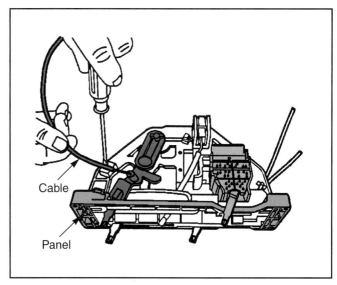

Figure 21-4. *To replace a cable, remove the fasteners and remove the cable. The cable fastener on one end may also be the cable adjustment device. (General Motors)*

Illumination Devices

Most panel illumination problems consist of burned out bulbs, blown fuses, or loss of control panel grounds. When the panel is only partially illuminated, the problem is almost always a burned out bulb. If the bulbs are difficult to reach, check the system fuse first. Then check for burned out bulbs and ground problems. **Figure 21-5** shows typical bulb placement.

If the panel uses LEDs, a loss of part of the panel display usually means the panel has failed internally and must be replaced. If no LEDs are illuminated, check the panel fuse and grounds before deciding the panel is defective.

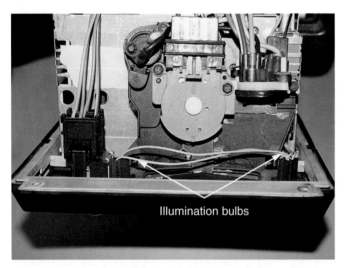

Illumination bulbs

Figure 21-5. *Location of the illumination bulbs on a manual control head. Most modern HVAC control panels have many electrical connectors. Most of them are held in place by tabs or other holding devices. Do not break these holding devices to remove the connector.*

Air Delivery and Control System Service

Once the problem has been found, most air delivery and control service is fairly easy. Most defective components are replaced instead of being repaired. Most components are relatively easy to reach. A few components can only be serviced after the dashboard and/or HVAC case components are removed.

⚠️ **Warning: Check the service manual for precautions when servicing vehicles equipped with air bags. Improper procedures may damage the air bag system or cause accidental deployment.**

Blend Door Cable Adjustments

All modern blend door cables are adjusted in one of two ways. Most cable mountings have an ***adjusting slot*** at one end, **Figure 21-6.** To adjust this type of cable, move the temperature lever to the full heat or full cold position. Then loosen the adjusting slot mounting bolt and push the blend door completely closed. Moving the blend door will cause the cable to slide forward or backward to change the cable length. Once the adjustment is made, tighten the slot mounting bolt and recheck blend door operation. The blend door should make a solid thump when the temperature lever is moved to the end of its travel.

Some older cables are adjusted by a ***turnbuckle,*** **Figure 21-7.** Turning the turnbuckle changes the cable length. To adjust this type of cable, move the temperature lever to the full heat or full cold position. Then turn the turnbuckle until a slight resistance is felt. This indicates the blend door is completely closed. Recheck blend door operation; you should hear a solid thump when the temperature lever is moved to the end of its travel. If it does not in either design, the door may be binding inside the blower case. You will have to remove and/or disassemble the blower case to correct the problem.

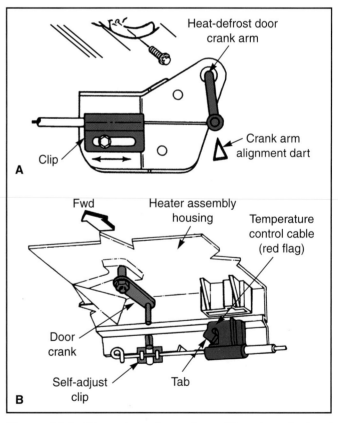

Figure 21-6. *These two views show different methods of adjusting a cable. A—The cable mounting bracket is moved to adjust cable length. B—The cable-to-door arm connection is moved to adjust the cable. (Ford, DaimlerChrysler)*

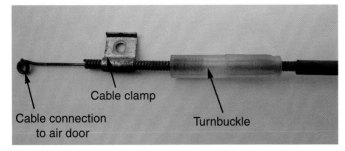

Figure 21-7. *All cable adjustment turnbuckles resemble the one shown here. The turnbuckle is rotated to change the length of the control cable.*

Removal and Replacement of Components

Unlike some other HVAC system components, air delivery and manual control components do not require removal of refrigerant or cooling system parts. The only exception to this is when some cases are removed. The placement of the heater core or evaporator may make refrigerant recovery or cooling system draining necessary. Replacement procedures are explained in the following paragraphs.

Cabin Filter

Manufacturers recommend that cabin filters be changed every 12,000–15,000 miles (19,000–24,000 km) or whenever the system airflow becomes sluggish. The cabin filter is usually easy to change, but information on its location may be difficult to obtain. Sometimes the location is noted in the owner's manual. Many parts catalogs list the filter's location.

If a cabin filter is located under the glove compartment, remove the kick panel covering the filter housing, loosen the housing, and remove the filter. Install a new filter in the reverse order of removal. Some vehicles have two filters installed in the housing. Both filters should be changed at the same time.

If the cabin filter is located behind the glove compartment, open the glove compartment and filter access doors. Two filters will be installed in the housing. Remove the top filter by pulling it forward. Pull the second filter upward into the space occupied by the first filter and then pull the filter out of the housing. Slide the first new filter into the place occupied by the top filter and then push it downward into place. Install the second new filter in the top position. Finish by closing the access door and the glove compartment door.

To replace a cabin air filter installed in the cowl, place the windshield wipers in the straight-up position. Then, open the hood and remove the right-side cowl cover. This will expose the cabin air filter. **Figure 21-8** shows the location of the cabin air filter on a popular vehicle. To replace the filter, pull on the removal tab to lift the filter from the recess. Install the new filter and make sure it is fully seated in the recess. Reverse the removal process to reinstall the cowl cover and then park the windshield wipers.

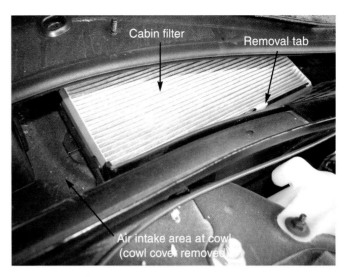

Figure 21-8. *Some cabin air filters are located under the cowl, between the windshield and the engine firewall. To remove this filter, remove the cowl cover fasteners and cover, and then pull the filter from the recess.*

Blower Motor

Blower motors are replaced rather than repaired. To begin removal, first locate the blower motor. Many blower motors are located under the hood where they are easily accessible, **Figure 21-9.** Others are located under side panels, between the firewall and inner fender, or under the dashboard. In almost all cases, the blower can be removed without removing the HVAC case.

 Note: The blower motor may be located in a difficult-to-reach spot on some vehicles. Check the service manual and any bulletins for the proper removal procedures.

Figure 21-9. *Some blowers are installed under the hood and are easy to reach. Others are under the dashboard and interior parts may have to be removed.*

Once the blower is located, make sure the ignition key is in the *off* position, then remove any parts necessary to gain access to the blower. Then remove the electrical connector, mounting screws, and then the blower from the case. **Figure 21-10** shows a blower motor being removed from a case.

Once the blower is removed, the blower wheel usually must be removed from the old motor and installed on the new motor, **Figure 21-11**. Some factory and aftermarket blower assemblies are provided with a new wheel already installed. Blower wheels commonly are held to the motor by a threaded nut or push-on clip.

 Note: Some blower wheel nuts have left-hand threads.

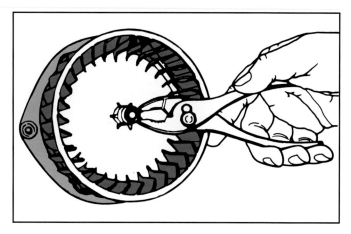

Figure 21-11. *Blower wheels are held to the motor shaft by a clip, as shown here or by a nut that tightens on the threaded blower shaft. (DaimlerChrysler)*

To remove the wheel, remove the nut or clip and pull the wheel from the shaft. If the wheel is hard to remove, hold the wheel and lightly tap the motor shaft with a rubber hammer. Install the wheel on the new motor and install the nut or clip; the new motor should come with a new nut or clip. Be sure to install any washers used with the original wheel.

To install the replacement blower, place it in position and install the mounting screws, making sure any ground wires are installed. Install the power wire and check blower operation. Then reinstall any other vehicle parts removed.

Blower Resistor Assembly

Most resistor assemblies are located in the case or ductwork under the hood. A few assemblies are installed under the vehicle dashboard. To remove the resistor assembly, unplug the electrical connector and remove the screws holding the assembly to the duct. Then remove the assembly from the duct, **Figure 21-12**. Clear any debris that might touch the resistor assembly. Place the new resistor in position and tighten the screws. Reinstall the electrical connector and check blower operation.

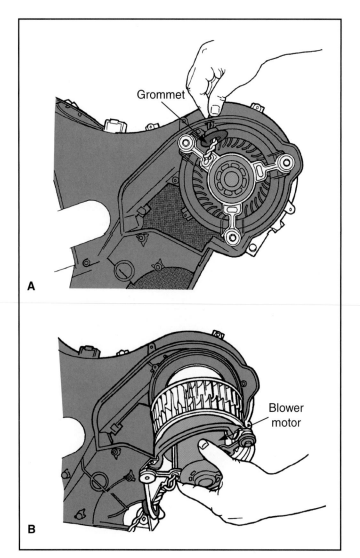

Figure 21-10. *Most blower motor and wheel installations resemble this setup. A—Remove the electrical connector. B—Remove the mounting bolts and pull the blower motor and squirrel cage free. (DaimlerChrysler)*

 Note: Like some blower motors, the resistor assembly is sometimes located in a hard-to-reach area. Check the service manual for proper removal procedure.

Blower Relays

Older blower relays are installed on the vehicle firewall or the underhood portion of the case, **Figure 21-13**. On late model vehicles, the relay is installed in one of the fuse and relay boxes, **Figure 21-14**. To replace a blower relay, first unplug the electrical connector. Then remove the screws holding the assembly. Install the new relay and reconnect the electrical connector. After replacing the relay, recheck HVAC system operation.

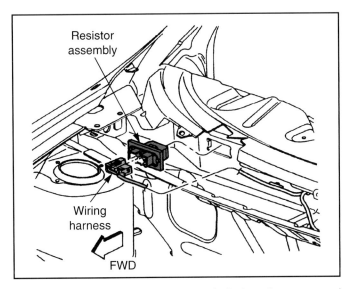

Figure 21-12. *Blower resistors are built into the case and cooled by airflow. They can usually be removed by unplugging the electrical connector and removing the mounting screws. (DaimlerChrysler)*

Vacuum Diaphragms, Solenoids, and Lines

Most vacuum operated units can be serviced without removing the HVAC case. However, most vacuum devices are located under the dashboard, requiring kick panel, ductwork, ashtray, glove compartment, and in some cases, center console removal to gain access.

Repairing Vacuum Lines

Sometimes cracked or leaking vacuum lines can be repaired instead of replacing the entire harness. Cut out the defective section of plastic line and splice in a section of rubber hose over the plastic ends, **Figure 21-15.** To repair a rubber hose, cut the hose and install a ***plastic connector***, **Figure 21-16.**

> **Caution: Exercise care when cutting the plastic hose as it kinks very easily.**

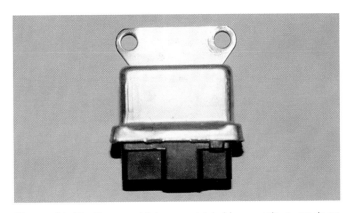

Figure 21-13. *Some vehicles use high blower relays, such as the one shown here. This type of relay is usually mounted on the vehicle's firewall.*

Figure 21-14. *Blower and other control relays are often installed in a combination fuse and relay compartment called an electrical center. Many vehicles have more than one electrical center.*

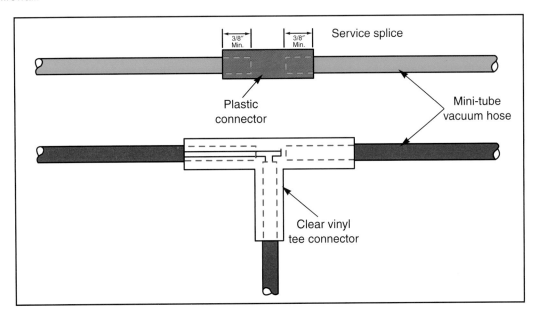

Figure 21-15. *Vacuum hoses can often be repaired instead of being replaced. This repair involves installing a larger hose over the damaged hose section. (Ford)*

Figure 21-16. *Other vacuum hoses can be fixed by the use of plastic connectors. Cut out the leaking or collapsed hose section and install the connector. (Gates)*

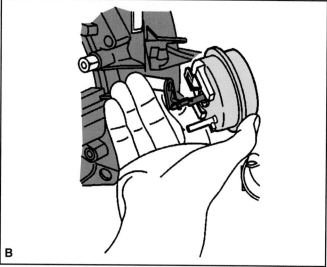

Figure 21-17. *A—One or two screws usually hold vacuum diaphragms to the case. B—Remove the vacuum hose first, then remove the screws and the clip holding the door arm to remove the diaphragm. (General Motors)*

Vacuum Diaphragms

Many vacuum diaphragms are located under the dashboard and various components may have to be removed to gain access. Diaphragms installed at the side (kick) panels may be removed after the panel trim is removed. Once the vacuum diaphragm has been located, remove the vacuum line and the air door to diaphragm linkage, **Figure 21-17.** Then remove the fasteners holding the diaphragm to the case and remove it. Place the new diaphragm in position and attach the air door linkage. Install the fasteners and reattach the vacuum line. Start the engine and check diaphragm operation, then reinstall other dashboard parts. Some diaphragms are located under the hood and are relatively easy to reach.

Vacuum Control Solenoids

Solenoids or solenoid assemblies are usually mounted under the dashboard. To remove a solenoid, remove dash and interior components as needed. Then disconnect the electrical and vacuum connectors. Remove the solenoid attaching fasteners and remove the solenoid from under the

dashboard, **Figure 21-18.** Install the new solenoid and tighten the fasteners, then reinstall the electrical and vacuum connectors. Start the engine and check solenoid and general HVAC system operation. Then reinstall any dashboard components removed.

Stepper Motors

Most stepper motors can be removed by locating them under the dashboard, then removing the electrical connector and the motor fasteners. **Figure 21-19** shows a typical motor removal procedure.

Vents, Ducts, Cases, and Air Doors

Since there are so many case and duct designs, replacement procedures will vary greatly. As a general rule, vents and ducts can usually be removed without taking the HVAC blower case apart. Servicing cases and blend doors usually requires major disassembly.

 Note: Be sure to follow all precautions regarding air bags when servicing ducts, cases, or air doors.

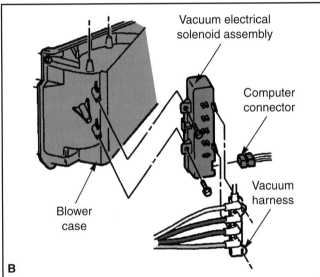

Figure 21-18. *A—Vacuum control solenoid packs are mounted under the dashboard. B—Remove the vacuum connector, electrical connector, and mounting screws to remove the solenoid pack. (General Motors)*

Vents

Clips are built into the dashboard to hold some types of vents. To remove these vents, insert a flat screwdriver into the space between the vent and dashboard, **Figure 21-20.** Inserting the screwdriver will release the vent, and it can be pulled forward to clear the clip. Repeat the process on the other side of the vent and withdraw the vent from the dashboard. Install the new vent by pressing it into place. The clips will engage the vent and hold it in place.

Other vents are attached with screws, **Figure 21-21.** Remove the screws and pull the vent from the dashboard. If the vent is connected to a flexible duct, pull the duct away from the vent. To install the vent, attach the flexible duct (if used) place the vent in position, and reinstall the screws.

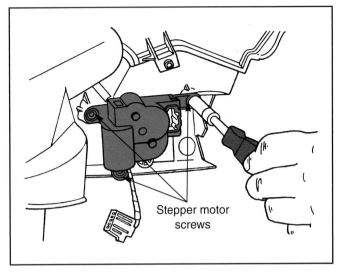

Figure 21-19. *Electrical door actuator motors are installed on the case, usually under the dashboard. Removal is accomplished by removing the electrical connector and mounting screws. Be sure to line up the motor and door drives before tightening the mounting screws. (DaimlerChrysler)*

Ducts

Clips, bolts, or sheet metal screws hold ducts to the case and dashboard. A few ducts are bolted to the case or under the dashboard. Rear air conditioning ducts are bolted to the floorpan or to the sides of the vehicle, depending on the design of the system. Some ducts are slip fit into other ductwork and have no separate fasteners. To remove a duct, remove any fasteners and pull the duct from under the dashboard. See **Figure 21-22.** To reinstall, carefully place the duct in position and replace the fasteners.

 Caution: Be sure not to pinch any wiring when removing or installing ducts.

Cases and Air Doors

Removing the case is usually a major job. Always consult the service manual if you are not familiar with the removal process. Remember from Chapters 12 and 19 there are two kinds of cases, the split case and the module. Both types of cases contain the heater core and evaporator. Most air doors can only be removed after the HVAC case is removed and partially or completely disassembled. Some blend doors can be replaced without changing the entire case. Other blend doors are not serviced separately, and the case must be replaced if a door is defective. Air doors mounted on the kick panels may be removed after the kick panel trim is removed.

To remove the case, the refrigerant may need to be reclaimed and/or the coolant may need to be drained. To remove the case, perform any refrigeration and cooling system depressurization and removal as needed. Then remove the heater hoses and refrigeration lines if necessary. Cap the refrigeration lines.

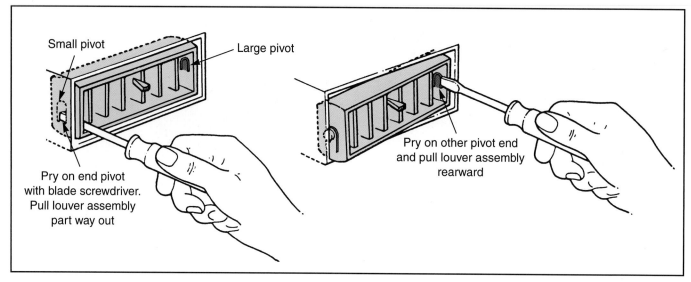

Figure 21-20. *Most dashboard vents are held in place with clips that can be depressed to remove the vent. (Ford)*

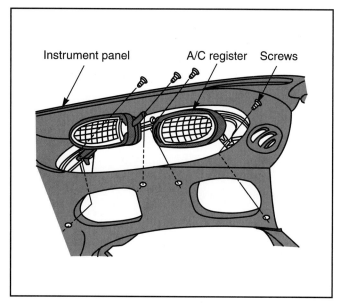

Figure 21-21. *Some vents are held in place by screws. Remove the screws to remove the vent. (Ford)*

Remove any ducts or other under dash components that might prevent case removal. If necessary, disarm the air bag system. Also check that all underhood components are clear of the case and fasteners. Then remove the fasteners holding the case to the firewall. See if the case is held by one or more studs pressed into the firewall. The studs help with alignment when the case is reinstalled. The case will be secured to the studs with nuts. If the case is a split type, it is not necessary to remove both halves of the case. Simply remove the side necessary for service. If the case is a modular type, remove the entire assembly.

Place the case on a clean bench and transfer parts to the new case as necessary. Service the air doors as needed before installing the vacuum diaphragms, **Figure 21-23.** Then place the new or repaired case in position over the studs. One or more alignment pins may be installed in the firewall to help with realignment. Once the case is in

position, install and tighten the fasteners. Reinstall all other HVAC and dashboard components. Install and rearm the air bag system. If the evaporator was removed, use new O-rings to reinstall the lines. If necessary, recharge the refrigeration system and refill the cooling system. Then check HVAC system operation.

Control Panels

HVAC control panels are attached to dashboards by many different methods. The technician should always consult the service manual before attempting to remove the control panel:

❑ Some control panels are attached to the front of the dashboard with screws. Remove the screws to remove the panel, **Figure 21-24.**

❑ Some dashboards are designed so trim pieces must be removed before the control panel can be removed, **Figure 21-25.**

❑ Some control panels require special tools for removal. Check the service manual.

❑ Some control panels can only be removed from the rear of the dashboard. Ashtrays, ducts, radios, and other components may require removal to gain access to the control panel fasteners.

❑ Most control panels have at least one support brace inside of the dashboard. Ducts and other HVAC components may need to be removed to gain access to the brace fasteners.

❑ To prevent damage, the electrical connectors, vacuum lines, and cables should be removed from the rear of the control panel as soon as they can be reached.

To begin panel removal, disconnect the battery negative cable. Placing the ignition switch in the *off* position may not be sufficient, since some dashboard components are powered at all times. Then remove the control knobs as necessary. Most knobs are pressed onto the lever or shaft. Clips or small Allen head set screws hold some knobs. Once the knobs are removed, remove the panel attaching fasteners by following service manual

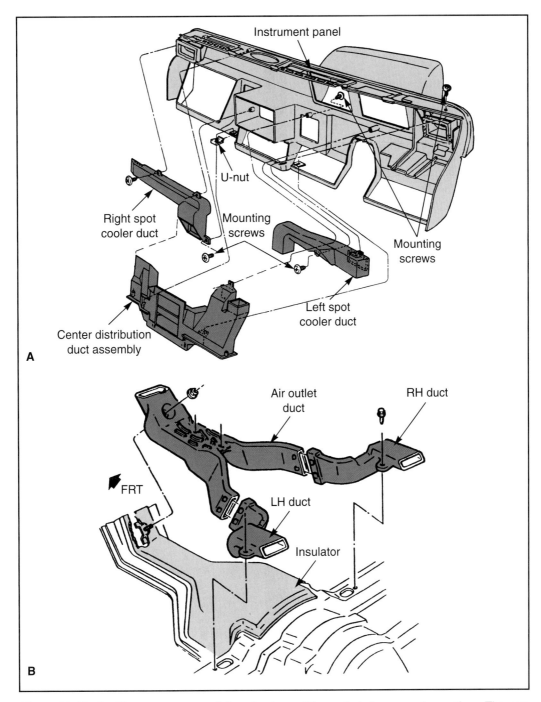

Figure 21-22. *A—There are many variations in air conditioner duct shapes and mountings. They are usually mounted to the dash with screws or bolts. B—Vent assembly for a vehicle with rear air conditioning. Always consult the service manual before beginning duct removal. (DaimlerChrysler, General Motors)*

instructions. As soon as they can be reached, remove the electrical and vacuum connectors. If used, remove the blend door cable. It is sometimes easier to detach the cable at the blend door and remove it with the control panel.

Control Panel Components

The illumination bulbs, when used, can usually be replaced from under the dashboard. Some older electro-mechanical panels have other components that can be replaced. The control panel is usually removed from the dashboard to service individual components. A few dashboards are arranged so switches and valves can be reached without removing the entire assembly.

To reinstall the control panel, put in position, and reconnect electrical and vacuum connectors. Install but do not tighten the panel fasteners. Install the blend door cable, if necessary. Then tighten the fasteners and reconnect the battery negative cable. Install the control knobs, start the engine and check HVAC system operation. If the HVAC system checks out okay, reinstall other dashboard components that were removed.

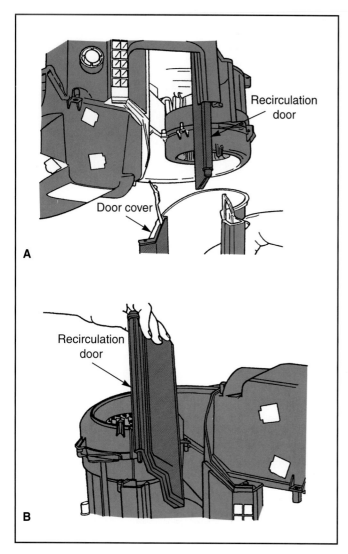

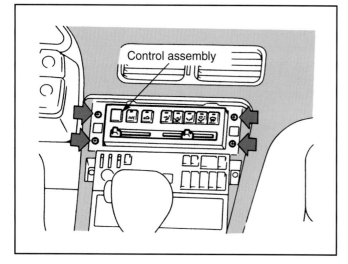

Figure 21-23. A—Disassembling a module case to replace doors. B—Removing the recirculation door, be sure all doors are in the correct position before resealing the case. (DaimlerChrysler)

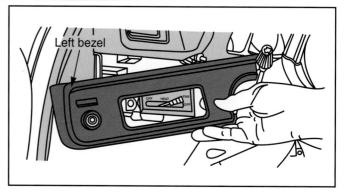

Figure 21-25. Some HVAC control panels can be removed after a trim piece is removed. (DaimlerChrysler)

Checking Afterblow Modules

An afterblow module is used to turn on the blower motor some time after the vehicle has been parked. The blower will run at high speed for about 5 minutes to dry the evaporator core. On some vehicles, the afterblow module turns on the blower as soon as the engine is turned off. Other modules cause the blower to come on as much as an hour after the vehicle is stopped. This allows as much moisture as possible to drip from the evaporator before it is dried by airflow.

An add-on afterblow module consists of a microprocessor timer connected in series with the blower motor. Some manufacturers have procedures for testing the afterblow module by grounding a test pin on the module harness. Carefully follow manufacturer's directions to make this test. To test the afterblow module on other vehicles, you must operate the air conditioner for several minutes, then stop the engine and wait until the proper time has elapsed and note whether the blower is energized. To avoid complaints, be sure that the vehicle driver knows how the afterblow module operates before releasing the vehicle. Afterblow module installation and service was discussed in Chapter 19.

(Control assembly figure)

Figure 21-24. HVAC control panel mounting screws may be visible from the front seat. Remove the screws to remove the panel. (Subaru)

Summary

Problems with the air delivery and HVAC control system are not as common as problems in the refrigeration system. They can also cause loss of heating and cooling as well as other complaints.

Common blower problems are shorted or open blower windings, a blown fuse, defective switch, or a defect in the blower resistor assembly or related relays. The blower can be checked with jumper wires. You should also check for a proper ground. If the blower operates in some speeds but not others, the problem is in the controls or resistors. Blower noises are caused by defective bearings or something caught in the blower wheel.

If the air control system works partially, the problem may be in a vacuum line, vacuum diaphragm, control device or a stepper motor when used. Diagnosis procedures vary depending on the type of door control system. A cable operates most blend doors. The cable can stick, break, or go out of adjustment. A damaged cable should be replaced. Electric motor operated blend doors can suffer from motor problems. The motor is usually serviced as a unit.

Manual control panels use vacuum switches, electrical switches, and cable levers to operate the HVAC system. Vacuum switches can leak or stick. Switches can fail to deliver current. Cable levers can bend or come loose. Burned out bulbs, blown fuses, or loss of control panel grounds can cause panel illumination problems.

Many air delivery and control system components can be reached from under the hood. Others can only be serviced after the dashboard and/or HVAC case components are removed. Some blend door cables are adjustable. Blower motors are replaced instead of repaired. Most blowers can be replaced without removing the HVAC case. The blower resistor assembly and blower relay can be changed easily.

Most vacuum operated units can be serviced without removing the HVAC case. Leaking or split vacuum lines can often be repaired. Vacuum diaphragms may be removed after any interfering parts have been removed. To remove a vacuum control solenoid, remove any needed dashboard components, then remove the solenoid.

Vents and ducts are usually easy to service. Cases and blend doors usually require major disassembly to remove them. Often the heater hoses and refrigeration lines must be removed to remove a case.

HVAC control panels are attached to dashboards by many different methods. Some are held by screws, some can only be removed after trim pieces are removed, and some control panels can only be removed from the rear of the dashboard. To remove the control panel, follow service manual instructions. Some control panel components can be serviced separately. After the panel has been replaced, recheck HVAC system operation. Some vehicles require the installation of an afterblow module to prevent evaporator mold formation.

Review Questions—Chapter 21

Please do not write in this text. Write your answers on a separate sheet of paper.

1. A blower motor mounted in a plastic case must have a separate _____ wire.

2. Blower resistors can be checked with an _____.

3. What is the most common duct problem?

4. A hissing noise from the control panel with the engine running indicates a _____.

5. Vacuum solenoid condition can be determined by checking _____.

6. An electrical switch can be bypassed with _____.

7. The two kinds of cable adjusting devices are the _____ and the _____.

8. What must be removed from the old blower motor and installed on the new motor?

9. Blower relays are almost always installed on the vehicle _____ or the underhood portion of the _____.

10. The first step in HVAC panel removal is to disconnect the _____.

ASE Certification-Type Questions

1. A blower motor does not operate in second speed. Technician A says the blower motor may be defective. Technician B says the resistor assembly may be defective. Who is right?
   (A) A only.
   (B) B only.
   (C) Both A and B.
   (D) Neither A nor B.

2. All of the following can cause partial operation of the air doors, except:
   (A) vacuum hose off at the intake manifold.
   (B) leaking rotary vacuum valve.
   (C) leaking vacuum diaphragm.
   (D) clogged vacuum restrictor.

3. All of the following are possible causes of blower noise, except:
   (A) incorrect blower wheel.
   (B) incorrect blower windings.
   (C) incorrect blower shaft.
   (D) debris in wheel.

4. The usual cause of poor heating or cooling problems is a stuck:
   (A) diverter door.
   (B) heater shutoff valve.
   (C) blend door.
   (D) blower wheel.

5. All of the following statements about diagnosing electrical switches are true, except:
   (A) electrical switches usually work or do not work.
   (B) always make sure the switch has power.
   (C) always make sure the unit being operated is not defective.
   (D) if the unit operates when the switch is bypassed, the switch is good.

6. Technician A says loss of some LEDs on the control panel means the panel has failed internally. Technician B says LEDs can be replaced individually. Who is right?
   (A) A only.
   (B) B only.
   (C) Both A and B.
   (D) Neither A nor B.

7. What is the most common way of attaching a blower wheel to the motor shaft?
   (A) Clip.
   (B) Nut.
   (C) Weld.
   (D) Press fit.

8. Which of the following vacuum system components is *most likely* to be repaired rather than replaced?
   (A) Line.
   (B) Diaphragm.
   (C) Solenoid.
   (D) Switch.

9. All of the following may need to be removed to get at the under dash HVAC components, *except:*
   (A) glove compartment
   (B) ashtray.
   (C) backseat.
   (D) radio.

10. Which of the following control panel components is *most likely* to be replaced instead of replacing the entire assembly?
    (A) Vacuum switch.
    (B) Blower switch.
    (C) Blend door lever.
    (D) Illumination bulb.

Chapter 22

Automatic Temperature Control System Service

After studying this chapter, you will be able to:
- ❏ Diagnose electronic temperature control system problems.
- ❏ Retrieve electronic temperature control system trouble codes.
- ❏ Diagnose mechanical temperature control system problems.
- ❏ Diagnose electromechanical temperature control system problems.
- ❏ Remove and replace control system components.

Technical Terms

Trouble code	Diagnostic routines	Test function
Self-diagnosis mode	Dedicated scan tools	Snapshot
Scan tool	Generic scan tools	Freeze frame
Data link connector (DLC)	Bi-directional	Reprogram
Keypad	OBD II diagnostic system	Recalibrating
Display screen	The Society of Automotive Engineers (SAE)	Trouble code
Menu	Power-on self-test (POST)	

Problems in an automatic temperature control system may seem almost impossible to correct at first. However, like any HVAC problem, they can be diagnosed and corrected if you use a combination of knowledge and logic. This chapter covers diagnosis and service of automatic temperature control systems. Modern systems are covered as well as older mechanical and electromechanical systems. Studying this chapter will add to the knowledge of HVAC control systems you have begun by studying the information in Chapter 14.

Automatic Temperature Control System Problems

The most common complaint about automatic temperature control systems is failure to react when the temperature changes. The temperature control may remain on one setting, the system may over or undercool the vehicle's interior, or may fail to switch modes when needed. Another common defect is failure to react to driver input, such as not changing fan speeds when requested. Since many small differences exist between different systems, always consult the proper service manual when troubleshooting any automatic temperature control.

When troubleshooting an automatic temperature control system, keep the seven step troubleshooting process in mind. Review the seven-step process in Chapter 15 if necessary. In addition, you should perform the following before troubleshooting the automatic control system.

❑ Make sure the basic refrigeration and heating components, blower, and air delivery systems are working.
❑ Retrieve trouble codes when applicable.
❑ Check service literature for common system problems and updates.

In the following sections, automatic temperature control system problems are grouped by type of system. Electronic systems are covered first, followed by older mechanical and electromechanical systems.

⚠️ **Warning: Check the service manual for precautions when servicing vehicles equipped with air bags. Improper procedures may damage the air bag system or cause accidental deployment.**

Electronic Temperature Control Systems

Common problems in the control system are caused by defects in the input sensors or the temperature control computer. Output devices covered in earlier chapters, such as the compressor clutch, blend door, and blower motor, operate in the same manner as on manual HVAC systems.

When an electronic control system fails to operate, the most common cause of trouble is a control head or temperature sensor defect. Since modern systems use three or four sensors, it may be difficult to determine which sensor is defective. For sensor diagnosis, it will be necessary to retrieve trouble codes and take ohm and volt readings at the sensor.

If the temperature control computer is defective, it can cause erratic or no system operation. Often the control panel will display erratic readings and/or fail to respond to driver input. If the basic refrigeration, heating, and air

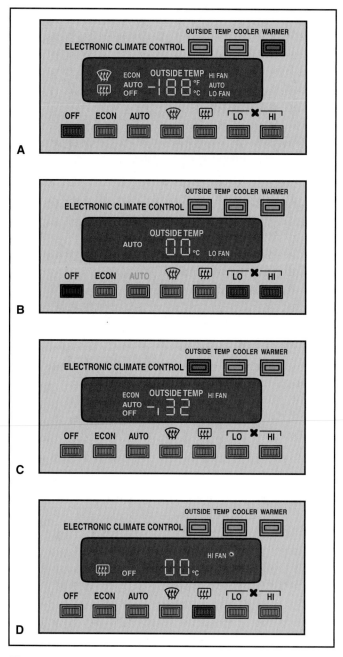

Figure 22-1. On some older vehicles pressing a certain series of buttons on the HVAC control panel accesses the automatic temperature control computer. The computer will display trouble codes and other information on the panel display. A—Entering self-diagnostics mode. B—Requesting change in parameters. C—Parameter data. D—Clearing codes. (General Motors)

delivery components are working properly, check the electrical and vacuum supply to the control module. If these check out ok, begin diagnosis of the control module. Often a defective temperature control computer will be unable to provide trouble codes. In this case, the only way to make sure the temperature control computer is the problem is to replace it with a known good unit.

Electronic Control System Self-Diagnosis

When an electronic temperature control system develops a problem, the temperature control computer stores information about the problem as a *trouble code.* Trouble codes identify either the general HVAC system problem or a defective component. The exact nature of the problem identified by the trouble code varies with the system. The computer may also indicate a problem by turning on a warning light on the HVAC control panel. On many modern vehicles, a scan tool must be used to retrieve trouble codes.

Reading Codes Using the HVAC Control Panel

To obtain trouble codes, the system must be placed in the *self-diagnosis mode.* Many electronic temperature control systems use the LEDs or other lights on the control panel to display trouble codes. Always consult the vehicle service manual for correct code information. The service manual will explain the correct retrieval method and the meaning of each trouble code.

Many electronic temperature controls can be placed in the self-diagnosis mode by pressing a certain series of buttons on the HVAC control panel. After the buttons are pressed, the control panel will display the trouble codes. **Figure 22-1** shows how a typical HVAC control panel

displays trouble codes. This system can display all vehicle trouble codes. The technician may have to separate the HVAC system codes from engine and other codes. Some systems allow the technician to access powertrain, antilock brake/traction control, air bag, and HVAC codes separately by pressing certain buttons on the HVAC control panel. A typical sequence for separating HVAC and other computer codes is shown in **Figure 22-2.**

The self-diagnostic system of some temperature control computers can perform a self-test on the output devices by operating the HVAC system in all modes. The vehicle service manual will give directions for performing the self-test. Compare the system operation against the manual chart to ensure the system is operating correctly.

Reading Codes Using a Scan Tool

To get the most out of the self-diagnostic capabilities of an electronic temperature control system, you should have a *scan tool,* **Figure 22-3.** Scan tools are commonly used to retrieve trouble codes. The codes will appear as numbers on the scan tool screen. Many scan tools can be used to perform other diagnostic activities, such as reading the input and output voltages of the compressor clutch, blower motor, relays, and output solenoids. These readings can be compared with factory specifications to determine whether or not the component is defective.

Some scan tools can override the control module and directly operate the compressor clutch, blower, or solenoids. Forcing a component to operate allows the technician to determine the condition of the component and related wiring. Most modern scan tools can be used to reprogram the temperature control system module. This procedure is explained later in this chapter.

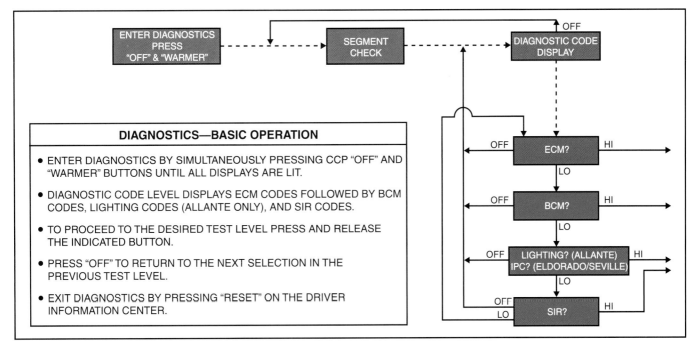

Figure 22-2. *This figure illustrates the steps in using the dashboard HVAC control panel to access the automatic temperature controls self-diagnostic system. (General Motors)*

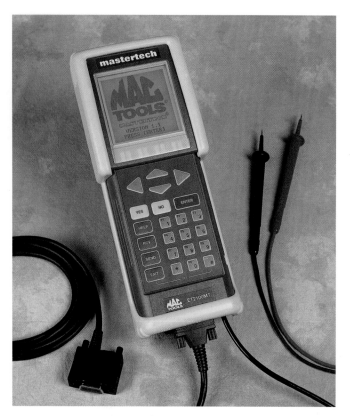

Figure 22-3. *Modern scan tools can be used to access diagnostic information, as well as reprogram some computers. (Mac Tools)*

Data Link Connectors

Most temperature control computers can also be accessed through the vehicle's data link connector. The **data link connector (DLC)** is the output terminal for the vehicle's self-diagnostic system. Vehicles that control the HVAC system through the ECM or BCM may require codes be accessed through the data link connector, rather than the HVAC panel. Plugging the scan tool into the DLC will automatically place the system in self-diagnosis mode. On some older vehicles, connecting two of the connector terminals with a jumper wire will place the system in self-diagnosis mode. This should only be done if specifically recommended by the manufacturer.

The data link connector is usually under the dashboard or center console. On vehicles built before 1996, it may be in the engine compartment, fuse box, or elsewhere, depending on the vehicle. Most vehicles have only one data link connector. This connector is used to access both engine and electronic temperature control codes. **Figure 22-4** shows some typical data link connector locations.

Scan Tool Construction

Scan tools are small and light enough to be handheld. Most scan tools are similar in appearance to the one shown in **Figure 22-5**. All scan tools have a **keypad** with a set of function keys used to select the various functions. Typical function keys are Enter and Exit keys, arrow keys,

and number keys. Many scan tools have specialized keys to access specific diagnosis features, or to connect to a remote computer or printer.

Some scan tools have indicator lights to tell you the scan tool is operating, the HVAC system mode, or an obvious system defect. The **display screen** is used to display trouble codes and other diagnostic results, and may also display a **menu.** The menu shows possible scan tool functions, often called **diagnostic routines.** The technician can pick one of these routines to begin using the scan tool. See **Figure 22-6.**

Cables and adapters provided with the scan tool allow it to be attached to one or more data link connectors. Some scan tools require adapters that plug into the data link connector to allow the scan tool to access the temperature control module. Some scan tools also have battery clips or a cigarette lighter adapter to allow them to use the vehicle battery as a power source. A few scan tools have an internal backup battery.

Types of Scan Tools

Some scan tools can be used with only one kind of electronic temperature control system. These are called **dedicated scan tools.** However, most newer scan tools can be used on many vehicles and are called *multi-system* or **generic scan tools.** These scan tools use specialized cartridges and adapters which can be changed to match the vehicle being worked on. Some scan tools can be updated by an interface with a personal computer or over a phone line to the manufacturer. Scan tools with this feature have an extra terminal that can be connected to a phone line.

Most scan tools are **bi-directional.** A bi-directional scan tool is able to retrieve information from the vehicle computer, and can send data into the computer. This ability to transfer data two ways allows the scan tool to operate system components for diagnosis, and to reprogram the module if updates are necessary.

After the 1996 model year, all vehicle manufacturers began to use the **OBD II diagnostic system.** All OBD II systems use the same 16-pin diagnostic link and trouble code format. Newer generic scan tools can retrieve trouble codes from any vehicle using the OBD II system. An automotive technical standards organization called **The Society of Automotive Engineers** or **SAE** has standardized trouble codes. Many post 1996 trouble codes will indicate the same problem, no matter what make of vehicle they are obtained from. Other trouble codes will be specific to the vehicle being tested. The technician should always consult the correct model and year service manual to determine what a particular code means.

Using a Scan Tool

The following sections explain how to use a scan tool to retrieve trouble codes and perform other diagnostic routines. Always refer to the scan tool manufacturer's instructions for exact procedures.

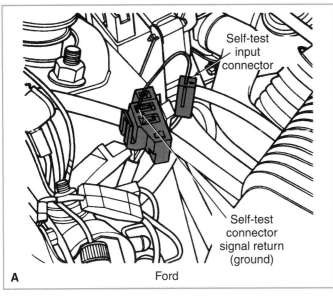

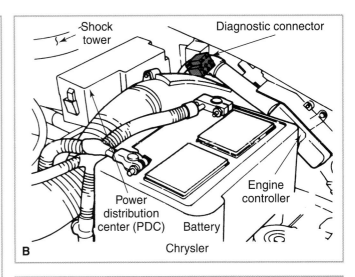

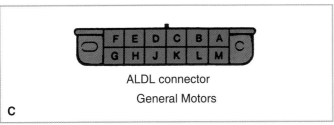

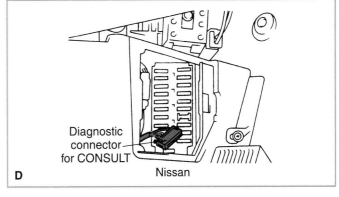

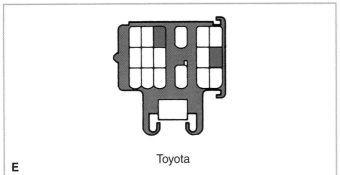

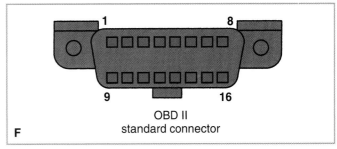

Figure 22-4. *Various data link connectors used on modern vehicles. A—Ford EEC connector. B—Chrysler diagnostic connector. C—General Motors ALDL. D—Nissan diagnostic connector. E—Toyota DLC. F—OBD II standardized connector. (Ford, DaimlerChrysler, General Motors, Nissan, Toyota)*

Figure 22-5. *Scan tools have keypads which can be used to select functions.*

Retrieving Trouble Codes

To retrieve trouble codes using a scan tool, first ensure the correct software cartridge is installed in the tool, if applicable. Next, locate the data link connector. Ensure the ignition is in the off position and connect the tool to the data link connector using the proper adapter. A typical connection is shown in **Figure 22-7.** If necessary, connect the scan tool power adapter after ensuring the battery is fully charged.

Note: Some vehicle systems may not function when the scan tool is connected to the DLC; check the tool and service manual for details.

```
                Climate Control

             Diagnostics Include:

      ATC, Manual A/C and Engine Cooling Items

             Turn Ignition Key On
```

Figure 22-6. *After the scan tool performs its power-on self-test, it can be used to retrieve trouble codes and perform other diagnostic tasks. (General Motors)*

Figure 22-7. *A—Various connectors allow the scan tool to be used on different vehicles. B—Scan tool plugged into the data link connector. Most connectors are located under the dashboard, but a few are under the hood.*

After the power adapter is connected, the scan tool will perform a ***power-on self-test (POST).*** This self-test checks the scan tool for problems or an incorrect cartridge. Most scan tools will then prompt you to enter the vehicle identification number (VIN) or otherwise identify the type of vehicle or system being worked on. Some scan tools can identify the vehicle automatically with little or no input from the technician. Once the vehicle information has

been entered, the display screen will prompt you to select the system to be diagnosed, in this case, the electronic temperature control system.

Next, the display screen will prompt you to retrieve trouble codes. If the ignition is not switched on, the screen may prompt you to turn the key to the *On* position. Once you select this option, the scan tool will retrieve and display the trouble codes. On older electronic temperature control systems, the trouble code numbers will not match. On later vehicles with the OBD II diagnostic system, the trouble codes are standardized. Once the trouble codes have been retrieved, test the indicated components for defects.

 Note: Always retrieve trouble codes before performing any other diagnostic routines.

Checking Sensor Input Readings

To check the condition of sensors, select the proper diagnostic routine from the scan tool menu. This may be called the **test function** or a similar name. Follow menu directions to check the voltage and/or waveforms produced by the input sensors, **Figure 22-8.** Compare the readings with manufacturer's specifications. The scan tool may have these specifications in its memory. Many of the latest scan tools have a **snapshot** or **freeze frame** mode, which will enable the control module to display system readings or waveforms at the time a malfunction occurred.

Operating Output Devices

Some scan tools allow the technician to diagnose output devices by bypassing the control module. Inputting commands through the tool will cause the device to operate. The scan tool will monitor operating voltage and determine whether the device is operating properly. Some scan tools can simulate actual road conditions to determine the operation and interaction of all system output devices. You should consult the scan tool and vehicle manufacturer service manuals for detailed information on these diagnostic routines.

Sensor Monitor	
Evap Temperature:	+36°F
In-car Temperature:	+68°F
Outside Temperature:	+76°F
Sunload:	4 volts
Mode Door:	2.1 volts
Temperature Door:	0.61 volts

Figure 22-8. *The scan tool will provide a menu of tests and functions. This screen allows the technician to monitor sensor input.*

Electronic Temperature Control System Testing

Testing of electronic temperature controls will be determined by the type of system you are working on. Newer systems have more test features, however, many of those cannot be accessed without a scan tool or other special equipment. Some of the tests can be accomplished by using a waveform meter.

The first and easiest test to perform is the performance and function tests, which are outlined in the service manual and discussed in Chapter 15 of this text. In conjunction with these tests, you should listen carefully for system operation. Roll the windows up, turn the ignition to key on, engine off, and with the blower motor disconnected, listen for stepper motor operation. You should not be able to hear the stepper motors; if they make noise, the motor is defective. Since most components in an automatic temperature control system require the removal of interior components, you can save some time by simply listening for stepper motor operation.

Testing Input Sensors

Most input sensors are thermistors. Thermistors allow the temperature control system to adjust temperature by monitoring the change in resistance values. This resistance can be measured with an ohmmeter, once the sensor is disconnected from the wiring harness. To check any thermistor sensor, you must know what the resistance

should be at a particular temperature. A temperature-resistance chart, such as the one shown in **Figure 22-9,** can be used to compare readings with known good values. If the sensor does not have the proper resistance across its temperature range, it should be replaced.

Testing Output Devices

Remember from Chapter 14 that many automatic temperature control output devices are the same ones as used on manual HVAC control systems. These devices can be tested by the same methods used in Chapter 21. Always consult the manufacturer's service manual since slight changes in procedures are often used to protect the electronic control system circuitry. The condition of most automatic temperature control output devices can be tested with a scan tool.

Mechanical Temperature Control Systems

Mechanically operated automatic temperature control systems are generally trouble free. The mechanical parts, however, can become worn or go out of adjustment. If a mechanical system fails to respond to temperature changes, the problem is usually in the transducer assembly or the power servo.

Before checking the automatic temperature control, make sure the basic HVAC components are in working order. Make sure electrical power and vacuum are available to the control units. Vacuum operated air doors can

Temperature-Resistance Chart					
°F	°C	Outside Temperature-Resistance		Inside Temperature-Resistance	
		Minimum Resistance K Ohms	Maximum Resistance K Ohms	Minimum Resistance K Ohms	Maximum Resistance K Ohms
14	-10	5.27	5.98	15.42	16.89
23	-5	4.03	4.52	11.76	12.80
32	0	3.11	3.44	9.03	9.75
41	5	2.43	2.66	7	7.53
50	10	1.91	2.07	5.47	5.85
59	15	1.51	1.62	4.31	4.58
68	20	1.21	1.28	3.41	3.61
77	25	0.97	1.02	2.72	2.86
86	30	0.78	0.82	2.17	2.29
95	35	0.63	0.67	1.74	1.85
104	40	0.51	0.55	1.41	1.50

Figure 22-9. *A temperature-resistance chart is useful when checking thermistors and other temperature sensors.*

be checked as explained in Chapter 21. If all the basic systems check ok, proceed to check the transducer and power servo.

 Note: The following procedures are general, designed to apply to all mechanical temperature control systems. If the service manual gives a different test procedure, or different vacuum specifications, use them instead.

Transducer

The transducer is both input sensor and control device. Therefore, a problem in the transducer can cause failure of the automatic temperature control. First check the ambient air tube that directs passenger compartment air to the transducer element. If the tube is disconnected or out of position, transducer operation will be erratic. If the tube is in place, use a vacuum pump to apply vacuum directly to the power servo.

 Caution: Do not exceed a vacuum of about 15 inches of mercury.

The servo should move in relation to changes in vacuum. If the power servo moves when operated by the vacuum pump, but does not respond correctly when attached to the transducer, try adjusting the cable as explained later in this chapter. If adjustment does not improve operation, replace the transducer.

Power Servo

The power servo contains a vacuum unit that operates the blend door and also operates some of the HVAC electrical devices. To test servo operation, "tee" in a vacuum gauge between the transducer and servo. Start the engine, turn the HVAC control to AUTO, and move the temperature lever while observing the power servo. When the temperature lever is moved to 85°F (30°C) the vacuum should lower to less than 2" (51 mm) of mercury. This vacuum reading varies between manufacturers and may be as low as .3" (8 mm) of mercury. The servo should move the blend door to the full heat position. When the temperature lever is moved to 65°F (18°C) the vacuum should rise to about 12-13" (305-330 mm). The power servo should move the blend door to the full cool position.

While checking vacuum, note whether the electrical and vacuum switches are operating according to service manual specifications. With the servo in the full cool position, the compressor clutch should engage and the blower should switch to one of the higher speeds. Cooled air should exit from the dashboard vents.

With the servo in the full heat position, the compressor clutch should disengage, and air should begin blowing out of the heater ducts. If the power servo does

not operate as indicated, it is defective. Some power servos can be adjusted. Others can only be fixed by replacing the entire unit.

Electromechanical Temperature Control Systems

Electromechanical temperature control systems operate like mechanical systems, but use temperature sensors and an amplifier (sometimes called a programmer) to modify transducer operation. In addition, the power servo electrical functions are taken over by the amplifier. Some of the same tests can be performed to the transducer and power servo as are done on a mechanical system. The transducer may be built into the amplifier. General sensor and amplifier testing procedures are given in the following paragraphs.

Sensors

To test the sensors, obtain an ohmmeter. Set the ohmmeter to the proper range (if necessary) and check each sensor. The sensors should have the specified resistance at a certain temperature. If the system uses a variable resistor to control temperature, it can be checked with an ohmmeter. Resistance readings should rise and fall smoothly as the resistor knob or thumb wheel is turned.

Amplifier

If all other components check ok, the amplifier may be at fault. Some amplifiers can be checked with an ohmmeter. However, ohmmeter checks may not uncover every possible amplifier defect. Substituting a known good unit is sometimes the only way to check the amplifier. A few older amplifiers have an adjusting screw. Mark the original position of the screw with a slight scratch on the case. Then start the HVAC system and turn the screw. If turning the screw does not affect operation, replace the amplifier.

Electronic Temperature Control System Service

Unlike the diagnosis portion, automatic temperature control system service is fairly easy. Most components are relatively easy to reach. A few components can only be reached by removing the dashboard or HVAC case.

 Note: Service of all basic units, such as the refrigeration system, heater, blower, and vacuum and electrical controls was covered in earlier chapters.

Once the problem has been located, electronic temperature control system service is limited to replacing parts. The only exception to this is computer reprogramming. The following sections are grouped by temperature control component.

It is possible to generate enough static electricity to destroy a computer simply by sliding across the vehicle seat. Always discharge any static electricity by touching a metal vehicle part before touching the computer or any associated parts.

 Caution: Stray static electricity (called *electrostatic discharge*) can severely damage electronic components. Guard against electrostatic discharge by following all discharge procedures. Static sensitive components will have a warning sticker like the one shown in Figure 22-10.

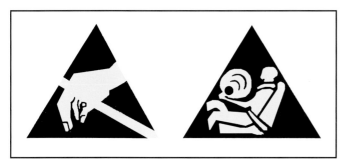

Figure 22-10. *These symbols indicate that special precautions must be taken to avoid electrostatic discharges and accidental activation of the air bag system. (General Motors)*

Computer Reprogramming

On some vehicles, a scan tool can be used to *reprogram* the temperature control computer. Reprogramming replaces the original system program with an updated program. The updated program is designed to correct any HVAC problems that have been reported from the field. After vehicles have been operated for a few years, unexpected problems with the original programming may surface. Reprogramming the original temperature control computer avoids the expense of installing a new computer.

 Note: On many newer vehicles, the temperature control computer is part of the powertrain or body/chassis ECM.

To reprogram the temperature control computer, the scan tool must have the new programming. The scan tool is usually loaded with the new information from a shop computer. The new programming is sent to the shop on a CD-ROM disc and loaded into the scan tool using the shop computer. The scan tool can then load

the information into the HVAC computer. On a few systems, the scan tool is used as a connector between the vehicle and another computer.

To reprogram the computer, load the new information into the scan tool and connect it to the vehicle data link connector. Then follow the menu directions to erase the old information and install new programming.

Control Head

Control head service is usually limited to replacement. To service the control head, remove dashboard or center console components as needed. Remove the control head. Installation is in the reverse order. Be sure to follow static discharge guidelines when handling the new control head.

In some cases, what seems like a major problem with an automatic control head may be caused by a poor connection between the control head and its wiring harness. You may want to try this before condemning an electronic control head. Remove the control head from the dash and carefully clean the contacts on the control head. Apply a chemical contact enhancer and reinstall. If this does not repair the problem, replace the control head

Temperature Control Computer

The first step in replacing the temperature control computer is to locate it on the vehicle. The computer is usually installed under the dashboard. Consult the service manual to locate the computer. Sometimes the computer is part of the HVAC control panel. Control panel removal was covered in Chapter 21.

To remove the computer, first disconnect the battery negative cable, **Figure 22-11.** Briefly touch the negative cable to the positive to discharge any system capacitors. Then remove the computer electrical connectors. Remove the screws holding the computer and remove the computer from the vehicle, **Figure 22-12.**

Install the new computer and tighten the mounting screws. Then reinstall the electrical connectors. Make sure the ignition key is in the off position, then reinstall the battery cable. Start the engine and check system operation.

Air Temperature and Sunload Sensors

Passenger compartment air temperature and sunload sensors are located in the dashboard. Ambient air temperature sensors are located either at the grill or in front of the condenser. To replace an interior temperature sensor, make sure the ignition switch is in the off position. Then remove the trim panel covering the sensor. Most sensor trim consists of small screen cover, which can be snapped off, or is held by plastic tabs. With the trim removed, locate and remove the sensor fastener(s). Most sensors are held in place with a clip, but a few are held into place with small screws. Once the sensor is removed, remove the electrical connector from the sensor pigtail, **Figure 22-13A.** Install

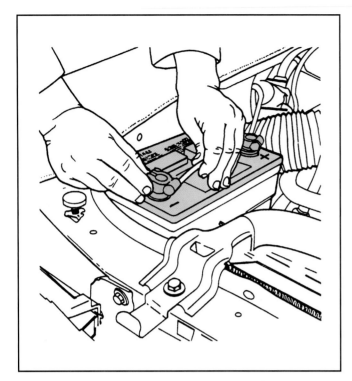

Figure 22-11. *Before servicing the automatic temperature control system, disconnect the battery. This will prevent damage to the electronic components. (General Motors and Ford)*

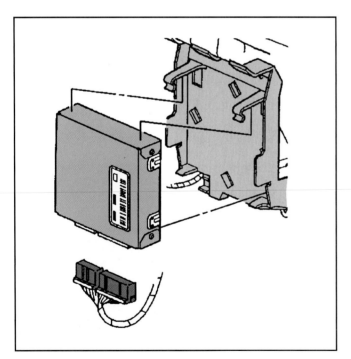

Figure 22-12. *Replacing an HVAC control computer. Some vehicles control the operation of the HVAC system with a body control computer (BCM) or power train control computer (PCM). Defective computers are replaced, not repaired. (General Motors)*

the electrical connector on the new sensor and position the sensor in the dashboard well. Install the fasteners and replace the cover. Start the engine and check sensor and HVAC system operation.

Sunload sensors are located at the top of the dashboard, near the edge of the windshield. They are usually not difficult to replace. Begin by removing the defrost vent cover, which usually houses the sunload sensor. Remove the sunload sensor and replace, **Figure 22-13B.**

Note: If the vehicle is equipped with an automatic headlamp system, be sure you are replacing the correct sensor. The photodiode for the automatic headlamp system and the sunload sensor are often installed close to each other and look somewhat similar.

To change an ambient air temperature sensor, locate the sensor, then remove the electrical pigtail and sensor fasteners, **Figure 22-14.** Remove the sensor and install the new sensor. Install the fasteners and electrical pigtail. Then start the engine and recheck system operation.

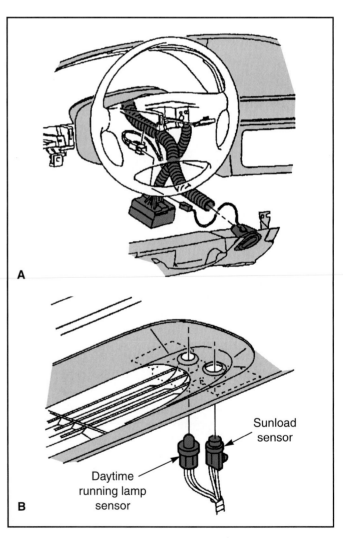

Figure 22-13. *To replace a passenger compartment temperature sensor, remove dashboard trim panels as necessary and remove the sensor. A—Interior temperature sensor. B—Sunload sensor. (General Motors)*

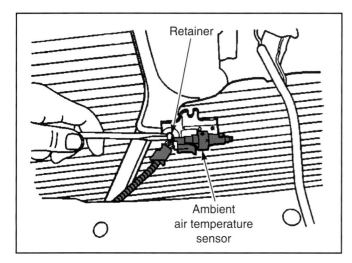

Figure 22-14. *Once the ambient air temperature sensor is located, remove the electrical connector and mounting hardware. (General Motors)*

Replacing Output Devices

Remember from Chapter 14 that automatic temperature control output devices are similar to those used on manual systems. Replacement procedures for most of these devices are similar to the procedures for manual systems. There are a few differences between automatic and manual device service procedures. These are discussed in the following sections.

Replacing Stepper Motors

Stepper motors frequently fail and must be replaced. Replacement of most stepper motors is fairly straightforward. Begin by removing kickpanels and other interior components as needed. Remove the electrical connector and screws or bolts holding the stepper motor to the case, **Figure 22-15.** Installation is the reverse of removal.

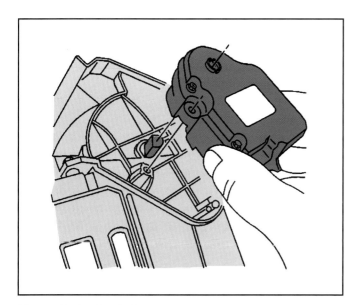

Figure 22-15. *Replacing stepper motor; be sure to recalibrate the motor, if needed. (General Motors)*

However, before reinstalling all the interior components, check to see if the stepper motor needs to be calibrated, which is covered in the next section.

Recalibrating Stepper Motors

On some electronic HVAC control systems, the electronic stepper motors must be positioned so that they are at the end of their travel before the system is re-energized. This is sometimes called **recalibrating** the motors. Always consult the service manual before proceeding with a motor recalibration. The general procedure to recalibrate a stepper motor begins with turning the ignition switch to the On position. Next, push the HVAC control panel AUTO and OFF buttons at the same time. Hold both buttons for at least five seconds. You may hear the motors operating as you hold the buttons, this is normal. After five seconds, the motors have been calibrated, and the HVAC system can be checked for proper operation

Replacing Power Module

Instead of a blower relay, some automatic temperature control systems control the fan and compressor clutch through a solid state power module. The power module is located in the ductwork downstream of the blower where it is cooled by airflow. The module is finned to dissipate as much heat as possible. **Figure 22-16** shows a typical location of a power module.

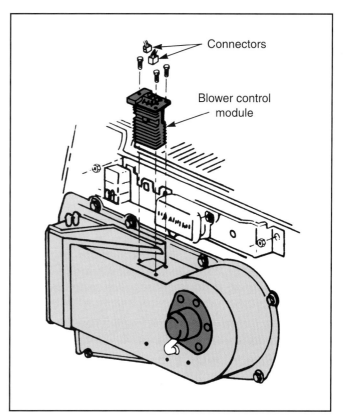

Figure 22-16. *Some automatic temperature control systems use power modules, which are usually located in the blower case. Replacement is fairly simple, however, some of the modules can be difficult to access. (General Motors)*

To remove the power module, first locate it under the dashboard or hood. Then disconnect the electrical leads and remove the fasteners holding the module to the duct. Then pull the module from the duct. Compare the old and new modules to ensure the new module is the correct replacement. Then place the new module in the duct and attach the fasteners. Reconnect the electrical leads and check HVAC system operation

Engine and Coolant Temperature Sensors

 Warning: Before removing any coolant sensors, review the cooling system safety precautions in Chapter 20.

Most engine and coolant temperature sensors are located on the engine. A few sensors are installed in the heater hoses. To remove the sensor, locate it on the engine and remove the electrical connector.

If the sensor extends into the cooling system, depressurize and drain the cooling system. Then use the correct tool to unscrew the sensor from the engine block. If the sensor is installed on a heater hose, remove the clamps and pull the sensor from the hose. Coat the new sensor with sealant if required.

 Caution: Some sensors must not be installed using sealant as this will affect the electrical flow between the sensor and engine block.

Install the new sensor and tighten it into place. Tighten the hose clamps if applicable. Refill the cooling system and bleed according to the procedures in Chapter 20. Then start the engine and make sure the sensor is operating properly.

 Note: After all repairs are made and the HVAC system has operated for about 10 minutes, re-enter the self-diagnosis mode and make sure no trouble codes have appeared.

Mechanical and Electromechanical System Service

These systems have fewer parts than electronic systems. Service procedures are identified in the following paragraphs.

Adjusting Transducer Cables

All mechanical temperature control transducer cables are adjustable. To adjust the cable, it will be necessary to remove dashboard parts to gain access to the transducer and control panel.

Early cables were adjusted by a turnbuckle. To adjust this type of cable, insert a gauge tool in the transducer. Then turn the turnbuckle until the temperature lever is in the middle position (usually 75°F or 24°C). Then remove the gauge tool.

Later mechanical system cables use an adjusting slot at the transducer. To adjust the cable, move the transducer cable arm to line up with the mark on the transducer body. Then loosen the adjusting slot mounting bolt and move the temperature lever to the middle position (75°F or 24°C). Then tighten the slot mounting bolt and recheck operation.

Replacing System Parts

Mechanical and electromechanical temperature controls are easy to replace by following the instructions in the following paragraphs. Remember to readjust the transducer cable whenever the cable is disturbed.

 Note: Mechanical and electromechanical control panels are replaced in the same way as manual control panels. Refer to manual control panel procedures in Chapter 21. Electromechanical sensors are replaced in the same manner as electronic sensors.

Transducers

To replace a transducer, remove the trim panel covering the transducer air inlet. This is usually a small louvered cover held by trim screws. Then remove the cable connection and aspirator duct. Remove the vacuum lines, and electrical connector if used. Remove the fasteners and remove the transducer. Install the new transducer and tighten the fasteners. Install the aspirator tube, vacuum lines, and cable, then adjust the cable. Reinstall the air inlet cover and recheck HVAC system operation.

Power Servo

Locate the power servo next to the blend door. Remove the electrical connectors, vacuum line, and blend door linkage. Then remove the fasteners holding the power servo to the case. Remove the power servo from the case. Place the new power servo in position and attach the blend door linkage. Install the fasteners and reattach the vacuum line and electrical connectors. Recheck HVAC system operation, and adjust servo position if needed. After all adjustments are made, reinstall other dashboard parts.

Amplifiers

To replace an amplifier, first make sure the ignition switch is in the off position. Then remove the amplifier electrical and vacuum connectors. Remove the screws holding the amplifier and remove it from the vehicle.

Install the new amplifier and tighten the mounting screws. If the amplifier is installed on the plastic case, make sure any ground wires used are reinstalled. Then reinstall the electrical and vacuum connectors. Start the engine and check system operation.

Summary

Automatic temperature control system problems can be fixed with the application of knowledge and logic. Common problems in electronic temperature control systems are defective input sensors and computers.

The temperature control computer stores information about system problem in the form of trouble codes, which identify the general HVAC system problem or a defective component. Many electronic temperature control systems use the HVAC control panel to display trouble codes. Pressing a series of buttons on the control panel causes the control panel to display trouble codes. The technician may have to separate HVAC system codes from other vehicle codes. Some temperature control computers can be accessed through the data link connector. Plugging a scan tool into the data link connector will automatically place the system in self-diagnosis mode. Trouble codes will appear as numbers on the scan tool screen.

Some scan tools also have indicator lights. Dedicated scan tools can be used with only one kind of electronic temperature control system. Multi-system and generic scan tools can be used to diagnose many manufacturer's control systems. All OBD II systems use the same 16-pin diagnostic link, and generic scan tools can retrieve trouble codes from any vehicle using the OBD II system. Scan tools can also be used to check sensors and output devices.

Mechanical temperature control systems are usually trouble free. If all of the basic HVAC systems are ok, check the transducer and power servo. Remember the transducer is both input sensor and control device. Electromechanical temperature systems operate like mechanical systems, but use temperature sensors and an amplifier.

Once the problem has been located, electronic temperature control system service is limited to replacing parts. Some computers can be reprogrammed with a scan tool. Always pay attention to electrostatic discharge problems when servicing any electronic part.

Mechanical and electromechanical systems have fewer parts than electronic systems. Sometimes the transducer cable must be adjusted. Parts replacement is relatively simple. Always recheck HVAC system operation after any service procedures.

Review Questions—Chapter 22

Please do not write in this text. Write your answers on a separate sheet of paper.

1. What should the technician check before troubleshooting the automatic control system?

2. The most common cause of electronic control system problems is a defective _____.

3. To obtain trouble codes, the system must be placed in the _____ mode.

4. A system self-test allows the self-diagnosis system to check the operation of _____.

5. Name three things scan tools can be used to do.

6. A scan tool that can send and receive data is a _____ type.
   (A) dedicated
   (B) generic
   (C) multi-system
   (D) bi-directional

7. An OBD II data link connector has a ____ -pin connector.

8. If a mechanical temperature control system has a problem, what two devices should the technician check?

9. On an electromechanical temperature control system the _____ may be part of the amplifier assembly.

10. Match the component with the service procedure *most likely* to be performed to service it.

    Ambient temperature sensor ___   (A) Adjust

    HVAC control computer ___   (B) Reprogram

    Sunload sensor ___   (C) Replace

    Transducer cable ___

    Passenger compartment
    temperature sensor ___

    Power module ___

ASE Certification-Type Questions

1. All of the following statements about electronic temperature control diagnosis are true, *except:*
   (A) the most common cause of trouble is a sensor defect.
   (B) a defective computer can cause erratic display readings.
   (C) trouble codes can always be retrieved from a defective computer.
   (D) sometimes the only way to diagnose the computer is to replace it with a known good unit.

2. Technician A says a sunload sensor is never used on a modern electronic temperature control system. Technician B says a transducer is never used on a modern electronic temperature control system. Who is right?

   (A) A only.
   (B) B only.
   (C) Both A and B.
   (D) Neither A nor B.

3. Which of the following procedures is *not* a method of retrieving HVAC system trouble codes?

   (A) Turning the key on, then off, three times.
   (B) Pressing a certain sequence of HVAC control panel buttons.
   (C) Using a dedicated scan tool.
   (D) Using a generic scan tool.

4. The modern scan tool contains all of the following, *except:*

   (A) keypad.
   (B) display screen.
   (C) reprogramming data.
   (D) indicator lights.

5. The scan tool performs a self-test when it is first energized. The scan tool is checking:

   (A) sensors.
   (B) the HVAC computer.
   (C) output devices.
   (D) itself.

6. If the technician uses a vacuum pump to apply vacuum to a power servo, the maximum vacuum should not exceed ___ inches of vacuum.

   (A) .3
   (B) 2
   (C) 15
   (D) 23

7. Electromechanical temperature control system sensors can be checked with an:

   (A) ohmmeter.
   (B) voltmeter.
   (C) ammeter.
   (D) powered test light.

8. Stray static electricity can damage the:

   (A) sensors.
   (B) HVAC computer.
   (C) output devices.
   (D) system fuses.

9. To adjust a mechanical temperature control system cable, the temperature lever should be placed at:

   (A) full hot.
   (B) full cold.
   (C) the middle of its travel.
   (D) It does not matter.

10. When replacing an electronic temperature control computer, which of the following steps should be taken *first*?

    (A) Touch the positive and negative battery cables together.
    (B) Touch a metal vehicle part.
    (C) Disconnect the computer electrical connector.
    (D) Remove the battery cable.

Chapter 23

Air Conditioning System Installation and Retrofitting

After studying this chapter, you will be able to:
- ❏ Install an aftermarket air conditioner in a vehicle.
- ❏ Retrofit an R-12 refrigeration system to R-134a.

Technical Terms

Air conditioning kit

Universal kits

Custom kits

Packing list

Grommet

Retrofitting

Label

Since factory air conditioning is very common, you will seldom be asked to install an aftermarket air conditioner. Eventually, however, an air conditioner installation job will turn up and you must be prepared. This chapter explains how to install an aftermarket air conditioner on vehicles that did not leave the factory with air conditioning. This chapter also covers procedures for converting R-12 refrigeration systems to R-134a. This chapter will draw on information covered throughout the text, so review the earlier chapters as needed.

Aftermarket Air Conditioners

The majority of vehicles made within the last ten years have factory-installed air conditioners. Only in relatively cold areas are many new vehicles sold without air conditioning. Anyone who wants a new or used vehicle with an air conditioner can usually find one.

A few vehicles are commonly found without factory air conditioning. They include small economy cars, lower priced pickup trucks, and some small sport utility vehicles. In many cases, the owner will want to install an aftermarket air conditioner. Many owners of classic or custom vehicles travel long distances to shows or other events and want to install an aftermarket air conditioner to make the trip more comfortable.

Aftermarket Air Conditioning Kits

The parts and instructions necessary to install an aftermarket air conditioner are called an *air conditioning kit.* Modern kits can be divided into two general classes:

❑ *Universal kits* can be installed in many different vehicles. These kits use an evaporator and case assembly as small as possible to fit under any dashboard.

❑ *Custom kits* are made to fit a certain vehicle. A typical custom kit will fit one series of vehicle. Many custom designs use the original dashboard controls and may be designed to deliver cooled air through the existing dashboard and defroster vents.

Modern Kits

Modern aftermarket air conditioning kits are much different from early units. Early units were installed under the dashboard, **Figure 23-1.** They simply cooled the air inside the passenger compartment. Almost all of these early units were universal kits. Modern air conditioning kits combine an evaporator, heater core, and defroster vents. These kits can give a car or truck the same performance as a factory air conditioning system. The air conditioner evaporator fits under the dash where it is barely visible, **Figure 23-2.** Many vehicle manufacturers produce aftermarket air conditioners for installation by the dealer. These units are almost identical to the factory installed units.

Some aftermarket suppliers sell the evaporator unit only. The customer can use the evaporator with a compressor and condenser designed for the vehicle. Most aftermarket air conditioning kits are available in R-12 and R-134a versions. A few manufacturers even supply electronically operated automatic temperature control systems.

Control Systems

Almost all aftermarket air conditioners control evaporator pressure with a cycling clutch and expansion valve. An evaporator temperature sensor controls the clutch. A thermostat modifies the operation of the temperature sensor to control evaporator temperature. Operation of this type of switch was discussed in Chapter 10. Blower speeds are controlled through a blower switch, high speed relay, and resistor block. Some systems have electrically operated solenoids to control the flow of vacuum to the heater door, heater shutoff valve, and defroster door.

Aftermarket Air Conditioning Considerations

An aftermarket air conditioner can be installed on just about any vehicle. The customer may be dissatisfied, however, if some questions are not answered first. Before installing an aftermarket air conditioner on the vehicle, you should consider the following factors.

Air Conditioner and Vehicle Compatibility

Although it is possible to install an air conditioner on almost any vehicle, it may not be possible to properly cool some vehicles. Large vans, convertibles, dark color vehicles, and vehicles with glass sunroofs or T-tops are often very hard to cool. Even the largest capacity aftermarket air conditioner may be only partially successful in cooling

Figure 23-1. *Older aftermarket air conditioners were installed under the dashboard as shown here. They took up interior space and were not integrated with the factory installed heating and ventilating systems.*

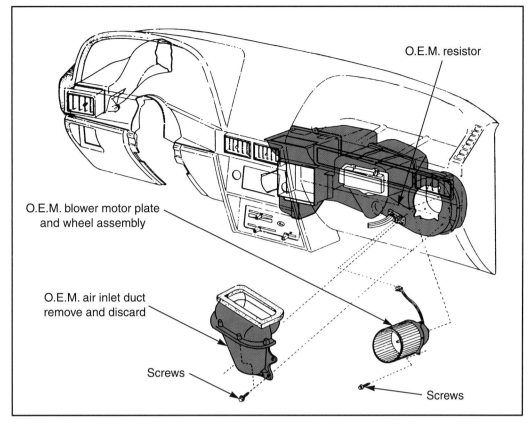

O.E.M. resistor

O.E.M. blower motor plate
and wheel assembly

O.E.M. air inlet duct
remove and discard

Screws

Screws

Figure 23-2. *Newer aftermarket air conditioners are designed to fit out of sight under the dashboard and to work with the original heater and defroster vents.*

these vehicles on very hot and sunny days. Make sure the customer understands these factors before beginning the installation.

Also, make sure the vehicle's engine and cooling system can handle the extra load of the compressor and heat from the condenser. Some vehicles are not designed for nonfactory systems and any warranty may be voided if a nonfactory air conditioning system is added.

Cooling System Modifications

The air conditioning system places an extra heat and horsepower load on the engine. This is especially true of smaller engines. If the cooling system is in poor condition, it must be brought up to its original condition. Even when the cooling system is in good condition, it may be necessary to install a larger radiator and auxiliary (extra) engine or transmission oil coolers. Often an auxiliary electric fan must be installed to prevent overheating.

Component Mounting

You should consider the following component mounting considerations before ordering the kit or beginning the installation process. On a smaller vehicle, the evaporator assembly may severely limit front seat space. Be sure the customer understands this before he or she approves the job.

The condenser is installed for maximum airflow and may take up most of the space in front of the radiator, making access to other front components difficult. Plan for this to ensure other parts can be reached for eventual service.

> **Caution: The front impact sensors for the air bag system on some vehicles are located just in front of the radiator. Be sure the condenser does not interfere with sensor location or operation.**

The compressor must be installed as close to the engine as possible to reduce belt length and vibration. In most cases the compressor will be installed on the top of the engine. Sometimes space is limited due to low hood clearance, and the compressor must be installed on the side of the engine.

Refrigerant lines should be mounted away from moving parts. Make sure the hoses will not be too close to any exhaust system parts. Ideally, all refrigerant hoses should be at least 3″ (7.5 mm) away from exhaust parts.

Compressor Compatibility

Most of the time you will buy a complete kit with all parts. It is possible, however, to buy the evaporator assembly only and obtain a compressor and condenser from another source. A common method is to use an OEM compressor and condenser. This has the advantage of giving the finished system a factory installed appearance. It is also easier to get the mounting brackets from a salvage yard vehicle with factory air conditioning.

The difficulty with this approach is the compressor and condenser may not be compatible with the evaporator. A compressor with too little capacity will not be able to keep up with the cooling load. A compressor with too much capacity will interfere with expansion valve operation, will be hard to control by cycling, and may result in excessive high side pressures. Replacing the drive pulley with a different size can sometimes correct compressor over or under capacity. The fitting sizes on the evaporator may not be the same as the factory evaporator, which would require the making of custom hoses and lines.

Older radial piston compressors may vibrate excessively if not solidly bracketed to the engine. A variable capacity compressor usually will not work with an aftermarket system. Check with the evaporator assembly manufacturer before deciding on a particular compressor pumping capacity.

Passenger Compartment Sealing

The air conditioner cannot cool properly when outside air is allowed to enter. Before beginning air conditioner installation, check the firewall and underside of the dashboard for any precut holes. These holes may be sealed with a rubber plug to keep outside air from entering the passenger compartment. Also check the condition of door and window seals. Also check for missing insulation at the firewall or floor. Any problems that are found should be corrected. If the air conditioner is being installed in a convertible, make sure the customer understands that cooling will not be as good as with the same system on a hard top sedan.

Installing an Aftermarket Air Conditioner

To install an aftermarket air conditioner, first determine the right kit for the vehicle. Use the criteria outlined earlier to determine which aftermarket air conditioner is best for the particular application. Also determine which refrigerant the system will use. R-134a is the easiest to obtain. Other refrigerants may be better, but may be more expensive or not readily available. The kit manufacturer can advise you as to the best type of air conditioning system to install. Once you have determined all of the factors, order the proper kit.

Unpack and Lay Out Parts

When the kit arrives, carefully unpack the components and make sure the correct kit has been sent, and all parts are included. The manufacturer will supply a *packing list* with the kit. The packing list shows all of the parts that should be in the kit. Compare the actual parts against the packing list, and contact the parts supplier or manufacturer immediately if any parts are missing.

 Note: Leave the sealing caps on all refrigeration system parts until they are ready to be installed.

Perform Preliminary Steps

Begin the actual installation by reading the installation instructions supplied with the kit. If it is necessary to preassemble any parts, for instance installing the expansion valve or hoses on the evaporator, take these steps before proceeding. Make sure hose and valve openings point in the proper direction before tightening. Do not forget to install insulation tape around the sensing bulb of the expansion valve, as shown in **Figure 23-3.**

Mark and Drill Holes

Before drilling any holes, position the evaporator and case assembly and make sure it is a good fit, **Figure 23-4.** Ensure the evaporator drain hole will be on the very bottom of the case when installed. Failure to ensure this will allow condensation to collect in the case. Also make sure the refrigerant hoses will not be too close to the engine or exhaust components. Then mark the mounting holes, refrigerant hole, and drain hole openings with a grease pencil. Carefully check out the area behind the potential drilling spots to ensure the drill bit does not damage other vehicle components. Always double-check the hole location before drilling.

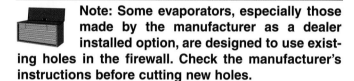 **Note: Some evaporators, especially those made by the manufacturer as a dealer installed option, are designed to use existing holes in the firewall. Check the manufacturer's instructions before cutting new holes.**

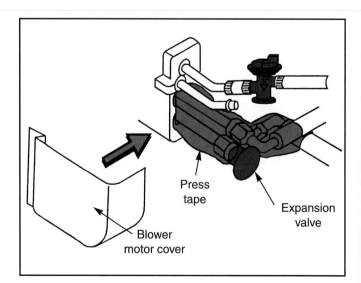

Figure 23-3. *Carefully insulate the expansion valve bulb before installing the evaporator unit. This will be much harder to do after the evaporator unit has been installed. (Vintage Air)*

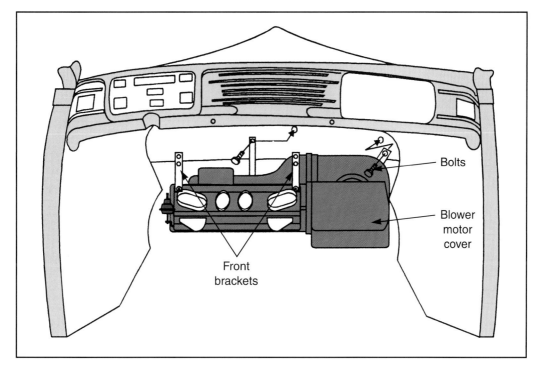

Figure 23-4. *Place the evaporator unit in position and make sure it fits properly. Only after you are sure the evaporator unit is in the proper position should you drill holes for the mounting brackets. (Vintage Air)*

Once all marking and checking steps have been completed, drill out the holes using the proper size bits. It may be necessary to use a hole cutter to make the openings for the refrigerant lines. Holes should not be any larger than necessary. If the unit includes a defroster, mark and cut holes for the defroster hoses and vents now. Grind the holes smooth and paint them to prevent corrosion.

Install Evaporator Assembly

Once all holes have been made, place the evaporator assembly in position under the dashboard and loosely install the mounting bolts. If necessary, thread the refrigerant hoses through the matching holes in the firewall. Install grommets around the hoses where they pass through the firewall. A **grommet** is a rubber ring that fits into the firewall hole and surrounds the hose. Grommets will prevent damage to the hose and seal the firewall opening.

Install Ducts as Necessary

If the system uses ducts for the defroster or auxiliary vents, they should be installed before the evaporator is tightened into place. See **Figure 23-5.** Once all parts are in place, tighten the evaporator assembly mounting fasteners.

Install Compressor

Install the compressor in the recommended location on the engine. If possible, use original equipment brackets, **Figure 23-6,** or the brackets provided with the new compressor. Other brackets must be obtained or made if the compressor will be installed in a non-stock location. Be sure the brackets hold the compressor securely in place.

If necessary, install the compressor drive pulley on the engine. Some vehicles may already be equipped with an extra pulley groove, and installing an extra pulley will not be necessary. After the compressor and drive pulley are installed, install and tighten the compressor drive belt.

Install Condenser

Begin condenser installation by installing any needed condenser to body brackets. Leave the fasteners hand tight to make placement easier. Then remove any under hood braces or shrouds to gain access to the front of the radiator. Place the condenser in front of the radiator, and make sure the condenser inlet and outlet fittings are accessible. Also make sure there is about 1″ (25.4 mm) between the condenser and radiator surfaces. Mark any place where bracket holes must be drilled. Drill holes as needed and install the condenser. Once the condenser is in place, tighten the fasteners. If an extra cooling fan is recommended, install it now then replace the other underhood components. To install the temperature control for the fan, see the next section.

Install Receiver-drier

After all other components are in place, install the receiver-drier. The receiver-drier is usually installed next to the condenser, but can be located at any point in the high-pressure line. Make sure the receiver-drier is installed in the upright position. If it is installed sideways or upside down, it will be less efficient in removing moisture and separating vapor from the liquid refrigerant.

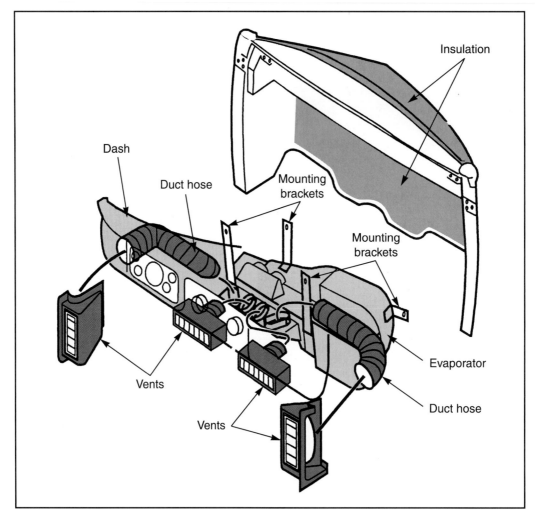

Figure 23-5. *After making sure that the evaporator unit will fit, install the ductwork and defroster vents as necessary. On a newer vehicle, it may be necessary to fit the evaporator unit ducts to the existing vehicle ductwork. (Vintage Air)*

Figure 23-6. *Try to find an original equipment bracket, or use a bracket supplied by the aftermarket air conditioner manufacturer. Do not use lightweight brackets or straps to hold the compressor is place. The compressor will vibrate and eventually break the mountings.*

Attach Lines

Install the refrigeration hoses. A typical hose routing sequence is shown in **Figure 23-7**. Procedures for installing refrigeration hoses are similar to replacing a hose on a factory installed system. Be sure to install O-rings where needed.

> **Caution: Route hoses carefully to keep them away from heat or moving parts. Install grommets around the hoses where they pass through the firewall. The grommets will prevent damage to the hose and seal the firewall opening.**

Install the heater hoses if the unit has a heater. Also install the heater shutoff valve if one is used. The heater shutoff valve must be installed in the proper flow direction. If necessary, drain the cooling system and install the heater hose fittings on the engine. If a thermostatically controlled cooling fan is being added, install the temperature control unit now while the cooling system is drained. Refill and bleed the cooling system as explained in Chapter 20.

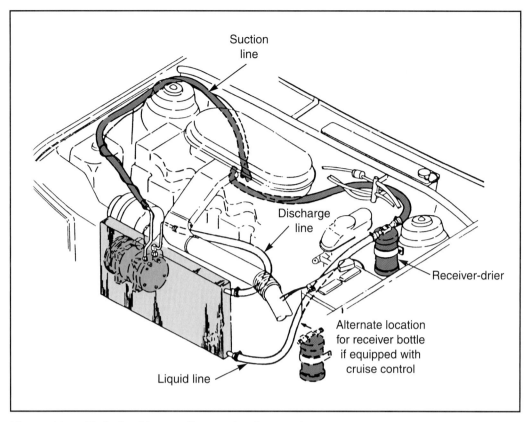

Figure 23-7. *Typical refrigerant line routing for an aftermarket air conditioner. As with factory installations, the lines should be at least 3 inches (7.5 mm) away from exhaust components and moving parts. (Vintage Air)*

Install Electrical and Vacuum Components

After all other parts have been installed, install the electrical components that have not already been installed. Typical electrical components are high blower relays, compressor relays, and the control panel. If these parts are installed in the evaporator assembly, ignore this step. Also install any vacuum components that are not part of the evaporator assembly.

After all the electrical and vacuum components are in place, connect the electrical wiring and vacuum hoses. A common aftermarket electrical schematic is shown in **Figure 23-8.** Always run the air conditioner controls through the ignition switch. After installation, check to ensure the control panel has power only when the ignition switch is on. One lead of the high blower relay may be run directly from a source of battery current. Be sure to install fuses and circuit breakers when needed. After all electrical connections have been made, turn the ignition switch on and make sure the blower runs in all speeds, and the compressor clutch engages.

The vacuum source should be the engine intake manifold. Do not tap into another vacuum accessory as this may affect component operation. If the engine is small or has a high performance camshaft, add a vacuum reservoir. The reservoir will hold enough vacuum to ensure the vacuum diaphragms can operate when the intake manifold vacuum is low. Typical vacuum reservoir can be obtained from a salvage yard, **Figure 23-9.**

Carefully route all wiring and hoses, using clips as needed to hold them in place and reduce movement. After all electrical wiring and vacuum hoses are installed, doublecheck to make sure they will not contact the exhaust system or moving parts.

Install Extra Cooling Fan

If the kit comes with an extra cooling fan, install it in a location that will not interfere with the fan(s) already installed. In some installations, you may need to install the fan on the opposite side of the radiator and condenser, and wire the fan to push air, rather than pull air.

To install the fan, find a location that will allow the fan to pull air through the condenser. If necessary, loosen the radiator support to allow space. Position the fan and mount to the condenser or radiator using the hardware that came with the fan kit. Connect the fan and switch control wiring. Connect the fan power to a fused connection that has power only when the ignition switch is in the On position.

 Note: Some technicians wire the extra cooling fan to come on only when the compressor clutch is engaged.

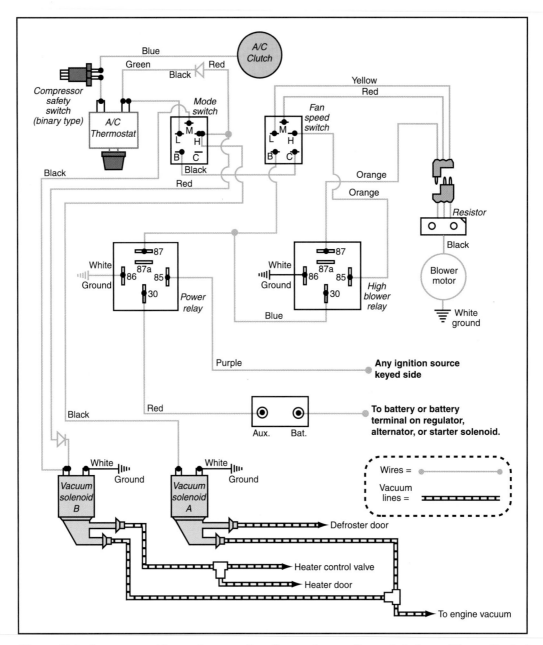

Figure 23-8. *A common wiring and vacuum line diagram for an aftermarket air conditioner. Most of these wires and lines are contained in harnesses and are simply plugged in. (Vintage Air)*

Evacuate and Charge System

If not done already, refill and bleed the cooling system. Then evacuate, leak check, and prepare to charge the refrigeration system. Refer to Chapter 16 if you need additional information. R-134a systems require a smaller charge of refrigerant than R-12 systems. Most R-134a systems use about 75% to 80% of the R-12 charge. The R-134a charge should never be more than 90% of the R-12 charge. Add the refrigerant as dictated by the type of equipment available. If you are installing another type of refrigerant, follow the refrigerant manufacturer's recommendations.

After charging the refrigeration system, start the engine, turn the air conditioner controls to maximum cooling, and carefully monitor pressures and system operation. Also make sure all radiator and condenser cooling fans energize and operate properly.

 Caution: Do not depend on the sight glass to determine refrigerant level. A sight glass is not a reliable refrigerant level indicator.

Road Test and Adjust Expansion Valve as Necessary

Road test the vehicle with a temperature gauge installed in the vent closest to the evaporator. As with a factory unit, the aftermarket air conditioner should be able to reduce interior temperatures by at least 30°F (17°C). Check all system modes, including heating and defrosting if equipped.

While road testing the vehicle, closely monitor the engine temperature indicator. If the engine shows signs of overheating, it may be necessary to install additional cooling fans or a larger radiator.

Figure 23-9. *To ensure the proper working of the vacuum devices, consider obtaining a vacuum reservoir from a salvage yard. Almost all vehicles with factory air conditioners use vacuum reservoirs.*

It may be necessary to adjust the thermostat to obtain maximum cooling. Many thermostats have an adjusting screw on the rear of the control knob assembly. The screw is usually turned clockwise to increase cooling, and counterclockwise to reduce cooling. If the thermostat is set too cold, the evaporator will ice up. To correct, turn the adjusting screw about 1/8 turn at a time, then recheck operation. If the thermostat is set too warm, the result will be a warm evaporator and rapid clutch cycling. Turn the adjusting screw about 1/8 turn to increase cooling. Then road test the vehicle to recheck operation.

R-12 to R-134a Retrofitting

Refrigerant *retrofitting* is becoming increasingly common. As R-12 becomes more expensive and harder to find, the HVAC technician will find retrofitting an increasing part of his or her business. Because of the decreasing stockpiles of R-12, new technicians will have to be familiar with retrofitting almost from the first day at work. This section will give you the basics of performing efficient and profitable retrofits. The procedures apply to R-12 to R-134a retrofits only. If you are retrofitting from R-12 to another type of refrigerant, consult the refrigerant manufacturer for details about their retrofit procedures.

⚠ **Warning: Remember it is illegal to mix refrigerants. A retrofit means complete replacement of the original refrigerant with another type. Topping off with a different refrigerant is not a retrofit, and is illegal.**

Whether to Retrofit

Although R-134a is cheaper and easier to obtain, some refrigeration systems should continue to use R-12. Typical reasons for staying with R-12 involve two kinds of systems:

❏ Systems where low side pressures cannot be readjusted, such as vehicles with non-adjustable POA valves. Non-adjustable POA valves, **Figure 23-10,** were used on many domestic and a few imported vehicles until the mid-1980s. Later vehicles with cycling clutches have adjustable pressure switches or the switch can be replaced with an R-134a compatible switch.

Figure 23-10. *Older cars with POA or other evaporator pressure control valves should not be converted to R-134a unless there is no alternative. The POA will not allow evaporator pressures to fall enough to get maximum efficiency from an R-134a system.*

❏ Systems with condensers too small to allow for the higher condensation pressure of R-134a. Many minivan condensers are too small to handle an R-134a conversion. One indication of this is system high-side pressures that are higher than normal with R-12. If the pressure is above 300 psi (2067 kPa) at idle, and the condenser fans are operating properly, the condenser is at its maximum capacity. Some manufacturers are now offering replacement condensers for R-134a retrofits. Check to see if one is available for the vehicle you are retrofitting.

If the vehicle refrigeration system meets either of the above conditions, they should be refilled with R-12 as long as R-12 can be obtained. For more information about various refrigerants, refer to Chapter 6.

Level of Retrofit

When retrofitting an air conditioning system, three factors must be considered: cost, climate, and components. These factors must be taken into account when retrofitting to get close to the same level of performance as R-12.

The cost of the retrofit must be figured in along with the cost of repairing the system's leaking or malfunctioning component(s). Depending on the age of the system, a retrofit can become more expensive than the actual repair. Ask the driver if he or she knows how much longer they plan to keep the vehicle; this can factor into the decision to perform the retrofit.

Much of the air conditioning system's performance is based on climate conditions and driver perception. For example, an owner of a vehicle driven on Sundays only in Minnesota may not expect the same level of system performance as the owner of a vehicle driven six days a week in Louisiana. Many customers choose to have the refrigerant only retrofit, despite the fact it may not provide maximum performance.

The third aspect of retrofit is existing components. Will the existing system components work with the retrofit? In some cases, additional components may have to be replaced along with the malfunctioning component(s) for the retrofit to be successful.

Refrigerant Only Retrofits

It is possible to perform a basic retrofit consisting of changing the service fittings and replacing the R-12 with R-134a. The retrofitted system will produce cold air. There are some factors, however, the technician must consider and the owner should be made aware of before performing a basic retrofit:

❏ The R-12 desiccant is not compatible with R-134a. The desiccant may not absorb all of the moisture in the system. The desiccant may break down and circulate through the system, clogging screens and orifices.

❏ The R-134a molecule is smaller than the R-12 molecule. If not detected, a slight leak in an R-12 system will become a severe leak when the system is converted to R-134a.

❏ The R-12 compressor oil will not work with R-134a. The oil will not mix properly with the refrigerant, causing it to collect in places other than the compressor. The compressor may starve for oil and wear out. The oil will eventually break down, causing severe system damage.

❏ If the system is not equipped with a high-pressure cutoff switch, a switch and wiring harness for the switch must be installed.

❏ Some vehicles need extra condenser airflow when converted to R-134a. Many minivans need an extra cooling fan. If the vehicle is a rear-wheel drive model with a fan clutch, the fan clutch may need replacing. If this is not done, the compressor will be placed under a strain and will fail prematurely. High-side pressures may become so high the pressure relief valve opens, losing refrigerant.

A basic retrofit may initially appear to work well. Eventually, the factors just listed will cause the system to fail, requiring expensive repairs. The only time a refrigerant-only retrofit makes sense is when the vehicle will

be scrapped within a few months, or when the owner intends to return for a complete retrofit within a short time.

Performing a Complete Retrofit

The procedures in the following paragraphs are a general guide to retrofitting a refrigeration system. There may be some specific steps that apply to certain systems. These will be stated in the factory service literature. You can retrofit an R-12 system to use R-134a or other approved refrigerant. However, in all cases, the new refrigerant must be EPA approved for use in motor vehicles and you must follow all guidelines on the use of proper fittings, labels, and hoses.

 Note: This procedure outlines an R-12 to R-134a retrofit. This procedure is not specific for any vehicle and should not be used as a substitute for the manufacturer's recommended procedure.

Preliminary Steps

Check the refrigeration system and other HVAC parts first, and determine what repairs, if any, are needed. If the system is still operable, run it at idle with the blower on high for 5 minutes. This will return the maximum amount of oil to the compressor for removal. Next, make a thorough leak check. Remember R-134a leaks out at a much faster rate than R-12. Then recover the R-12 refrigerant into a dedicated recycling machine.

Thoroughly inspect the engine cooling system according to the procedures explained in Chapter 20. If the cooling system has any defects, they should be corrected before proceeding. Pay particular attention to the cooling fans. Correct any fan problems, and decide whether or not the system will require an extra cooling fan. The vehicle manufacturer can advise you about the need for an extra fan.

 Note: In areas where high temperature/ high humidity days are frequent, it is a good idea to install an extra cooling fan even if the retrofit does not require one. Fan kits can be bought separately.

Complete Other Refrigeration System Repairs

Replace parts as necessary using R-134a compatible parts. Complete all repairs before beginning the retrofit process. If a hose must be replaced, use a quality barrier hose. If any fittings with O-rings have been removed, replace with R-134a compatible O-rings. Original R-12 hoses and O-rings will be oil soaked, and do not need replacing if they are not defective and have not been removed for other service.

If major parts are being replaced, add the correct amount of R-134a compatible PAG or ester oil instead of the original mineral oil. Flush the system and change the orifice tube or expansion valve screen as necessary. Some manufacturers recommend installing a filter, **Figure 23-11**. The filter may be installed permanently, or long enough to collect debris after a compressor failure.

Figure 23-11. *Install an inline filter if there is any doubt as to the amount of debris in the system. A filter should be installed whenever the system has suffered a debris producing failure, even if it is not being retrofitted.*

Retrofitting with R-134a Compatible Parts

Most of the parts in this section can be changed with a small amount of labor. The only exception is a compressor that must be removed to replace the oil. These procedures are outlined in the following paragraphs.

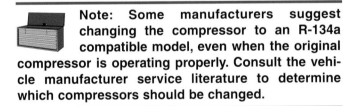

Note: Some manufacturers suggest changing the compressor to an R-134a compatible model, even when the original compressor is operating properly. Consult the vehicle manufacturer service literature to determine which compressors should be changed.

Replace Refrigerant Oil

It is impossible to remove all the original R-12 mineral oil, but you must remove as much as possible. Oil removal was discussed in Chapter 16. If the compressor has a drain plug, remove and allow as much oil as possible to drain out. Applying sight pressure to the refrigeration system with nitrogen will help to force the oil out. When as much oil as possible has been removed, reinstall the drain plug using a new R-134a compatible O-ring.

If the compressor has no drain plug, remove it and drain the oil through the intake and discharge ports. After most of the oil has drained out, turn the compressor clutch plate through several revolutions to ensure oil has not entered the cylinders. Then reinstall the compressor and add the correct amount of PAG or ester oil as recommended.

Compressors

Any compressor that is not in good shape should be replaced as part of a retrofit. If the compressor must be replaced, use a compatible compressor. If the compressor appears to be in good shape, try to keep it. The internal components of a compressor broken in with R-12 have a thin film of metal chloride, which acts as a good antiwear agent, even after the R-134a conversion.

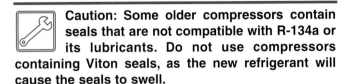

Caution: Some older compressors contain seals that are not compatible with R-134a or its lubricants. Do not use compressors containing Viton seals, as the new refrigerant will cause the seals to swell.

Replace Desiccant

Next, install a new receiver-drier or accumulator containing R-134a compatible desiccant. Desiccant XH-7 is used with R-134a only, while desiccant XH-9 works with both kinds. Desiccant XH-5 is for use with R-12 only.

Replace Fittings

Replace the original R-12 fittings with the proper R-134a fittings, **Figure 23-12**. There are two common methods of replacing fittings, using the existing fitting and a separate saddle clamp fitting. Installation procedures for each type are explained on the next page.

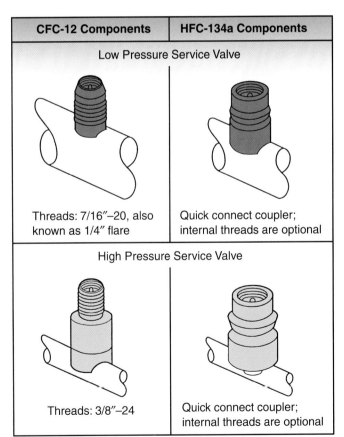

Figure 23-12. *This chart shows the relationship of R-134a and R-12 fittings. Do not mix fittings! (Gates)*

Figure 23-13 illustrates an R-134a fitting that threads over the existing R-12 fitting. To install this type of fitting, remove the valve core from the existing R-12 fitting. Then coat the threads of the male fitting with Loctite or other thread locking compound to prevent removal. Then thread the R-134a fitting onto the R-12 fitting.

Another type of fitting is the saddle clamp fitting. To install a saddle clamp fitting, locate a straight section of metal line on both the high and low sides. Install the two halves of the clamp over the first line and tighten it into place. Make sure any sealing devices are also installed. Then thread the piercing tool into the fitting until it punctures a hole in the line. Remove the piercing tool and install the valve core. Then coat the cap of the original R-12 fitting with Loctite or other thread locking compound to prevent removal.

Figure 23-13. *R-134a retrofit fittings are available and can be installed over the factory R-12 fittings. Use Loctite when installing the fittings to prevent them from being removed.*

Replace Rubber Components

Some manufacturers call for replacing certain seals and O-rings when the system is retrofitted. Typical rubber parts are high-side fitting gaskets and O-rings. Change these parts at this time, being sure to use R-134a compatible parts. Many retrofit O-rings are a different color than the original O-ring.

Adjust or Replace Pressure/Temperature Switches

Install or change the low-side pressure cycling switch if the old switch is not adjustable. If the switch is adjustable, it will need to be set after the system is operating. If the

 Note: You may need to change the evaporator temperature sensor to avoid false trouble codes on some automatic temperature control systems.

vehicle has an adjustable POA valve, set it to about 4 psi (28 kPa) below the R-12 point.

Install High-Pressure Switch

Check to see if the system is equipped with a high-pressure switch. If the system is not equipped with a high-pressure cutout switch, you are required to install one as part of a retrofit to R-134a or any other refrigerant. Most newer systems are equipped with high-pressure switches, making this step unnecessary in most cases.

 Warning: This is a mandatory step in the retrofit, required by law. Do not skip this step simply to save money on the job.

The easiest way to install a pressure switch is to use an accumulator (if the system has one) equipped with a extra Schrader valve port. The other method is to use a tee fitting to install the switch. If necessary, install a saddle port and install the switch. Follow the manufacturer's instructions for wiring the switch connector into the vehicle's HVAC control system.

Add Cooling Fan/Replace Condenser

Some refrigeration systems will not operate properly on R-134a unless an extra fan is added. The fan pulls more air through the condenser at idle. Install the extra fan as was explained in the aftermarket air conditioner section earlier.

A few refrigeration systems have condensers that are too small to handle the higher condensation pressures of R-134a. The manufacturer will call for replacing the condenser with a more efficient model. As stated earlier, some manufacturers are offering replacement condensers for R-134a retrofits. Basic condenser replacement procedures are the same as those described in Chapter 19. If the replacement condenser is larger or differently shaped, revised installation instructions will be provided by the condenser manufacturer.

Evacuate and Charge System.

After all new parts are in place, attach the gauge manifold or refrigeration machine hoses and evacuate the system for at least 45 minutes. The longer evacuation time is needed to remove water as PAG oil is very hygroscopic. After evacuating the system, thoroughly leak test by turning off the pump and closing the manifold or service machine valves. Allow the system to sit for at least 20 minutes. The low side gauge should not move. If it does, locate and correct leaks.

If the system has no leaks, proceed with charging. Remember, less refrigerant is needed with R-134a. Do not add more than 90% of the R-12 charge. As an example, a system designed to contain 36 oz. (1.02 kg) of R-12 will probably work well with 27 oz. (.756 kg) of R-134a, and should not be charged with more than 32 oz. (.896 kg).

Start System and Make Adjustments

Start the engine and check for leaks. Then adjust the pressure switch to cycle the clutch off at approximately 21 psi (145 kPa). The clutch should cycle back on at about 40 psi (276 kPa). If the system uses a POA valve instead of a cycling clutch, check the valve controls pressure to about 21 psi (145 kPa). Readjust the POA valve if applicable. If pressures indicate more refrigerant is needed, add more in small amounts, no more than 90% of the original R-12 charge.

Check Operation

Make a road test first to stabilize pressures and distribute oil. Then recheck pressures. High-side pressures should be no higher than about 115% of R-12 pressures. For instance, if the R-12 high-side pressure was 250 psi (1723 kPa) at 70°F (21°C) ambient temperature and the engine idling, it should be no more than 288 psi (1984 kPa) with R-134a under the same conditions. Finally, check vent temperatures. Most vent temperatures will be virtually the same as they were with R-12. On a few vehicles, the vent temperature may be 2-4°F (1-2°C) higher than R-12 readings. A small variation such as this in most cases will not be noticeable.

Turn off the engine and recheck for leaks. If none are found and the system is operating properly, proceed with labeling.

Install Label

On all vehicles retrofitted to accept R-134a or another refrigerant, a detailed *label* must be applied in a conspicuous location in the engine compartment, **Figure 23-14.** The label is color-coded, depending on the refrigerant, **Figure 23-15.** On vehicles retrofitted to accept R-134a or any other refrigerant, the label must show the following:

❏ The name and address of the shop or technician performing the retrofit.

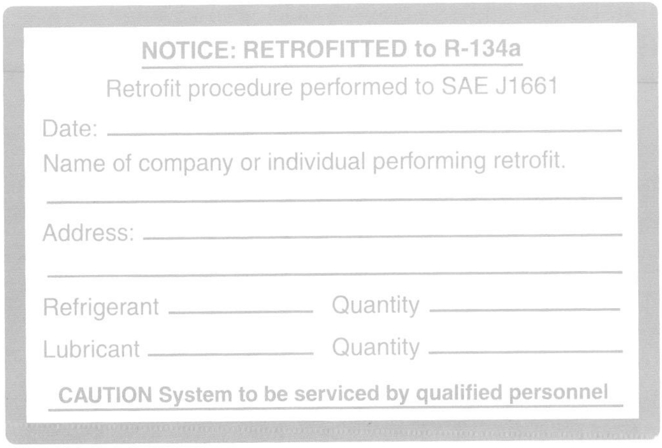

Figure 23-14. *Before finishing the retrofitting job, always install a retrofit label. The label should be placed where it can easily be seen by the next person to work on the vehicle. You must fill out the label completely.*

❏ The date of the retrofit.
❏ The refrigerant trade name and charge amount. In some areas, you may need to add the refrigerant's ASHRAE numerical designator.
❏ The type, manufacturer, and amount of lubricant.
❏ The phrase "ozone depleter" if the refrigerant contains an ozone-depleting substance.

Mark on the label or elsewhere, the high- and low-side pressure gauge readings, as well as how much refrigerant was charged into the system.

Refrigerant	Label Background Color
HFC-134a	Sky Blue
CFC-12	White
Freeze 12	Yellow
Free Zone/RB-276	Light Green
Hot Shot	Medium Blue
GHG-X4	Red
R-406A	Black
GHG-X5	Orange
FRIGC FR-12	Gray

Figure 23-15. *Label background colors for each type of refrigerant. Only refrigerants approved by the EPA's SNAP program are on this list. (EPA)*

Summary

Vehicles with factory air conditioning are widely available, but the technician will occasionally be asked to install an aftermarket air conditioner. Usual candidates for aftermarket air conditioning are small cars, pickup trucks, sport utility vehicles, and classic or custom vehicles. The parts and instructions necessary to install an aftermarket air conditioner are called an air conditioning kit. The two kinds of kits are the universal kit, which can be installed in many kinds of vehicle and the custom kits made to fit a certain vehicle. Modern air conditioning kits may combine an evaporator, heater core, and defroster vents. Aftermarket air conditioning kits are available in R-12 and R-134a versions.

Before agreeing to install an aftermarket air conditioner on the vehicle, the technician should consider air conditioner and vehicle compatibility, whether the cooling system requires modification, component mounting, compressor compatibility, and body sealing.

Begin aftermarket air conditioning by determining the right kit to be used. Check all parts against the packing list before starting, and leave the sealing caps on all refrigeration system parts until they are ready to be installed.

Mark and drill holes, then install the evaporator assembly. Install ducts as needed, then tighten the evaporator mounting brackets. Next, install the compressor on the engine. If necessary, install the compressor drive pulley on the engine, then install the compressor drive belt.

Drill holes as needed and install the condenser brackets. Install an extra cooling fan if needed, then install the receiver-drier and connect the refrigeration hoses. Then install the heater hoses and heater shutoff valve if a heater is used. If necessary, drain the cooling system and install the heater hose fittings and fan control unit on the engine. Then refill and bleed the cooling system. Install the electrical and vacuum components that have not already been installed. Connect all wiring and hoses.

Evacuate, leak check, and charge the new refrigeration system. R-134a systems require a smaller charge of refrigerant than R-12 systems. Next, start the engine, turn the air conditioner controls to maximum cooling, and monitor pressures and system operation. Road test the vehicle with a temperature gauge installed in the vent closest to the evaporator. Adjust the thermostat if necessary.

R-12 to R-134a retrofits are commonly performed and will be more common as R-12 becomes harder to get. The only vehicles that should not be retrofitted are those with non-adjustable low-side pressure controls or low capacity condensers.

A retrofit in which only the refrigerant is changed will produce cold air. However, the system will eventually fail because the original oil, desiccant, and rubber components are not compatible with R-134a, and because high-side pressures may become excessive. Whenever possible, perform a complete retrofit.

Before beginning a complete retrofit, check the refrigeration and HVAC systems and determine what, if any, other repairs are needed. Complete other refrigeration system repairs as necessary, using R-134a compatible parts. Then change other parts to R-134a compatible parts as necessary. This includes oil, desiccant, service fittings, some rubber parts, and possibly pressure switches.

Finally, evacuate, leak test, and charge the system. Then start the system and make adjustments as necessary. Road test, then recheck pressures and vent temperatures.

Review Questions—Chapter 23

Please do not write in this text. Write your answers on a separate sheet of paper.

1. An aftermarket air conditioning kit that can be used in many kinds of vehicle is a _____ type.

2. An aftermarket air conditioning unit uses which of the following?
   (A) Expansion valve.
   (B) POA valve.
   (C) Variable capacity compressor.
   (D) All of the above.

3. The thermostatic switch on an aftermarket air conditioner can be adjusted to vary the _____ temperature.

4. Vehicles that should not be retrofitted to R-134a are those with _____ low-side pressure controls or _____ condensers.

5. Which of the following R-12 parts is *not* compatible with R-134a?
   (A) Compressor oil.
   (B) Desiccant.
   (C) POA valves.
   (D) All of the above.

6. Evacuate a retrofitted system for at least ___ minutes.

7. The clutch should _____ at approximately 21 psi (145 kPa).
   (A) cycle on
   (B) cycle off
   (C) ice up
   (D) cause the belt to squeal

8. An R-134a system will usually operate well on about _____ of the original R-12 charge.
   (A) 25%
   (B) 50%
   (C) 75%
   (D) 115%

9. Condenser pressures on an R-134a retrofit will usually be ___ they were with R-12.
   (A) higher than
   (B) lower than
   (C) the same as
   (D) None of the above.

10. Vent temperatures on an R-134a retrofit will usually be ___ they were with R-12.
    (A) higher than
    (B) lower than
    (C) the same as
    (D) None of the above.

ASE Certification-Type Questions

1. Technician A says few vehicles are manufactured without factory air conditioning. Technician B says only new vehicle dealers install custom aftermarket air conditioners. Who is right?
   (A) A only.
   (B) B only.
   (C) Both A and B.
   (D) Neither A nor B.

2. Before ordering the aftermarket air conditioning kit, the technician must consider all of the following variations, *except:*
   (A) the type of refrigerant.
   (B) the type of pressure control.
   (C) the type of vehicle.
   (D) whether the compressor and condenser will be supplied with the kit.

3. Technician A says a low capacity compressor may cause cooling problems. Technician B says a condenser that is too small may cause cooling problems. Who is right?
   (A) A only.
   (B) B only.
   (C) Both A and B.
   (D) Neither A nor B.

4. All of the following statements about properly installing an aftermarket air conditioner evaporator are true, *except:*
   (A) drill all holes, then position the evaporator assembly and make sure it fits.
   (B) ensure the evaporator drain hole will be on the bottom of the case.
   (C) ensure the refrigerant hoses will not be too close to the engine or exhaust components.
   (D) carefully check out the area behind all potential drilling spots.

5. Placement of the receiver-drier in an aftermarket air conditioner installation is being discussed. Technician A says the receiver-drier can be located at any point in the high pressure line. Technician B says the receiver-drier can be installed in any position. Who is right?
   (A) A only.
   (B) B only.
   (C) Both A and B.
   (D) Neither A nor B.

6. Which of the following procedures is *most likely* to result in a satisfactory retrofit?
   (A) Removing all R-12 and installing R-134a.
   (B) Topping off an R-12 system with R-134a.
   (C) Changing the compressor and charging with R-134a.
   (D) Changing oil and desiccant and charging with R-134a.

7. Which of the following procedures violates EPA regulations?
   (A) Removing R-12 from a system, evacuating, and charging with R-134a.
   (B) Changing system oil and desiccant and charging with R-134a.
   (C) Failing to install a high pressure cutoff switch on a system not equipped with a high-pressure switch from the factory.
   (D) Changing the compressor, adding PAG oil, and charging with R-134a.

8. All of the following statements about R-134a compatible desiccant are true, *except:*
   (A) the desiccant should always be changed if the refrigerant is changed.
   (B) desiccant XH-5 is used with R-12 only.
   (C) desiccant XH-7 is used with R-134a only
   (D) desiccant XH-9 is used with R-134a only.

9. All of the following statements about changing pressure and temperature switches during an R-134a retrofit are true, *except:*
   (A) if the old pressure cycling switch is not adjustable, leave it in place.
   (B) an adjustable pressure cycling switch should be set after the system is operating.
   (C) a few POA valves are adjustable.
   (D) the evaporator temperature sensor may set false trouble codes if not replaced.

10. Technician A says the retrofitted system should be evacuated for at least 2 hours to remove all residual R-12. Technician B says the amount of R-134a charged into the system should never be more than 90% of the R-12 amount. Who is right?
    (A) A only.
    (B) B only.
    (C) Both A and B.
    (D) Neither A nor B.

ASE certified technicians have the training to properly diagnose and service vehicle air conditioning and heating systems. You must be certified in recovery and recycling in order to purchase R-12 refrigerant and other chemicals. (Goodyear)

Chapter 24

ASE Certification

After studying this chapter, you will be able to:
- ❑ Explain why technician certification is beneficial to technicians and vehicle owners.
- ❑ Explain the process of registering for ASE tests.
- ❑ Explain how to take the ASE tests.
- ❑ Identify typical ASE test questions.
- ❑ Identify the format of the ASE test results.
- ❑ Explain how the ASE test results are used.
- ❑ Discuss certification programs for refrigerant recovery and recycling.

Technical Terms

National Institute for Automotive Service Excellence

ASE

Standardized tests

Certified

Master Technician

ACT

Registration Booklet

Pass/fail letter

Test score report

Certificate in evidence of competence

In this chapter, the purpose and organization of the National Institute for Automotive Service Excellence (ASE) will be explained. This chapter also covers ASE certification, the advantages, and in the case of air conditioning recovery and recycling, the need to become certified. In this chapter are directions for applying for and taking the ASE tests. When you have finished studying this chapter, you will know the purposes of ASE, the design of the ASE tests, how the test results are delivered, and the purpose of the test results.

Reasons for ASE Tests

The concept of setting standards of excellence for skilled jobs is not new. In ancient times, metalworkers, weavers, potters, and other artisans were expected to conform to set standards of product quality. In many cases, this need for standards resulted in the establishment of associations of skilled workers who set standards and enforced rules of conduct. Ancient civilizations had such associations, and many medieval industries were regulated by guilds. Many modern American labor unions are descended from early associations of skilled workers. Certification processes for aircraft mechanics, aerospace workers, and electronics technicians have existed since the beginnings of these industries.

However, this has not always been true of the automotive industry. Automobile manufacturing and repair began as a fragmented industry, made up of many small vehicle manufacturers, and thousands of small repair shops. Although the number of vehicle manufacturers decreased, the number of repair facilities continued to grow in number, and also in variety. Due to its fragmented, decentralized nature, standards for the automotive repair industry were difficult to establish. For over 50 years, there was no unified set of standards of automotive repair knowledge or experience. Anyone could claim to be an automotive mechanic, no matter how unqualified, often resulting in unneeded or improperly done repair work. As a result, a large segment of the public came to regard mechanics as unintelligent, dishonest, or both.

This situation changed in 1972, when the **National Institute for Automotive Service Excellence,** now called **ASE,** was established to provide a certification process for automobile technicians. ASE is a non-profit organization formed to encourage and promote high standards of automotive service and repair. ASE does this by providing a series of written tests on various subjects in the automotive repair, truck repair, collision repair/refinishing, school bus, and engine machinist areas. These tests are called **standardized tests,** which means the same test in a particular subject is given to everyone throughout the United States. Any person passing one of these tests, and meeting certain experience requirements, is **certified** (officially recognized as meeting all standards) in the subject covered by that test. If a technician can pass all the tests in a given area he or she is certified as a **Master Technician** in that skill area.

The purpose of the ASE certification test program is to identify skilled and knowledgeable technicians. Periodic recertification provides an incentive for updating skills, and also provides guidelines for keeping up with current technology. The test program allows potential employers and the driving public to identify good technicians, and helps the technician advance his or her career. The program is not mandatory on a national level, but many repair shops now hire only ASE certified technicians. Close to 500,000 persons are now ASE certified in one or more areas.

The ASE certification program has now been extended to Canada and Brazil. The tests in these countries are similar to those given in the United States. Select tests written in French and Portuguese are offered in the respective countries. In the near future, ASE plans to offer certification tests in Mexico. Other countries may become involved in the ASE certification program in the near future.

Other activities ASE is involved in are encouraging the development of effective training programs, and conducting research on the best methods of performing instruction, and publicizing the advantages of technician certification. ASE is managed by a board made up of persons from the automotive and truck service industries, motor vehicle manufacturers, state and federal government agencies, schools and other educational groups, and consumer associations.

The advantages the ASE certification program has brought to the automotive industry include increased respect and trust of automotive technicians, at least of those who are ASE certified. This has resulted in better pay and working conditions for technicians, and increased standing in the community. Thanks to ASE, automotive technicians are taking their place next to other skilled artisans.

Applying for the ASE Tests

You do not have to be employed in the automotive service industry to apply for and take any ASE test. However, to become certified, the applicant must have two years experience working as an automobile or truck technician. In some cases, training programs or courses, an apprenticeship program, or time spent performing similar work may be substituted for all or part of the work experience.

ASE tests are given twice each year, in the Spring and Fall. Tests are usually held during a two-week period at night during the workweek. The actual test administration is performed by **ACT,** a non-profit organization experienced in administering standardized tests.

The tests are given at designated test centers in over 300 places in the United States. If necessary, special test centers can be set up in remote locations. However, there must be a certain minimum of potential test takers before a special test center can be set up.

To apply for the ASE tests, begin by obtaining an application form like the one shown in **Figure 24-1**. To obtain the most current application form, contact ASE at the following address:

National Institute for Automotive Service Excellence
13505 Dulles Technology Drive, Suite 2
Herndon, VA 20171-3421

ASE will send the proper form, enclosed in a **Registration Booklet** explaining how to complete the form. When you get the form, carefully fill it out, recording all needed information. You may apply to take as many tests as are being given, or fewer tests, or only one test if desired. Work experience, or any substitutes for work experience, should also be included, according to the instructions in the latest Registration Booklet. If there is any doubt about what should be placed in a particular space,

consult the Registration Booklet. Be sure to determine the closest test center, and record its number in the appropriate space. Most test centers are located at local colleges and schools. You can also register at ASE's Internet site, *http://www.asecert.org,* **Figure 24-2.**

When you send in the application, you must include a check, money order, or credit card number to cover all necessary fees. A fee is charged to register for the test series, and a separate fee is charged for each test to be taken. See the latest Registration Booklet for the current fee structure. In some cases, your employer may pay the registration and test fees. Check with your employer before sending in your application. If you will need to take the ASE tests in a language other than English, indicate this on the application form. Test booklets in some certification areas are available in Spanish, French, and Portuguese.

Figure 24-1. *A sample form for registering for ASE certification tests. Be sure you fill in all pertinent information required. Include payment for all applicable test fees. (ASE)*

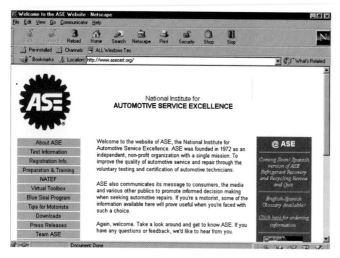

Figure 24-2. *You can register for ASE tests or find out more information about ASE over the information superhighway at ASE's Internet site. (ASE)*

To be accepted for either the Spring or Fall ASE tests, your application and payment must arrive at ASE headquarters at least one month before the test date(s). To ensure you can take the test at the location of your choice, send in your application as early as possible.

After sending the application and fees, you will receive an admission ticket to the test center. This should arrive by mail within two weeks of sending the application. See **Figure 24-3.** If your admission ticket has not arrived, and it is less than two weeks until the test date(s), contact ASE using the phone number given in the latest ASE Registration Booklet. If the desired test center is filled when ASE receives your application, you will be directed to report to the nearest center that has an opening. If it is not possible for you to go to the alternate test center, contact ACT immediately, using the phone number given in the Registration Booklet.

Preparing for the ASE Tests

ASE tests are designed to measure your knowledge of three things:

❑ Basic information on how automotive systems and components work.

❑ Diagnosis and testing of systems and components.

❑ Repair of automotive systems and components.

Therefore, you should study the basic principles of how automotive systems work, and also study the latest information about diagnosis and repair steps. Good sources of this material are your textbook, service manuals, factory training material, periodicals such as Motor or Motor Age magazines, and service bulletins.

Remember, ASE tests are designed to test you knowledge of correct diagnosis and repair procedures. Do not assume the way you have always done something is the correct way.

Taking the ASE Tests

Be sure to bring your admission ticket with you when reporting to the test center. When you arrive at the test center, you will be asked to produce the admission ticket and a driver's license or other photographic identification. In addition to these items, bring some extra Number 2 pencils. Although pencils will be made available at the test center, some extra pencils may save you time if the original pencil breaks.

After you enter the test center and are seated, listen to and follow all instructions given by the test administrators. During the actual test, carefully read all test questions before making a decision as to the proper answer. ASE tests consist of multiple choice questions with four possible answers.

Types of ASE Test Questions

Each ASE test will contain between 40 and 80 test questions, depending on the subject to be tested. All test questions are multiple-choice, with four possible answers. These types of multiple-choice questions are similar to the multiple-choice questions used in this textbook. The next section discusses the types of ASE questions you are likely to encounter.

One-part Question

One-part questions require you to answer a single question:

1. Which of the following refrigerants are approved for retrofitting by most vehicle manufacturers?

    (A) R-22.

    (B) R-134a.

    (C) Blends containing R-12.

    (D) Blends containing R-22.

Notice the question calls for the best answer out of all of the possibilities. R-22 and all blends (A, C, and D) are not approved by vehicle makers for retrofitting. Only R-134a is manufacturer approved. Therefore, "B" is correct.

Two-part Question

Two-part questions used in the ASE tests involves two technicians named Technician A and Technician B. Each technician makes a statement. The statements usually address one type of system. You are asked to determine whether each of the statements is true:

1. Technician A says a refrigeration system with a variable displacement compressor does not use a cycling

National Institute for Automotive Service Excellence

ACT, P.O. Box 4007, Iowa City, Iowa 52243, Phone: (319) 337-1433   017910042 T

Admission Ticket

Test Center to which you are assigned:

A

REGULAR TESTS (Late arrivals may not be admitted.)		
DATE	REPORTING TIME	TEST(S)
11/14	7:00 PM	A5, A7

RECERTIFICATION TESTS (Late arrivals may not be admitted.)

TEST CODE KEY

A1 Auto: Engine Repair	M1 Machinist: Cylinder Head Specialist	T6 Med/Hvy Truck: Elec./Electronic Systems	P2 Parts: Automobile Parts Specialist
A2 Auto: Automatic Trans/Transaxle	M2 Machinist: Cylinder Block Specialist	T8 Med/Hvy Truck: Preventive Main. Inspec.	F1 Alt. Fuels: Lt. Veh. Comprsd. Nat. Gas
A3 Auto: Manual Drive Train & Axles	M3 Machinist: Assembly Specialist	B2 Coll.: Painting & Refinishing	L1 Adv. Level: Adv. Engine Perf. Spec.
A4 Auto: Suspension & Steering	T1 Med/Hvy Truck: Gasoline Engines	B3 Coll.: Non-structural Analysis	S1 School Bus: Body Sys. & Spec. Equip.
A5 Auto: Brakes	T2 Med/Hvy Truck: Diesel Engines	B4 Coll.: Structural Analysis	S4 School Bus: Brakes
A6 Auto: Electrical/Electronic Systems	T3 Med/Hvy Truck: Drive Train	B5 Coll.: Mechanical & Elec. Components	S5 School Bus: Suspension & Steering
A7 Auto: Heating & Air Conditioning	T4 Med/Hvy Truck: Brakes	B6 Coll.: Damage Analysis & Estimating	S6 School Bus: Elec./Electronic Systems
A8 Auto: Engine Performance	T5 Med/Hvy Truck: Suspension & Steering	P1 Parts: Med/Hvy Truck Parts Specialist	

See Notes and Ticketing Rules on reverse side. An asterisk (∗) indicates your certification in these areas is expiring.

SPECIAL MESSAGES

-REVIEW ALL INFORMATION ON THIS TICKET. CALL IMMEDIATELY TO REPORT AN
ERROR OR IF YOU HAVE QUESTIONS.
-IF YOU MISS ANY EXAMS, FOLLOW THE REFUND INSTRUCTIONS ON THE BACK OF THIS SHEET.
THE REFUND DEADLINE IS
-YOU HAVE BEEN ASSIGNED TO AN ALTERNATE TEST CENTER. THE CENTER
ORIGINALLY REQUESTED IS FULL.
8010-IL/LOCAL 150 IS LOCATED ON JOLIET AVE, THREE DOORS W. OF LAGRANGE RD ON
SOUTH SIDE OF THE ST. ENTER THROUGH BACK DOOR. NO ALCOHOL ON PREMISES.

MATCHING INFORMATION: The information printed in blocks B and C at the right was obtained from your registration form. It will be used to match your registration information and your test information. Therefore, the information at the right must be copied EXACTLY (even if it is in error) onto your answer booklet on the day of the test. If the information is not copied exactly as shown, it may cause a delay in reporting your test results to you.

IF THERE ARE ERRORS: If there are any errors or if any information is missing in block A above or in blocks B and C at the right, you must contact ACT immediately. DO NOT SEND THIS ADMISSION TICKET TO ACT TO MAKE SUCH CORRECTIONS.

Check your tests and test center to be sure they are what you requested. If either is incorrect, call 319/337-1433 immediately. Tests cannot be changed at the test center. **ON THE DAY OF THE TEST,** be sure to bring this admission ticket, positive identification, several sharpened No. 2 pencils, and a watch if you wish to pace yourself.

B FIRST FIVE LETTERS OF LAST NAME

C SOCIAL SECURITY NUMBER OR ACT IDENTIFICATION NUMBER

SIDE 1

Figure 24-3. *The Admission Ticket will arrive via mail. If your ticket does not arrive within two weeks after sending in your registration form, contact ASE immediately.*

clutch. Technician B says a refrigeration system with an axial piston compressor does not use a cycling clutch. Who is right?

(A) A only.
(B) B only.
(C) Both A and B.
(D) Neither A nor B.

Note both statements can be true, both can be false, or only one of them can be false. In this case, the statement of Technician A is correct since a variable displacement compressor does not need a cycling clutch. However, the statement of Technician B is incorrect, since the many axial piston compressors are used with a cycling clutch. Therefore, the correct answer is "A".

A variation of the two-part question sets up a situation or states a subject under discussion. Again, Technician A and Technician B each make a statement and you are asked to determine whether each of the statements is true:

1. An air conditioner stops cooling, accompanied by low airflow, after operating for 10 minutes. When the system is turned off for a few minutes, it begins cooling again, but then stops cooling after a few minutes. Technician A says this is normal operation on hot days. Technician B says the evaporator may be icing up. Who is right?

(A) A only.
(B) B only.
(C) Both A and B.
(D) Neither A nor B.

In this case, you must read the statement carefully to determine the exact nature of the situation. You can then determine whether the statements made by Technicians A and B are correct. In this case, Technician A's statement is incorrect, since the air conditioner should continue to produce cool air at all times. Technician B's statement is correct, since evaporator icing can produce the exact symptoms described in the opening sentence. Therefore, the correct answer is "B".

Negative Questions

Some questions are called negative questions. These questions ask you to identify the incorrect statement from four possible choices. They will usually have the word "except" in the question. An example of a negative question is given below.

1. A blend door may be operated by all of the following, *except:*

(A) vacuum diaphragm.
(B) electric motor.
(C) cable and linkage.
(D) thermostatic spring.

Vacuum, electric motors, or a cable can operate the blend door. Therefore A, B, and C cannot be the correct answer. The blend door is never operated by a thermostatic spring. Therefore, the correct answer is "D".

A variation of the negative question will use either the words "most" or "least," such as the one below:

1. Air blows out of the defroster vents in any HVAC mode. Which of the following defects is the *least likely* cause?

(A) Defective heater-defroster door diaphragm.
(B) Vacuum hose loose at intake manifold.
(C) Recirculation door stuck.
(D) Disconnected vacuum control valve.

A defective heater-defroster door diaphragm (A) could cause air to exit from the defroster at all times. A loose manifold vacuum hose (B) could keep the vacuum doors from operating properly. A disconnected vacuum control valve (C) would also keep the vacuum doors from operating properly. The least likely cause of air exiting from defroster vents would be a stuck recirculation door, which has no effect on the defroster airflow. Therefore, the answer is "C".

Incomplete Sentence Questions

Some test questions are incomplete sentences, with one of the four possible answers correctly completing the sentence. An example of an incomplete sentence question is given here.

1. A 400 psi (2756 kPa) reading on the refrigeration system high side could be caused by a _____.

(A) restricted condenser
(B) low refrigerant charge
(C) missing compressor drive belt
(D) worn out compressor

Once again the question calls for the best answer. All of the possibilities except a restricted condenser would cause low pressures. Therefore "A" is correct.

The ASE Heating and Air Conditioning Test

The ASE certification in heating and air conditioning (A7) is one of the most important certifications to obtain. While there are several organizations that provide certification in various aspects of air conditioning service, ASE covers most of the areas in heating and air conditioning service. ASE certification is the one most frequently recognized by governmental agencies, as well as the motoring public.

The ASE heating and air conditioning test contains approximately 50 questions. The categories in this test include:

❑ A/C System Diagnosis and Repair.
❑ Refrigeration System Component Diagnosis and Repair.
❑ Heating and Engine Cooling Systems Diagnosis and Repair.
❑ Operating Systems & Related Controls Diagnosis and Repair.
❑ Refrigerant Recovery, Recycling, and Handling.

The section on refrigeration system component diagnosis and repair is subdivided into two sections covering the compressor, clutch, evaporator, condenser, and other related components. The section on operating systems & related controls diagnosis and repair is subdivided into three areas covering electrical, vacuum/mechanical, and automatic and semiautomatic HVAC systems.

After completing all the questions on the test, recheck your answers one time to ensure you did not miss anything that would change an answer, or that you did not make a careless error on the answer sheet. In most cases, rechecking your answers more than once is unnecessary and may lead you to change correct answers to incorrect ones. The time allowed for each test is usually about four hours. However, you may leave after completing your last test and handing in all test material.

Test Results

Test grading takes from six to eight weeks. After this time, you will receive a *pass/fail letter* from ASE. A typical pass/fail letter is shown in **Figure 24-4.** This letter will tell

**National Institute for
AUTOMOTIVE SERVICE EXCELLENCE**

December 20, XXXX   032527

John Smith
123 Main Street
Edens, Il 60000

Dear ASE Test Taker:

Listed below are the results of your November XXXX ASE Tests. You will soon be receiving a more detailed report.

If your test result is "Pass", and if you have fulfilled the two-year "hands-on" experience requirement, you will receive a certificate and credential cards for the tests you passed.

If your test result is "More Preparation Needed", you did not attain a passing score. Check your detailed score report when it arrives. This information may help you prepare for your next attempt.

If you do not receive your detailed report within the next three weeks, please call.

Thank you for participating in the ASE program.

| A5 | BRAKES | PASS |
| A7 | HEATING & AIR CONDITIONING | PASS |

123-45-6789

13505 Dulles Technology Drive · Herndon, Virginia 22071-3415 · (703) 713-3800

Figure 24-4. *A confidential pass/fail letter will be sent to you shortly after your ASE tests. (ASE)*

you only whether or not you have passed each test. This letter will be followed in about two weeks with a ***test score report.*** The test score report is a confidential report of your performance on the tests. The report will list the number of questions that must be answered correctly to pass the test, and the number of questions you have answered correctly. The test questions are also subdivided into the general areas to help you to determine any weak areas that require more study. A typical test diagnostic report is shown in **Figure 24-5.**

If you passed one or more tests, included with the report is a ***certificate in evidence of competence.*** This certificate lists all of the areas in which you are certified. In addition, a pocket card and a wallet card are provided. Like the certificate, they list all of the areas you are certified in. Also included is an order form for shoulder patches, wall plates, and other ASE promotional material, **Figure 24-6.**

Figure 24-6. *Auto repair businesses will advertise the fact they employ certified technicians. It is to your advantage to become ASE certified.*

Note: If you did not indicate you have two years of automotive experience on the test application, you will not receive a certificate in evidence of competence. After you have met the experience requirement, you should provide ASE with the necessary information to receive your certificate.

ASE takes the position that all ASE test results are confidential information, and provides them only to the person who took the test. This is done to protect your privacy. The only test information ASE will release is to confirm to an employer that you are certified in a particular area. Test results will be mailed to your home address and will not be provided to anyone else. This is true even if your employer has paid the test fees. If you wish your employer to know exactly how you performed on the tests, you must provide him or her with a copy of your test results.

If you fail a certification test, you can retake it again as many times as you would like. However, you (or your employer) must pay all of the applicable registration and test fees again. You should study all available information in the areas where you did poorly. A copy of the ASE Preparation Guide may be helpful to sharpen your skills in these areas. The ASE Preparation Guide is free, and can be obtained by filling out the coupon at the back of the Registration Booklet.

Recertification Tests

Once you have passed the certification test in any area, you must be recertified every five years. This assures your certification remains current, and is proof

Your score is *40* (Passed) The total score needed to pass A7 is *33* out of 50		
Test A7 Heating and Air Conditioning **Content Area**	Number of Questions Answered Correctly	Total Number of Questions
A/C System diagnosis & Repair	8	12
Refrigeration System Component	9	10
Heating & Engine Cooling Systems	4	5
Operating Sys & Related Controls	12	16
Refrig Recovery, Recyc & Handling	7	7
Total Test	40	50

Figure 24-5. *A diagnostic report will follow the pass/fail letter. This allows you to see how you performed in each section and, if needed, which areas you need further study. (ASE)*

you have kept up with current technology. The process of applying to take the recertification tests is similar to other certification tests. Use the same form and enclose the proper recertification test fees. If you allow your certification to lapse, you must take the regular certification test(s) to regain your certification.

Certification in Refrigerant Recovery and Recycling

One of the regulations outlined in the Clean Air Act is the *mandatory requirement* that all technicians servicing mobile vehicle air conditioning systems, be certified in refrigerant recovery and recycling. While passing the ASE heating and air conditioning certification test will give you better recognition to potential employers, it does not currently satisfy the recovery and recycling certification requirements of the EPA.

Note: Certification in refrigerant recovery and recycling is *not* the equivalent of technical certification (passing the ASE heating and air conditioning test). It is only deemed a certification in the eyes of the EPA.

To satisfy EPA requirements for technicians who perform air conditioning service on mobile vehicles, ASE and several other organizations have developed separate tests in refrigerant recovery and recycling. These tests are an open book, and can be done at work, home, or other convenient locations. There is usually no registration fee, however, a small fee is charged for processing the test. Certification allows you not only to work on air conditioning systems, but gives you the ability to purchase R-12 refrigerant and other ozone-depleting chemicals. Future regulations may restrict the sale of R-134a only to certified technicians.

Note: Certification requirements vary from area to area, so check local regulations. You may be required to obtain additional certifications, licensing, or permits in order to service air conditioning systems and/or purchase refrigerants.

Summary

The automotive service industry was one of the few major industries that did not have testing and certification programs. This caused a lack of professionalism in the automobile industry, often leading to poor or unneeded repairs, and decreased status and pay for automobile technicians. The National Institute for Automotive Service Excellence, or ASE, was started in 1972 to improve status of the automotive service industry. ASE tests and certifies automotive technicians in major areas of automotive repair. This has increased the skill level of technicians, resulting in better service, and increased benefits for technicians.

ASE tests are given two times each year, in the Spring and Fall. Anyone can register to take the tests by filling out the proper registration form and paying the proper registration and test fees. The registrant must also select the test center he or she would like to go to. To be considered for certification, the registrant must have two years of hands-on experience as an automotive technician. Proof of this should also be included with the registration form. About three weeks after applying for the test, the technician will receive a test entry ticket, which he or she must bring to the test center.

The actual test questions will test your knowledge of general system operation, problem diagnosis, and repair techniques. All of the questions are multiple-choice questions with four possible answers. The questions must be read carefully. The entire test should be gone over one time only to catch careless mistakes.

Test results will arrive within six to eight weeks after the test session. A letter will be sent telling the technician whether he or she has passed the tests. This is followed by a detailed report showing the test scores for all areas of the tests taken. Results are confidential, and will be sent only to the home address of the person who took the test. If a test was passed, and the experience requirement has been met, the technician will be certified for five years. Anyone who fails a test can take it again in the next session. Tests can be retaken as many times as necessary. Recertification tests are taken at the end of the five year certification period.

Review Questions—Chapter 24

Please do not write in this text. Write your answers on a separate sheet of paper.

1. Even though the number of automobile manufacturers has decreased it has been hard to come up with a unified set of automotive technician standards since there are so many _____.

2. Before 1972, what automotive certification process was used?

3. How many times are ASE tests given each year?

4. An ASE certified Master Technician has passed all of the tests in the _____ or _____ areas.

5. What three major areas of automotive knowledge are the ASE tests designed to measure?

6. *True or False?* ASE sets technician pay scales in all eight certification areas.

7. *True or False?* ASE provides test results to whoever paid for the tests.

8. *True or False?* ASE advertises the benefits of certification.

9. *True or False?* Recertification tests must be taken every five years.

10. Name some sources of study material for the ASE tests.

ASE Certification-Type Questions

1. All of the following are ASE activities, *except:*
   (A) encouraging the development of effective training programs.
   (B) conducting research on the best methods of performing instruction.
   (C) publicizing the advantages of technician certification.
   (D) negotiating pay and benefits with large employers.

2. Technician A says ASE encourages high standards of automotive service and repair by establishing a series of instructional courses. Technician B says ASE encourages high standards of automotive service and repair by providing a series of standardized tests. Who is right?
   (A) A only.
   (B) B only.
   (C) Both A and B.
   (D) Neither A nor B.

3. All of the following statements about the ASE tests are true, *except:*
   (A) the same test is given to everybody in a particular state.
   (B) these tests are called standardized tests.
   (C) the number of questions varies with the area being tested.
   (D) the tests are always given at official test centers.

4. Which of the following is an advantage ASE certification has helped to bring to automotive technicians?
   (A) Longer working hours.
   (B) Increased use of the salary pay system.
   (C) Increased use of the commission pay system.
   (D) Increased respect.

5. Bring all of the following to the ASE test center, *except:*
   (A) two #2 pencils.
   (B) any needed study materials.
   (C) your admission ticket.
   (D) photographic identification.

6. Technician A says a technician can retake a certification test up to three times. Technician B says certified technicians must take a recertification test every five years. Who is right?
   (A) A only.
   (B) B only.
   (C) Both A and B.
   (D) Neither A nor B.

7. ASE test questions are always _____ types.
   (A) true-false
   (B) two-part
   (C) multiple-choice
   (D) negative

8. ASE tests are designed to measure your knowledge of all of the following, *except:*
   (A) basic information on how automotive systems and components work.
   (B) customer relations and dealing with difficult customers.
   (C) diagnosis and testing of automotive systems and components.
   (D) repairing automotive systems and components.

9. ASE provides test results to _____.
   (A) the technician who took the test
   (B) whoever paid for the test
   (C) the technician's employer
   (D) the Environmental Protection Agency

10. When the word "least" is used in a question, what should you should look for among the possible answers?
    (A) The answer that is probably the best choice.
    (B) The answer that is probably the worst choice.
    (C) The answer that is somewhere between the best and worst choice.
    (D) There is no way to tell without actually reading the question.

Chapter 25

Career Preparation

After studying this chapter, you will be able to:
- ❏ Identify three classifications of automotive technicians.
- ❏ Identify the major sources of employment in the automotive industry.
- ❏ Identify advancement possibilities for automotive technicians.
- ❏ Explain how to fill out a job application.
- ❏ Explain how to conduct oneself during a job interview.

Technical Terms

Helper

Apprentice

Certified technicians

New vehicle dealers

Used car superstore

Chain and department stores

Independent repair shops

Entrepreneur

Shop supervisor

This chapter is an overview of career opportunities in the automotive service industry. It discusses types of automotive technicians, and what kind of work they perform. This chapter also includes information on the types of repair outlets, and the type of work, working conditions, and pay scales the beginning technician can expect to find in each place. Also included are some of the ways you can locate jobs and become employed in the auto repair industry. It includes information on the types of available automotive related jobs. Studying this chapter will help you to find and get a job in the automotive service industry.

Automotive Servicing

The business of servicing and repairing cars and trucks has provided employment for many millions of people over the last 100 years. It will continue to provide good employment opportunities for many years to come. Like any career, it has its drawbacks, but it also has its rewards.

Most persons in the auto service business work long hours, and the diagnosis and repair procedures can be mentally taxing, physically hard, and often hot and dirty. Automotive service has never been a prestigious career, although this is changing as vehicles become more complex, and technicians better trained. The technician often has to deal with difficult, condescending, and sometimes dishonest vehicle owners.

The advantages of the auto service business are the opportunity to work with your hands, much less confinement than with many other professions, and the enjoyment of taking something broken, and making it work again. Auto repair salaries are usually competitive with those for similar jobs, and it is a secure profession where the good technician can always find work. To ensure you stay employable, always seek to learn new things, and become ASE certified in as many areas as possible.

Levels of Automotive Service Positions

Although the public tends to classify all automotive technicians as "mechanics," there are many types of auto service professionals. The types of auto service professionals range from the helper who changes oil or performs other simple tasks; through apprentices who remove and install parts; to the certified technician, capable of diagnosing and repairing various automotive systems. Although these levels are unofficial, they tend to hold true throughout the automotive repair industry. It would be possible to further break these levels down into more sublevels, but these three will adequately cover the general skill classifications. These levels are explained in more detail in the following paragraphs.

Helpers

The *helper,* **Figure 25-1,** performs the easier types of service and maintenance, such as installing and balancing tires, changing engine oil and filters, and installing batteries. The skills required of the helper are low, and the pay will be less than the other levels. However, the helper position is a good way for many people to start. In fact, many technicians started out doing this kind of automotive service when they were in their teens.

Figure 25-1. *Helpers usually perform minor tasks around the shop, such as servicing equipment, cleaning oil and chemical spills, and helping the other technicians.*

Apprentices

The *apprentice,* **Figure 25-2,** installs parts on vehicles and performs more complex repairs than the helper. Most of the parts installed are suspension components such as shock absorbers or struts; mufflers and other exhaust system components; and possibly brake master cylinders, alternators, and voltage regulators. They, along with helpers, work under the guidance of certified technicians. Apprentices seldom do any more complicated repair work, and do not generally diagnose vehicle problems. Apprentices are paid more than helpers, but less than certified technicians. Many apprentices take the opportunity to improve their knowledge and skills, and eventually become certified technicians.

Certified Technicians

Certified technicians, **Figure 25-3,** are at the top level, and have the skills to prove it. Most modern technicians are ASE certified in at least one automotive area, and many are certified in all car or truck areas. The certified technician is able to make diagnosis and repairs in every area he or she is certified in, and can perform many other service jobs. The certified technician is at the top of the pay scale also.

Figure 25-2. *Apprentices are generally service persons who perform routine maintenance and light repairs, such as evacuating and recharging air conditioning systems.*

Types of Auto Service Facilities

There are many types of auto service facilities where the technician can work. The traditional place to get started in automotive repair, the corner service station, is all but gone, replaced by self-service stations/convenience stores. However, many opportunities to repair vehicles still exist. Even the smallest community has many types of automotive repair facilities.

New Vehicle Dealers

All **new vehicle dealers** must have large service departments to meet the warranty service requirements of the vehicle manufacturer. These service departments are usually well equipped, with all special testers, tools, and service literature needed to service the manufacturer's vehicles. Dealership service departments are also equipped with lifts, parts cleaners, hydraulic presses, brake lathes, electronic test equipment, and other equipment for efficiently servicing vehicles. However, the technician has to provide his or her own hand and air tools. Dealers stock all of the most common parts, and are usually tied into a factory parts network, which allows them to quickly obtain any part. See **Figure 25-4.**

Pay scales at most dealerships are based on flat rate hours. The rate per hour is competitive between dealerships in the same area, and usually higher than local industry in general. The number of hours the technician is paid for depends on what work comes in and how fast he or she can complete it. If you can work fast, and enough work comes in, the pay can be excellent. Most modern dealers offer some sort of benefits package.

Dealership working conditions are relatively good, and most of the vehicles are new or well cared for older models. Since the dealer must fix any part of the vehicle, the technician can perform a large variety of work.

Figure 25-3. *An ASE Certified Technician has the knowledge and experience to perform most repairs. (Jack Klasey)*

Figure 25-4. *New vehicle dealers offer excellent working conditions and benefits. However, you do have to deal with all problems on the vehicle, and there is not a very good opportunity for specialization. (Land Rover)*

Although many dealer service departments have technicians who work only in specific areas, the trend is toward training all technicians to handle any type of work. Most repairs will be on the same make of vehicle, although many large dealerships handle more than one make. Most modern dealerships are tied into manufacturer hot lines. These hot lines are used to access factory diagnosis and repair information, which makes it easier to troubleshoot and correct problems.

The disadvantages of dealership employment are the lack of salary guarantees, low pay rates for warranty repairs, and fast paced, often hectic, working conditions. If you welcome the challenge of being paid by the job, and do not mind working under deadlines, a dealership may be the ideal employer. Also check out the local large truck dealerships. Although the work is much heavier, the pay is usually somewhat higher, and working conditions are not as fast paced.

Used Car Superstore

A recent addition to the list of places for technicians to find work is the **used car superstore.** Used car lots are not new to the automotive industry, however, they were not known as good places to work. Used car superstores are national chains that service and sell all makes and models of vehicles. Typically, the work at these chains includes most of the repairs performed at new vehicle dealers. You may also be responsible for inspecting and, if needed, performing repairs on vehicles coming in for sale. Pay at used car superstores is very good and benefits are excellent, with plenty of opportunity for additional training and advancement.

Chain and Department Store Auto Service Centers

Many national **chain and department stores** have auto service centers where various types of automotive repairs are performed. Companies with these types of auto service centers include Sears, Montgomery Ward, and K-Mart. These centers often hire technicians for entry-level jobs and may have several classifications above entry-level. Technicians in most of these shops are paid a salary, plus commission for work performed. Pay scales for the various classifications of work are competitive, and most companies offer generous benefit packages. One advantage of working for large companies such as these is the chance of advancement into other areas, such as sales or management.

One disadvantage of working at the average auto center is the lack of variety. Most auto centers concentrate on a few types of repairs, such as brake work and alignment, and turn down most other repairs. The work can become monotonous due to the lack of variety. Although the job pressure is usually less than at dealerships, customers still expect their vehicles to be repaired in a reasonable period of time. However, if you would enjoy the opportunity to work on only one or two areas of automotive repair, this type of job may be ideal for you.

Tire, Muffler, and Other Specialty Shops

Tire, muffler, and other specialty shops can offer good working conditions and good pay. Most of these shops concentrate on their major specialty, with a few other types of repair, such as alignments and other suspension and steering service. Examples of these types of shops are those run by major tire manufacturers, Midas and Mieneke muffler shops, AAMCO transmission shops, and the auto centers of various tire manufacturers. These shops usually offer a base salary, plus commission for work performed. Pay scales are generally competitive. A disadvantage of specialty shops is the lack of variety. Since they concentrate on a few types of repairs, working at these shops can become monotonous.

Many of these shops are franchise operations, and the demands of the franchise can create problems. If the prime purpose of the shop is to sell tires, for instance, the technician who was hired to do steering and suspension repairs may be forced to spend time installing tires. This can be annoying to technicians who want to be doing the job they were hired for. However, if this does not bother you, the specialty shop could be a good work situation.

Independent Repair Shops

There are millions of **independent repair shops,** **Figure 25-5.** As places to work, they range from excellent to terrible. Many shops are run by competent and fair-minded managers, and have first rate equipment and good working conditions. Other independent shops have almost no equipment, low pay rates, and extremely poor, even dangerous working conditions. The prospective employee should carefully check all aspects of the shop environment before agreeing to work there. Technicians at independent shops are usually paid on a salary plus commission basis.

There are two major classifications of independent repair shop, the general repair shop and the specialty shop. The general repair shop takes in most types of work, and offers a variety of jobs. General repair shops may avoid jobs that require special equipment, such as automatic transmissions, alignment, or driveability service. However, they will usually take a variety of other repair work on different makes and types of vehicles. This can be a good place to work if you like to be involved in many different types of diagnosis and repair.

Specialty shops usually confine their repair work to one area of repair, such as transmissions, tune-up, brake repair, or alignments. They are fully equipped to handle all aspects of their particular specialty. These shops may occasionally take in other minor repair work when business in their specialty is slow. These shops can be ideal places to work if you want to concentrate on one or two areas of repair.

Figure 25-5. *General repair shops handle a variety of work. This makes the job more interesting.*

Fleet Agencies

Many companies have a fleet of vehicles, such as rental car companies, often hire technicians to maintain their vehicles. Fleet technicians usually perform only routine maintenance and minor repairs, since most companies operating fleet vehicles usually purchase new ones after 2-3 years. In most fleets, if a vehicle has a problem requiring major repair, the vehicle is sent to a dealership or specialty repair shop. Fleet technicians are usually paid a straight salary, with very good to excellent benefits. Since most vehicle fleets are fairly large, there is not as much pressure to get a vehicle finished within a set time.

Government Agencies

Many local, state, and federal government agencies maintain their own vehicles. Government operated repair shops can be good places to work. Pay is straight salary, on a per hour or per week basis. Pay is usually set by law, with no commission. Although civil service pay scales are lower than private industry, the benefits are usually excellent. Pay raises, while relatively small, are regular. Most government shops work a 35 or 40 hour week, and have the same holidays as other government agencies.

The working conditions in most of these shops are good, without the stress of deadlines, or having to deal with customers. Often, technicians at government agency shops perform the same work as a technician at a dealership. You may also be called upon to install special equipment in vehicles, such as radios and lights. Hiring procedures are more involved than they are with other auto repair shops. Prospective employees must take civil service examinations, which often have little or nothing to do with automotive subjects. Some government agencies require a certain level of education, thorough background checks, lists of former employers or other references, and may require the employee be a registered voter.

If you think you would be interested in working for a government owned repair shop, contact your local state employment agency for the addresses of local, state, and federal employment offices in your area.

Self-Employment

Many persons dream of going into business for themselves. This can be a good and profitable option for the good technician. However, in addition to mechanical and diagnostic ability, the person with his or her own business must have a certain type of personality to be successful. This type of person must be able to shoulder responsibilities, handle problems, and look for practical ways to increase business and make a profit. This type of person must maintain a clear idea of what plans, both long and short term, need to be made. A person like this is often called an ***entrepreneur,*** in other words, a person who has the energy and skill to build something out of nothing.

When you have your own business, all of the responsibility for repairs, parts ordering, bookkeeping, debt collection, and a million other problems are yours. Starting your own shop requires a large investment in tools, equipment, and working space. If the money must be borrowed, you will be responsible for paying it back. However, many people enjoy the feeling of independence, of not having to answer to an employer. If you have the personality to deal with the problems, you may enjoy the feeling of being your own boss.

Another possible method of self-employment is to obtain a franchise from a national service chain. A franchise operation removes some of the headaches of being in business for yourself. Many muffler, tire, transmission, tune-up, and other nationally recognized businesses have local owners. They enjoy the advantages of the franchise affiliation, including national advertising, reliable and reasonable parts supplies, and employee benefit programs. Disadvantages include high franchise fees and start-up costs, lack of local advertising, and some loss of control of shop operations to the national headquarters.

Other Opportunities in the Auto Service Industry

Many other opportunities are available to the automotive technician. These jobs still involve the servicing of vehicles, without some of the physical work. If you like cars and trucks, but are unsure whether you want to make a career of repairing them, one of these jobs may be for you.

Shop Supervisor or Service Manager

The most likely automotive promotion you will be offered is *shop supervisor,* **Figure 25-6.** Many repair facilities are large enough to require one or more supervisors or service managers. If you move into management from the shop floor, your salary will increase, and you will be in a cleaner, less physically demanding position. Many technicians enjoy the management position because it lets them in on the fun part of service, troubleshooting, without getting dirty or greasy by actually making the repairs.

The disadvantage of a move to management is you will no longer be dealing with the logical principles of troubleshooting and repair. Instead, you will deal with the illogical and arbitrary personalities of people. Both customers and technicians will have problems and attitudes you will have to deal with. Unlike a vehicle problem, these problems require considerable personality and tact. Sometimes the manager has to compromise, which can be hard for the person who is used to being right and saying so.

The paperwork load is large for any manager, and may not be something a former technician can get used to. Record keeping requires a good bit of desk, and usually, computer time. Automotive record keeping is like balancing a checkbook and writing a term paper every few days. If you do not care to deal with people or keep records, a career in management may not be for you.

Salesperson or Service Advisor

Many people enjoy the challenge of selling. The salesperson or service advisor performs a vital service, since repairs will not be performed unless the owner is sold on the necessity of having them done. Salespersons are not necessary in many small independent shops, but are often an important part of dealership service departments, department store service centers, and specialty repair shops.

The salesperson may enjoy a large income, and be directly responsible for a large amount of business in the shop. However, selling is a people oriented job, and takes a lot of persuasive ability and diplomacy. If you are not interested in dealing with the public, you would probably not be happy in a sales job.

Parts Person

One often overlooked area of the automotive service business is the process of supplying parts. It is as vital as any other area of auto service. There are many types of parts outlets, including dealership parts departments, independent parts stores, parts departments in retail stores, and combination parts and service outlets. All of these parts outlets meet the needs of technicians and shops, as well as the needs of the do-it-yourselfer. The many different makes of vehicles, as well as the complexity of the every modern vehicle, means a large number of parts must be kept in stock and located quickly when needed.

Parts persons are trained in the methods of keeping the supply of parts flowing through the system until they reach the ultimate endpoint, the vehicle. Parts must be carefully checked into the parts department, and stored so they can be found again. When a specific part is needed, it must be located and brought to the person requesting it. If more of the same parts are needed, they must be

Figure 25-6. *Shop supervisors usually deal with people and the everyday operation of the shop, however, they can be called upon to perform diagnostic work.*

ordered. If the part is not in stock, it must be special ordered. This can be a challenging job.

The job of the parts person appeals to many people. The job of the parts person does not pay as much as some other areas of automotive service, but rates are comparable with other jobs with the same skill level. If this type of work appeals to you, it may be a good job choice. There is an ASE test for parts specialists, and you may want to consider taking it to enhance your employability.

Getting a Job

There are many automotive jobs available at all times. The problem is connecting with the job when it is available. There are essentially two hurdles to getting a job: finding a job opening, and successfully applying for the job.

Figure 25-7. *Some shops advertise openings on their store-front marquee.*

Figure 25-8. *State employment agencies have many job listings not normally advertised.*

Finding Job Openings

Before applying for a job, you must know about a suitable opening. A good place to start is with your instructors. They often have contacts in the local automobile industry, and may be able to recommend you to a local company. Another good place to begin your search is the classified section of your local newspaper. Automotive jobs are often advertised in newspapers, especially those for automotive dealerships, specialty and franchise shops, and independent repair shops. Sometimes, shops will simply advertise the need for technicians by posting a "Help Wanted" sign on their street signage, **Figure 25-7.**

Also visit your local state employment agency or job service, **Figure 25-8.** Most of these agencies keep records of job openings, usually throughout the state, and may be able to connect to a data bank of nationwide job openings. Some private employment agencies specialize in automotive placement. If there is an agency of this type in your area, arrange for an interview with one of their recruiters.

Visit local repair shops where you are interested in working. Sometimes these shops have an opening they have not advertised. If you are interested in working for a chain or department store, most of these stores have a personnel department where you can fill out an application. Even if no jobs are available, your application will be placed on file in the event a job becomes available in the future.

Applying for the Job

No matter how good your qualifications, you will not get a job if you make a poor first impression on the job interviewer. If you do not impress your potential employer as competent and dedicated, you will not get the job.

Start creating a good impression when you fill out the employment application. Type or neatly print when filling out the application. Complete all blanks, and completely explain any blanks in your education or previous employment history. List all of your educational qualifications, including those that may not apply directly to the automotive industry. **Figure 25-9** shows a typical employment application.

When you are called for an interview, try to arrange for a morning interview, since your potential employer is most likely to be in a positive mood in the morning and less likely to be overwhelmed by last minute problems in the shop. Dress neatly, and arrive on time or a little early. When introduced to the interviewer, make an effort to repeat and remember their name. Speak clearly when answering the interviewer's questions. Do not smoke or chew gum during the interview. State your qualifications for the job without bragging or belittling your accomplishments. At the conclusion of the interview, thank the interviewer for his or her time. If you do not hear from the interviewer in a few days, it is permissible to make a brief and polite follow-up call.

Ice Cold Auto Air   **Employment Application**

213 Credibility Street Edens, IL 60000

Date: _____ Social Security Number: _____

Name: _____ Age: _____ Sex: _____

Address: _____

Phone: _____ United States citizen? _____ Can you furnish proof? _____

Employment Desired

Position: _____ Date you can start: _____ Expected salary: _____

Are you currently employed? _____ May we inquire of your present employer? _____

Education

Circle the number for the highest level completed:

High School   Trade/Technical School   Community/Junior College   University

1 2 3 4   1 2   1 2   1 2 3 4

Other: _____

Specialized Training or Certifications: _____
_____

Employment Record

Current/Last employer: _____ From: _____ To: _____

Address: _____ Phone: _____

Salary: _____ Job description: _____

Reason for leaving: _____

Previous employer: _____ From: _____ To: _____

Address: _____ Phone: _____

Salary: _____ Job description: _____

Reason for leaving: _____

Previous employer: _____ From: _____ To: _____

Address: _____ Phone: _____

Salary: _____ Job description: _____

Reason for leaving: _____

Figure 25-9. *Typical job application. Most job applications are longer than this one.*

Summary

The automotive service industry provides employment for many people, and will continue to do so. Automotive service has some disadvantages, such as long hours; hard work, both mental and physical; lack of status; and difficulties in dealing with the public. Advantages include interesting work, the security of a guaranteed career, and the enjoyment of diagnosing and correcting problems. Always stay employable by learning new things and taking the ASE tests.

The three general classes of technicians are the helper, the apprentice, and the certified technician. The helper does the simplest tasks, such as tire changing and lubrication. Many helpers move up into the other classes after a short time. The apprentice installs new parts, such as shock absorbers and strut assemblies, and sometimes moves into brake repair and wheel alignment. The certified technician performs the most complex diagnosis and repair jobs on vehicles, and makes the most money.

There are many places to work as an automotive technician. Among the most popular are new car and truck dealers, auto centers affiliated with department or chain stores, specialty shops, independent repair shops, and government agencies. Some people prefer to have their own businesses, either as independent owners, or as part of a franchise system.

Other opportunities in the automotive service field include moving into management as a foreman or service manager, or into sales as a service advisor or specialty sales person. Another often overlooked employment possibility is in the automotive parts business.

To obtain a job in the automotive business, first locate possible job openings. Try the local newspaper and state job service, and visit repair shops in your area. Most department or other large stores with attached auto service centers have personnel departments where you can fill out a job application. To get a job, you must make a good impression, no matter how qualified you are. Fill out all job applications carefully and neatly, listing you qualifications honestly. When invited to a job interview, dress neatly, arrive on time, and be courteous. Answer all questions without over- or understating your abilities and experience. Follow up the interview with a brief phone call within a few days.

Review Questions—Chapter 25

Please do not write in this text. Write your answers on a separate sheet of paper.

1. An advantage of the automotive service field is the opportunity to work with your _____.

2. Before beginning any needed work, get an ok from the vehicle _____.

3. *True or False?* Working on vehicles can be hot and dirty.

4. *True or False?* The helper will be called on to do the most complex jobs.

5. *True or False?* New vehicle dealers must have well-equipped service departments.

6. *True or False?* A lot of the work at a new vehicle dealer will be warranty work.

7. *True or False?* Pay at government garages is based on a weekly salary.

8. Pay at most dealerships is based on _____ hours.

9. An independent shop or franchise is a good way to have your own _____.

10. A good place to start looking for a job is at your local state _____.

ASE Certification-Type Questions

1. In what sort of shop would the technicians be most likely to work only 35 hours weekly?
   (A) Dealership service departments.
   (B) Government agencies.
   (C) Independent shops.
   (D) Franchise shops.

2. Technician A says one advantage of working in the automotive service field is that diagnosis and repair is often interesting. Technician B says one advantage of working in the automotive service field is the respect of vehicle owners. Who is right?
   (A) A only.
   (B) B only.
   (C) Both A and B.
   (D) Neither A nor B.

3. Which of the following automotive service jobs will most likely be done by a helper?
   (A) HVAC noise diagnosis.
   (B) Retrofitting to R–134a.
   (C) Changing a compressor.
   (D) Draining and refilling a cooling system.

4. All of the following statements about working for a new vehicle dealer shop are true, *except:*
   (A) most of the vehicles serviced are new or almost new.
   (B) the opportunity exists to become very familiar with one make of vehicle.
   (C) the technician is usually paid a salary.
   (D) new vehicle warranty repairs are commonly performed.

5. Which of the following is a business where the automotive technician is *most likely* to get experience in many kinds of repairs?

   (A) New vehicle dealers.

   (B) Brake shops.

   (C) Government garages.

   (D) Chain stores

6. All of the following statements about working for a local, state, or federal government operated repair shop are true, *except:*

   (A) raises are small but regular.

   (B) deadlines are rare.

   (C) working conditions are often hectic.

   (D) benefits are equivalent to other civil service positions.

7. At which of the following places is the automotive technician *most likely* to move up to a management position?

   (A) The same shop where he/she is working.

   (B) An automotive parts store.

   (C) Another shop in the same area.

   (D) A totally different business.

8. Which of the following is *not* a job duty of the parts specialist?

   (A) Obtaining parts for technicians.

   (B) Selling parts to the general public.

   (C) Telling the technicians how to install parts.

   (D) Special ordering parts.

9. Employment applications are being discussed. Technician A says you should fill out the application in a sloppy manner to let the owner know you are not a paper kind of person. Technician B says filling out the application is a good time to start making a good impression. Who is right?

   (A) A only.

   (B) B only.

   (C) Both A and B.

   (D) Neither A nor B.

10. All of the following help to make a good impression during the job interview, *except:*

   (A) dress in dirty work clothes.

   (B) arrive a little early.

   (C) do not smoke.

   (C) make an effort to remember the interviewer's name.

Appendix A

Rear Window and Mirror Defrosters

After completing this appendix, you will be able to:
- ❑ Explain the operation of rear window and mirror defrosters.
- ❑ Identify the components of rear window and mirror defrosters.
- ❑ Diagnose rear window and mirror defroster problems.
- ❑ Repair or replace rear window and mirror defroster components.

Technical Terms

Rear window defroster	Heating grid
Mirror heating	Electrical bus
Resistance heater	Defroster relay

Many modern vehicles are equipped with devices to defrost the glass of rear windows and outside rearview mirrors. The purpose of any type of glass defroster is to heat the glass, causing frost or dew to evaporate. HVAC technicians are often called upon to repair these defrosters. This appendix explains how these defrosters operate, identifies their major components, and explains how to diagnose and repair them.

Defroster Components and Operation

While the operation of glass defrosters is relatively simple, several components are used to protect the glass and the vehicle electrical system. Individual components and how they work together are discussed in the following sections.

Rear Window Defroster and Mirror Heating Systems

The **rear window defroster** and **mirror heating** systems consist of several components working together with one purpose, to heat the glass surface. If all components are operating properly, the glass or mirror will be quickly defrosted with minimum electrical demand.

As you remember from Chapter 4, every conductor has some resistance to the flow of electricity. Some materials, called resistors, are designed to have a specific amount of electrical resistance, which creates heat. In many resistor applications, heat is an unwanted byproduct. However in the glass heating resistor, heat is deliberately created and put to work. This type of resistor is called a **resistance heater.** The **heating grid** consists of a series of flat (or ribbon) resistor wires applied to the surface of the glass. The grid wires are glued or baked onto

the glass surface. **Figure A-1** shows a typical defroster grid applied to the rear window of an automobile. The end wire connecting all of the horizontal wires is usually called an **electrical bus.**

Current flows through the grid wires, heating the glass. Heating the glass causes frost to melt and evaporate. Dew can also be evaporated quickly on damp mornings.

Control Switch

The defroster control switch is mounted on the dash, close to or on the HVAC control panel, **Figure A-2A.** Some switches are built into one of the HVAC controls, **Figure A-2B.** The defroster switch receives 12-volt power when the ignition switch is in the On position only. Most defroster control switches have an indicator light.

Defroster Relay

To protect the dashboard switch from overheating, and to automatically turn the defroster off after a few minutes, a **defroster relay** is used. Defroster relays can carry high amounts of current. The relay is energized by the dashboard switch and directs current to the window and mirror grids. On older vehicles, the relay was an electromechanical unit similar to a high blower relay. It was usually located under the dashboard, away from the switch assembly. Some late-model vehicles use a solid state electronic relay, sometimes called a module. Many newer vehicles have the switch and relay built into a single solid state unit.

Most relays contain an internal timer. The timer will shut off current to the grid after a few minutes. Most defroster systems are set to turn off after 5 to 10 minutes. If the glass is not fully defrosted, the driver must again turn on the defroster control switch. If the driver turns on the switch again, the defrost timer will keep the system on for one-half the normal amount of time.

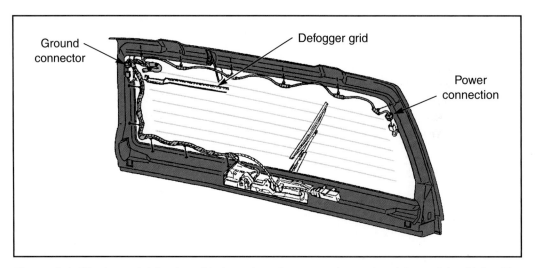

Figure A-1. *The type of defroster grid shown here is commonly used on late-model vehicles. The grid material has a relatively high amount of electrical resistance. Electricity passing through the grid causes it to heat up. This heats the glass and moisture is evaporated. (General Motors)*

Figure A-2. *Glass defroster switches are always located near the HVAC system controls. A—Some defroster switches are built into one of the control knobs. B—Other switches are on the control head or close to the control head, mounted in the dash.*

Defroster Operation

Both electromechanical and electronic defrosters operate in the same manner. The driver turns the defroster switch on. The switch energizes the relay, which directs current to the defroster grids. After a set time, the relay turns off, de-energizing the grid. If necessary, the driver can restart the defrosting sequence by turning the switch back on. **Figure A-3** is an electrical schematic of a rear window and mirror defroster.

Older Rear Window Defoggers

Some older rear window defoggers consisted of a blower that directed air against the rear window. However, these defoggers were not effective until the interior of the vehicle was heated by the HVAC system and could not remove frost. The blower control switch was mounted on the dashboard, and a relay was usually not used. Power was supplied to the switch only when the ignition switch

was on. Turning the dashboard switch to the On position activated the blower motor. The motor operated until the driver turned the switch to the Off position or turned off the ignition.

Defroster Diagnosis

To diagnose the defroster, turn the switch to the On position. The indicator light should come on. If the light does not come on, go to the next paragraph. If the light does come on, determine whether the defroster grid(s) are drawing current. Attach a voltmeter to the vehicle battery terminals. Turn the ignition switch to the On position (engine and defroster off) and observe the voltage. Then turn the defroster on and observe the voltage. Voltage should drop at least 1 volt when the defroster is turned on. If the voltage drop is small, or there is no change in voltage, the defroster is not operating. If there is any doubt about whether the system is operating, feel the glass at a grid wire, and at the glass away from the grid. The glass should be warmer at the grid wire. The same procedure can be used to check heated mirrors. If the rear glass grid does not become hot, make a close visual inspection of the grid wires. A broken wire may be seen as a gap or discolored spot.

> **Note: If only one grid of a multiple grid rear window and mirror heating system is not working, the problem is likely in that portion of the grid circuit and not the main control switch or relay.**

If the grids are not heating, begin by checking the system fuses or circuit breakers. Many systems use a separate circuit breaker or breakers to control the heavy current flow into the grid. Some systems use separate relays for the rear window and mirror grids. If the fuses and circuit breakers are ok, use a voltmeter or non-powered test light to ensure electricity is entering the switch. If no voltage is present at the switch and the fuses and relays are good, the problem may be in the wire harness. If voltage is available at the switch input terminal but not the output terminal, the switch is defective. Also check the indicator bulb is not burned out.

If power is leaving the switch and the relay does not operate, the relay may be defective. If possible, bypass the relay and determine whether the grid is working. Do not leave the jumper wires in place any longer than necessary to determine whether the grid is heating up.

>
> **Caution: Disconnect the relay before attempting to bypass it. If this is not done, the relay may be damaged.**

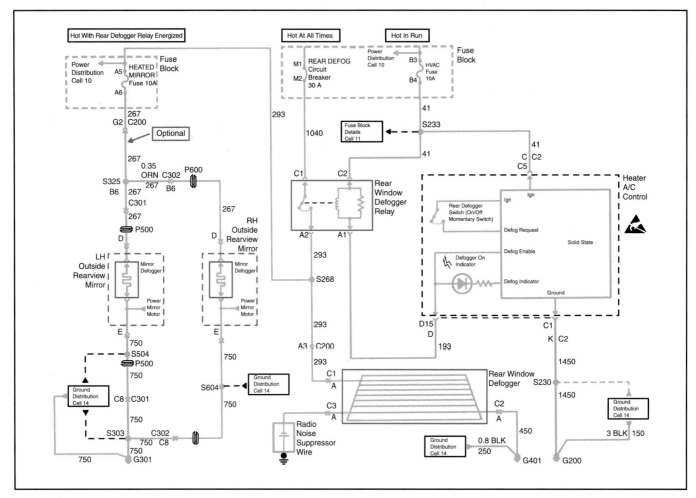

Figure A-3. *This electrical diagram shows how current flows through a heated rear window and mirror defogger system. The main contact set in the relay is closed by an electromagnet energized by the dashboard control. Current then flows through the timer contacts. (General Motors)*

If power appears to be entering the grid, but the grid is not heating, use a voltmeter to check the grid wires, **Figure A-4.** A wire showing less than 10-14 volts is defective. A non-powered test light can also be used to quickly check the grid. Some grids can be checked at their electrical connectors. If the grid is defective, it can be repaired or replaced. Other defective parts are simply replaced.

If the vehicle is equipped with an older blower type defogger, check the defogger motor for voltage when the ignition and defogger switches are on. If no voltage is present, check the switch and fuse as necessary. If electricity is reaching the motor, it is burned out or stuck. Sometimes a stuck motor in a defogger that has not been used for some time can be freed up by turning it by hand. Be sure power to the motor is disconnected before trying to turn it by hand.

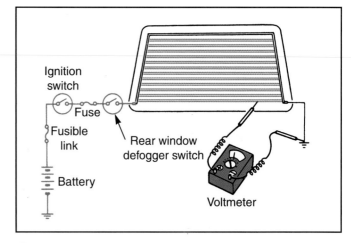

Figure A-4. *An ohmmeter can be used to check for a broken grid wire. An analog or digital ohmmeter can be used. Infinite resistance indicates that the wire has broken. (Ford)*

Defroster Service

Replacement of defroster control switches and relays is relatively easy. However, grid repair requires careful work procedures to avoid damaging the glass. The following sections explain how to service glass defroster components.

Replacing Switches and Relays

Most switches are part of the HVAC control panel, and replacement of the switch is similar to replacing other HVAC controls. Begin by removing the control panel fasteners and pulling the panel from the dashboard. Then disconnect the switch electrical connector and remove the fasteners holding the switch to the panel. Install the new switch and reconnect the electrical connector. Then reinstall the HVAC control panel and recheck defroster operation.

If the relay is not part of the switch assembly, locate it under the dashboard. Next, remove the electrical connectors and fasteners. Compare the old and new relays and then install the new relay. Reinstall the electrical connectors and fasteners and recheck defroster operation.

Repairing the Window Grid

Grids usually fail because they develop an open section, which prevents current flow. An electrically conductive compound can be applied (painted) to the broken section of a grid wire to restore conductivity. The grid repair material may be supplied as part of a kit.

To repair the wire break, begin by thoroughly cleaning the area around the break with a cleaner recommended by the manufacturer. This cleaner may be included in the grid repair kit. Then tape the area around the broken spot with masking tape, **Figure A-5.** Next, mix the conductive compound as instructed and carefully paint it onto the broken grid section. See **Figure A-5.** Allow the compound to dry and apply additional coats as required. After the compound has been applied, allow it to dry thoroughly, then remove the masking tape. Some manufacturers suggest drying the compound with a heat source, **Figure A-6.**

If a grid lead-in wire has become broken where it enters the grid bus, soldering the lead will sometimes repair it, **Figure A-7.** If the grid is too badly damaged to be repaired, the entire back glass must be replaced.

Mirror Grid

Mirror grids are not repairable. Some manufacturers call for replacing the entire mirror when the heater is defective, **Figure A-8.** Other manufacturers provide replacement heated mirror glass. To replace this type of mirror glass, break the defective glass as shown in **Figure A-9A.** Breaking the glass exposes the mounting screw.

 Warning: Wear safety glasses when breaking mirror glass.

Next, remove the mounting screw, unplug the electrical connector, and remove the assembly, **Figure A-9B.** Then install the new glass, being sure to install the electrical connector. If the replacement mirror glass is glued in place, secure it firmly with rubber bands, **Figure A-9C.** The bands should be allowed to remain in place for at least 24 hours.

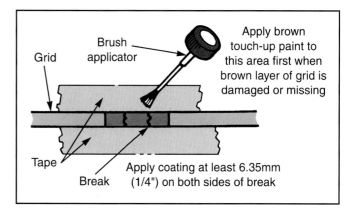

Figure A-5. *After being mixed in a separate container, conductive compound is painted over the broken spot in the grid. Note how the area around the grid wire has been taped with masking tape to make a neat repair. (Ford)*

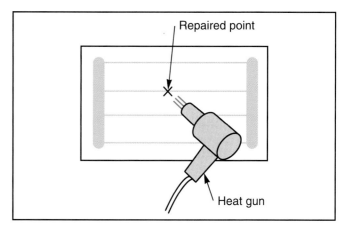

Figure A-6. *To speed drying of the conductive compound, gently heat the repaired area with a heat gun or hair dryer. (Nissan)*

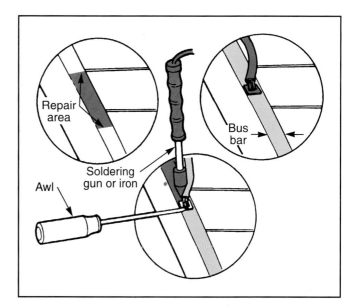

Figure A-7. *Some grid connectors can be repaired by soldering. (Ford)*

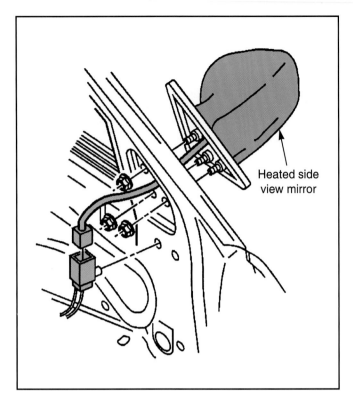

Figure A-8. *The heater grid of many vehicles is not repairable, and the entire mirror will have to be replaced if the grid wires break. (General Motors)*

Summary

Modern vehicles are often equipped with devices that defrost the rear window and outside rearview mirror glass. These defrosters heat the glass to evaporate frost or dew. The defroster grid wires are glued or baked onto the glass surface. Current flows through the grid wires, heating the glass. Defrosters are controlled through a dashboard switch and high current relay. Defroster controls usually have an indicator light. The relay has a timer that removes power from the grid after a set interval.

If operating properly, the grid should warm the glass when the switch is turned on. If the defroster does not operate, check the fuses, relays, and grids. Make a visual inspection of the grid for breaks. Switches and relays can be bypassed or checked with a voltmeter. If the grid does not become hot, make a close visual inspection of the grid wires. Blower type defroster motors can be checked for voltage when the ignition and defroster switches are on.

Defroster control switches and relays can be replaced by accessing the switch and removing the electrical connectors and fasteners. Replacement is the reverse of removal. Applying a conductive coating to the broken section will repair the rear window grid. Mirror grids are not repairable and should be replaced. Blower defrosters can be replaced by accessing them through the vehicle trunk.

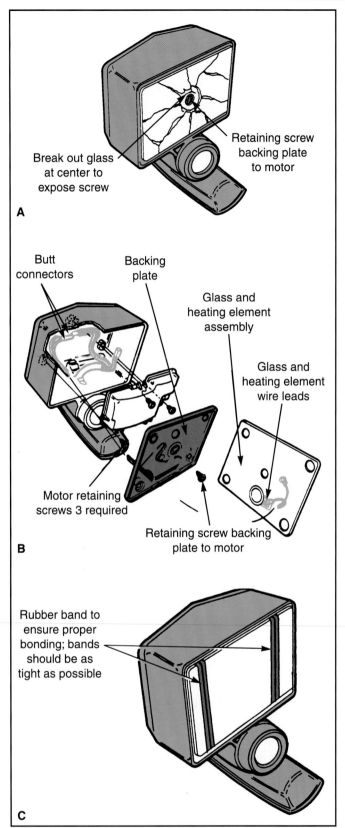

Figure A-9. *A—Some manufacturers provide for replacing the mirror and grid portion of a heated rear view mirror. To do this the glass must be broken to expose the attaching screw. B—After removing the old glass and grid, install the new grid and tighten the attaching screw. After this step, the replacement mirror must be glued to the grid assembly. C—Use rubber bands to hold the repaired mirror together for at least 24 hours. (Ford)*

Review Questions—Appendix A

Please do not write in this text. Write your answers on a separate sheet of paper.

1. The purpose of any type of glass defroster is to _____ the glass, causing frost or dew to _____.

2. Resistance to the flow of electricity causes _____ in a conductor.

3. The defroster switch may be built into one of the _____ controls.

4. The defroster will almost always have a dashboard mounted _____.

5. The internal timer in the defroster relay will perform what action after 5 or 10 minutes of defroster operation?

6. Older rear window defoggers consisted of a motor that operated a _____.

7. A grid wire showing below _____ volts is defective.

8. The first step in repairing a broken grid is to thoroughly _____ the area around the broken section.

9. Some manufacturers suggest drying the grid repair compound by applying _____.

10. After a new heated mirror is glued into place, the _____ should be left on for 24 hours.

ASE Certification-Type Questions

1. Technician A says the defroster relay may be an electromechanical unit. Technician B says the defroster relay may be an electronic unit. Who is right?
   (A) A only.
   (B) B only.
   (C) Both A and B.
   (D) Neither A nor B.

2. A voltmeter is attached to the battery of a vehicle equipped with a rear window defroster. The ignition switch is on and the engine is not running. When the defroster switch is turned to the on position, voltage drops by 2 volts. Which of the following is the *most likely* cause of the problem?
   (A) The relay is defective.
   (B) The grid wires are broken.
   (C) The indicator light is shorted.
   (D) The system is operating normally.

3. The most common defroster defect is a:
   (A) broken grid wire.
   (B) stuck relay.
   (C) blown fuse.
   (D) broken dashboard switch.

4. A defective window grid can be repaired by all of the following procedures, *except:*
   (A) replacing the entire rear window.
   (B) painting a conductive solution on the broken grid section.
   (C) scraping the old grid from the window and applying a new grid.
   (D) resoldering the bus connector.

5. Technician A says a defective heated mirror should always be replaced as a unit. Technician B says to remove a defective heated mirror it is sometimes necessary to break the old glass to access a screw. Who is right?
   (A) A only.
   (B) B only.
   (C) Both A and B.
   (D) Neither A nor B.

6. After a grid type rear window defroster has been operating for five minutes, the glass at the grid wires feels warmer to the touch than other areas of the glass. Technician A says this indicates abnormal current draw. Technician B says that this indicates normal defroster system operation. Who is right?
   (A) A only.
   (B) B only.
   (C) Both A and B.
   (D) Neither A nor B.

7. If only one part of a defroster grid is not working, which of the following problems is *most likely* to be the cause?
   (A) A defective control switch.
   (B) Low alternator output.
   (C) A broken grid wire.
   (D) A defective timer relay.

8. The defroster electrical bus is located on or near the:
   (A) broken grid wire.
   (B) blower motor.
   (C) timer relay.
   (D) system circuit breaker.

9. A discolored spot can be seen on one of the rear window defroster grid wires. Technician A says that the wire may be broken. Technician B says that the problem at the discolored spot may be confirmed by using an ohmmeter. Who is right?

   (A) A only.
   (B) B only.
   (C) Both A and B.
   (D) Neither A nor B.

10. A heated rear view mirror glass does not remove moisture. The dashboard indicator light is on. All of the following are possible causes, *except:*

   (A) a broken grid wire.
   (B) a disconnected mirror electrical lead.
   (C) a timer defect.
   (D) a blown system fuse.

Useful Tables

CONVERSION CHART

METRIC/U.S. CUSTOMARY UNIT EQUIVALENTS

Multiply:	by:	to get:	Multiply:	by:	to get:
ACCELERATION					
feet/sec^2	x 0.3048	= meters/sec^2 (m/s^2)	x 3.281	= feet/sec^2	
inches/sec^2	x 0.0254	= meters/sec^2 (m/s^2)	x 39.37	= inches/sec^2	
ENERGY OR WORK (watt–second = joule = newton–meter)					
foot–pounds	x 1.3558	= joules (J)	x 0.7376	= foot–pounds	
calories	x 4.187	= joules (J)	x 0.2388	= calories	
Btu	x 1055	= joules (J)	x 0.000948	= Btu	
watt–hours	x 3600	= joules (J)	x 0.0002778	= watt–hours	
kilowatt–hrs.	x 3.600	= megajoules (MJ)	x 0.2778	= kilowatt–hrs	
FUEL ECONOMY AND FUEL CONSUMPTION					
miles/gal	x 0.42514	= kilometers/liter (km/L)	x 2.3522	= miles/gal	
Note:					
235.2/(mi/gal) = liters/100km					
235.2/(liters/100 km) = mi/gal					
LIGHT					
footcandles	x 10.76	= lumens/meter^2 (lm/m^2)	x 0.0929	= footcandles	
PRESSURE OR STRESS (newton/sq meter = pascal)					
inches Hg(60 °F)	x 3.377	= kilopascals (kPa)	x 0.2961	= inches Hg	
pounds/sq in	x 6.895	= kilopascals (kPa)	x 0.145	= pounds/sq in	
inches H$_2$O(60 °F)	x 0.2488	= kilopascals (kPa)	x 4.0193	= inches H$_2$O	
bars	x 100	= kilopascals (kPa)	x 0.01	= bars	
pounds/sq ft	x 47.88	= pascals (Pa)	x 0.02088	= pounds/sq ft	
POWER					
horsepower	x 0.746	= kilowatts (kW)	x 1.34	= horsepower	
ft–lbf/min	x 0.0226	= watts (W)	x 44.25	= ft–lbf/min	
TORQUE					
pounds–inches	x 0.11298	= newton–meters (N-m)	x 8.851	= pound–inches	
pound–feet	x 1.3558	= newton–meters (N-m)	x 0.7376	= pound–feet	
VELOCITY					
miles/hour	x 1.6093	= kilometers/hour (km/h)	x 0.6214	= miles/hour	
feet/sec	x 0.3048	= meters/sec (m/s)	x 3.281	= feet/sec	
kilometers/hr	x 0.27778	= meters/sec (m/s)	x 3.600	= kilometers/hr	
miles/hour	x 0.4470	= meters/sec (m/s)	x 2.237	= miles/hour	

COMMON METRIC PREFIXES

mega	(M)	= 1 000 000	or 10^6	centi	(c)	= 0.01	or 10^{-2}	
kilo	(k)	= 1 000	or 10^3	milli	(m)	= 0.001	or 10^{-3}	
hecto	(h)	= 100	or 10^2	micro	(μ)	= 0.000 001	or 10^{-6}	

METRIC/U.S. CUSTOMARY UNIT EQUIVALENTS

Multiply:	by:	to get:		by:	to get:
LINEAR					
inches	x 25.4	= millimeters (mm)		x 0.03937	= inches
feet	x 0.3048	= meters (m)		x 3.281	= feet
yards	x 0.9144	= meters (m)		x 1.0936	= yards
miles	x 1.6093	= kilometers (km)		x 0.6214	= miles
inches	x 2.54	= centimeters (cm)		x 0.3937	= inches
microinches	x 0.0254	= micrometers (μm)		x 39.37	= microinches
AREA					
inches^2	x 645.16	= millimeters^2(mm^2)		x 0.00155	= inches^2
inches^2	x 6.452	= centimeters^2(cm^2)		x 0.155	= inches^2
feet^2	x 0.0929	= meters^2(m^2)		x 10.764	= feet^2
yards^2	x 0.8361	= meters^2(m^2)		x 1.196	= yards^2
acres^2	x 0.4047	= hectares (10^{4}m^2)			
			ha	x 2.471	= acres
miles^2	x 2.590	= kilometers^2 (km^2)		x 0.3861	= miles^2
VOLUME					
inches^3	x 16387	= millimeters^3 (mm^3)		x 0.000061	= inches^3
inches^3	x 16.387	= centimeters^3 (cm^3)		x 0.06102	= inches^3
inches^3	x 0.01639	= liters (L)		x 61.024	= inches^3
quarts	x 0.94635	= liters (L)		x 1.0567	= quarts
gallons	x 3.7854	= liters (L)		x 0.2642	= gallons
feet^3	x 28.317	= liters (L)		x 0.03531	= feet^3
feet^3	x 0.02832	= meters^3 (m^3)		x 35.315	= feet^3
fluid oz	x 29.57	= milliliters (mL)		x 0.03381	= fluid oz
yards^3	x 0.7646	= meters^3 (m^3)		x 1.3080	= yards^3
teaspoons	x 4.929	= milliliters (mL)		x 0.2029	= teaspoons
cups	x 0.2366	= liters (L)		x 4.227	= cups
MASS					
ounces (av)	x 28.35	= grams (g)		x 0.03527	= ounces (av)
pounds (av)	x 0.4536	= kilograms (kg)		x 2.2046	= pounds (av)
tons (2000 lb)	x 907.18	= kilograms (kg)		x 0.001102	= tons (2000 lb)
tons (2000 lb)	x 0.90718	= metric tons (t)		x 1.1023	= tons (2000 lb)
FORCE					
ounces—f (av)	x 0.278	= newtons (N)		x 3.597	= ounces—f (av)
pounds—f (av)	x 4.448	= newtons (N)		x 0.2248	= pounds—f (av)
kilograms—f	x 9.807	= newtons (N)		x 0.10197	= kilograms—f

TEMPERATURE

°F -40 0 [32] 40 80 [98.6] 120 160 200 [212] 240 280 320 °F

°C -40 -20 0 20 40 60 80 100 120 140 160 °C

°Celsius = 0.556 (°F – 32)　　°F = (1.8 °C) + 32

TAP/DRILL CHART

COARSE STANDARD THREAD (N.C.) Formerly U.S. Standard Thread					FINE STANDARD THREAD (N.F.) Formerly S.A.E. Thread				
Sizes	Threads Per Inch	Outside Diameter at Screw	Tap Drill Sizes	Decimal Equivalent of Drill	Sizes	Threads Per Inch	Outside Diameter at Screw	Tap Drill Sizes	Decimal Equivalent of Drill
1	64	.073	53	0.0595	0	80	.060	^{3}/$_{64}$	0.0469
2	56	.086	50	0.0700	1	72	.073	53	0.0595
3	48	.099	47	0.0785	2	64	.086	50	0.0700
4	40	.112	43	0.0890	3	56	.099	45	0.0820
5	40	.125	38	0.1015	4	48	.112	42	0.0935
6	32	.138	36	0.1065	5	44	.125	37	0.1040
8	32	.164	29	0.1360	6	40	.138	33	0.1130
10	24	.190	25	0.1495	8	36	.164	29	0.1360
12	24	.216	16	0.1770	10	32	.190	21	0.1590
^{1}/$_4$	20	.250	7	0.2010	12	28	.216	14	0.1820
^{5}/$_{16}$	18	.3125	F	0.2570	^{1}/$_4$	28	.250	3	0.2130
^{3}/$_8$	16	.375	^{5}/$_{16}$	0.3125	^{5}/$_{16}$	24	.3125	I	0.2720
^{7}/$_{16}$	14	.4375	U	0.3680	^{3}/$_8$	24	.375	Q	0.3320
^{1}/$_2$	13	.500	^{27}/$_{64}$	0.4219	^{7}/$_{16}$	20	.4375	^{25}/$_{64}$	0.3906
^{9}/$_{16}$	12	.5625	^{31}/$_{64}$	0.4843	^{1}/$_2$	20	.500	^{29}/$_{64}$	0.4531
^{5}/$_8$	11	.625	^{17}/$_{32}$	0.5312	^{9}/$_{16}$	18	.5625	0.5062	0.5062
^{3}/$_4$	10	.750	^{21}/$_{32}$	0.6562	^{5}/$_8$	18	.625	0.5687	0.5687
^{7}/$_8$	9	.875	^{49}/$_{64}$	0.7656	^{3}/$_4$	16	.750	^{11}/$_{16}$	0.6875
1	8	1.000	^{7}/$_8$	0.875	^{7}/$_8$	14	.875	0.8020	0.8020
1^1/$_8$	7	1.125	^{63}/$_{64}$	0.9843	1	14	1.000	0.9274	0.9274
1^1/$_4$	7	1.250	^{17}/$_{64}$	1.1093	1^1/$_8$	12	1.125	1^3/$_{64}$	1.0468
					1^1/$_4$	12	1.250	1^{11}/$_{64}$	1.1718

BOLT TORQUING CHART

METRIC STANDARD

Grade of Bolt	5D	.8G	10K	12K	Size of Socket or Wrench Opening
Min. Tensile Strength	71,160 P.S.I.	113,800 P.S.I.	142,200 P.S.I.	170,679 P.S.I.	
Grade Markings on Head	5D	8G	10K	12K	
Metric					**Metric**
Bolt Dia.	U.S. Dec Equiv.	Foot Pounds			Bolt Head
6mm	.2362	5	6	8	10	10mm
8mm	.3150	10	16	22	27	14mm
10mm	.3937	19	31	40	49	17mm
12mm	.4720	34	54	70	86	19mm
14mm	.5512	55	89	117	137	22mm
16mm	.6299	83	132	175	208	24mm
18mm	.709	111	182	236	283	27mm
22mm	.8661	182	284	394	464	32mm

SAE STANDARD/FOOT POUNDS

Grade of Bolt	SAE 1 & 2	SAE 5	SAE 6	SAE 8	Size of Socket or Wrench Opening	
Min. Tensile Strength	64,000 P.S.I.	105,000 P.S.I.	133,000 P.S.I.	150,000 P.S.I.		
Markings on Head	●	◆	✚	✖		
Metric	**U.S. Standard**				**U.S. Regular**	
Bolt Head	Bolt Dia.	Foot Pounds			Bolt Head	Nut
10mm	1/4	5	7	10	10.5	3/8	7/16
14mm	5/16	9	14	19	22	1/2	9/16
17mm	3/8	15	25	34	37	9/16	5/8
19mm	7/16	24	40	55	60	5/8	3/4
22mm	1/2	37	60	85	92	3/4	13/16
24mm	9/16	53	88	120	132	7/8	7/8
27mm	5/8	74	120	167	180	15/16	1.
32mm	3/4	120	200	280	296	1-1/8	1-1/8

Inches of Mercury/Micron Equivalents

Inches of Mercury	Microns
0.00	759,968
5.00	535,000
9.81	525,526
16.02	355,092
20.80	233,680
24.12	149,352
26.36	92,456
27.83	55,118
28.75	31,750
29.00	25,400
29.10	22,860
29.20	20,320
29.30	17,780
29.40	15,240
29.50	12,700
29.60	10,160
29.70	7,620
29.82	4,572
29.90	2,540
29.95	1,270
29.99	254
29.995	127

SOME COMMON ABBREVIATIONS

U.S CUSTOMARY		METRIC	
UNIT	ABBREVIATION	UNIT	ABBREVIATION
inch	in.	kilometer	km
feet	ft.	hectometer	hm
yard	yd.	dekameter	dam
mile	mi.	meter	m
grain	gr.	decimeter	dm
ounce	oz.	centimeter	cm
pound	lb.	millimeter	mm
teaspoon	tsp.	cubic centimeter	cm^3
tablespoon	tbsp.	kilogram	kg
fluid ounce	fl. oz.	hectogram	hg
cup	c.	dekagram	dag
pint	pt.	gram	g
quart	qt.	decigram	dg
gallon	gal.	centigram	cg
cubic inch	in^3	milligram	mg
cubic foot	ft^3	kiloliter	kl
cubic yard	yd^3	hectoliter	hl
square inch	in^2	dekaliter	dl
square foot	ft^2	liter	L
square yard	yd^2	centiliter	cl
square mile	mi^2	milliliter	ml
Fahrenheit	F°	square kilometer	km^2
barrel	bbl.	hectare	ha
fluid dram	fl. dr.	are	a
board foot	bd. ft.	centare	ca
rod	rd.	tonne	t
dram	dr.	Celsius	C°
bushel	bu.		

Acknowledgments

The production of a textbook of this type is not possible without assistance from the automotive industry. In procuring and compiling the materials for this text, the industry has been extremely helpful. The author would like to thank the following firms and individuals for their assistance in the preparation of **Auto Heating & Air Conditioning Technology.**

AC-Delco; Acura Automobile Division; Alfa-Romeo, Inc.; Allied-Signal Inc.; Audi of America; Autolite; Automotive Diagnostics; Balco Inc.; Barret Company; Beauron Industries, Inc.; Bee Line Company; Black and Decker; BMW of North America; Bower/BCA; Bridgestone U.S.A.; British-Leyland; Buick Motor Division; Cadillac Motor Division; Central Tools; Champion; Chevrolet Motor Division; Chief Automotive Systems; Chief Industries; Clayton Manufacturing; Continental Teves; Cooper Industries; CR Industries; Daihatsu America Inc.; DaimlerChrysler; Dana Publication Services; Dayton; Deere and Company; Delphi Corporation; Dorman Products, Inc.; Douglas Components Corporation; Dover Corporation; EMI-Tech; Federal-Mogul Corporation; Ferrari North America; John Fluke Manufacturing Company; FMC Corporation; Ford Motor Company; Gates Rubber Company; General Motors Corporation, Service Technology Group; GMC Trucks; Goodyear Tire and Rubber Company; Graymills Corporation; H and M Dreyer, Inc.; Honda Motor Company; Hyundai Motor Corporation; ICI Americas; Imperial-Eastman; Infiniti North America; International Mobile Air Conditioning Association; Jaguar Cars; Kent-Moore Tool Group; Kia Motors; Kocs, Wesson, & Associates Inc.; Land Rover of North America; Lexus Automobile Division; Lisle Corporation; Loctite Corporation; Mac Tools; Maremont Corporation; Mazda Motor of America; Mercedes-Benz of North America; Mitsubishi Motors of America; Mobile Air Conditioning Society; Modine Manufacturing; Mohawk Resources; Moog Automotive; NAPA; National Institute for Automotive Service Excellence; National Oceanic & Atmospheric Administration; Niehoff Company; Nissan North America; Oldsmobile Motor Division; OTC Division of SPX Corp.; Paprl Automotive Industries; PBS Corporation; Pontiac Motor Division; Porsche Cars of North America; Prestolite; Raytek; Robinair Div. of SPX Corp.; RTI Technologies, Inc.; Saab Cars USA; Sanden International; Saturn Corporation; SCS/Frigette; SFK Industries; SK Hand Tools; Snap-On Tools; Subaru of America; Sun Electric Corporation; Technical Chemical Co; TIF Instruments; Toyota Motor Sales; Trumark; TRW Inc.; U.S. Environmental Protection Agency; Vaco Tool Company; Vetronix Corporation; Volkswagen of America; Volvo Corporation of North America; Walker Company; Warner Company; White Industries.

The author would also like to thank the following persons and organizations who provided vehicles, parts, and test equipment for the photographs as well as other items used throughout this text.

David Bryson, Billy Benson, Sears Auto Center, Greenville, SC; Pete Chase, Protech Motor Sports, Greenville, SC; Steve Dunn, Auto Air, Greenville, SC; James Daley; Danny McCown, Century Lincoln Mercury/BMW, Reuben Byrd, Robert Holbert, Lynn Anders, Cline Hose and Hydraulics, Greenville, SC; Ronald Parker, Larry Neighbors, A-1 Auto, Greenville, SC; Herbert and Dorothy MacMillian; John Edwards, Sandi Ridge, Karl Martus, Glenn Gaines, Raymond Rhoads, Steve Benge, Auto Zone #0163, Greenville, SC; Palmetto Chapter, Pontiac-Oakland Club International; Jeff Nags; Richard Rackow, Moraine Valley Community College, Palos Hills, IL; Oliver Scheurmann; C.W. Weeks; Natchitoches Junior College, Natchitoches, LA; Michael Wood.

"Portions of materials contained herein have been reprinted with permission of General Motors Corporation, Service Technology Group."

"This publication contains material that is reproduced and distributed under a license from Ford Global Technologies, Inc., a Subsidiary of Ford Motor Company. No further reproduction or distribution of the material is allowed without the express written permission from Ford Global Technologies, Inc."

Glossary

A

Abbreviations: Letters or letter combinations that stand for words. Used extensively in the automotive industry.

Absolute zero: The temperature point in which an object is considered to have no heat. Absolute zero is approximately -459°F (-273°C).

Absorption: State where moisture enters a desiccant and is held there.

Accumulator: Metal cylinder used to store vaporized refrigerant from the evaporator.

Actuator: Any output device controlled by a computer.

Adsorption: State where moisture sticks to the surface of the desiccant.

Afterblow module: Electronic module used to operate the blower motor for a short period of time after the vehicle has been shut off. Designed to dry off the evaporator surface to prevent mold formation.

Air bleed: Valve installed at the high points on the engine. Used to bleed air from the cooling system.

Air bypass door: Door used during maximum air conditioning operation to direct additional cooled air around the blend door. This aids cooling under extremely hot conditions.

Air doors: Movable panels inside a blower case. Used to alter the direction and temperature of the air flowing through the ventilation system.

Air ducts: See *Register.*

Air gap: Amount of space between the compressor clutch plate and the compressor pulley.

Air-cooled engine: Any engine which uses air as the primary cooling medium. Few automobiles use air-cooled engines.

ALDL: Abbreviation for Assembly Line Diagnostic Link. Also referred to as a *data link connector.*

Alkalinity: The acidity level of a fluid measured by its pH level.

Alkylbenzene (AB): Refrigerant oil used in stationary air conditioning systems; should not be used in mobile applications.

Allen wrench: A hexagonal wrench, which is usually "L" shaped, designed to fit into a hexagonal hole.

Alternating current (ac): An electrical current that moves in one direction and then the other.

Ambient air: The outside air surrounding a vehicle or person.

Ambient air temperature sensor: Thermistor placed outside the vehicle, usually in the front grill or near the condenser, to measure outside air temperature.

American Society of Refrigerating and Air Conditioning Engineers (ASHRAE): Organization that classifies and sets standards for marking and handling refrigerants.

Ammeter: Instrument used to measure the flow of electric current in a circuit in amperes. Normally connected in series in a circuit.

Ampere: The unit of measurement for the flow of electric current.

Analog: A signal that continually changes in strength.

Anodized: Electrically deposited coating used to reduce wear on aluminum parts.

Antifreeze: See *Coolant.*

Antifriction bearing: A bearing that uses balls or rollers between a journal and bearing surface to decrease friction.

ASE: Abbreviation for National Institute For Automotive Service Excellence.

Atmospheric pressure: The pressure exerted by the earth's atmosphere on all objects. Measured with reference to the pressure at sea level, which is around 14.7 psi (101 kPa).

Automatic temperature control system: HVAC control system that uses electronic controls to operate vacuum servos or electronic stepper motors to control mode and blend door positions.

Auxiliary vacuum pump: A small engine or electrically driven pump used to produce additional vacuum for vehicle accessory systems. Usually found on luxury cars.

AWG: Abbreviation for American Wire Gage, which is a standard measure of wire size. The smaller the wire, the larger the number.

Axial compressor: Compressor design that places the pistons lengthwise to the compressor shaft.

Axial load: Sideways load on a bearing.

Azeotropic blend: Class of blended refrigerants that will not separate once blended. Numbered in the R-500 classification.

B

Ball bearing: A bearing consisting of an inner and outer hardened steel race separated by a series of hardened steel balls.

Barrier hose: A refrigerant hose that has an inner liner which helps prevent refrigerant leakage.

Belt tension gauge: Tool used to measure the tension of an installed V-belt or serpentine belt.

Belt tensioner: Spring loaded pulley which is turned to install and remove serpentine belts and keeps tension on the belt.

Belt-driven fan: Cooling system fan driven by engine power through one or more belts.

Bi-level: Operating mode that allows air to blow from more than one vent level, usually from the floor and defrost vents.

Biocide: Chemical disinfectant used to clean mildew and mold from evaporator cores and blower cases.

Blend door: A movable partition in the blower case that regulates air temperature according to the demands of the vehicle's occupants.

Blend refrigerant: A refrigerant made from two or more refrigerants. Depending on its stability, the blended refrigerant can be classified as either azeotropic or zeotropic.

Block valve: See *H-block*.

Blower motor: Electric motor that provides air movement in the blower case.

Blowgun: Tool used to spray liquids under shop air pressure.

Body control module (BCM): On-board computer responsible for controlling such functions as interior climate, radio, instrument cluster readings, and on some vehicles, cellular telephones. May also interact with the engine's electronic control unit.

Bulkhead: The rear wall behind the engine. Also referred to as the *firewall*.

Bus: Pathway for data inside a computer. Can also be used to refer to a circuit used to connect two on-board computers.

C

Capacity control valve: Valve which varies the pumping ability of the compressor. This eliminates the need to cycle the compressor clutch on and off.

Capillary tube: See *Sensing bulb*.

Captured O-ring: Fitting that has a special depression designed to accept and hold an O-ring.

Carbon monoxide: A colorless, odorless gas that forms as a result of unburned fuel. Can be deadly if allowed to build up in an enclosed area.

Carcinogen: A substance that can cause cancer.

Center of gravity: The point on an object on which it could be balanced.

Central processing unit (CPU): Microprocessor inside a computer that receives sensor information, compares the input with information stored in memory, performs calculations, and makes output decisions.

Centrifugal fan clutch: Clutch assembly comprised of two internal sets of ridges that move in relation to each other. The space between the sets of ridges is filled with a thick silicone fluid compound.

Centrifugal force: A force which tends to keep a rotating object away from the center of rotation.

CFC-12: See *R-12*.

Chamfer: To bevel an edge on an object at the edge of a hole.

Change of state: The switching of the physical condition of a substance.

Charging scale: See *Refrigerant scale*.

Charging station: See *Refrigerant service center*.

Chatter: Rapid clicking or knocking noise typical of a loose or damaged compressor clutch.

Check valve: A valve that permits flow in only one direction.

Chlorine (Cl): Gaseous element used to manufacture some refrigerants.

Chlorofluorocarbon (CFC): Name for refrigerants, propellants, and chemicals that contain chlorine, carbon, fluorine, and sometimes hydrogen. Capable of causing damage to the ozone layer.

Circuit breaker: Circuit protection device consisting of a contact point set attached to a bimetallic strip. The bimetallic strip will heat and bend as current flows. Unlike a fuse, it does not blow out.

Class 2 serial communication: Electronic data transfer medium used on newer vehicles. Operates by toggling the line voltage from 0-7 volts, with 0 being the rest voltage, and by varying the pulse width.

Clean Air Act: Document that enforces the provisions of the Montreal Protocol in the United States.

Closed loop flushing: Refrigerant or cooling system flushing method where the flushing agent and any debris is captured by a tank or waste receptacle.

Clutch cycling: Condition where the compressor clutch frequently engages and disengages. Condition usually caused by low refrigerant charge.

Coefficient of friction: The amount of friction produced as two materials are moved against each other. Coefficient of friction is calculated by dividing the force needed to push a load across a given surface.

Combination machine: Refrigerant service center designed to service more than one type of refrigerant, usually R-134a and R-12.

Combustion leak tester: Chemical used to determine the presence of exhaust gas in engine coolant.

Compression fitting: Refrigerant line fitting consisting of a nut with internal threads tightened against a seat with external threads. An O-ring is included in the fitting for additional sealing.

Compressor: The "pump" for the refrigeration system. Compresses and moves refrigerant and oil throughout the refrigeration system.

Compressor clutch: The means of engaging and disengaging the compressor from the engine.

Condenser: Heat exchanger used to remove heat from the refrigerant. Mounted beside or in front of the radiator.

Conduction: Heat transfer by direct contact between two objects.

Conductor: Any material or substance that provides a path for electricity to flow. Examples of good conductors are copper and aluminum.

Control head: The control panel for the heating and air conditioning system.

Control loop: A continuous series of operations in a system.

Control module: A computer that controls the operation of a system based on sensor inputs. Usually less complex than the computer that controls the engine and other vehicle systems.

Convection: The transfer of heat by air.

Coolant: A liquid mixture of ethylene glycol, propylene glycol, or OAT solution, other chemicals, and water. Used in the cooling system.

Coolant passages: See *Water jacket.*

Coolant pump: See *Water pump.*

Coolant recovery system: Part of closed cooling system which keeps air out of the cooling system. Also see *Reservoir tank.*

Cooling fins: Metal fins cast into the engine components of air-cooled engines. Used to dissipate heat.

Cowl: Area just above the bulkhead, and before the windshield. Entry point for air going to the blower case.

Crimp fitting: An intentional pinch placed to hold a fitting to a hose. Two types are used in refrigerant hoses, barb and beadlock.

Cross-contamination: Contamination with a substance not designated for that system, for example an R-134 service center accidentally charged with R-12.

Crossflow: Type of radiator that allows coolant to flow in a horizontal direction.

Current: The number of electrons flowing past any point in the circuit.

Cycling clutch orifice tube (CCOT) system: Air conditioning system which uses a fixed orifice tube. System controls refrigerant flow by using a cycling compressor clutch.

D

De minimis: Latin phrase meaning *minor.*

Defrost: Air conditioning mode that allows air to blow on the windows. Refrigeration system is usually engaged during this mode. Also known as *Defog.*

Dehumidify: Process of removing humidity or moisture from the air.

Density: The relative mass of an object in a given volume.

Desiccant: Drying agent used in every refrigeration system to minimize possible damage from moisture.

Diaphragm: A flexible rubber sheet that divides two sides of a chamber.

Dichlorodifluoromethane (CCl_2F_2): Chemical name and formula for refrigerant R-12.

Diode: Semiconductor device that allows current to flow in one direction.

Direct current (dc): An electrical current that moves in only one direction.

Diverter door: One or more movable panels inside the blower case that change airflow direction.

Documentation: Process of writing on the repair order a vehicle's problem, what caused the problem, and what was done to correct the problem.

Double-ended piston: Compressor piston that has heads on both ends.

Downflow: Type of radiator that allows coolant to flow vertically.

Downstream: Used to describe the location of a component after another component in the direction of flow.

Drop-in refrigerant: Term used to mean a refrigerant substitute that can be added to a partially charged refrigeration system with no modifications. Currently, there are no drop-in refrigerants.

Dual Zone: Type of air conditioning system having controls that allow the driver and front seat passenger to control air temperature and flow for their particular side.

Ductwork: Channels used to move air from the blower case into the vehicle.

Dye: Chemical marker added to a closed system to locate a leak. Often used along with a black light.

E

Eccentric: A circle within a circle that has a different shape and center.

Electric compressor: Type of compressor installed on some SULEV and zero emissions vehicles. Compressor is driven by electricity rather than engine power.

Electric fans: Engine cooling fans operated by electrical power.

Electrochemical degradation (ECD): Damage to the inside of heater and radiator hoses caused by a reaction between the chemicals in the coolant and the metals in the engine and radiator.

Electromagnet: A magnet that is produced by placing a coil of wire around a steel or iron bar. When current flows through the wire, the bar becomes magnetized.

Electromagnetic interference (EMI): Electronic noise caused by voltage spikes and stray magnetic fields created by defects in such circuits as the ignition and charging systems.

Electron: A negatively charged particle that makes up part of an atom.

Electronic leak detector: An extremely sensitive tool designed to detect refrigerants.

Electronically erasable programmable read only memory (EEPROM): A type of microprocessor whose programming can be changed by special electronic equipment that "burns in" the new programming.

Endplate: See *Wobble plate*.

Energy module: Electronic module which operates the air bag system; also stores extra power should the battery be disconnected in an accident.

Engine oil cooler: Tube or plate device connected to the engine's lubrication system. Designed to remove excess heat from the engine oil. May use either engine coolant or air as the medium of transfer.

Environmental Protection Agency (EPA): Agency of the United States government charged with enforcing laws against pollution and environmental destruction.

Erasable programmable read only memory (EPROM): A type of microprocessor whose programming can be altered only by erasing it with special equipment and reprogramming.

Ethylene glycol: Chemical used in the manufacture of antifreeze. Term used to classify type of refrigerant.

Evacuate: Process of removing moisture from a refrigeration system through the use of a vacuum pump.

Evacuation pump: Electric or air powered pump used to evacuate air from the refrigeration system.

Evaporation: The changing of a liquid to a gas.

Evaporator: Heat exchanger which removes heat from the air entering the vehicle.

Evaporator case: Plastic or metal case that houses the evaporator and sometimes the heater core. Also called the *blower case*.

Evaporator pressure regulator (EPR): Switch that allows the control of evaporator pressure by monitoring its temperature. Sometimes called an *evaporator temperature sensor*.

Evaporator water drain: Hole in the evaporator case that allows condensation to drain out.

Expansion valve: See *Thermostatic expansion valve*.

F

Fan clutch: Temperature controlled device that allows a belt-driven fan to freewheel when the engine is cold or the vehicle is moving at highway speeds.

Fiber optic: A path for electricity or data transmission in which light acts as the carrier.

Firewall: See *Bulkhead*.

Fixed orifice tube (FOT): Calibrated restriction in the refrigeration system which allows a metered amount of refrigerant to enter the evaporator. Used with cycling clutch air conditioning systems. Usually just referred to as an *orifice tube*.

Flare-nut wrench: Sometimes called a *tubing wrench* or *line wrench*. Tool used to remove tube fittings used on all brake line connections.

Flex fan: Belt-driven fan which flexes as it turns faster to pull more air through the radiator. Designed to be used without a fan clutch.

Flexible hose: Radiator hose designed to be installed on most vehicles.

Flushing: The removal of contaminants from a refrigeration or cooling system.

Formed hose: Hose designed for use on a particular vehicle or engine.

Freon: Name originally given to R-12. Now used as a generic name to describe any refrigerant.

Friction bearing: A bearing made of babbitt or bronze with a smooth surface.

Functional test: Test used to determine proper system operation at different control panel settings.

Fuse: Circuit protection device made of a soft metal that melts when excess current flows through it, before the current can damage other components or circuit wiring.

Fuse block: Location for fuses, circuit breakers, and fusible links. Sometimes referred to as a *junction block*.

Fusible link: A special calibrated wire installed in a circuit. Will allow an overload condition for short periods. A constant overload will melt the wire and break the circuit.

Fusible plug: A bolt installed in some older compressors. Designed to melt if the pressure becomes too high.

G

Galvanize: To coat a metal with a molten alloy mixture of lead and tin. Used to prevent corrosion.

Gasket: A material used to prevent leaks between two stationary parts.

Gauge manifold: A tool used to monitor air conditioning system pressure. Also can be used to recover, evacuate, and add refrigerant.

Ground: The part of a circuit connected to a body terminal or the battery.

H

Halide flame detector: Torch-type leak detector. A change in the color of the flame indicates a refrigerant leak.

Hand valve: Valve used to control the flow of refrigerant through the passages of a gauge manifold.

Hard code: Trouble code set by an on-going problem that still exists in the vehicle.

Harness: Collection of wiring for one or more circuits.

Hazardous waste: Any chemical or material that has one or more characteristics that makes it hazardous to health, life, and/or the environment.

H-block: Expansion valve design shaped into a block form resembling the letter H. Frequently used on DaimlerChrysler vehicles.

HCFC-22: See *R-22*.

Header: Tank or enclosure on the end of a radiator, condenser, or evaporator core.

Heat dissipation: Heat removed from a friction surface by direct transfer to the surrounding air.

Heat exchanger: Any device that allows for the transfer of heat from a medium to the surrounding air.

Heat load: The total effect of heat and humidity of the ambient air.

Heat mode: Vent mode which brings warmed air into the vehicle.

Heater case: Separate case which houses the heater core.

Heater core: Heat exchanger used to transfer heat from coolant to air coming in the vehicle.

Heater door: Door used on air-cooled engines to redirect heated air into the passenger compartment.

Heater hose: Small diameter hose used to transfer coolant from the engine to the heater core to warm the vehicle interior.

Heater shutoff valve: Valve installed in one of the heater hoses and stops the flow of coolant as needed.

Heater-defroster door: Mode door used to direct heated air from the heater core to the floor and defrost registers.

Heat-shrink tubing: Plastic tube used to insulate electrical solder joints.

HFC-134a: See *R-134a.*

Hg: Abbreviation for *mercury.*

High blower relay: Relay used to allow full current flow to bypass the resistor assembly.

High efficiency particulate air (HEPA) vacuum: Type of vacuum cleaner that can trap extremely small particles, including asbestos particles.

High-pressure cutoff switch: Electrical or computer-controlled switch that cuts off the compressor before the pressure reaches the relief valve opening point.

High-pressure relief valve: Spring-loaded valve installed on older R-12 systems to relieve excess refrigeration system pressure.

Hose clamp: Screw or spring metal clamp used to hold cooling system hoses.

Hose cutters: Tool that is used to cut off old radiator and heater hoses. Also cutting tool used in hose fabrication.

Humidity: The amount of water vapor in the surrounding air.

Hydrochlorofluorocarbon (HCFC): Class of refrigerants comprised of methane or ethane in combination with fluorine or other halogen. Example is R-22.

Hydrofluorocarbons (HFC): Class of refrigerants that contain hydrogen and no chlorine. Example is R-134a.

Hygroscopic: Able to absorb water.

I

Idler pulley: Simple pulley installed at a point where a long belt run would create vibration.

Independent case: Single piece blower case which fits under the dash.

Independent vent: Any air inlet independent of the HVAC system.

Induction: The imparting of electricity by magnetic fields when a wire moves through a magnetic field.

Inertia: Force which tends to keep stationary objects from moving and keeps moving objects in motion.

Inlet screen: Screen built into an expansion valve, used to prevent debris from clogging the valve.

Inline filter: A filter installed in the refrigeration system to remove debris that may remain after flushing.

Input sensors: Any sensor that provides information to a computer.

Inside temperature sensor: Thermistor used to measure the interior temperature. This reading is compared to the ambient air temperature for improved climate control system operation.

Insulators: Any material or substance which resists the flow of electricity. Glass and plastic are examples of good insulators.

Integrated circuit (IC): Electronic semiconductor device containing thousands of circuits, used in computers and other electronic components.

Interior air filter: A conventional filter, electrostatically charged in some cases that removes dust, pollen, and in sometimes, odors from the incoming air.

Intermittent code: Computer diagnostic code that does not return immediately after it has been cleared.

Intermittent problem: System malfunction that occurs infrequently, usually when certain conditions are met, such as temperature, humidity, vehicle operation, or in response to tests performed by a computer.

J

Jumper wire: A wire used to make a temporary electrical connection.

K

Kinetic friction: Sometimes called *sliding* or *dynamic friction.* Friction that slows a moving object by converting momentum to heat.

L

Latent heat: Heat hidden in a substance, such as a liquid or gas.

Leak detector: Means used to detect leakage from the refrigeration or cooling system. Can be electronic, dye, or other means.

Liquid-cooled engine: An engine cooled by antifreeze and water.

Long-life coolants: Engine coolants designed with low corrosive potential. Long-life coolants will last far beyond the useful life of conventional coolants. Also see *Organic acid technology (OAT)*.

Low-pressure cutoff switch: Electronic switch that monitors for excessive low pressure. Will cause the compressor clutch to disengage, and prevent clutch engagement, in some cases, if low refrigerant pressure is detected.

M

Malfunction indicator light (MIL): Amber-colored light in the instrument cluster used to indicate that a problem exists in a vehicle's computer control system. Generalized term used for any instrument cluster light used to indicate a problem in a system.

Manifold hose: Service hose used to connect a gauge manifold or service center to a vehicle.

Material Safety Data Sheet (MSDS): Information on a chemical or material that must be provided by the material's manufacturer. Lists potential health risks and proper handling procedures.

Microprocessor: Small electronic circuit used in computers and other systems.

Mineral oil: Petroleum based oil is used in R-12 systems. Usually clear to light yellow in color. Also used with some replacement refrigerants. However, it should never be used in an R-134a system.

Mirror defrosters: Electric grid installed in side view mirror glass to defrost the mirror in icy weather.

Miscible: Term used to describe any oil that will mix well with refrigerant.

Mode door: See *Diverter door*.

Molded radiator hose: Neoprene hose that is shaped to fit a certain vehicle or engine application.

Momentum: The combination of the vehicle's weight and speed. Often referred to as *kinetic energy*.

Montreal Protocol: An international agreement which calls for the reduction and elimination of substances, including refrigerants, which contribute to ozone layer depletion.

Muffler: Portion of a refrigerant hose that reduces noises caused by pulsation of the refrigerant leaving the compressor.

Muffler hose: Hose assembly that connects the compressor to the refrigeration system.

Multiflow condensers: Condenser that directs refrigerant through a small number of tubes laid in rows.

Multimeter: An electrical test meter that can be used to test for voltage, current, or resistance.

Multiple-pass evaporator: Evaporator designed so refrigerant enters through the top and passes three or four of the plates before being redirected upward through the next set of plates.

Multiplexing: A method of using one communications path to carry two or more signals simultaneously.

N

NORM: Air conditioning mode that allows air from the outside to enter the blower case. It is then cooled and transferred to the vehicle's interior.

Neck: The fill opening in a radiator, reservoir, or cooling system.

Nitrogen pressurizing: Leak detection method that calls for the pumping of pressurized nitrogen into the refrigeration system.

Noncondensible gas: Air, nitrogen, or any other nonrefrigerant gas in the refrigeration system.

Non-powered test light: Tool that is used to check for the presence of voltage.

O

OBD II: On-Board Diagnostics—Generation Two. Protocol adapted by vehicle manufacturers for standardization of diagnostic trouble codes and automotive terminology.

Ohm: Unit of measurement for resistance to the flow of electric current in a given unit or circuit.

Ohmmeter: An electrical instrument used to measure the amount of resistance in a given unit or circuit.

Ohm's law: Formula for computing unknown voltage, resistance, or current in a circuit by using two known factors to find the unknown value.

Oil return hole: Small hole in the bottom of the pickup tube in an accumulator. Allows refrigerant oil to return to the compressor.

One pound can: Industry term used to describe a small can of refrigerant, however, the can almost never contains a pound of refrigerant.

Open circuit: A circuit that is broken or disconnected.

Open loop flushing: Refrigerant or cooling system flushing method that allows the flushing medium and any loose debris to escape to the air or ground.

Organic acid technology (OAT): Type of coolants based on a blend of two or more organic acids with long-life properties. Formulated without silicates, phosphates, and other minerals and substances that can form deposits in the cooling system.

Orifice tube: See *Fixed orifice tube*.

O-ring: A rubber or neoprene ring which serves the same function as a gasket in a smaller space.

Overcooling: Condition where the engine is running cooler than normal.

Overheating: Condition where the engine is running hotter than normal.

Ozone layer: Invisible barrier in the Earth's stratosphere that filters out harmful ultraviolet light and radiation from the sun.

P

Pascal's law: Law of hydraulics which states the pressure in a closed hydraulic system is the same everywhere in the system.

Performance test: Used to check the refrigeration and heating system components for proper pressures and temperatures.

Permanent magnet: A magnet capable of retaining its magnetic properties over a very long period of time.

Petcock: A drain located in one of the radiator tanks.

pH level: Measure of a fluid's acidity.

Phosgene gas: A poisonous gas that is formed when refrigerant is burned in an open flame.

Photovoltaic cell: Semiconductor device that produces a small electric current when exposed to sunlight.

Pilot operated absolute (POA) valve: Refrigerant control valve that uses a control piston which seals the passage between the evaporator and compressor.

Piston compressor: Compressor that uses two or more pistons to intake, compress, and move refrigerant.

Pitch: Angle of the blades on an engine cooling fan.

Plate and fin evaporator: An evaporator consisting of a set of flat aluminum plates between two aluminum end tanks.

Plug-in connector: Electrical junction connector used in computer control and other circuits.

Polyalkylene glycol (PAG): A synthetic (non-petroleum) oil similar in chemical makeup to cooling system antifreeze. PAG oil is usually light blue in color. It is intended for R-134a systems, and cannot be used in an R-12 system.

Polyol ester (POE): An alcohol-based oil that is compatible with small amounts of PAG or mineral oil. POE is usually clear with no tint. It has a slight odor similar to brake fluid. Can be used with either R-134a or R-12, which makes it ideal for retrofitting.

Power module: Electrical module that is part of an automatic control system. Used to control blower speed.

Powered test light: Tool that uses a small battery to send power through a circuit. Used to check for continuity.

Press-fit: Condition of fit between two parts that requires pressure to force the two parts together.

Pressure cap: Radiator cap. Raises the pressure of the cooling system, which raises the boiling point of the coolant.

Pressure tester: Tool used to test the cooling system for leaks by applying pressure to the system.

Pressure-temperature relationship: The relationship of pressure to change of state for a substance.

Primary wiring: Small insulated wires which serve the low voltage needs of the ignition and vehicle systems.

Programmable read only memory (PROM): A semiconductor chip that contains instructions that are permanently encoded into the chip. Instructions contain base operating information for how a system's components should operate under various conditions.

Programmer: Electromechanical control device used in some semi-automatic control systems.

Propylene glycol: Chemical base for so-called "environmentally safe" coolants.

Proton: A positively charged particle that makes up part of an atom.

Pulley ratio: The difference in diameter between a drive and driven pulley.

Pulse width: The length of time an output device is operated by a computer.

Purge modes: One of several modes in an automatic control system. Used to remove excess moisture and moisture laden air from the air distribution system.

Purging: The removal of air or noncondensible gas from a system.

Push-on service fitting: Refrigeration service fittings that allow for a manifold hose to be pushed on. Used on R-134a and some blend refrigerant systems.

Q

Quick disconnect hose: Heater hose with male and female connectors that push together and snap to form a tight connection.

R

R-11: Trichlorofluoromethane (CCl_3F), a refrigerant that was used as a flushing agent. No longer produced, as it is classified as a CFC.

R-12: Dichlorodifluoromethane (CCl_2F_2), refrigerant used in automotive air conditioning systems for many years. It is no longer produced and is being phased out of older systems.

R-22: Chlorodifluoromethane ($CHClF_2$), a hydrochlorofluorocarbon refrigerant used in many blend refrigerants.

R-134a: Tetrafluoroethane (CF_3CH_2F), a hydrofluorocarbon refrigerant, which contains no chlorine atom. Used in all air conditioning systems in new cars built after 1993.

Radial compressor: Compressor that has its pistons placed around the compressor shaft.

Radiator: Heat exchanger used to remove heat from engine coolant.

Radio frequency interference (RFI): Electronic noise caused by stray signals from electronic transmitters such as police and CB radios.

Ram air: Air that enters the engine or vehicle due to vehicle movement.

Random access memory (RAM): Type of computer memory that is volatile; stores temporary information.

Read only memory (ROM): Type of nonvolatile computer memory; stores basic operating information for the computer and vehicle.

Rear window defroster: Electrical grid used to defrost and defog the rear window.

Receiver: Also referred to as an *emitter* or *responder*. Point where a vibration is heard or felt.

Receiver-drier: A tank that stores condensed refrigerant and a small amount of vaporized refrigerant. The receiver-drier is always located at the condenser outlet.

Recirculation door: Door used to close off outside air and allow air from the vehicle interior to be recirculated.

Recovery tank: Reservoir for coolant overflow when the cooling system is under pressure.

Redundant control: Steering wheel or remote mounted controls for the heating and air conditioning system.

Reed valves: Small flap valves inside a compressor, used to seal the compressor pistons from the incoming and outgoing refrigerant.

Reference voltage: A known voltage (can vary from 0.5-5V) that is sent to a sensor by a computer. The changes in sensor resistance will change the voltage, which is read by the computer as a change in temperature, airflow, etc.

Refrigerant containment switch: See *High-pressure cutoff switch.*

Refrigerant identifier: Device used to check refrigerant for type and purity before a service center is connected.

Refrigerant oil: Oil added to the refrigeration system to lubricate the compressor. See *Mineral oil, Polyalkylene glycol,* and *Polyol ester.*

Refrigerant recovery: Process of discharging refrigerant from an air conditioning system.

Refrigerant recycling: The reuse of refrigerant that came from the system or another vehicle.

Refrigerant scale: Electronic scale assembly used to accurately meter a set amount of refrigerant. Term also applied to conventional scales used to measure refrigerant tanks during charging.

Refrigerant service center: Device which combines a refrigerant pump, storage cylinder, and gauge manifold. Used to recover, recycle, and recharge refrigerant.

Refrigeration system: Portion of the air conditioning system that performs the task of chilling the air.

Refrigeration temperature sensor: Measures the temperature of the refrigerant in the system. The control computer uses temperature inputs to determine the pressures in the refrigeration system as part of malfunction monitoring.

Register: Air conditioning vents in the dash.

Relay: Switching device which uses a magnetic field that closes one or more sets of electrical contacts, causing electrical flow in a circuit.

Resistance: The opposition of atoms in a conductor to the flow of electrons.

Resistor pack: Series of coiled wires of different gage sizes, used to control blower motor speed.

Retrofitting: The process of converting an R-12 refrigeration system to operate on another type of refrigerant. R-134 is usually the refrigerant of choice when retrofitting.

Reverse fill: Coolant fill technique that involves filling the cooling system through the upper radiator hose first. Used to minimize the amount of air trapped in the cooling system.

Reverse flushing: Flush technique that cleans a system in the opposite direction of flow.

S

SAE: Abbreviation for Society of Automotive Engineers.

Scan tool: Diagnostic tool used to read computer sensor and output device values and states of operation.

Schematic: Diagram of an electrical circuit.

Schrader valve: Valve shaped like a tire valve core, used in R-12 systems and as connection points for switches.

Screw-on service fittings: Refrigerant hose fittings used to connect to R-12 and other systems. Also used to connect between tanks on refrigerant service centers.

Scroll compressor: A compressor that uses two lengths of flat metal, formed into spiral shapes and placed together. Sometimes called a spiral or *orbital compressor.*

Sealing washer: Combination seal and washer used to connect a muffler hose to the compressor.

Seat heater: Electric heater built into the base frame of the interior seats.

Semi-automatic temperature control: Control system that uses both electronics and mechanical devices to control HVAC system operation.

Semiconductor: Device that depending on its operational state can be either a conductor or insulator.

Sensible heat: Heat that can be felt, or sensed.

Sensing bulb: Expansion valve tube used to monitor temperature. It has a bulb-shaped end filled with a refrigerant gas and is connected to the evaporator.

Serial data: Sensor and actuator information that is shared by the various vehicle on-board computers.

Serpentine belt: Accessory drive belt used on newer vehicles. Used to drive all the engine-mounted accessories.

Serpentine evaporator: A single-pass evaporator consisting of a flat tube coiled to allow refrigerant to pass through the entire tube.

Service fittings: Push-on or Schrader valve fittings used to service the refrigeration system.

Set temperature: The temperature requested by the driver and passengers.

Shroud: Panels used to force air to move in a particular direction.

Sight glass: Clear glass installed at the top of a receiver-drier. Used to quick-check the charge status of the refrigeration system.

Significant New Alternatives Policy (SNAP): Program initiated by the U.S. Environmental Protection Agency to review and approve or reject refrigerant alternatives to those containing CFCs.

Single-pass evaporator: Evaporator made from a single piece of tubing folded into a series of coils.

Slugging: The migration of liquid refrigerant to the inlet side of the compressor.

SNAP list: Listing of refrigerants approved by the U.S. Environmental Protection Agency for use in automotive air conditioning systems.

Snap ring: A split ring snapped in a groove to hold a bearing, thrust washer, gear, etc., in place.

Snow ingestion mode: Purge mode used in some automatic control systems to clear snow from the blower case.

Soap solution: Leak detection method. Uses a soapy solution spread over the surfaces of the refrigeration system.

Soft code: Also referred to as an *intermittent code.* Trouble code that is set by a problem that occurs occasionally, or only once.

Solenoid: Switching device that uses a magnetic field to perform a mechanical task, such as opening or closing blend doors.

Solvent: A liquid used to dissolve or thin another material.

Special tools: Custom tools designed to perform a specific task, such as adjusting or disassembling a system.

Specific gravity: A relative weight of a given volume of a specific material as compared to an equal volume of water.

Split blower case: Two-piece blower case that can be disassembled without removing the entire case from the vehicle.

Split crankcase: Compressor body that can be disassembled into two or more pieces.

Spring lock coupling: Sometimes called the *quick connect coupling.* Refrigerant system fitting found on Ford, DaimlerChrysler, and some Asian and European made vehicles.

Squirrel cage blower: A round fan used in air conditioning systems to move air; resembles the wheel in a squirrel cage.

Static friction: Friction produced by a holding action that keeps a stationary object in place.

Stepper motor: Electronic motor, usually computer-controlled, that can be started, stopped, and moved into different positions. Used on blower case doors in newer vehicles.

Suction throttling valve (STV): A spring-loaded shutoff valve used on older vehicles to control evaporator pressure.

Sunload: The amount of sunlight exposure on the vehicle.

Sunload sensor: Photovoltaic cell used to measure the amount of sunlight exposure.

Superheat: Additional heat above the temperature needed to vaporize a substance.

Superheat switch: Switch used on older vehicles with thermal limiters. Designed to close if the low side pressure went below a preset value.

Swash plate: See *Wobble plate.*

T

Technical service bulletins (TSB): Information published by vehicle manufacturers in response to vehicle conditions and problems that may not be diagnosed by normal methods.

Temperature control computer: Microcomputer used to operate the climate control system.

Temperature gauge: Analog or digital gauge used to measure the temperature of the air coming from the air conditioning system registers.

Test strips: Chemically treated strips of paper or plastic used to measure coolant freeze point and pH level.

Tetrafluoroethane (CF_3CH_2F): Chemical name and formula for R-134a.

Thermal limiter: Fuse type device used on older air conditioning systems to prevent high system pressures.

Thermal lockout switch: Switch designed to shut off coolant flow to the heater core when the coolant is too cold to heat the passenger compartment.

Thermistor: Sensors made of resistor material that responds to temperature changes. Unlike other kinds of resistors, thermistor resistance goes down as its temperature increases.

Thermostat: Temperature sensitive valve used to control coolant flow through the engine.

Thermostat monitor: Internal computer timer that measures the amount of time needed for the thermostat to open.

Thermostatic expansion valve (TXV): A variable opening refrigerant flow control device. It is sometimes called an *expansion valve* or an *X valve.*

Transfer path: Also referred to as a *conduit.* The path a vibration or noise takes to the receiver.

Transistor: Semiconductor electronic device that operates as a switch.

Transmission oil cooler: Tube or plate heat exchanger mounted inside the radiator tank. Used to cool automatic transmission fluid.

Transmitter: Also referred to as the *source.* The point of origin of a vibration.

Trinary cutoff switch: A combination low and high pressure switch. It will cut off current to the compressor if refrigerant pressure becomes too high or too low.

Trouble code: Numeric or alpha-numeric designator identifying the general location of a problem or defective component.

Tube and fin evaporator: Single pass evaporator with double or triple rows of coils.

Turnbuckle: Movable adjustment point on a temperature cable in a manual system.

Two-speed fan: Electric fan that uses a set of two or three relays to control fan speed.

U

Universal asynchronous receive and transmit (UART): Electronic communications medium used between the ECM, off-board diagnostic equipment, and other control modules. UART is a data line that varies voltage between 0-5 volts at a fixed pulse width rate.

Upstream: Used to describe the location of a component before another component in the direction of flow.

V

Vacuum: Negative pressure; pressure in an enclosed area that is lower than atmospheric.

Vacuum gauge: Diagnostic instrument that measures the amount of vacuum in a system or generated by the engine.

Vacuum pump: A motorized or air driven pump used to evacuate air and noncondensible gases from the refrigeration system.

Vacuum servo: Diaphragm or solenoid operated by engine vacuum.

Valves-in-receiver (VIR): A single control valve assembly housing the expansion valve, POA valve, and receiver-drier in one unit.

Variable displacement orifice tube (VDOT): Refrigeration system that uses a fixed orifice tube and a variable displacement compressor.

Variable orifice valve: Orifice tube that has an adjustable opening. Used in some retrofits.

V-belts: Accessory drive belt used on older vehicles. Usually only used to drive 1-2 components.

Vehicle control module (VCM): On-board computer which controls ABS/TCS functions as well as engine and powertrain operation.

Vehicle identification number (VIN): Individual series of letters and numbers assigned to a vehicle by the manufacturer at the factory.

Vehicle speed sensor (VSS): Sensor placed in the transmission/transaxle or the rear axle assembly. Used by the engine's ECM to monitor vehicle speed.

Vent: An opening in the kickpanels of older cars used to bring in outside air.

Vent mode: Operating mode which circulates outside air in the vehicle without conditioning, either heating or cooling.

Volt: Unit of measurement of electrical pressure or force that will move a current of one ampere through a resistance of one ohm.

Voltage drop: A lowering of circuit voltage due to excessive lengths of wire, undersize wire, or through a resistance.

Voltage: Electrical pressure, created by the difference in the number of electrons between two terminals.

Voltmeter: Instrument used to measure voltage in a given circuit.

W

Water jacket: Passages cast inside an engine which allows coolant to circulate around the engine's moving components.

Water pump: Engine driven pump used to circulate coolant through the cooling system. Also knows as a *coolant pump.*

Waveform: Pattern created by electrical current.

Wire: Copper or aluminum conduit used to carry electricity.

Wire gage: System used to classify the size of wire, the smaller the number, the larger the wire.

Wind chill: Temperature caused by the combination of air moving over skin.

Wobble plate: The movable plate on a compressor clutch assembly.

Z

Zeotropic blend: Type of blended refrigerant that can separate into its various component refrigerants. Most alternative refrigerants fall into this classification and are in the R-400 family of refrigerants.

Index

C

D

E

S

T

U

V

W

Z